Android
学习精要

高洪岩 编著

清华大学出版社
北 京

内 容 简 介

本书是作者在北大青鸟多年教授软件开发课程的经验总结。本书以丰富的实例、完整的代码解说、清晰的操作步骤，言简意赅，直达 Android 开发核心要点为目标，让读者用最短的时间掌握 Android 开发技能。主要内容包括 Activity 对象的使用及其生命周期，各种自定义对话框的使用，多种创建 View 视图对象的方式，常用 5 大布局对象的使用，Android 控件的使用，使用 Intent 对象进行隐式和显式的调用，通知 Notification 的使用，Activity 对象常用 flag 标记的使用，持久化技术 ContentProvider、SharedPreferences 和 SQLite、File IO 的使用，Android 中 Service 服务技术的使用，定时服务 AlarmManager 的使用，串行化 Parcelable 接口的使用，详细的 AIDL 使用案例，Handler 对象使用的知识点，HTTP 协议结合 JSON 和 XML 技术与服务器通信，加强与 Internet 的数据交互，Android 控件的美化，Fragment 对象的使用等。

本书完全使用实例代码演示的方式教学，紧跟 Android 的技术潮流，适合 Android 初学者、Android 开发人员使用，也可以用作培训机构和大专院校的教学参考书。

本书封面贴有清华大学出版社防伪标签，无标签者不得销售。
版权所有，侵权必究。侵权举报电话：010-62782989　13701121933

图书在版编目（CIP）数据

Android 学习精要 / 高洪岩编著. —北京：清华大学出版社，2012.9
ISBN 978-7-302-29264-7

Ⅰ. ①A… Ⅱ. ①高… Ⅲ. ①移动终端－应用程序－程序设计 Ⅳ. ①TN929.53

中国版本图书馆 CIP 数据核字（2012）第 151602 号

责任编辑：	王金柱
封面设计：	王　翔
责任校对：	闫秀华
责任印制：	张雪娇

出版发行：清华大学出版社
网　　址：http://www.tup.com.cn, http://www.wqbook.com
地　　址：北京清华大学学研大厦 A 座　　邮　编：100084
社 总 机：010-62770175　　邮　购：010-62786544
投稿与读者服务：010-62776969, c-service@tup.tsinghua.edu.cn
质 量 反 馈：010-62772015, zhiliang@tup.tsinghua.edu.cn
印　刷　者：清华大学印刷厂
装　订　者：三河市溧源装订厂
经　　销：全国新华书店
开　　本：190mm×260mm　　印　张：38　　字　数：973 千字
　　　　　（附光盘 1 张）
版　　次：2012 年 9 月第 1 版　　印　次：2012 年 9 月第 1 次印刷
印　　数：1～4000
定　　价：79.00 元

产品编号：047271-01

前 言

本书是笔者在北大青鸟讲解软件开发课程的教学笔记，期间历经多期学员，并在教学过程中不断改进和完善，应该说，这是一本来自于教学实践和开发者需求的 Android 图书，全书对 Android 常用的开发技术和控件进行了细致地讲解，并使用实例和代码演示的方法教学，读者在学习过程中，只须对照实例代码边学边练，就可以很快掌握 Android 开发技术，应对软件开发公司的用人需求。

笔者在学习 Android 之前先入手了一个 HTC G7 型号的手机，目的是想看看 Android 这个操作系统到底有哪些功能，是什么样的界面，有哪些选项，只有熟练地掌握了操作系统，才能更加深刻地认识 Android。不可否认的是，Android 是一个操作系统，它具有操作系统所有的特点，包括在学习开发时的复杂度，它所含的知识点太多，而且这些知识点都具有很"新颖"的特点，在其他的操作系统上并没有出现，这就增加了学习上的成本，但是只要坚持下来并勤于动手实践，相信读者很快就会成为开发高手。

本书十大知识点：

（1）Android 技术入门，包括 Android 的体系结构，Eclipse 中项目的结构安排，Activity 对象的使用及非常重要的生命周期，以及常用的各种对话框样式的使用。

（2）View 与 ViewGroup 组件在 Android 技术中的具体应用，比如创建 View 视图对象的多种方式，在 Android 中常用的 5 大布局对象的使用及其相关的注意事项等。

（3）在 Android 控件的使用相关知识点中，使用了大量的篇幅，因为在开发 Android 的过程中始终离不开控件，它是学习 Android 必须要经过的一个步骤。

（4）Intent 对象的使用，包括隐式和显式调用，Intent 匹配的过程，通知 Notification 的使用，Activity 对象不同的启动方式，以及 Activity 对象常用 flag 标记的使用等。

（5）ContentProvider、SharedPreferences、SQLite 及 File IO 持久化技术在 Android 中的使用，这也是学习 Android 技术的重点所在。

（6）启动 Service 服务的两种方式，定时服务 AlarmManager 的使用，串行化 Parcelable 接口的使用，详细的 AIDL 使用案例，以及非常重要的 Handler 对象使用的知识点。

（7）使用 HTTP 协议结合 JSON 和 XML 技术实现 Android 客户端和远程服务器之间进行数据交互，并且详细地介绍了不同类型 JSON 字符串的转换和解析。

（8）Android 4 大核心技术之间的无缝调用，这是学习 Android 的进阶知识点。

（9）常用 Android 控件的美化，具体效果请参看对应章节的截图。

（10）Fragment 对象的使用，使开发 Android 实现了模块化，分工更加明确，这也是开发 Android 平板电脑必须掌握的技术。

掌握了这 10 大知识点后，完全可以掌握 Android 常用的开发技术，并且建立起系统的知识体系，对未来的深入学习和实际开发打下良好基础。

本书的知识体系经过多次真实教学过程中的改良，书中的实例更易被读者消化吸收，更以"所见即所得"的方式展示了代码的运行结果，读者只需跟着实例练习，应该很快就能掌握，这也是本书的一大特点。

在此非常感谢支持撰写本书的前辈，包括公司的领导、QQ上尚未谋面的好友以及笔者的家人，感谢这样一个良好的环境给笔者力量完成本书。真心的希望本书能作为读者学习Android路上的辅路石，减少读者在学习Android技术上的时间和成本。

高洪岩

2012年6月于北京东三环

目 录

第1章 初识 Android 1
1.1 Android 平台概述 1
1.2 Android 平台体系 2
1.2.1 Linux Kernel 内核层 2
1.2.2 系统运行库 Libraries 和 Android Runtime 层 2
1.2.3 Application Framework 应用程序框架层 3
1.2.4 Application 应用程序层 3
1.3 Android 开发环境配置 4
1.4 在 Eclipse 环境配置 Android SDK 及创建 AVD 7
1.5 在 Eclipse 中创建 Android 第一个项目并运行 10
1.6 在 Eclipse 中创建 Android 项目结构 14
1.6.1 Runme.java 主程序文件 15
1.6.2 R.java 资源索引文件 16
1.6.3 main.xml 界面布局文件 17
1.6.4 AndroidManifest.xml 应用程序配置文件 19
1.6.5 R.java 文件的自动索引 19
1.6.6 AndroidManifest.xml 文件相关的知识点 20
1.6.7 main.xml 界面布局文件 24
1.7 Log 类中的方法使用 25
1.7.1 通用日志方法 26
1.7.2 getStackTraceString 方法的使用 34
1.7.3 v()、e()、i()、v()和 w()方法的区别与 isLoggable 方法的使用 35
1.8 文件夹 res 中更多的资源类型 39
1.9 常用资源的读取操作 41
1.10 Activity 的生命周期 45
1.10.1 实现 onCreate()->onStart()->onResume()->onPause()->onResume 52
1.10.2 实现 onCreate()->onStart()->onResume()->onPause()->onStop()-> onRestart()->onStart() 56
1.10.3 实现 onCreate()->onStart()->onResume()->onPause()-> onStop()->onDestroy() 60
1.10.4 应用程序列表时的生命周期情况 64
1.10.5 AVD 横竖屏切换时的生命周期情况 65
1.10.6 onSaveInstanceState()和 onRestoreInstanceState()回调方法的使用 67
1.11 LinearLayout 布局对齐方式和 Dialog 提示的使用 69
1.11.1 使用自定义对话框实现登录功能（对话框与 Activity 通信） 73

1.11.2　AlertDialog 对话框的使用 78
　　1.11.3　ProgressDialog 对话框的使用 88
　　1.11.4　对话框中的内容是列表条目的情况并取消后退按钮 90
　　1.11.5　使用自定义 XML 布局义件填充 AlertDialog 对话框的另外一种方法 ... 93
　　1.11.6　实现自动关闭对话框 94
　　1.11.7　toast 提示的使用 96
　　1.11.8　设置 Dialog 对话框的尺寸 99
　　1.11.9　PopupWindow 对话框 100
　1.12　抽象类 Window 与布局分析工具 Hierarchy View 101
　1.13　控制控件位置和大小的常用属性 105
　1.14　设置应用程序背景图片 106

第 2 章　View 与 ViewGroup 类和控件事件 108

　2.1　View 和 ViewGroup 类的概述 108
　2.2　View 类的构造函数 110
　　2.2.1　View(Context context)构造方法的使用 110
　　2.2.2　View(Context context, AttributeSet attrs)构造方法的使用 112
　2.3　View 单线程模型特性与在非 UI 线程中更新界面异常的实验 116
　2.4　动态创建 View 和 ViewGroup 控件 118
　　2.4.1　第一种创建控件的办法 118
　　2.4.2　第二种创建控件的办法 120
　　2.4.3　第三种创建控件的办法 123
　2.5　界面布局的空间分配与权重 124
　2.6　常用布局 127
　　2.6.1　RelativeLayout 相对布局实验 127
　　2.6.2　TableLayout 布局的使用 133
　　2.6.3　FrameLayout 布局的使用 139
　　2.6.4　AbsoluteLayout 布局的实验 139
　　2.6.5　用程序来实现 margin 的实验 140
　2.7　控件事件 141

第 3 章　Android 的 UI 控件 148

　3.1　UI 控件与 Adapter 和 ListView 对象 148
　3.2　Adapter 接口 149
　3.3　ListAdapter 接口 150
　3.4　ListView 对象 151
　3.5　ArrayAdapter 对象 152
　3.6　AnalogClock 和 DigitalClock 控件 152
　3.7　AutoCompleteTextView 控件的使用与 XML 数据源 154

3.8	Button 控件	157
3.9	CheckBox 控件	159
3.10	CheckedTextView 控件	160
3.11	Chronometer 控件	165
3.12	DatePicker 和 TimePicker 控件	167
3.13	EditText 控件	173
3.14	Gallery 控件和 ImageSwitcher 控件	179
3.15	TextView 控件	184
3.16	ImageView 和 ImageButton 控件	188
3.17	MultiAutoCompleteTextView 控件	190
3.18	ProgressBar 控件	191
3.19	RadioGroup 与 RadioButton 控件	192
3.20	RatingBar 控件	194
3.21	SeekBar 控件	196
3.22	ListView 对象和 Spinner 控件	197
	3.22.1　Spinner 控件初步使用	197
	3.22.2　在 ListView 控件中显示文本列表功能	201
	3.22.3　在 ListView 控件中使用多选 checkedbox 控件	203
	3.22.4　在 ListView 控件中使用单选 radioButton 控件	206
	3.22.5　在 ListView 中自定义布局内容	208
	3.22.6　在 ListView 中添加及删除条目	213
	3.22.7　在 ListView 中使用带图标的自定义布局	214
3.23	VideoView 控件	219
3.24	SimpleAdapter 对象	221
3.25	WebView 对象	223
3.26	控件的显示与隐藏	225
3.27	GridView 对象	226
	3.27.1　GridView 中放置文字	226
	3.27.2　在 GridView 中放置图片	227
	3.27.3　在 GridView 中放置图片和文字	230
3.28	菜单 Menu 控件之选项菜单	234
	3.28.1　创建选项菜单	234
	3.28.2　为菜单加多选和单选功能	237
3.29	菜单 Menu 控件之子菜单	240
3.30	菜单 Menu 控件之上下文菜单	243
3.31	ScrollView 垂直滚动视图和 HorizontalScrollView 水平滚动视图	250
3.32	DatePickerDialog 和 TimePickerDialog 对话框	252
3.33	TextView 控件小示例继续讨论	254
3.34	ToggleButton 对话框	256

3.35 ListActivity 对象 ... 258
3.36 TabHost 标签页控件 ... 259
3.37 控件显示内容的国际化 i18n ... 261
3.38 Color 颜色的操作 ... 262
3.39 draw9Patch 工具的使用 ... 264
3.40 以 9 格图片资源作为 Button 背景 ... 267
3.41 使用 selector 改变按钮状态 ... 269

第 4 章 Intent 对象 ... 271

4.1 Intent 对象必备技能 ... 271
 4.1.1 指定 componentName 组件名称与显式调用 ... 271
 4.1.2 指定 Action 动作名称与隐式调用 ... 273
 4.1.3 指定 Action 的动作名称和 Data 数据 ... 280
 4.1.4 两个 Activity 之间传递 Extra 字符串和 Extra 实体对象的实验 ... 282
 4.1.5 category 类型的使用 ... 289
 4.1.6 data 标签的使用 ... 295
4.2 创建 Dialog 式的 Activity 登录实例 ... 297
4.3 显式启动其他应用程序的 Activity ... 301
4.4 发送文本短信的简单示例 ... 302
4.5 Notification 通知的使用 ... 304
 4.5.1 Notification 通知的初入 ... 304
 4.5.2 自动隐藏状态条的图标 ... 306
 4.5.3 每个通知对象拥有自己的 Intent 对象 ... 306
 4.5.4 设置状态栏中通知的数量显示 ... 308
 4.5.5 取消通知 ... 309
 4.5.6 设置振动模式和发出提示音和 LED 灯 ... 310
 4.5.7 自定义通知布局内容 ... 312
 4.5.8 Notification.FLAG_INSISTENT 和 Notification.FLAG_ONGOING_EVENT 的使用 ... 313
4.6 Activity 的 4 种启动方式 ... 314
 4.6.1 standard 模式 ... 315
 4.6.2 singleTop 模式 ... 317
 4.6.3 singleTask 模式 ... 323
 4.6.4 singleInstance 模式 ... 327
4.7 Activity 常用 flag 标记的学习 ... 329
 4.7.1 FLAG_ACTIVITY_CLEAR_TOP 标记 ... 330
 4.7.2 FLAG_ACTIVITY_CLEAR_WHEN_TASK_RESET 标记 ... 333
 4.7.3 FLAG_ACTIVITY_EXCLUDE_FROM_RECENTS 标记 ... 336
 4.7.4 FLAG_ACTIVITY_FORWARD_RESULT 标记 ... 337

第 5 章 ContentProvider、SharedPreferences 和 SQLite 持久化存储 351

- 4.7.5 FLAG_ACTIVITY_NEW_TASK 标记 340
- 4.7.6 FLAG_ACTIVITY_NO_ANIMATION 标记 343
- 4.7.7 FLAG_ACTIVITY_NO_HISTORY 标记 343
- 4.7.8 FLAG_ACTIVITY_NO_USER_ACTION 标记 345
- 4.7.9 FLAG_ACTIVITY_REORDER_TO_FRONT 标记 348

- 5.1 在 Android 中使用 File 对象实现文件基本操作 351
- 5.2 在 Android 中使用 Android 平台自带对象实现文件的基本操作 354
  - 5.2.1 使用 openFileOutput 和 openFileInput 读写文件 354
  - 5.2.2 读取 assets 目录中的文件 357
  - 5.2.3 读取 res/raw 文件夹中已经存在的 TXT 和 PNG 文件 358
  - 5.2.4 读取 res/xml 文件夹中已经存在的 XML 文件 361
  - 5.2.5 操作 SD 卡中的文件 363
- 5.3 Linux 中的文件操作权限 364
- 5.4 SharedPreferences 的读写权限实验 365
- 5.5 Uri 对象的匹配 368
- 5.6 ContentProvider 对象的初步使用 369
- 5.7 SQLite 数据库的使用 375
  - 5.7.1 使用 Navicat_for_SQLite 工具创建 SQLite 数据库及表 375
  - 5.7.2 使用 SQLiteDatabase 对象的常用方法操作数据库 378
  - 5.7.3 封装数据库操作类 396
  - 5.7.4 使用 DBOperate 对象将数据表中的数据显示在 ListView 中 405
- 5.8 ContentProvider 对象的使用 407
  - 5.8.1 创建数据提供者 ContentProvider 对象 407
  - 5.8.2 创建 ContentProvider 对象的使用者 412
  - 5.8.3 调用 ContentProvider 对象的应用运行效果 414
- 5.9 Application 全局数据存储对象的使用 417

第 6 章 Broadcast、Service 服务及 Handle 对象 420

- 6.1 使用 Broadcast 的种类 420
  - 6.1.1 多 BroadcastReceiver 同时匹配 Intent 的情况 420
  - 6.1.2 用广播实现程序开机运行的效果 422
  - 6.1.3 sendStickyBroadcast 函数的使用 423
- 6.2 Service 服务 424
  - 6.2.1 用 startService 启动 Service 方式与生命周期 426
  - 6.2.2 用 bindService 启动 Service 的方式与生命周期 431
  - 6.2.3 回调函数 onRebind()的调用时机 435
  - 6.2.4 ServiceConnection 对象的 onServiceDisconnected()方法调用时机 439

6.3 Service 相关示例及知识点 439
6.3.1 定时服务 AlarmManager 的使用 439
6.3.2 判断 Service 是否在运行中 442
6.3.3 方法 onStartCommand 的返回值实验 443
6.3.4 Parcelable 接口串行化的使用 446
6.3.5 使用 AIDL 技术跨进程传递 Parcelable 对象 449
6.4 Handle 对象的使用 461
6.4.1 Handler 对象的初步使用 461
6.4.2 postDelayed 方法和 removeCallbacks 方法的使用 465
6.4.3 post 方法的使用 468
6.4.4 postAtTime 方法的使用 470
6.4.5 在线程对象的 run 方法中实例化 Handler 对象的注意事项 470
6.4.6 以异步方式打开网络图片 473
6.5 Appwidget 小部件的使用 476
6.5.1 初入 Appwidget 小部件 477
6.5.2 Appwidget 的生命周期 480
6.5.3 Appwidget 的隔时刷新界面的效果——使用 AlarmManager 482
6.6 章节 AsyncTask 对象的使用 484
6.6.1 初入 AsyncTask 485
6.6.2 使用 AsyncTask 更新 UI 的示例 487
6.6.3 使用 AsyncTask 时外界无参数与其进行交互的情况 489

第 7 章 HTTP 交互、JSON 和 XML 490
7.1 JSON 介绍 490
7.1.1 Gson 框架与 JSON 字符串交换数据示例 490
7.1.2 在 Android 中通过 HTTP 协议用 JSON 与 Web 项目通信 495
7.2 在 Android 中通过 HTTP 协议访问 TXT 文件和 PIC 图片 498
7.3 用 java 语言 DOM 解析 XML 502

第 8 章 Activity 活动、Service 服务和 Broadcast 广播彼此调用实验 505
8.1 Activity->BroadCaseReceiver->Activity 实验 505
8.2 Activity->Service(startService)->Activity 实验 507
8.3 Activity->BroadCaseReceiver->Service(startService)实验 508
8.4 Activity->Service(startService)-> BroadCaseReceiver 实验 510
8.5 Activity->BroadCaseReceiver->Service(bindService)实验 511

第 9 章 UI 控件的美化与动画 515
9.1 style 的使用 515
9.1.1 style 的概述与定义 516

9.1.2 style 的使用与继承 518
9.2 文字颜色 selector 状态列表 519
9.2.1 文字颜色 selector 的概述与定义 519
9.2.2 文字颜色 selector 的使用 520
9.3 背景图片 selector 状态列表 521
9.3.1 背景图片 selector 状态列表 521
9.3.2 用 selector 状态列表美化 Button、CheckBox、RadioButton 和 EditText 常用控件 522
9.3.3 美化 Option 选项面板 528
9.3.4 美化 ListView 控件 531
9.3.5 美化 TabHost 控件 534
9.3.6 美化 RadioGroup 组件 539
9.3.7 美化 ExpandableListView 组件 542
9.4 动画 548
9.4.1 alpha 透明动画演示 549
9.4.2 scale 缩放动画演示 550
9.4.3 translate 移动动画演示 552
9.4.4 rotate 旋转动画演示 553
9.4.5 动画中 Interpolators 的使用 554
9.4.6 动画的混合应用演示 554

第 10 章 Fragment 对象的使用 557
10.1 Fragment 对象简介 557
10.2 Fragment 对象生命周期与事务 557
10.2.1 Fragment 对象生命周期 558
10.2.2 Fragment 对象的事务 563
10.3 Fragment 对象使用案例 569
10.3.1 Fragment 对象的初步使用与 inflate 方法参数的解析 569
10.3.2 FragmentActivity 与 Fragment 对象交互 572
10.3.3 Fragment 对象之间的交互 576
10.3.4 在 DialogFragment 对象中使用 onCreateView 回调函数生成对话框 578
10.3.5 将 DialogFragment 对象放入 back stack 后退栈中 580
10.3.6 在 DialogFragment 对象中使用 onCreateDialog 回调函数生成对话框 583
10.3.7 切换 Fragment 添加动画效果 584
10.3.8 Fragment 的显示和隐藏 587
10.3.9 ListFragment 对象的使用 588
10.3.10 Fragment 对象的分页处理方式 1 589
10.3.11 Fragment 对象的分页处理方式 2 592
10.3.12 使用 Fragment 对象实现 TabHost 样式的分页及滑动 593

第 1 章　初识 Android

作为本书的第一章不会像其他技术教程一样，以浓重的文字来对 Android 做全面的历史演化理论概述，而是以实例的方式来介绍 Android 开发的入门步骤与基本的控件使用，开门见山，永远是快速学习一门技术最好的方式。

本章应该着重掌握如下知识点：
- Android 技术的框架结构
- 理解在使用 ADT 插件的 Eclipse 中 Android 的项目结构
- Activity 的生命周期
- 对话框的使用。本章对话框的使用方式较多，常用于各种数据展示的情况

1.1　Android 平台概述

看到本教程的读者都是带着一点点对 Android 的了解而来，Android 主要的使用场合就是移动通信领域，也就是 Android 开发出来的软件在手机或移动电脑上运行，随时随地可以很方便地处理数据和管理数据。

先来看看学习 Android 需要掌握的一些基本知识。

（1）Android 是由 Google 公司进行设计与开发的移动通信平台。

（2）Android 平台基于 Linux 操作系统，所以内存的分配、线程的调度以及作业的执行都由底层 Linux 操作系统来进行处理。

（3）Android 平台是开放源代码的。

（4）Android 单词的中文翻译是"机器人"的意思。

（5）Android 平台被很多知名的通信运营商支持，所以 Android 逐渐成为了一个通信平台的标准。

（6）Android 相当于一个操作系统，在这个操作系统上可以运行任何 Android 支持的软件。Android 相当于 Windows XP，所以安装有 Android 系统的手机叫做"智能手机"。

（7）Android 支持多任务环境。

（8）现阶段支持 Android 平台比较好的手机制造商有 HTC、三星等。

（9）基于 WebKit 引擎的浏览器。

（10）具有 2D 和 3D 软件开发能力，其中 3D 图形库基于 OpenGL ES 1.0 标准。

（11）数据存储使用 SQLite，它是一个文件型数据库。

（12）支持常用的图形及视频格式（MPEG4、H.264、MP3、AAC、AMR、JPG、PNG、GIF）。

1.2 Android 平台体系

任何的技术都有其独有的体系结构，Android 也不例外，图 1.1 显示出了完整的 Android 技术体系。

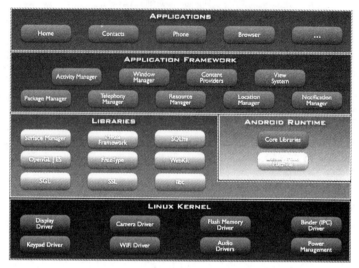

图 1.1 Android 平台的体系结构

从图 1.1 中可以看到，Android 的体系结构分为 4 层，从高到低分别是：

（1）Applications 应用层。
（2）Application Framework 应用程序框架层。
（3）Libraries 和 Android Runtime 层。
（4）Linux Kernel 内核层。

这 4 个层的功能将会在下面进行详细介绍。

1.2.1 Linux Kernel 内核层

Android 使用基于 Linux version 2.6 版本的内核，所以安全和内存管理、进程处理、网络传输、驱动模型等这些核心功能的处理都在内核完成，也就是说 Android 程序员不需要知道过多的底层实现，只须以透明的方式进行基于 Android 的软件开发，具体的程序运行细节由内核来进行管理。

1.2.2 系统运行库 Libraries 和 Android Runtime 层

在本层中，分为两个部分，一个是 Libraries 运行库层，另一个是 Android Runtime 层。

（1）Libraries 运行库层

Android 平台系统底层包含不同功能的组件库，这些组件库都是由 C/C++语言来进行编写，从而可以正确地运行在 Linux version 2.6 版本上，开发者通过 Android 应用程序框架可以很容易地来使用这些功能组件。

以下就是常用的功能组件库 Libraries 的介绍。

- Media Libraries 媒体库：基于 PacketVideo's OpenCORE 库，使用 OpenCORE 库可以播放和录制声音和视频，并且支持大部分的常用媒体格式和静态图片，例如 MPEG4、H.264、MP3、AAC、AMR、JPG 和 PNG 等。
- Surface Manager 显示管理库：负责显示的子系统可以与 2D 或 3D 技术进行整合。
- SGL 库：基于 2D 的图形图像处理引擎。
- FreeType 库：位图及矢量字体支持。
- SQLite 库：非常流行的关系型数据库引擎，支持几乎所有的应用程序。

（2）Android Runtime 层

有了 Libraries 运行库层还不行，因为还没有程序运行环境的支撑，所以就需要 Android Runtime 组件，它的主要功能为：

Android Runtime 使用 Dalvik virtual machine 虚拟机来运行 Android 程序。在 Dalvik virtual machine 虚拟机中每一个 Android 程序都有其自己的实例和应用程序进程。Dalvik virtual machine 虚拟机只运行 Dalvik Executable（.dex）格式的文件，与 JVM 运行 class 文件不同。.dex 文件是经过 CPU 优化，并且尽量地被设计成占用内存最少的文件格式。

Dalvik Executable（.dex）文件创建的过程：首先，*.java 生成*.class 文件；然后*.class 文件再生成*.dex 文件。

Android 应用程序的扩展名为.apk，使用 AAPT（即 Android Asset Packaging Tool 工具）生成，使用前需要将 APK 安装到 AVD 或真机设备中才可以运行应用程序。

Dalvik VM 依赖于 Linux 内核实现内存的管理及线程的分配等任务。在 Android 中每一个应用程序都有一个 Dalvik 虚拟机为其进行服务。

1.2.3　Application Framework 应用程序框架层

在 Linux 内核和 Dalvik virtual machine 虚拟机以及 Android Runtime 运行环境基础上，就可以创建一些通用组件了，这些组件是用 Java 语言来进行编写，这些代码组成了 Android 的 SDK，SDK 中的主要功能有：

- 包含界面控件 lists、grids、text boxes、buttons 等。
- 提供应用程序之间互相访问数据的功能，即 Content Provider 技术。
- 允许应用程序之间访问文件，例如 MP3、图形图像和音视频文件等。
- Notification Manager 通知管理，允许应用程序在 status bar 状态栏中显示一些相关的提示信息。
- Activity Manager 活动管理，主要目的是管理 Activity 的生命周期，并且提供一个通用的后台活动栈（navigation backstack），可以使 Activity 活动进行回退或销毁。

1.2.4　Application 应用程序层

本层是 Android 程序员通过使用 Android SDK 开发出来的软件，例如发短信的程序、数据管理 ERP 的程序和电子商务程序等。

1.3 Android 开发环境配置

 在配置 Android 开发环境之前一定要把 JDK 安装到当前的计算机中，如果使用的是 EXE 安装版的 JDK，则在安装时会自动配置环境变量。

如果使用的是绿色版的 JDK，则必须在电脑中配置环境变量。值得庆幸的是，Android 平台也可以在 Eclipse 开发工具中得到支持，并且需要下载必要的安装组件，如图 1.2 所示。

但在这里需要注意的是，文件 Android-sdk_r09-windows.zip 并不是真正的 Android 的 SDK，而是一个管理 SDK 的软件，所以还得在线下载指定版本的 Android SDK。另外到目前为止，Google 并没有发布离线 SDK 包。

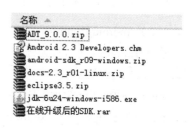

图 1.2 开发 Android 必要的软件

01 将 Android-sdk_r09-windows.zip 文件解压，然后执行 SDK Manager.exe 文件，如图 1.3 所示。

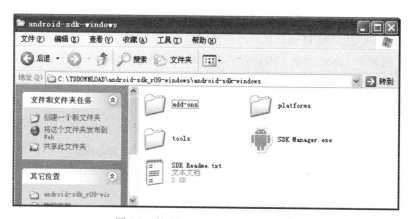

图 1.3 执行 SDK Manager.exe 文件

02 弹出如图 1.4 所示的对话框，由于不需要下载全部的 SDK，只需要下载指定版本的 SDK，所以直接单击 Cancel 按钮关闭对话框。

03 在图 1.5 所示下载指定版本的 Android SDK 界面中单击 "Available packages" 选项，然后选中图中指定版本的 SDK 下载，下载的内容分别是 SDK 和 Samples 示例。

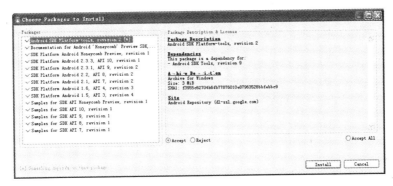

图 1.4　下载所有版本的 Android SDK

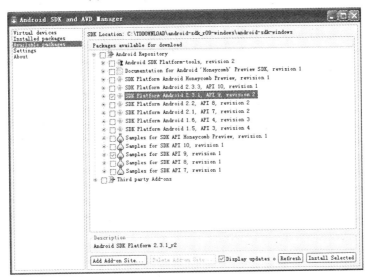

图 1.5　下载指定版本的 Android SDK

04 选中指定版本的 Android SDK 和 Samples 后单击右下角的"Install Selected"按钮进行下载并且安装，出现确认安装界面，如图 1.6 所示。

图 1.6　确认安装 Android SDK 和 Samples 界面

05 单击 Install 按钮后 SDK Manager.exe 软件开始下载，如图 1.7 所示。

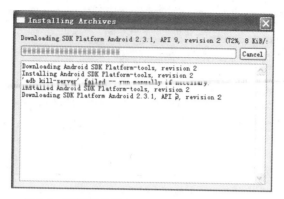

图 1.7　开始下载指定 Android SDK 和 Samples

06 下载结束后询问是否重启 ADB（Android Debug Bridge）服务，如图 1.8 所示。

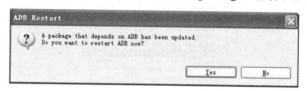

图 1.8　重启 ADB 服务

07 单击 Yes 按钮，重启 ADB 服务。重启结束后出现如图 1.9 所示的界面，证明 SDK 和 Samples 下载成功。

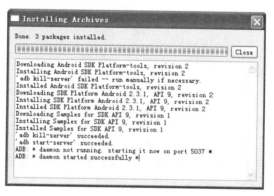

图 1.9　下载成功

08 单击 Close 按钮关闭下载界面。出现如图 1.10 所示的界面，单击 "Installed packages" 选项，出现安装成功后的 SDK 列表。

图 1.10　安装成功后的 SDK

到这步已经下载成功了 SDK 和 Samples 示例，下载结束后会在图 1.11 Android-sdk-windows\platforms 目录下创建指定 SDK 版本的目录，Android-sdk-windows\platforms\Android-9 的目录内容如图 1.12 所示。

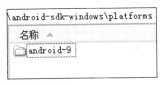

图 1.11　创建指定 SDK 版本的目录

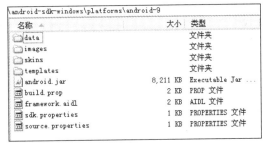

图 1.12　Android-9 的目录内容

1.4　在 Eclipse 环境配置 Android SDK 及创建 AVD

01　解压 eclipse3.5.zip 文件。

02　解压 ADT_9.0.0.zip 文件，ADT 是 Android development Tools，是在 Eclipse 下开发 Android 的插件。解压后出现的文件列表如图 1.13 所示。将这些文件复制到 Eclipse 安装目录中，以使 Eclipse 和 ADT 进行整合。

03　运行 eclipse.exe 文件。

04　如果在 Eclipse 中出现 Android 按钮，则证明 Eclipse 与 Android 整合成功，如图 1.14 所示。

图 1.13　ADT_9.0.0.zip 文件解压后的文件列表

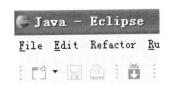

图 1.14　出现 Android 按钮

05　这时仅仅 Eclipse 与 ADT 整合成功，但 Eclipse 还从未与 Android SDK 进行整合，所以单击 Eclipse 的 Window 菜单的子菜单 Preferences 中的 Android 节点配置 SDK 所存在的路径，如图 1.15 所示。

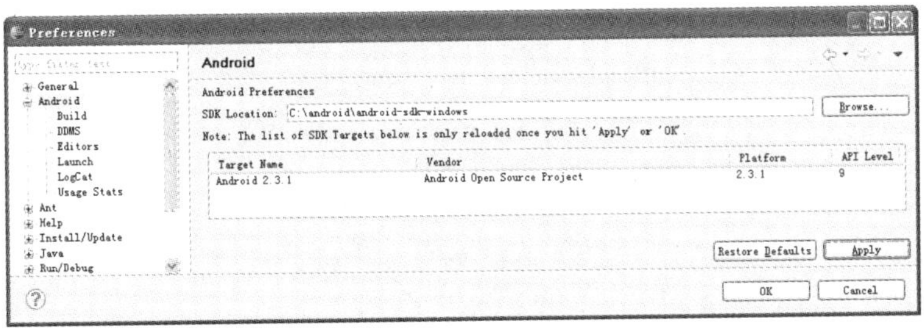

图 1.15　配置 Android 的 SDK

配置路径完成后单击右下角的 Apply 按钮，出现了当前使用 Android SDK 的版本号是 2.3.1。单击"OK"按钮完成配置。

06 第（5）步虽然把 Eclipse 与 Android SDK 进行了整合，但由于 Android 是运行在手机上的，所以还需要配置一个虚拟/模拟的运行环境，这个环境叫做 AVD（Android virtual Device）。

单击 Window 菜单的"Android SDK and AVD Manager"命令，如图 1.16 所示。

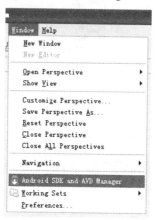

图 1.16　单击 Android SDK and AVD Manager 命令

弹出 Android 的 AVD 配置界面，如图 1.17 所示。

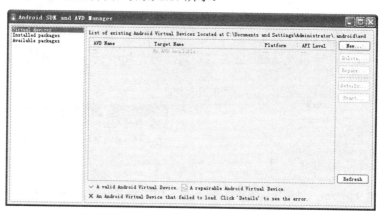

图 1.17　出现配置 Android AVD 界面

在图 1.17 中单击右上角的"New..."按钮，新建一个 AVD 设备，弹出如图 1.18 所示的对话框。

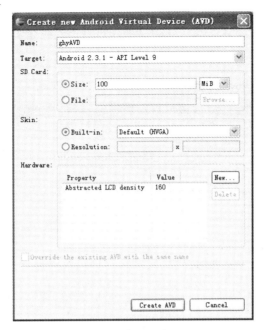

图 1.18　详细配置 AVD

在图 1.18 所示的对话框中需要配置 3 个选项。

- Name：这个 AVD 设备的名称。
- Target：这个 AVD 设备使用的 Android SDK 版本是什么。
- Size：要创建的 SD 卡的大小。

配置完成后单击"Create AVD"按钮立即创建这个 AVD 设备。

07 成功创建的"ghyAVD"在列表中显示出来，如图 1.19 所示。

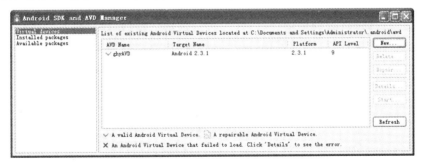

图 1.19　ghyAVD 创建成功

到这一步，Eclipse 关联 ADT、配置关联 Android SDK，以及创建 AVD 的步骤就结束了，下一步就到了使用 Eclipse 真正地创建一个 Android Project 的时候了。

1.5 在 Eclipse 中创建 Android 第一个项目并运行

01 单击图 1.20 中的 Other 命令。

图 1.20 单击 Other 命令

02 选中要创建的项目类型是 Android，如图 1.21 所示。

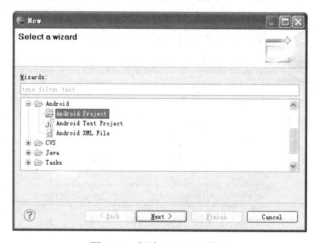

图 1.21 创建 Android 项目

在图 1.21 中单击 Next 按钮继续创建。

03 弹出一个详细配置 Android 项目的对话框，如图 1.22 所示。

图 1.22　详细配置 Android 项目

在图 1.22 中需要配置以下 4 个选项。

- Project Name：项目名称。
- Application name：应用程序名称，此选项非常重要，里面写的值就是在 Android 真机上显示应用程序图标下方的名称，可以允许为中文。
- Package name：设置源代码所存放的包名。

配置 Activity 活动的名称，也就是 Android 应用起始的 Activity 对象，也就是启动界面对象。

04 配置完成后单击 Next 按钮继续，出现如图 1.23 所示的对话框。

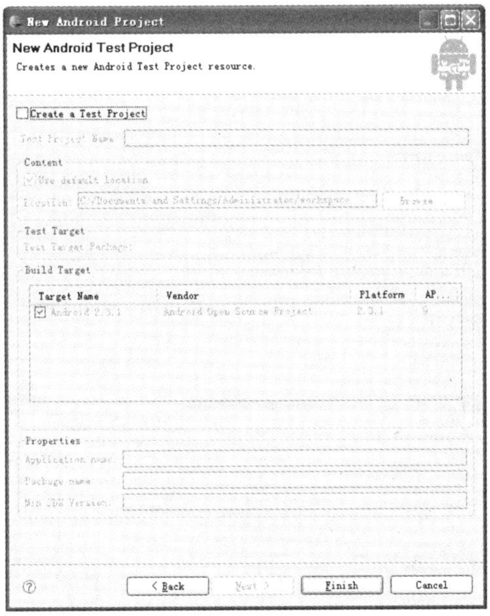

图 1.23　是否创建测试 Test 模块

不需要创建，所以"Create a Test Project"不打勾，直接单击 Finish 按钮完成 Android 项目的创建和配置。

05 创建完 Android 的项目结构如图 1.24 所示。

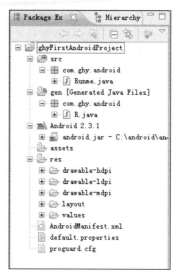

图 1.24　Android 项目的结构

先不需要了解每一个节点的具体功能。

06 右键单击 Android 项目"ghyFirstAndroidProject"，按图 1.25 所示执行"Android Application"菜单命令。

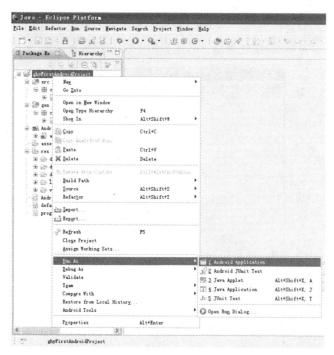

图 1.25　单击 Android Application 菜单

弹出启动状态界面如图 1.26 所示。

图 1.26　启动 Android

07　经过漫长的等待，终于出现项目的运行界面，如图 1.27 所示。

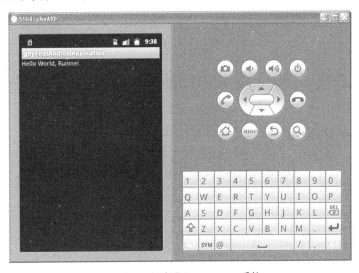

图 1.27　成功进入 Android 系统

并且在图 1.27 中看到打印出了默认的字符串"Hello World Runme!"。

现在的状态是直接在 AVD 中运行第一个项目，再来看看 Android 系统中是否成功安装了我们的第一个应用程序。

单击 图标，界面变换，出现如图 1.28 所示的界面。然后在图 1.28 中单击 按钮，来查看一下当前 Android 系统中已经被成功安装的应用程序，应用程序列表如图 1.29 所示。

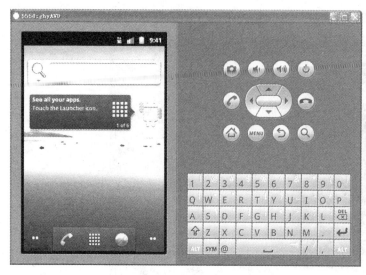

图 1.28　Android 系统默认主页　　　　　图 1.29　应用程序列表

从图 1.29 中可以看到应用程序"ghyFirstAndroidApplication",实践证明,当前的 Android 程序已经被正确地部署到 AVD 设备中。

到此,创建 Android 项目和运行 Android 程序结束。

1.6　在 Eclipse 中创建 Android 项目结构

前面创建的项目 ghyFirstAndroidProject 的文件结构如图 1.30 所示。

图 1.30　项目 ghyFirstAndroidProject 的文件结构

名称为 src 的节点定义了当前项目可以存放 java 源代码的位置，而 res 则是存放系统中所有资源的目录，assets 目录存放原始格式的文件，例如音视频文件等，这个目录中的资源不会被 R.java 文件索引，只能以流的形式进行读取，关于 R.java 文件的作用后面的章节有介绍。

在本节中将分为 7 个知识点讲解 Android 项目的文件结构及文件与文件之间的关系。

1.6.1 Runme.java 主程序文件

Android 主程序文件是在 com.ghy.android 包中的 Runme.java 文件，这是一个 Activity 对象，如图 1.31 所示。

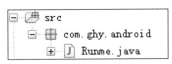

图 1.31　Runme.java 主程序文件的位置

其中 Runme.java 文件的代码如下：

```
public class Runme extends Activity {
    /** Called when the activity is first created. */
    @Override
    public void onCreate(Bundle savedInstanceState) {
        super.onCreate(savedInstanceState);
        setContentView(R.layout.main);
    }
}
```

从代码中可以看到 Runme 类继承自 Activity 父类，并且实现 onCreate() 方法，onCreate 方法有些类似构造方法，是自动执行的，onCreate 方法的知识与 Activity 生命周期有关，后面有更详细的介绍。

一个 Activity 就是一个用户界面，所以 Runme.java 文件就是这个 Android 项目用户界面的启动文件，当前项目也仅仅只有一个 Activity 文件。

代码："super.onCreate(savedInstanceState);"的功能是保存用户的界面状态，比如内存不够的情况下，系统将要销毁 Activity 对象，用户再次进入这个 Activity 对象想要把以前的输入状态恢复，这个方法就是处理这个功能的。

而代码："setContentView(R.layout.main);"的功能是设置启动用户界面布局文件唯一的 int 类型的标识，这样做可以将业务逻辑的代码与布局文件进行分离，更有助于软件的分层性设计，这点非常类似于 Flex 技术，也就是在 Android 中设计界面的文件就是一个 XML 文件，这个 XML 文件决定了界面的布局和样式等信息，而一个 Activity 类就是操作这个 XML 文件上组件的对象，通过 Activity 类就可以操作 XML 文件中的控件。

既然 Android 中的界面布局用 XML 文件，那为什么使用如下的代码来关联界面 XML 布局文件？

R.layout.main

上述代码中 main 为什么是 int 类型的呢？

下面再来继续看看 R.java 文件的代码。

1.6.2 R.java 资源索引文件

R.java 文件的位置如图 1.32 所示。

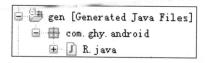

图 1.32 R.java 文件的位置

R.java 文件的代码如下：

```
package com.ghy.Android;

public final class R {
    public static final class attr {
    }
    public static final class drawable {
        public static final int icon=0x7f020000;
    }
    public static final class layout {
        public static final int main=0x7f030000;
    }
    public static final class string {
        public static final int app_name=0x7f040001;
        public static final int hello=0x7f040000;
    }
}
```

从 R.java 文件中可以看到，里面定义了项目所有资源的索引，这些资源的类型有 attr（属性）、drawable（图形）、layout（布局）和 string（字符串）资源等，其中就包括代码：

```
public static final class layout {
    public static final int main=0x7f030000;
}
```

在 layout 内置类中定义了一个常量 main，含义是一定在项目 layout 目录下存在一个名称为 main.xml 的布局文件，所以给这个 main.xml 文件定义一个唯一标识并且是自动生成的值 0x7f030000。说明在 res 目录中的所有资源必须在 R.java 文件中进行"备案"，生成每个资源的唯一索引。

目录 layout 下的 main.xml 文件如图 1.33 所示。此时，类之间的引用关系如图 1.34 所示。

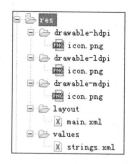

图 1.33 layout 目录存在 main.xml 文件

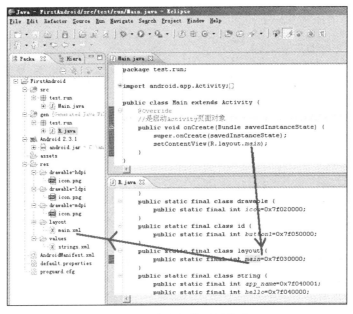

图 1.34 类之间的引用关系

从 R.java 文件中可以看到自动生成了内置类,内置类中的常量值是整个项目所有资源 res 的索引。

1.6.3 main.xml 界面布局文件

文件 main.xml 是定义界面布局的 XML 配置文件,代码如下:

```
<?xml version="1.0" encoding="utf-8"?>
<LinearLayout xmlns:android="http://schemas.android.com/apk/res/android"
    android:orientation="vertical"
    android:layout_width="fill_parent"
    android:layout_height="fill_parent"
    >
<TextView
    android:layout_width="fill_parent"
    android:layout_height="wrap_content"
    android:text="@string/hello"
    />
</LinearLayout>
```

从 main.xml 配置文件中可以看到,有一个<TextView>控件用来显示一段文本。

代码:"android:layout_width="fill_parent""的含义是这个 TextView 控件的宽度,其和父容器的宽度一样。

而代码:" android:layout_height="wrap_content""的含义是这个 TextView 控件的高度,它随着内容的多少而变化。

文本的内容定义在下述代码中：

android:text="@string/hello"

其中代码@string/hello 的功能是读取项目 values 目录下的 strings.xml 配置文件的 hello 节点值，把它当做文本内容。

文件 main.xml 和 strings.xml 的引用关系如图 1.35 所示。

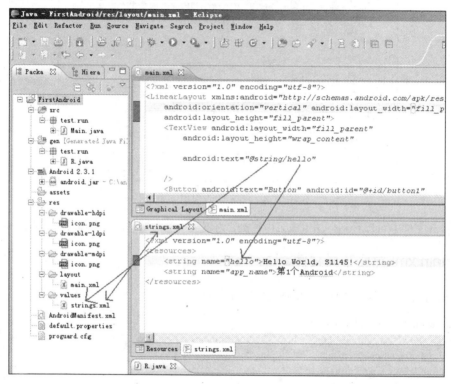

图 1.35　main.xml 和 strings.xml 的引用关系

目录 values 中的 strings.xml 文件的内容如下：

```
<?xml version="1.0" encoding="utf-8"?>
<resources>
    <string name="hello">Hello World, Runme!</string>
    <string name="app_name">ghyFirstAndroidApplication</string>
</resources>
```

文件 strings.xml 的位置如图 1.36 所示。

通过上面的 3 个知识点可以发现，Android 项目中的文件是彼此相互引用相互调用的关系，将不同功能放入不同的文件中，有利于 Android 项目结构的分层。

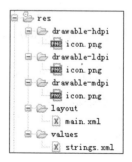

图 1.36　strings.xml 文件的位置

1.6.4 AndroidManifest.xml 应用程序配置文件

文件 AndroidManifest.xml 的功能是定义整个应用程序的公共信息，有点类似于 Web 项目中的 web.xml 文件，起整个项目配置的作用。

在 AndroidManifest.xml 这个文件中可以定义应用程序的名称、定义应用程序当前版本号等公共信息，示例代码如下：

```xml
<?xml version="1.0" encoding="utf-8"?>
<manifest xmlns:android="http://schemas.android.com/apk/res/android"
        package="com.ghy.Android"
        android:versionCode="1"
        android:versionName="1.0">
    <application android:icon="@drawable/icon" android:label="@string/app_name">
        <activity android:name=".Runme"
                android:label="@string/app_name">
            <intent-filter>
                <action android:name="android.intent.action.MAIN" />
                <category android:name="android.intent.category.LAUNCHER" />
            </intent-filter>
        </activity>

    </application>

</manifest>
```

又看到比较熟悉的代码：

```
android:label="@string/app_name"
```

也就是一定要在 strings.xml 文件中有一个 app_name 节点，代码如下：

```xml
<?xml version="1.0" encoding="utf-8"?>
<resources>
    <string name="hello">Hello World, Runme!</string>
    <string name="app_name">ghyFirstAndroidApplication</string>
</resources>
```

又是一个不同功能的文件相互引用的情况，关于该文件更详细的介绍请参看后面的内容。

1.6.5 R.java 文件的自动索引

上面已经介绍了 R.java 文件是项目资源的索引，并且里面的值是自动生成的，一起来做一个实验。

将文件 gaohongyan.png 复制到项目的 res 目录下时，R.java 文件就自动生成了 gaohongyan.png 文件的索引，如图 1.37 所示。

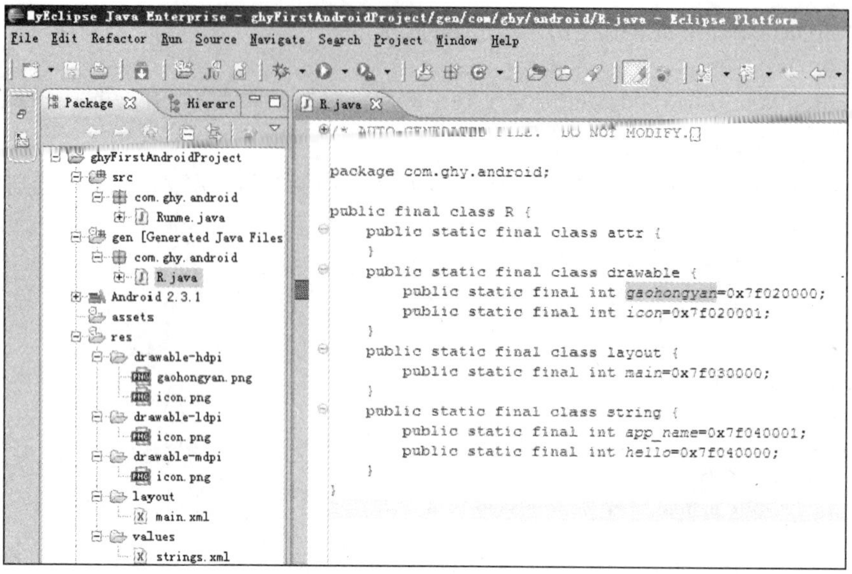

图 1.37　自动生成 gaohongyan.png 文件的索引

新生成的常量 gaohongyan 和 png 图片的主文件名称一样，并且自动赋于一个索引值，通过这个实验可以看到，R.java 文件是不需要程序员手动维护的，全是 ADT 自动生成代码的结果，由于 R.java 不需要程序员手动编辑，所以 R.java 文件是只读的。

1.6.6　AndroidManifest.xml 文件相关的知识点

AndroidManifest.xml 文件的示例代码如下：

```xml
<?xml version="1.0" encoding="utf-8"?>
<manifest xmlns:android="http://schemas.android.com/apk/res/android"
    package="com.ghy.android" android:versionCode="1"
    android:versionName="1.0">
    <application android:icon="@drawable/gaohongyan"
        android:label="@string/app_name">
        <activity android:name=".Runme"
            android:label="@string/app_name">
            <intent-filter>
                <action android:name="android.intent.action.MAIN" />
                <category
                    android:name="android.intent.category.LAUNCHER" />
            </intent-filter>
        </activity>
    </application>
</manifest>
```

上述代码中标签的功能如表 1.1 所示。

表 1.1　AndroidManfest.xml 文件的代码标签的属性及功能

标签名称	属性及功能
manifest	功能：AndroidManifest.xml 文件的根标签，包含定义应用程序基本相关信息 属性： xmlns:android：定义 android 对象的命名空间 package：定义应用程序的源代码在 src 中的路径 android:versionCode：一个内部的版本号管理机制，这个值不是给用户看的，而是给 Android 设备识别，以确定是否有新的版本要升级 android:versionName：这个值是给用户看到的版本号信息，可以来自 strings.xml 文件中的一个文本资源
application	功能：定义应用程序的基本组件信息，例如 icon、label、permission、process、taskAffinity 和 allowTaskReparenting 属性： android:icon：定义应用程序使用的图标 android:label：应用程序的名字
| activity | 功能：定义应用程序界面，是 android.app.Activity 类的子类，所有的界面 Activity 都必须使用 \<activity\> 元素进行定义，不然系统不能识别这些 Activity，并且不能在系统中显示出来<br>属性：<br>android:name：定义 Activity 类的名称，也就是当前应用程序中的 Runme.java 文件，代码如下：<br><br>```java
package com.ghy.Android;

import android.app.Activity;
import android.os.Bundle;

public class Runme extends Activity {
    /** Called when the activity is first created. */
    @Override
    public void onCreate(Bundle savedInstanceState) {
        super.onCreate(savedInstanceState);
        setContentView(R.layout.main);
    }
}
```<br><br>android:label：当前显示出来 Activity 类的标题文本内容，如果这个值没有被设置，则使用 \<application\> 标签的 android:label 属性值作为代替。另外如果 application 和 activity 标签中都有 android:label 属性，则优先使用 activity 标签中的 android:label 属性 |
| intent-filter | 功能：定义 Activity、service 或后台通知广播 broadcast 程序的意图过滤器，通过使用 intent-filter 过滤器可以对使用 Intent 请求的对象进行匹配，这样更有助于模块的解耦，intent 过滤器的功能有些类似于江河中的鱼网，符合鱼网捕捉鱼的大小就被捕捉到 |
```

（续表）

标签名称	属性及功能
action	功能：定义 action 名字 属性： android:name 属性值"android.intent.action.MAIN"定义当前的 Activity 是应用程序的启动界面。它也是<intent-filter>的子标签，参与匹配 Intent 请求的对象
category	功能：定义 category 类型 属性： android:name 属性值"android.intent.category.LAUNCHER"定义当前的应用程序在 Android 设备的应用程序列表中显示

下面的代码用于更改应用程序的图标：

```
<application android:icon="@drawable/gaohongyan"
    android:label="@string/app_name">
```

程序运行结果如图 1.38 所示。

图 1.38 更改图标的应用程序

在图 1.38 中可以看到应用程序名称"ghyFirstAnd..."的图标被更改了。

在 AndroidManifest.xml 文件中配置 Activity 对象是学习 Android 比较重要的知识点，在这里可以分为 3 种情况来进行 Activity 对象在 AndroidManifest.xml 文件的配置。

配置 Activity 对象可以使用<activity>标签的 android:name 属性和<manifest>标签的 package 属性，当然也可以选择性地使用<activity>标签的属性来进行组合配置。

（1）第一种情况：只写类名

需要在<activity>标签中的 android:name 属性来设置 Activity 对象的类名称，但类的路径必须是<manifest>标签中 package 包路径所指的路径范围，设置配置的内容如图 1.39 所示。

图 1.39　第一种配置情况

（2）第二种情况：指定完整的类路径

在 AndroidManifest.xml 文件中的<activity>标签的 android:name 属性设置完整的类路径也可以实现配置 Activity 对象。配置的代码如图 1.40 所示。

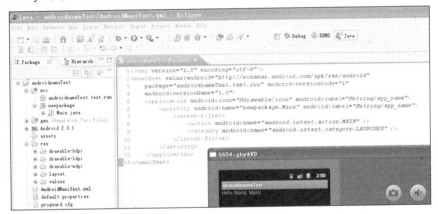

图 1.40　第二种配置情况

（3）第三种情况：多级包

Activity 对象可以在多级包的结构中进行配置，配置代码如图 1.41 所示。

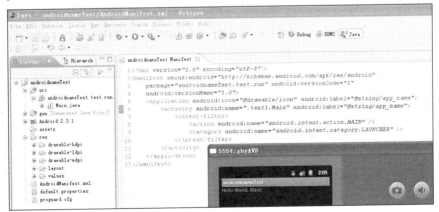

图 1.41　第三种配置情况

1.6.7　main.xml 界面布局文件

Main.xml 界面布局文件的示例代码如下：

```xml
<?xml version="1.0" encoding="utf-8"?>
<LinearLayout xmlns:android="http://schemas.android.com/apk/res/android"
    android:orientation="vertical" android:layout_width="fill_parent"
    android:layout_height="fill_parent">
    <TextView android:layout_width="fill_parent"
        android:layout_height="wrap_content" android:text="@string/hello" />
</LinearLayout>
```

上述代码中的标签属性及功能如表 1.2 所示。

表 1.2　main.xml 文件代码标签的属性及功能

标签名称	属性及功能

LinearLayout	功能：定义当前 Activity 的界面布局样式，以线性的方式从上到下排列界面中的组件，也可以从左到右进行控件的布局，也就是以一行一列的表格组织控件
	属性：
	xmlns:android：命名空间的名称
	android:orientation：设置布局中组件的排列方式，有两种值：HORIZONTAL 和 VERTICAL，不写属性默认的值是 HORIZONTAL，下面的代码也将<TextView>控件的宽度属性重新赋值为：
	`<TextView android:layout_width="wrap_content">`
	代表宽度随着内容的多少而变化
	完整示例代码如下：
	`<?xml version="1.0" encoding="utf-8"?>`
	`<LinearLayout xmlns:android="http://schemas.android.com/apk/res/android"`
	　　`android:layout_width="fill_parent" android:layout_height="fill_parent">`
	　　`<TextView android:layout_width="wrap_content"`
	　　　　`android:layout_height="wrap_content" android:text="@string/hello" />`
	　　`<TextView android:layout_width="wrap_content"`
	　　　　`android:layout_height="wrap_content" android:text="@string/hello" />`
	`</LinearLayout>`
	运行效果如下：
	（图：y268android　Hello World, Main!Hello World, Main!）
	android:layout_width 属性的值为 fill_parent，含义是当前的布局的宽度填充整个屏幕，如果值是 wrap_content 则代表宽度随着内容的多少而变化
	android:layout_height 属性的值为 fill_parent，含义是当前的布局的高度填充整个屏幕，如果值是 wrap_content 则代表高宽随着内容的多少而变化

标签名称	属性及功能

TextView	功能：定义一个显示文本的标签
	属性：
	android:layout_width 属性的值为 fill_parent，含义是当前的文本标签的宽度和父元素一样宽
	android:layout_height 属性的值为 wrap_content，含义是当前的文本标签的高度随着里面文本的高度而变化
	android:text 属性的值为@string/hello，含义是在 R.java 文件中找到 hello 变量，代码如下：
	　　public static final class string {
	　　　　public static final int *app_name*=0x7f040001;
	　　　　public static final int *hello*=0x7f040000;
	　　}
	再通过这个 hello 变量找到 strings.xml 文件中的 hello 节点中的文本，配置如下：
	<?xml version="1.0" encoding="utf-8"?>
	<resources>
	　　<string name="hello">Hello World, Runme!</string>
	　　<string name="app_name">ghyFirstAndroidApplication</string>
	</resources>

可以更改 strings.xml 文件中 hello 节点的文本，加入\r\n 回车的效果，运行结果如图 1.42 所示。

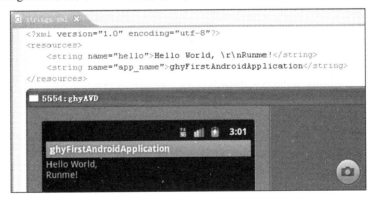

图 1.42　文本支持\r\n 回车

1.7　Log 类中的方法使用

不管开发什么样的应用程序，程序中的信息打印都对程序开发和调试阶段起到非常重要的作用，所以本节将会非常详细地介绍 Log 类中的每一个方法的使用，以使读者能非常熟练地将日志功能应用到实际的开发项目中。

1.7.1 通用日志方法

在 Android 开发中，日志输出主要使用的是 Log 类的静态方法来实现。表 1.3 所示就是 Log 类的方法列表。

表 1.3 Log 类的方法列表

ID	方法名声明	功能
1	static int v(String tag, String msg)	用 VERBOSE 方式来进行消息的打印
2	static int v(String tag, String msg, Throwable tr)	用 VERBOSE 方式来进行消息的打印并且输出异常信息
3	static int d(String tag, String msg)	用 DEBUG 方式来进行消息的打印
4	static int d(String tag, String msg, Throwable tr)	用 DEBUG 方式来进行消息的打印并且输出异常信息
5	static int i(String tag, String msg)	用 INFO 方式来进行消息的打印
6	static int i(String tag, String msg, Throwable tr)	用 INFO 方式来进行消息的打印并且输出异常信息
7	static int w(String tag, Throwable tr)	用 WARN 方式来进行异常对象的打印
8	static int w(String tag, String msg)	用 WARN 方式来进行消息的打印
9	static int w(String tag, String msg, Throwable tr)	用 WARN 方式来进行消息的打印并且输出异常信息
10	static int e(String tag, String msg)	用 ERROR 方式来进行消息的打印
11	static int e(String tag, String msg, Throwable tr)	用 ERROR 方式来进行消息的打印并且输出异常信息
12	static String getStackTraceString(Throwable tr)	取得异常对象的堆栈信息
13	static boolean isLoggable(String tag, int level)	指定标签是否可以在指定日志等级中进行输出
14	static int println(int priority, String tag, String msg)	将指定的日志消息按指定的日志等级进行输出
15	static int wtf(String tag, Throwable tr)	打印非常严重并且永远不会发生的错误异常
16	static int wtf(String tag, String msg)	打印非常严重并且永远不会发生的错误信息
17	static int wtf(String tag, String msg, Throwable tr)	打印非常严重并且永远不会发生的错误信息和异常

从上面的表格中可以看到，ID 列从 1 到 11 的方法声明的形式基本都是相同的，都是采用如下的格式：

static int 方法名称(String tag, String msg)
static int 方法名称(String tag, String msg, Throwable tr)

重点：参数 tag 的作用是消息字符串唯一标识，通常传入的值是当前 Activity 类的名称，msg 参数的值是具体的消息内容，而参数 tr 是具体出错的 Throwable 异常类的对象。

一起来做一个实验，新建名称为 LogTest 的 Android 项目，然后在 Index.java 文件中设计如下代码：

```java
public class Index extends Activity {

    private static final String TAG = "IndexActivity";

    @Override
    public void onCreate(Bundle savedInstanceState) {
        super.onCreate(savedInstanceState);
        setContentView(R.layout.main);
        Log.v(TAG, "v 方法打印出信息");
    }
}
```

在上面的代码段中，onCreate()方法是自动被系统所调用的，是自动运行的。其中使用了 Log 类的 v 方法来进行消息的打印，运行项目后在 Console 面板中并没有看到打印出来的消息 "v 方法打印出信息"。其实使用 Log 类打印出来的消息不是显示在 Console 面板中，而是显示在 ADT 插件中名称为 LogCat 的面板，单击 Eclipse 中的 Window 菜单中的 Show View 菜单中的 Other 子菜单，将 Show View 窗口显示出来，并且选中 Android 节点下的 LogCat 面板。如图 1.43 所示。

图 1.43 显示 LogCat 面板

单击 "LogCat" 选项后在 Eclipse 界面中显示出了 LogCat 面板，如图 1.44 所示。

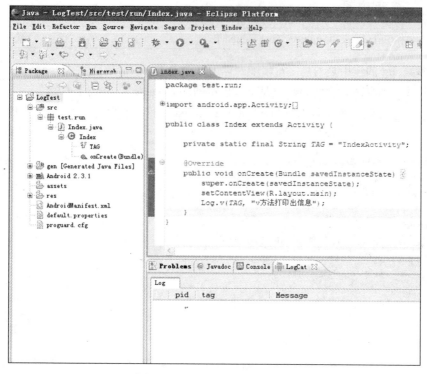

图 1.44 LogCat 面板显示出来

下面的关键步骤是启动并运行这个 LogTest 项目，来看看 LogCat 面板显示出来了哪些启动信息，但很可惜的是，直到 Android 虚拟机运行出来显示了虚拟机的界面，而 LogCat 面板中也没有任何的消息，如图 1.45 所示。

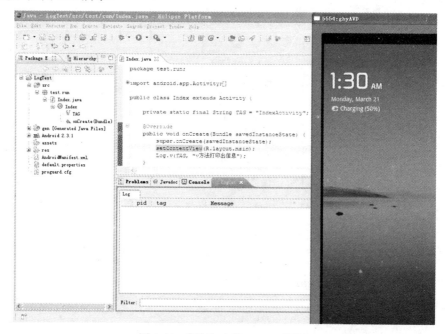

图 1.45 空空如也的 LogCat 面板

这到底是怎么回事呢？这样的情况是因为当前的 ADT 没有选中激活的 AVD 设备，造成没有监视结果，从而没有在 LogCat 面板中显示出任何信息，解决的办法其实很简单，切换 Eclipse 的透视图，变成 DDMS(Dalvik Debug Monitor Service)透视图，如图 1.46 所示。

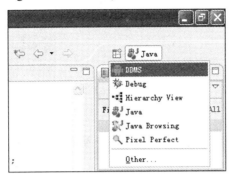

图 1.46　切换到 DDMS 透视图

在 DDMS 透视图界面中选中"Devices"面板中的 AVD 设备 emulator-5554，则自动在 LogCat 面板中显示出了相关的打印信息，如图 1.47 所示。

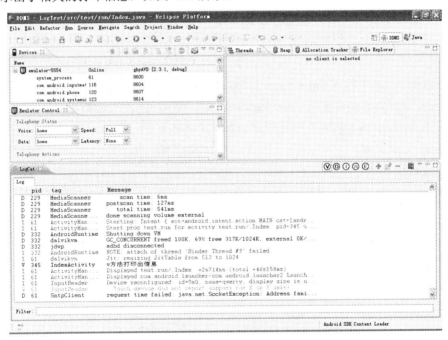

图 1.47　选中激活的 AVD 设备以显示 LogCat 中的信息

细心的读者会发现，LogCat 面板中显示出来的信息数据量从右边的滚动条就可以看出来，相当的多，如果这样，想快速地找到自己的信息是相当的麻烦和繁琐，那怎么办呢？很简单，创建一个消息过滤器 Message Filter。

单击 LogCat 面板右边的三角按钮▽，弹出菜单，选中"Create Filter"命令就可以了，如图 1.48 所示。

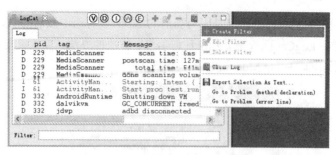

图 1.48　选中 Create Filter 子菜单

单击"Create Filter"命令后弹出如图 1.49 所示的对话框。

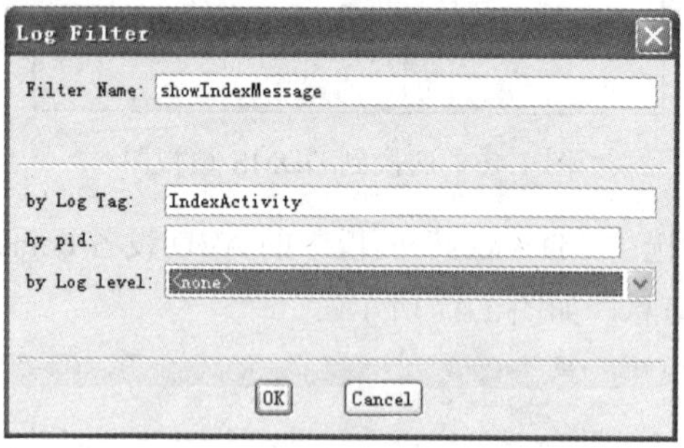

图 1.49　配置消息过滤器

在图 1.49 的对话框中输入 Filter Name 过滤器的名称：showIndexMessage，这个值可以任意写，但写的一定要有意义。还要设置 by Log Tag 的值，这个值的含义是查看指定标签名对应的消息，在这里输入 IndexActivity，因为在刚才创建的名称为 Index 的 Activity 中常量 TAG 的值就是"IndexActivity"。代码如下：

```
public class Index extends Activity {
    private static final String TAG = "IndexActivity";
```

设置完毕后单击 OK 按钮完成消息过滤器的创建，这时程序员自己的消息就被过滤出来并且显示在 LogCat 面板的子标签页中，如图 1.50 所示。

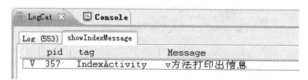

图 1.50　显示出自己的消息

这里需要注意的是，喜欢"重复运行项目"习惯的程序员，在不改变当前代码的情况下，鼠标右击 LogTest 项目并且选中"Run As"菜单中的"Android Application"命令，在 LogCat 面板中的 showIndexMessage 子标签是不重复打印信息的，但改动一下源代码如下：

```
public class Index extends Activity {
    private static final String TAG = "IndexActivity";
```

```
    @Override
    public void onCreate(Bundle savedInstanceState) {
        super.onCreate(savedInstanceState);
        setContentView(R.layout.main);
        Log.v(TAG, "v 方法打印出信息 new");
    }
}
```

从代码中可以看到仅仅改变了消息的内容，这时再用鼠标右击 LogTest 项目并且选中"Run As"菜单中的"Android Application"命令，在 LogCat 面板中的 showIndexMessage 子标签中显示出来最新的消息信息，如图 1.51 所示。

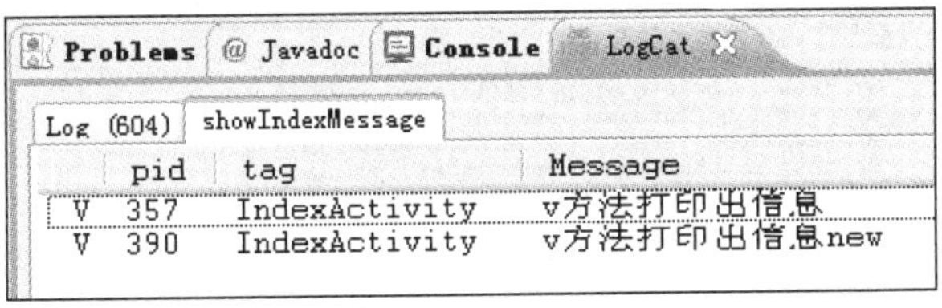

图 1.51 显示最新版的消息信息

通过这个实验不难发现，ADT 插件部属 Android 项目时要进行 Eclipse 中项目与 AVD 中应用程序代码差异的判断，如果断定 Eclipse 中的 java 代码和 AVD 中的 java 代码完全一致，也就是代码没有被改动过，则不进行进一步的部署，可以避免 AVD 设备对相同项目的重复运行。

到这一步，我们的第一个 Log 实验："Log.v(TAG, "v 方法打印出信息 new");"在经过若干个操作步骤后就结束了，也就是如下的格式：

| static int 方法名称(String tag, String msg) |
| static int 方法名称(String tag, String msg, Throwable tr) |

第 1 种写法 static int 方法名称(String tag, String msg)已经成功实现，可以看到 Eclipse 结合 ADT 插件来进行 Android 应用程序的开发是相当的方便，LogCat 也对大数据量的消息过滤得到了很好的支持。

下面再继续实现第 2 个写法：static int 方法名称(String tag, String msg, Throwable tr)。
改动代码，变成如下形式：

```
public class Index extends Activity {

    private static final String TAG = "IndexActivity";

    @Override
    public void onCreate(Bundle savedInstanceState) {
        super.onCreate(savedInstanceState);
        setContentView(R.layout.main);
        try {
            Integer.parseInt("a");
        } catch (NumberFormatException nfe) {
```

```
            Log.v(TAG, "v 方法打印出带异常的信息", nfe);
        }
    }
}
```

在代码中可以看到，使用 Log 对象的重载方法 v 带有 3 个参数的形式，运行这个项目，在 LogCat 面板中显示出来详细的出错信息，如图 1.52 所示。

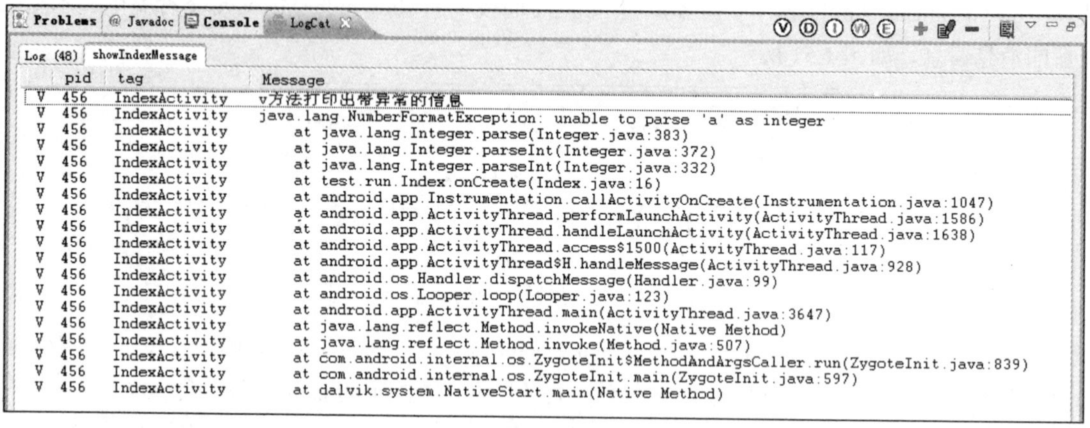

图 1.52 异常的 Log.v 方法

从图 1.52 中可以看到，ADT 很简单地打印出了详细的类型转换出错信息，并且将异常对象的堆栈信息也打印到 LogCat 面板中，使程序员通过查看堆栈信息可以快速定位出错的代码。

介绍到这，如下的两种格式相信读者已经掌握了使用规则：

| static int 方法名称(String tag, String msg) |
| static int 方法名称(String tag, String msg, Throwable tr) |

上面的示例使用的是 Log.v()方法来进行信息的打印，如果掌握了 v 方法的使用，也就掌握了 d 方法、i 方法、w 方法和 e 方法的使用，即表 1.4 中带阴影的方法的使用方式都是一致的，读者可以通过 Log.v()方法举一反三去应用。

表 1.4 方法名的声明

ID	方法名声明
1	static int v(String tag, String msg)
2	static int v(String tag, String msg, Throwable tr)
3	static int d(String tag, String msg)
4	static int d(String tag, String msg, Throwable tr)
5	static int i(String tag, String msg)
6	static int i(String tag, String msg, Throwable tr)
7	static int w(String tag, Throwable tr)

（续表）

ID	方法名声明
8	static int w(String tag, String msg)
9	static int w(String tag, String msg, Throwable tr)
10	static int e(String tag, String msg)
11	static int e(String tag, String msg, Throwable tr)
12	static String getStackTraceString(Throwable tr)
13	static boolean isLoggable(String tag, int level)
14	static int println(int priority, String tag, String msg)
15	static int wtf(String tag, Throwable tr)
16	static int wtf(String tag, String msg)
17	static int wtf(String tag, String msg, Throwable tr)

提示：表格 ID 序号为 7 的方法 static int w(String tag, Throwable tr)中只有两个参数，一个是 TAG 标志，另外一个是异常类的实例对象，和 Log.v()方法的使用非常相似，只是少了一个 msg 信息内容参数。示例代码如下：

```java
public class Index extends Activity {

    private static final String TAG = "IndexActivity";

    @Override
    public void onCreate(Bundle savedInstanceState) {
        super.onCreate(savedInstanceState);
        setContentView(R.layout.main);
        try {
            Integer.parseInt("a");
        } catch (NumberFormatException nfe) {
            Log.w(TAG, nfe);

        }

    }
}
```

程序运行后的结果如图 1.53 所示。

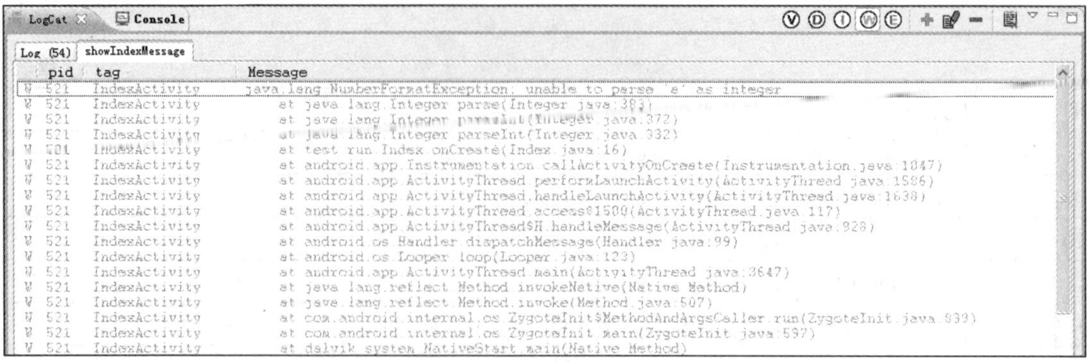

图 1.53 Log. w(String tag, Throwable tr)打印出来的信息

从图 1.53 中打印出来的信息的字体颜色可以看出，Log.w()方法打印出来的信息颜色是橙色，而 Log.v()方法打印出来的信息颜色是黑色。

1.7.2 getStackTraceString 方法的使用

方法 Log.v()、Log.d()、Log.i()、Log.w()和 Log.e()都是将信息打印到 LogCat 中，有时候需要将出错的信息插入到数据库或一个自定义的日志文件中，那么这种情况就需要将出错的信息以字符串的形式返回来，也就是使用 static String getStackTraceString(Throwable tr)方法的时候了，示例代码如下：

```java
public class Index extends Activity {

    private static final String TAG = "IndexActivity";

    @Override
    public void onCreate(Bundle savedInstanceState) {
        super.onCreate(savedInstanceState);
        setContentView(R.layout.main);
        try {
            Integer.parseInt("a");
        } catch (NumberFormatException nfe) {
            Log.w(TAG, nfe);
            String exceptionString = Log.getStackTraceString(nfe);
            Log.v(TAG, "getStackTraceString=================="
                    + exceptionString);

        }

    }
}
```

程序运行后在 LogCat 输出信息，如图 1.54 所示。

初识Android 第1章

图1.54 成功返回堆栈出错信息

程序顺利取出异常对象的堆栈信息，程序员可以将这个信息存入数据库以备项目运行时查看系统运行流程的日志。

1.7.3 v()、e()、i()、v()和w()方法的区别与isLoggable方法的使用

v()、e()、i()、v()和w()方法的区别仅仅是在LogCat打印出来的日志字体颜色不同和每个方法代表日志的严重等级不一样，从严重等级最小到最大分别是：v()、d()、i()、w()、e()。

先来实现不同颜色的效果，将Index.java 的文件代码更改如下：

```
public class Index extends Activity {

    private static final String TAG = "IndexActivity";

    @Override
    public void onCreate(Bundle savedInstanceState) {
        super.onCreate(savedInstanceState);
        setContentView(R.layout.main);

        Log.v("v----------v", "我是 VERBOSE   我是黑色的");
        Log.d("d----------d", "我是 DEBUG     我是蓝色的");
        Log.i("i----------i", "我是 INFO      我是绿色的");
        Log.w("w----------w", "我是 WARN      我是橙色的");
        Log.e("e----------e", "我是 ERROR     我是红色的");
```

        }
    }

运行一下,出现如图1.55所示的效果。

图1.55　i()、d()、e()、w()、e()方法打印不同颜色

再来实现日志等级的实验,这个实验非常有趣。

所谓的日志等级就是程序员可以设置一个日志 TAG 的等级,这样就可以以更改配置文件的方式来确定某一个日志是否进行输出,比如如下代码:

```java
public class Index extends Activity {

    private static final String TAG = "IndexActivity";

    @Override
    public void onCreate(Bundle savedInstanceState) {
        super.onCreate(savedInstanceState);
        setContentView(R.layout.main);

        Log.v(TAG, "" + Log.isLoggable(TAG, Log.VERBOSE));
        Log.d(TAG, "" + Log.isLoggable(TAG, Log.DEBUG));
        Log.i(TAG, "" + Log.isLoggable(TAG, Log.INFO));
        Log.w(TAG, "" + Log.isLoggable(TAG, Log.WARN));
        Log.e(TAG, "" + Log.isLoggable(TAG, Log.ERROR));

    }
}
```

程序运行后在 LogCat 面板中显示打印出来的信息,如图1.56所示。

图1.56　默认 INFO 级别

从图1.56中打印出来的信息可以看到,默认允许进行日志输出的等级是 INFO,INFO 之上的等级包括 INFO,还有 WARN 和 ERROR 都是允许进行日志的输出的,从图1.56中的第1列就可以看到日志级别的简写为 V,D,I,W,E。而且在 Android 的规范中也明确表示,如果调用 Log 类中的任意输出日志的方法之前都要进行一下日志输出等级的判断,还有 VERBOSE 类型的日志也就是使用 v()方法只允许在开发中存在,永远不能编译进应用程序,而 DEBUG 类型的日志也就是使用 d()方法在程序发布阶段应该从代码中删除,而 WARN(w()方法)、INFO(i()方法)和 ERROR(e()方法)可以一直保留在程序的代码中。

如果在程序的开发阶段，应用程序中有如下使用 Log.w()方法的代码：

```
public class Index extends Activity {

    private static final String TAG = "IndexActivity";

    @Override
    public void onCreate(Bundle savedInstanceState) {
        super.onCreate(savedInstanceState);
        setContentView(R.layout.main);

        Log.w(TAG, "a" + "b" + "c" + "d" + "e");

    }

}
```

则在 LogCat 也会打印出来，因为打印的方法并不与系统日志等级有关，不管什么方法都会输出打印日志，如图 1.57 所示。

```
W  466    IndexActivity              abcde
```

图 1.57　打印 WARN 日志

这时假设程序员对 Index.java 类的设计已经完毕，没有任何的出错 DEBUG，这个 Index.java 类是一个程序的成品，不需要程序员再进行二次改动和代码的维护，但运行这个 Index.java 文件时还在 LogCat 中打印出日志信息，这样频繁的 IO 操作对应用系统的运行效率起到了非常慢的结果，这时就可以定制这个 TAG 标记的日志输出等级，那么如何更改呢？很简单，在桌面创建一个名称为 local.prop 文件，内容为：

```
log.tag.IndexActivity=WARN
```

上面代码的功能是定义名称为 IndexActivity 的 TAG 的日志等级变成 WARN 级别，那么还需要改动 Index.java 文件中的代码，加入判断当前日志等级的功能，代码如下：

```
public class Main extends Activity {
    private final static String TAG = "IndexActivity";

    @Override
    public void onCreate(Bundle savedInstanceState) {
        super.onCreate(savedInstanceState);
        setContentView(R.layout.main);

        Log.v(TAG, "" + Log.isLoggable(TAG, Log.VERBOSE));
        Log.d(TAG, "" + Log.isLoggable(TAG, Log.DEBUG));
        Log.i(TAG, "" + Log.isLoggable(TAG, Log.INFO));
        Log.w(TAG, "" + Log.isLoggable(TAG, Log.WARN));
        Log.e(TAG, "" + Log.isLoggable(TAG, Log.ERROR));

        if (Log.isLoggable(TAG, Log.WARN)) {
```

```
            Log.e(TAG, "当前的日志等级为WARN");
        }
    }
}
```

并且还要将桌面的那个 local.prop 文件添加进 Android 模拟器的 data 文件夹中。选中 data 文件夹，如图 1.58 所示，然后依下面的步骤操作。

添加完成 local.prop 文件的模拟器内容如图 1.59 所示。

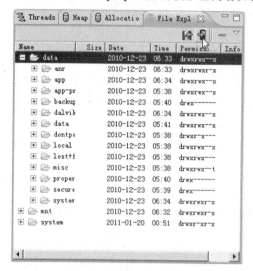

图 1.58　单击添加 local.prop 文件的按钮　　　图 1.59　添加完 local.prop 文件模拟器 data 目录

最重要的步骤就是将当前的模拟器关掉，然后重启模拟器，以将 data 目录中的 local.prop 文件读入内存，在 LogCat 面板中打印的效果如图 1.60 所示。

图 1.60　标签 IndexActivity 不支持 INFO、DEBUG 和 VERBOSE 日志级的打印

介绍到这，Log 类就剩以下 4 个方法还未学习。

14	static int println(int priority, String tag, String msg)
15	static int wtf(String tag, Throwable tr)
16	static int wtf(String tag, String msg)
17	static int wtf(String tag, String msg, Throwable tr)

这 4 个方法中也都是与日志打印有关，其中 println()方法是将指定的 TAG 对应的消息用指定

日志等级进行输出，wtf()方法也是将信息以日志的方式进行输出，与前面日志方法含义不同的是，wtf()方法打印的是一些应该从未发生过的异常，或永远不会发生的异常，wtf()方法与前面介绍过的日志方法参数类型一致，所以在此不再用代码的方式举例。最后看一下 static int println(int priority, String tag, String msg)方法的使用，代码如下：

```java
public class Index extends Activity {

    private static final String TAG = "IndexActivity";

    @Override
    public void onCreate(Bundle savedInstanceState) {
        super.onCreate(savedInstanceState);
        setContentView(R.layout.main);

        Log.println(Log.VERBOSE, TAG, "出错了，颜色是黑色");
        Log.println(Log.DEBUG, TAG, "出错了，颜色是蓝色");
        Log.println(Log.INFO, TAG, "出错了，颜色是绿色");
        Log.println(Log.WARN, TAG, "出错了，颜色是橙色");
        Log.println(Log.ERROR, TAG, "出错了，颜色是红色");

    }
}
```

程序运行效果如图 1.61 所示。

```
V  488    IndexActivity        出错了，颜色是黑色
D  488    IndexActivity        出错了，颜色是蓝色
I  488    IndexActivity        出错了，颜色是绿色
W  488    IndexActivity        出错了，颜色是橙色
E  488    IndexActivity        出错了，颜色是红色
```

图 1.61　println()方法打印出来的效果

1.8　文件夹 res 中更多的资源类型

文件夹 res 中的 XML 文件可以分成很多类型，每一种类型也使用不同的 XML 文件名称及放入不同名称的文件夹中，具体细节如表 1.5 所示。

表 1.5　XML 文件列表

XML 文件类型	存放文件夹	建议 XML 文件名	使用示例
字符串	res/values	strings.xml	<resources> 　　<string name="hello">：）</string> </resources>

（续表）

XML 文件类型	存放文件夹	建议 XML 文件名	使用示例

字符串数组	res/values	arrays.xml	`<resources>`
			　　`<string-array name="myStringArray">`
			　　　　`<item>A</item>`
			　　　　`<item>B</item>`
			　　　　`<item>C</item>`
			　　　　`<item>D</item>`
			　　`</string-array>`
			`</resources>`

颜色	res/values	color.xml	`<resources>`
			　　`<color name="myColor1">#ff0000</color>`
			　　`<color name="myColor2">#00ff00</color>`
			`</resources>`

大小	res/values	dimens.xml	`<resources>`
			　　`<dimen name="mydimens1">13px</dimen>`
			　　`<dimen name="mydimens2">13pt</dimen>`
			`</resources>`

布局文件	res/layout	自定义 xml 文件名	根据不同布局使用不同的标签
被编译的 xml	res/xml	自定义 xml 文件名	自定义标签名称
原始 xml	res/raw	自定义 xml 文件名	自定义标签名称

外观样式	res/values	styles.xml	`<resources>`
			　　`<style name="ghyStyle1">`
			　　　　`<item name="android:background">#ff0000</item>`
			　　`</style>`
			`</resources>`

图片文件	res/drawable-hdpi res/drawable-ldpi res/drawable-mdpi	png、jpg、gif、bmp 等图片格式	
动画	res/anim	自定义 xml 文件	`<set>`、`<alpha>`、`<scale>`、`<translate>`、`<rotate>`

(续表)

XML 文件类型	存放文件夹	建议 XML 文件名	使用示例

带颜色的简单 drawable 图形	res/values	drawables.xml	\<resources\>
			  \<drawable name="myDrawable1"\>#00ffff\</drawable\>
			  \<drawable name="myDrawable2"\>#ffffff\</drawable\>
			\</resources\>

1.9 常用资源的读取操作

提前注意一下：在 Android 中各种资源文件是不允许被放入任意名称的文件夹中的，一定要放入指定名称的文件夹中，但不排除 Android 版本更新的情况。

前面介绍了下述代码：

```
<TextView android:layout_width="fill_parent"
    android:layout_height="wrap_content" android:text="@string/hello" />
```

其中标签\<TextView\>的属性 android:text 的值是来自于当前项目 R.java 文件中名称为 string 的静态类中的 hello 常量所指向资源 strings.xml 文件同名的 XML 节点，对应关系如图 1.62 所示。

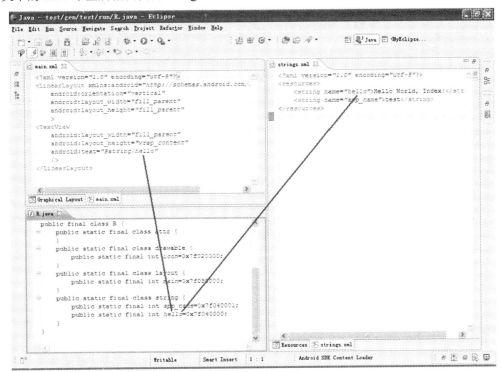

图 1.62 各个文件的引用关系

那如果想自定义一个类似 strings.xml 文件来放自己的资源文本该如何操作呢？很简单，创建它，一切都是自动化的。

在 values 目录下创建一个名称为 ghyText.xml 的文件并且编辑其内容，如图 1.63 所示。

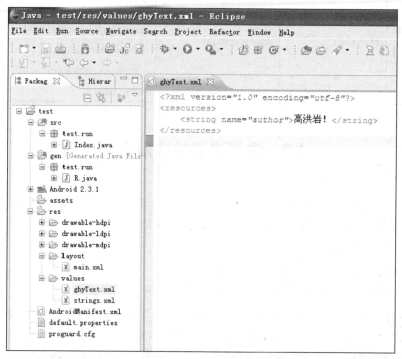

图 1.63　创建 ghyText.xml 文件添加文本资源内容

保存这个文件后，则自动在 R.java 文件中生成这个 author 文本资源的索引，如图 1.64 所示。

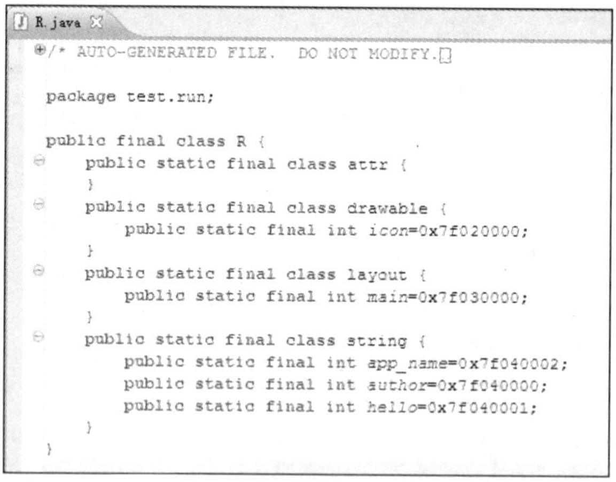

图 1.64　自动生成 author 文本资源的索引

再把 layout 目录中的 main.xml 文件的<TextView>标签的 android:text 属性代码改成如图 1.65 所示。

```xml
<?xml version="1.0" encoding="utf-8"?>
<LinearLayout xmlns:android="http://schemas.android.com/apk/res/android"
    android:orientation="vertical" android:layout_width="fill_parent"
    android:layout_height="fill_parent">
    <TextView android:layout_width="fill_parent"
        android:layout_height="wrap_content"
        android:text="@string/author" />
</LinearLayout>
```

图 1.65　使用新文本资源中的文本内容

再运行这个项目，在虚拟机中看到了使用最新文本资源 ghyText.xml 中的 author 节点的文本，如图 1.66 所示。

图 1.66　应用最新文本资源文件示例成功

这是使用<TextView>标签的 android:text 属性自动读出 ghyText.xml 文件中的内容，那如何实现用代码读取呢？很简单！新建名称为 readResource 的 Android 项目，创建 XML 文件，全部的 XML 资源文件及代码如图 1.67 所示。

图 1.67　全部 6 个资源 XML 文件及代码

文件 Main.java 的代码如下：

```java
public class Main extends Activity {
    private TextView textView2;
    private TextView textView3;
    private TextView textView4;

    @Override
    public void onCreate(Bundle savedInstanceState) {
        super.onCreate(savedInstanceState);
        setContentView(R.layout.main);

        // 操作颜色 color-半透明
        textView2 = (TextView) this.findViewById(R.id.textView2);
        int ghyColor = this.getResources().getColor(R.color.ghyColor);
        textView2.setBackgroundColor(ghyColor);
        Log.v("ghyColor=", "" + ghyColor);

        // 操作字符串 string-不带格式
        Log.v("ghyString=", ""
                + this.getResources().getString(R.string.ghyString));

        // 操作大小 dimens
        textView3 = (TextView) this.findViewById(R.id.textView3);
        float ghyDimens = this.getResources().getDimension(R.dimen.ghyDimens);
        textView3.setTextSize(ghyDimens);
        Log.v("ghyDimens=", "" + ghyDimens);

        // 操作 drawable
        Drawable ghyDrawable = this.getResources().getDrawable(
                R.drawable.ghyDrawable);
        textView4 = (TextView) this.findViewById(R.id.textView4);
        textView4.setBackgroundDrawable(ghyDrawable);

        // 操作字符串数组 array
        String[] ghyArray = this.getResources()
                .getStringArray(R.array.ghyArray);
        for (int i = 0; i < ghyArray.length; i++) {
            Log.v("!", "" + ghyArray[i]);
        }

    }
}
```

程序运行后的效果及 LogCat 打印日志如图 1.68 所示。

图 1.68　运行效果及 LogCat 的日志内容

1.10　Activity 的生命周期

每一种技术都有其生命周期，就像 Java EE 技术中的 Servlet 生命周期就分为 4 步，即实例化、初始化、服务和销毁，当然 Android 系统中的"窗体对象"Activity 也不例外，也有其自己的生命周期过程。

那么在 Android 中，进程的生命周期大多数时候是由系统管理的，但由于手机应用的一些特殊性，所以我们需要更多的去关注各个 Android Component 控件运行时的生命周期模型，其实所谓手机应用的特殊性主要是指以下两点：

（1）手机应用的大多数情况下只能在手机上看到一个程序的一个界面，用户除了通过程序界面上的功能按钮来在不同的窗体间切换，还可以通过 Back 键和 Home 键来返回上一个窗口，而用户使用 Back 或者 Home 键的时机是非常不确定的，任何时候用户都可以使用 Home 或 Back 来强行切换当前的界面。

（2）往往手机上一些特殊的事件发生也会强制地改变当前用户所处的操作状态，例如无论任何情况，在手机来电时，系统都会优先显示电话接听界面等这类的情况。

Activity 有 4 种本质区别的状态：

（1）Activity 在屏幕的最上方，称为活动状态或激活状态

（2）如果 Activity 失去焦点，但依然可见（比如弹出一个非全屏半透明的对话框）称为暂停状态（Paused）。

（3）如果 Activity 被另外一个 Activity 完全覆盖遮挡掉，称为停止状态（Stopped）

（4）如果 Activity 是 Paused 或 Stopped 状态时，由于内存不够等情况下系统可以随时销毁这些 Activity，称为销毁状态。

本小节将要创建具有两个 Activity 对象的 Android 项目，创建这个项目的目的是先对 Android 切换界面（切换 Activity 对象）进行一个热身，了解一下 Android 如何切换页面。

项目名称为 TwoActivity，创建项目的详细信息如图 1.69 所示。

图 1.69 创建 TwoActivity 明细

在图 1.69 中单击 Finish 按钮完成 TwoActivity 项目的创建，在项目中只有 1 个名称为 Index 的 Activity 对象，将这个名称为 Index.java 的 Activity 对象所对应的布局文件 main.xml 的代码更改如下：

```xml
<?xml version="1.0" encoding="utf-8"?>
<LinearLayout xmlns:android="http://schemas.android.com/apk/res/android"
    android:orientation="vertical" android:layout_width="fill_parent"
    android:layout_height="fill_parent">
    <TextView android:layout_width="fill_parent"
        android:layout_height="wrap_content" android:text="this is one page!" />
</LinearLayout>
```

为了有交互的效果，还要在 main.xml 文件中加入一个切换到第 2 个 Activity 的按钮，这时可以在 ADT 提供的"界面设计模式"进行添加，单击 main.xml 文件下方的"Graphical Layout 标签页"切换到设计模式下，标签页位置如图 1.70 所示。

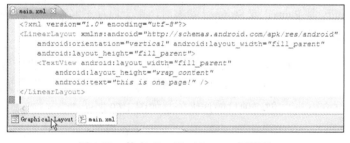

图 1.70 单击 Graphical Layout 标签页

切换到设计模式后显示出 Android 常用的界面控件，如图 1.71 所示。

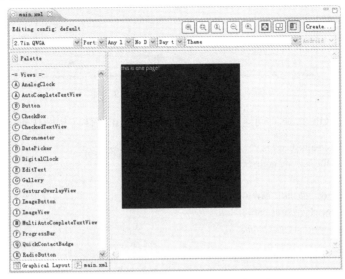

图 1.71 Android 支持的界面控件列表

这时可以用鼠标选中左边的 Button 按钮，然后再将这个 Button 按钮拖拽到右边设计界面"this is on page!"文本的下方，完成后效果如图 1.72 所示。

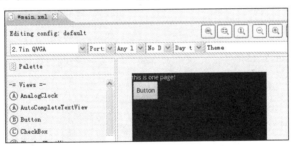

图 1.72 成功添加 Button 控件

更改 main.xml 文件中新添加这个 Button 的 id 属性值为@+id/button1_1，代码如下：

```
<?xml version="1.0" encoding="utf-8"?>
<LinearLayout xmlns:android="http://schemas.android.com/apk/res/android"
    android:orientation="vertical" android:layout_width="fill_parent"
    android:layout_height="fill_parent">
    <TextView android:layout_width="fill_parent"
        android:layout_height="wrap_content" android:text="this is one page!" />
    <Button android:text="Button" android:id="@+id/button1_1"
        android:layout_width="wrap_content" android:layout_height="wrap_content"></Button>
</LinearLayout>
```

由于本示例要实现的是两个 Activity 对象，则必须再创建 1 个 Activity 类，通常笔者不喜欢用菜单的方式来新建 1 个 Class，再设置这个 Class 的父类是 Activity，为了方便，经常使用的是 COPY 复制法，复制 Index.java 文件，再粘贴，重命名 java 类文件名称，变成 Index2.java，如图 1.73 所示。

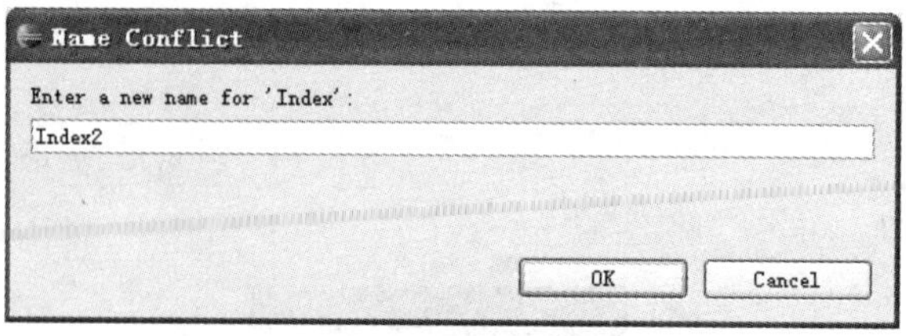

图 1.73 重命名 Index2.java 文件名

单击图 1.73 中的 OK 按钮后创建一个名称为 Index2.java 文件，初始代码如下：

```java
public class Index2 extends Activity {
    /** Called when the activity is first created. */
    @Override
    public void onCreate(Bundle savedInstanceState) {
        super.onCreate(savedInstanceState);
        setContentView(R.layout.main);
    }
}
```

其实在 Index2.java 文件中代码是错误的，因为 Index2.java 还在使用 main.xml 布局文件，如代码是：

```java
setContentView(R.layout.main);
```

更改代码如下：

```java
public class Index2 extends Activity {
    @Override
    public void onCreate(Bundle savedInstanceState) {
        super.onCreate(savedInstanceState);
        setContentView(R.layout.index2);
    }
}
```

由于 Index2.java 文件指定使用 index2.xml 这个名称的布局文件，所以还要使用"复制法"在 layout 目录下再创建一个名称为 index2.xml 文件，并且改动 index2.xml 文件的代码如下：

```xml
<?xml version="1.0" encoding="utf-8"?>
<LinearLayout xmlns:android="http://schemas.android.com/apk/res/android"
    android:orientation="vertical" android:layout_width="fill_parent"
    android:layout_height="fill_parent">
    <TextView android:layout_width="fill_parent"
        android:layout_height="wrap_content" android:text="this is two page!" />
    <Button android:text="Button" android:id="@+id/button2_1"
        android:layout_width="wrap_content" android:layout_height="wrap_content"></Button>
</LinearLayout>
```

改动了两处：

（1）android:text 的属性值变成"this is two page!"，标明当前的 Activity 界面是 index2.xml。

（2）<Button>标签的 android:id 属性改成"@+id/button2_1",标明当前按钮是 index2.xml 界面的 1 个 button,符号@+id 代表/符号后面的字符 button2_1 要在 R.java 文件中进行注册 id 资源,这时在 R.java 文件中生成这两个 Button 按钮的索引,代码如图 1.74 所示。

图 1.74　生成两个 Button 的索引

改动 Index.java 文件的代码如下：

```
public class Index extends Activity {
    /** Called when the activity is first created. */
    @Override
    public void onCreate(Bundle savedInstanceState) {
        super.onCreate(savedInstanceState);
        setContentView(R.layout.main);

        Button button1_1 = (Button) this.findViewById(R.id.button1_1);

        button1_1.setOnClickListener(new OnClickListener() {

            public void onClick(View arg0) {

                Intent intentRef = new Intent();
                intentRef.setClass(Index.this, Index2.class);
                startActivity(intentRef);
                Index.this.finish();
            }
        });
    }
}
```

方法 findViewById()是根据控件的 id 来找到对象,而 Button 对象的方法 setOnClickListener 是设置按钮单击事件的监听器。

类 Intent 的作用是使两个 Activity 对象之间能互相切换,也就是换界面的效果,方法 setClass()第 1 个参数指的是起始 Activity 对象,第 2 个参数是欲到达目的地的 Activity 对象。传递 Intent 对象使用的是 startActivity()方法。

改动 Index2.java 文件代码如下：

```java
public class Index2 extends Activity {
    /** Called when the activity is first created. */
    @Override
    public void onCreate(Bundle savedInstanceState) {
        super.onCreate(savedInstanceState);

        setContentView(R.layout.index2);// *******重点
        Button button2_1 = (Button) this.findViewById(R.id.button2_1);

        button2_1.setOnClickListener(new OnClickListener() {

            public void onClick(View arg0) {

                Intent intentRef = new Intent();
                intentRef.setClass(Index2.this, Index.class);
                startActivity(intentRef);
                Index2.this.finish();
            }
        });

    }
}
```

关键代码是：

```
setContentView(R.layout.index2);
```

用于将 1 个 Activity 文件和 1 个 Layout 布局文件进行关联，还需要在 AndroidManifest.xml 文件中手动改动代码，加入注册 Index2.java 文件的功能，也就是项目中所有的 Activity 对象都必须在这个配置文件中进行注册，完整代码如下：

```xml
<?xml version="1.0" encoding="utf-8"?>
<manifest xmlns:android="http://schemas.android.com/apk/res/android"
      package="test.run" android:versionCode="1" android:versionName="1.0">


    <application android:icon="@drawable/icon" android:label="@string/app_name">
        <activity android:name=".Index" android:label="@string/app_name">
            <intent-filter>
                <action android:name="android.intent.action.MAIN" />
                <category android:name="android.intent.category.LAUNCHER" />
            </intent-filter>
        </activity>

        <activity android:name=".Index2" android:label="@string/app_name">
        </activity>
```

```
</application>
</manifest>
```

运行这个项目，显示初始界面，如图 1.75 所示。

单击 main.xml 界面中的 Button 按钮切换到 index2.xml 界面，如图 1.76 所示。

图 1.75　显示 Index 界面

图 1.76　切换到 Index2 界面

再单击 index2.xml 界面中的 Button 按钮，则切换到 main.xml 界面布局。

到此我们已经将测试 Activity 对象生命周期的基本环境和关键代码介绍完毕，并且完全可以在 Android 虚拟机中实现切换 Activity 界面的操作了，切换 Activity 界面的操作就是掌握 Android 生命周期的基础。

下面开始进入测试生命周期。先看如图 1.77 所示，在这张图中我们要根据图中的步骤重现主要的生命周期过程。

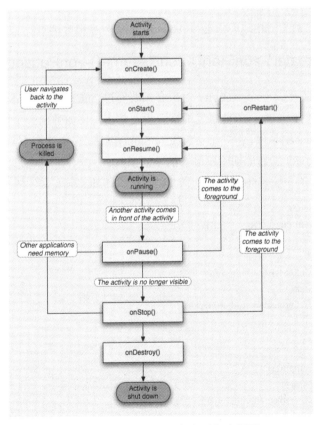
图 1.77　Activity 对象生命周期流程图

从图 1.77 中可以看到有 7 个事件：

- onCreate()：当 Activity 第一次被创建时调用，可以在这个方法中绑定数据或创建其他的视图控件，其中应该注意的问题是，覆写 onCreate()方法时尽量将当前的 Activity 状态保存进系统，以备以后再使用这个 Activity 时保存以前界面的状态，保存状态的代码如下：

```
public void onCreate(Bundle savedInstanceState) {
super.onCreate(savedInstanceState);
```

- onStart()：当 Activity 变为用户可见之前调用。
- onResume()：当 Activity 可以与用户交互之前调用，也就是 Activity 对象到达 Activity 栈的顶部即将成为前台进程时被调用。
- onPause()：当系统调用其他 Activity 对象时调用，可以在这个方法中将当前 Activity 对象没有保存的数据保存到持久化对象中，也可以在这个方法中结束比较耗费 CPU 时间的操作，比如动画之类的。用这个方法写的代码要尽量效率高一些，如果这个方法没有执行完，新的 Activity 对象将不会显示出来，会影响客户的体验性，也就是新的 Activity 对象必须要等待 onPause()方法执行完毕后再显示出来。大多数情况下，在 onPause()方法中关闭 onResume()中打开的资源。
- onStop()：当 Activity 不可视时调用。
- onDestroy()：当销毁 Activity 对象时调用。
- onRestart()：当处于 onStop()状态的 Activity 又变为可视时调用。

1.10.1 实现 onCreate()->onStart()->onResume()->onPause()->onResume

从图 1.77 中可以看到，一个 Activity 对象必须要经过 onCreate()、onStart()、onResume()这 3 个生命周期，其中第 4 步 onPause()被触发的情况也非常多，比如弹出一个以 Activity 作为对话框的示例，就会触发 onPause()事件，但并不会触发 onStop()事件，在本示例中实现全部 7 个生命周期，目的是全面地监控 Activity 生命周期阶段触发的不同函数。

下面在 Index.java 类中添加生命周期回调方法，更改 Index.java 文件的代码如下：

```
public class Index extends Activity {
    private static final String TAG = "IndexActivity";

    @Override
    protected void onStart() {
        super.onStart();
        Log.v(TAG, "protected void onStart()");
    }

    @Override
    protected void onPause() {
        super.onPause();
        Log.v(TAG, "protected void onPause()");
    }

    @Override
    protected void onResume() {
```

```java
        super.onResume();
        Log.v(TAG, "protected void onResume()");
    }

    @Override
    protected void onStop() {
        super.onStop();
        Log.v(TAG, "protected void onStop()");
    }

    @Override
    protected void onDestroy() {
        super.onDestroy();
        Log.v(TAG, "protected void onDestroy()");
    }

    @Override
    protected void onRestart() {
        super.onRestart();
        Log.v(TAG, "protected void onRestart()");
    }

    /** Called when the activity is first created. */
    @Override
    public void onCreate(Bundle savedInstanceState) {
        super.onCreate(savedInstanceState);
        setContentView(R.layout.main);

        Log.v(TAG, "protected void onCreate()");

        Button button1_1 = (Button) this.findViewById(R.id.button1_1);

        button1_1.setOnClickListener(new OnClickListener() {

            public void onClick(View arg0) {

                Intent intentRef = new Intent();
                intentRef.setClass(Index.this, Index2.class);
                startActivity(intentRef);
            }
        });

    }
}
```

还要更改 AndroidManifest.xml 文件中定义 Index2.java 的选项，更改代码如下：

```xml
<?xml version="1.0" encoding="utf-8"?>
<manifest xmlns:android="http://schemas.android.com/apk/res/android"
    package="test.run" android:versionCode="1" android:versionName="1.0">
```

```xml
<application android:icon="@drawable/icon" android:label="@string/app_name">
    <activity android:name=".Index" android:label="@string/app_name">
        <intent-filter>
            <action android:name="android.intent.action.MAIN" />
            <category android:name="android.intent.category.LAUNCHER" />
        </intent-filter>
    </activity>

    <activity android:name=".Index2" android:label="@string/app_name"
        android:theme="@android:style/Theme.Dialog">
    </activity>

</application>
</manifest>
```

加入 android:theme="@android:style/Theme.Dialog"属性代表 index2.xml 布局文件是一个对话框，不会添充满屏，而 main.xml 布局界面还会以背景的方式在后面显示出来。

相应的，还要更改 Index2.java 文件的代码如下：

```java
public class Index2 extends Activity {
    private static final String TAG = "Index2Activity";

    @Override
    protected void onStart() {
        super.onStart();
        Log.v(TAG, "protected void onStart()");
    }

    @Override
    protected void onPause() {
        super.onPause();
        Log.v(TAG, "protected void onPause()");
    }

    @Override
    protected void onResume() {
        super.onResume();
        Log.v(TAG, "protected void onResume()");
    }

    @Override
    protected void onStop() {
        super.onStop();
        Log.v(TAG, "protected void onStop()");
    }
```

```java
@Override
protected void onDestroy() {
    super.onDestroy();
    Log.v(TAG, "protected void onDestroy()");
}

@Override
protected void onRestart() {
    super.onRestart();
    Log.v(TAG, "protected void onRestart()");
}

/** Called when the activity is first created. */
@Override
public void onCreate(Bundle savedInstanceState) {
    super.onCreate(savedInstanceState);
    Log.v(TAG, "protected void onCreate()");

    setContentView(R.layout.index2);// *******重点
    Button button2_1 = (Button) this.findViewById(R.id.button2_1);

    button2_1.setOnClickListener(new OnClickListener() {

        public void onClick(View arg0) {
            Index2.this.finish();
        }
    });

  }
}
```

代码 Index2.**this**.finish();的功能是将当前的 Activity 对象关闭。

运行这个 Android 项目，在 LogCat 面板中显示如图 1.78 所示的结果。

```
D  437  jdwp            Got wake-up signal, bailing out
D  437  dalvikvm        Debugger has detached; object r
I  445  jdwp            Ignoring second debugger -- acc
V  445  IndexActivity   protected void onCreate()
V  445  IndexActivity   protected void onStart()
V  445  IndexActivity   protected void onResume()
I  61   ActivityManager Displayed test.run/.Index: +1s5
D  136  dalvikvm        GC_EXPLICIT freed 139K, 52% fre
D  244  dalvikvm        GC_EXPLICIT freed 6K, 54% free
```

图 1.78　执行了 Index.java 文件的 3 个生命周期函数

方法 onPause()没有执行！如何触发它呢？很简单，单击 main.xml 布局文件中的第一个 Button 按钮，以对话框显示出 index2.xml，界面就触发了，如图 1.79 所示。

单击 Button 后在控制台打印出了标签名称为 IndexActivity 的 onPause()方法被调用的信息，如图 1.80 所示。

图 1.79　单击 main.xml 布局中的 Button 按钮　　　图 1.80　执行 Index.java 的 onPause 方法

并且在 Android 虚拟机中出现了一个对话框，如图 1.81 所示。当单击 index2.xml 布局中的 Button 按钮时，再返回 main.xml 界面，在 LogCat 打印出了如图 1.82 所示的结果。

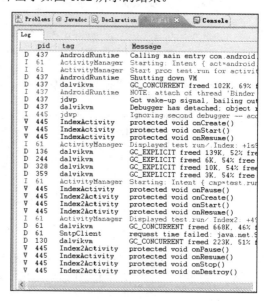

图 1.81　单击 index2.xml 布局文件的 Button 按钮　　　图 1.82　重回 main.xml 时执行的生命周期过程

从图 1.82 中可以看到，系统触发了 Index.java 文件的 onResume()方法。

本示例是以 main.xml 作为主界面，而 index2.xml 作为对话框界面的生命周期过程，其中 main.xml 文件不要从内存中删除，也就是在 Index.java 文件中不要有 Index.**this**.finish();代码。另外 main.xml 布局文件也一直在屏幕上显示，但仅仅是以背景的方式进行显示。

1.10.2　实现 onCreate()->onStart()->onResume()->onPause()->onStop()-> onRestart()->onStart()

在上一小节中实现的仅仅是如下的过程：

onCreate()->onStart()->onResume()->onPause()->onResume

从图 1.77 中可以看到,其实 onStop()事件也有一个分支,那就是 onRestart()方法。触发 onStop()方法的时机是 Activity 不再显示的时候,当执行 onStop()方法再显示的时候将会触发 onRestart()方法,然后再从 onStart()方法按生命周期顺序执行下去。

在 main.xml 界面中加入一个新的 Button2 按钮,新代码如下:

```xml
<?xml version="1.0" encoding="utf-8"?>
<LinearLayout xmlns:android="http://schemas.android.com/apk/res/android"
    android:orientation="vertical" android:layout_width="fill_parent"
    android:layout_height="fill_parent">
    <TextView android:layout_width="fill_parent"
        android:layout_height="wrap_content" android:text="this is one page!" />
    <Button android:text="Button" android:id="@+id/button1_1"
        android:layout_width="wrap_content" android:layout_height="wrap_content"></Button>
    <Button android:text="Button2" android:id="@+id/button1_2"
        android:layout_width="wrap_content" android:layout_height="wrap_content"></Button>
</LinearLayout>
```

在 Index.java 文件中的 onCreate()方法中追加如下代码:

```java
Button button1_2 = (Button) this.findViewById(R.id.button1_2);
button1_2.setOnClickListener(new OnClickListener() {

    public void onClick(View arg0) {
        Intent intentRef = new Intent();
        intentRef.setClass(Index.this, Index3.class);
        startActivity(intentRef);
    }
});
```

还要新建名称为 Index3.java 的 Activity 对象,代码如下:

```java
public class Index3 extends Activity {
    private static final String TAG = "Index3Activity";

    @Override
    protected void onStart() {
        super.onStart();
        Log.v(TAG, "protected void onStart()");
    }

    @Override
    protected void onPause() {
        super.onPause();
        Log.v(TAG, "protected void onPause()");
    }

    @Override
    protected void onResume() {
```

```java
        super.onResume();
        Log.v(TAG, "protected void onResume()");
    }

    @Override
    protected void onStop() {
        super.onStop();
        Log.v(TAG, "protected void onStop()");
    }

    @Override
    protected void onDestroy() {
        super.onDestroy();
        Log.v(TAG, "protected void onDestroy()");
    }

    @Override
    protected void onRestart() {
        super.onRestart();
        Log.v(TAG, "protected void onRestart()");
    }

    /** Called when the activity is first created. */
    @Override
    public void onCreate(Bundle savedInstanceState) {
        super.onCreate(savedInstanceState);
        Log.v(TAG, "protected void onCreate()");

        setContentView(R.layout.index3);// *******重点
        Button button3_1 = (Button) this.findViewById(R.id.button3_1);

        button3_1.setOnClickListener(new OnClickListener() {

            public void onClick(View arg0) {
                Index3.this.finish();
            }
        });

    }
}
```

创建 Index3.java 对应的布局文件 index3.xml，代码如下：

```xml
<?xml version="1.0" encoding="utf-8"?>
<LinearLayout xmlns:android="http://schemas.android.com/apk/res/android"
    android:orientation="vertical" android:layout_width="fill_parent"
    android:layout_height="fill_parent">
    <TextView android:layout_width="fill_parent"
        android:layout_height="wrap_content" android:text="this is three page!" />
    <Button android:text="Button" android:id="@+id/button3_1"
```

            android:layout_width="*wrap_content*" android:layout_height="*wrap_content*"></Button>
</LinearLayout>

在 AndroidManifest.xml 文件中关联名称为 Index3 的 Activity 对象，代码如下：

```
<?xml version="1.0" encoding="utf-8"?>
<manifest xmlns:android="http://schemas.android.com/apk/res/android"
    package="test.run" android:versionCode="1" android:versionName="1.0">

    <application android:icon="@drawable/icon" android:label="@string/app_name">
        <activity android:name=".Index" android:label="@string/app_name">
            <intent-filter>
                <action android:name="android.intent.action.MAIN" />
                <category android:name="android.intent.category.LAUNCHER" />
            </intent-filter>
        </activity>

        <activity android:name=".Index2" android:label="@string/app_name"
            android:theme="@android:style/Theme.Dialog">
        </activity>

        <activity android:name=".Index3" android:label="@string/app_name">
        </activity>

    </application>
</manifest>
```

运行程序，在 LogCat 打印出了 Index.java 文件必须执行的 3 个事件，如图 1.83 所示。单击 main.xml 布局界面的第 2 个按钮，结果如图 1.84 所示。

图 1.83 Index.java 文件必须要执行的 3 个步骤　　　图 1.84 单击第 2 个按钮

在控制台打印了 onPause() 和 onStop()，结果如图 1.85 所示。单击 index3.xml 布局界面上的按钮，结果如图 1.86 所示。

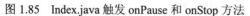

图 1.85　Index.java 触发 onPause 和 onStop 方法　　图 1.86　单击 index3.xml 布局界面的按钮

在控制台打印出了 onRestart()方法，如图 1.87 所示。

图 1.87　切换到 onRestart()方法

1.10.3　实现 onCreate()->onStart()->onResume()->onPause()-> onStop()->onDestroy()

上面的实验，就差 onDestroy()方法没有被调用了，更改 main.xml 文件加入第 3 个 Button 按钮，代码如下：

```
<?xml version="1.0" encoding="utf-8"?>
<LinearLayout xmlns:android="http://schemas.android.com/apk/res/android"
    android:orientation="vertical" android:layout_width="fill_parent"
    android:layout_height="fill_parent">
    <TextView android:layout_width="fill_parent"
        android:layout_height="wrap_content" android:text="this is one page!" />
    <Button android:text="Button" android:id="@+id/button1_1"
        android:layout_width="wrap_content" android:layout_height="wrap_content"></Button>
    <Button android:text="Button2" android:id="@+id/button1_2"
        android:layout_width="wrap_content" android:layout_height="wrap_content"></Button>
    <Button android:text="Button3" android:id="@+id/button1_3"
        android:layout_width="wrap_content" android:layout_height="wrap_content"></Button>
```

```
</LinearLayout>
```

在 Index.java 文件中的 onCreate()方法中追加如下代码：

```java
Button button1_3 = (Button) this.findViewById(R.id.button1_3);
button1_3.setOnClickListener(new OnClickListener() {
    public void onClick(View arg0) {
        Intent intentRef = new Intent();
        intentRef.setClass(Index.this, Index4.class);
        startActivity(intentRef);
        Index.this.finish();
    }
});
```

还要创建名称为 Index4.java 的 Activity，代码如下：

```java
public class Index4 extends Activity {
    private static final String TAG = "Index4Activity";

    @Override
    protected void onStart() {
        super.onStart();
        Log.v(TAG, "protected void onStart()");
    }

    @Override
    protected void onPause() {
        super.onPause();
        Log.v(TAG, "protected void onPause()");
    }

    @Override
    protected void onResume() {
        super.onResume();
        Log.v(TAG, "protected void onResume()");
    }

    @Override
    protected void onStop() {
        super.onStop();
        Log.v(TAG, "protected void onStop()");
    }

    @Override
    protected void onDestroy() {
        super.onDestroy();
        Log.v(TAG, "protected void onDestroy()");
    }

    @Override
```

```java
    protected void onRestart() {
        super.onRestart();
        Log.v(TAG, "protected void onRestart()");
    }

    /** Called when the activity is first created. */
    @Override
    public void onCreate(Bundle savedInstanceState) {
        super.onCreate(savedInstanceState);
        Log.v(TAG, "protected void onCreate()");

        setContentView(R.layout.index4);// *******重点

        Button button4_1 = (Button) this.findViewById(R.id.button4_1);
        button4_1.setOnClickListener(new OnClickListener() {
            public void onClick(View arg0) {
                Intent intentRef = new Intent();
                intentRef.setClass(Index4.this, Index.class);
                startActivity(intentRef);
                Index4.this.finish();
            }
        });

    }
}
```

创建 Index4.java 对应的布局文件 index4.xml，代码如下：

```xml
<?xml version="1.0" encoding="utf-8"?>
<LinearLayout xmlns:android="http://schemas.android.com/apk/res/android"
    android:orientation="vertical" android:layout_width="fill_parent"
    android:layout_height="fill_parent">
    <TextView android:layout_width="fill_parent"
        android:layout_height="wrap_content" android:text="this is three page!" />
    <Button android:text="Button" android:id="@+id/button4_1"
        android:layout_width="wrap_content" android:layout_height="wrap_content"></Button>
</LinearLayout>
```

在文件 AndroidManifest.xml 中注册 Index4.java，代码如下：

```xml
<?xml version="1.0" encoding="utf-8"?>
<manifest xmlns:android="http://schemas.android.com/apk/res/android"
    package="test.run" android:versionCode="1" android:versionName="1.0">


    <application android:icon="@drawable/icon" android:label="@string/app_name">
        <activity android:name=".Index" android:label="@string/app_name">
            <intent-filter>
                <action android:name="android.intent.action.MAIN" />
                <category android:name="android.intent.category.LAUNCHER" />
            </intent-filter>
        </activity>
```

```
            <activity android:name=".Index2" android:label="@string/app_name"
                android:theme="@android:style/Theme.Dialog">
            </activity>

            <activity android:name=".Index3" android:label="@string/app_name">
            </activity>

            <activity android:name=".Index4" android:label="@string/app_name">
            </activity>

        </application>
</manifest>
```

运行项目，在 LogCat 打印出 3 个基本事件信息，如图 1.88 所示。

单击 Index.java 界面的第 3 个按钮，结果如图 1.89 所示。

I	514	jdwp	Ignoring second debugger -- accepti
V	514	IndexActivity	protected void onCreate()
V	514	IndexActivity	protected void onStart()
V	514	IndexActivity	protected void onResume()
I	61	ActivityManager	Displayed test.run/.Index: +1s497ms

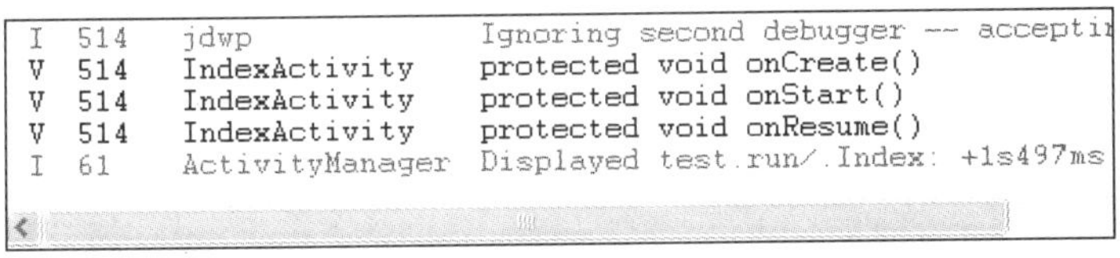

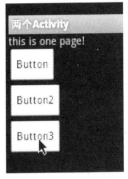

图 1.88　Index.java 文件执行了 3 个基本事件函数

图 1.89　单击第 3 个按钮

在 LogCat 中打印出了 Index.java 文件的 3 个函数 onPause()、onStop()和 onDestroy()，如图 1.90 所示。

D	505	dalvikvm	Debugger has detached; object regis
I	514	jdwp	Ignoring second debugger -- accepti
V	514	IndexActivity	protected void onCreate()
V	514	IndexActivity	protected void onStart()
V	514	IndexActivity	protected void onResume()
I	61	ActivityManager	Displayed test.run/.Index: +1s497ms
D	136	dalvikvm	GC_EXPLICIT freed 128K, 52% free 28
D	244	dalvikvm	GC_EXPLICIT freed 7K, 54% free 2547
D	328	dalvikvm	GC_EXPLICIT freed 8K, 54% free 2597
D	359	dalvikvm	GC_EXPLICIT freed 2K, 54% free 2538
I	61	ActivityManager	Starting: Intent { cmp=test.run/.In
V	514	IndexActivity	protected void onPause()
V	514	Index4Activity	protected void onCreate()
V	514	Index4Activity	protected void onStart()
V	514	Index4Activity	protected void onResume()
I	61	ActivityManager	Displayed test.run/.Index4: +414ms
V	514	IndexActivity	protected void onStop()
V	514	IndexActivity	protected void onDestroy()

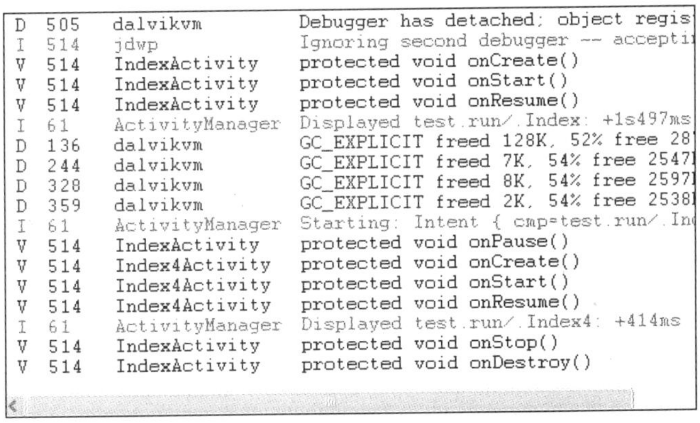

图 1.90　Index.java 对应的 Activity 对象销毁了

到此，Activity 常见的生命周期就通过上面的 3 个示例重现完毕。

1.10.4 应用程序列表时的生命周期情况

下面的示例将要实现一个程序运行时，单击 Home 按键后的生命周期过程的演示，新建名称为 returnApplicationList 的 Android 项目，更改 Main.java 的代码如下：

```java
public class Main extends Activity {

    @Override
    protected void onDestroy() {
        super.onDestroy();
        Log.e("Main", "onDestroy");
    }

    @Override
    protected void onPause() {
        super.onPause();
        Log.e("Main", "onPause");
    }

    @Override
    protected void onRestart() {
        super.onRestart();
        Log.e("Main", "onRestart");
    }

    @Override
    protected void onResume() {
        super.onResume();
        Log.e("Main", "onResume");
    }

    @Override
    protected void onStart() {
        super.onStart();
        Log.e("Main", "onStart");
    }

    @Override
    protected void onStop() {
        super.onStop();
        Log.e("Main", "onStop");
    }

    @Override
    public void onCreate(Bundle savedInstanceState) {
        super.onCreate(savedInstanceState);
        setContentView(R.layout.main);
        Log.e("Main", "onCreate");
    }
}
```

程序初始运行时出现如图 1.91 所示的效果。

图 1.91　程序初始运行时的生命周期与界面

这时单击 AVD 界面中的 按钮回到 AVD 的桌面，出现如图 1.92 所示的结果。

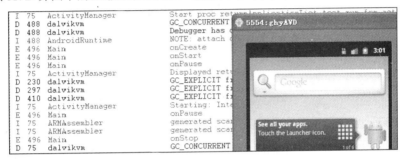

图 1.92　去往桌面的生命周期与界面

从图 1.92 中可以看到，当前的 Activity 变成 onStop 停止状态了，这时单击应用程序列表按钮，如图 1.93 所示。

图 1.93　单击应用程序列表按钮

单击当前的应用程序名称 再一次进入这个 Activity 界面，出现如图 1.94 所示的结果。

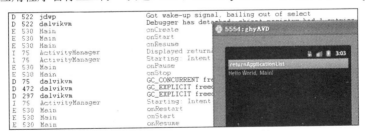

图 1.94　返回 Activity 生命周期的情况

从图 1.94 中可以看到，Activity 并没有新建一个新的实例，只是从 onStop 状态转到了 onRestart->onStart->OnResume 状态。

1.10.5　AVD 横竖屏切换时的生命周期情况

重新运行 returnApplicationList 项目，在 LogCat 中打印出相关的信息，界面效果如图 1.95 所示。

图 1.95 初始运行效果

单击 AVD 设备确保它是获得焦点的,然后按下 Ctrl+F12 键,这时由原来的竖屏变成了横屏,而 Activity 的生命周期的变化效果如图 1.96 所示。

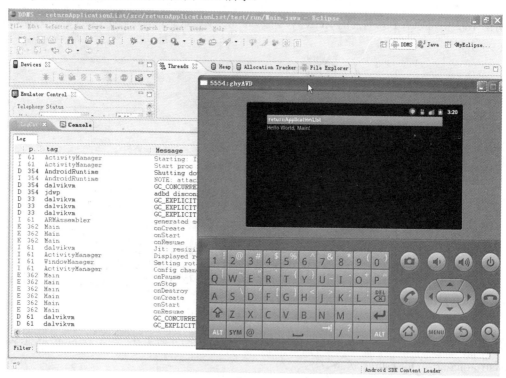

图 1.96 切到横屏时的生命周期变化

从图 1.96 中可以看到,Activity 由竖屏切换到横屏时是把原来的 Activity 销毁掉,再重新创建一个新的 Activity 对象。

在 Android 中判断屏幕是横屏还是竖屏由以下代码来进行，此代码在项目 screenOrientation 中：

```java
public class Main extends Activity {
    private TextView TextView1;

    @Override
    public void onCreate(Bundle savedInstanceState) {
        super.onCreate(savedInstanceState);
        setContentView(R.layout.main);
        TextView1 = (TextView) this.findViewById(R.id.textView1);

        if (this.getResources().getConfiguration().orientation == this
                .getResources().getConfiguration().ORIENTATION_LANDSCAPE) {
            TextView1.setText("水平");
        }
        if (this.getResources().getConfiguration().orientation == this
                .getResources().getConfiguration().ORIENTATION_PORTRAIT) {
            TextView1.setText("垂直");
        }
    }
}
```

1.10.6　onSaveInstanceState()和 onRestoreInstanceState()回调方法的使用

回调函数 onSaveInstanceState()可以让 Activity 在"某些特殊情况下"销毁前获得保存信息的机会，但需要注意的是这个函数不是什么时候都会被调用，所以"某些特殊情况下"是前题。那么这个"特殊情况"是指什么呢？可以这样理解，不是手机用户主动销毁 Activity 的情况，比如屏幕翻转时就是这种情况，还有如果从 ActivityA 启动 ActivityB 后，ActivityB 在 Activity 栈中位于 ActivityA 的前方，此时系统由于内存不足，肯定要销毁不在前台显示的 ActivityA，这时 ActivityA 就可以通过 onSaveInstanceState()函数保存临时的状态信息，使得将来用户返回到 ActivityA 时能通过 onCreate()或者 onRestoreInstanceState()函数恢复界面的状态。

另外需要说明的是，不要将 onSaveInstanceState()方法和 Activity 生命周期回调函数如 onPause()或 onStop()混为一谈，onPause()在 Activtiy 被放置到背景或者自行销毁时总会被调用，onStop()在 Activity 被销毁时调用，而 onSaveInstanceState()是在一个非人为因素操作下销毁 Activity 时才被调用。

一个会调用 onPause()和 onStop()但不触发 onSaveInstanceState()的例子是当用户从 ActivityB 返回到 ActivityA 时就没有必要调用 ActivityB 的 onSaveInstanceState ()，此时的 ActivityB 实例永远不会被恢复，因此系统不会调用 ActivityB 的 onSaveInstanceState()。

还有另外一种情况，就是一个调用 onPause()但不调用 onSaveInstanceState()的例子是当 ActivityB 启动并处在 ActivityA 的前端时，如果在 ActivityB 的整个生命周期里 ActivityA 的用户界面状态都没有被系统破坏的话，也就是系统从未因为特殊原因销毁 ActivityA 时，系统是不会调用 ActivityA 的 onSaveInstanceState()的。如果被调用，这个方法会在 onStop()前被触发，但系统并不保证是否在 onPause()之前或者之后触发。

要注意的是，onSaveInstanceState()方法和 onRestoreInstanceState()方法不一定是成对的被调用，

比如用户按下 Home 键时，Activity 的 onSaveInstanceState()被调用，但重新进入项目显示刚才的 Activity 时，onRestoreInstanceState()不被调用。那到底有哪些因素导致 onSaveInstanceState()方法被调用呢？下面通过示例来演示一下。

新建名称为 twoState 的 Android 项目来测试下面的这些示例。

（1）按下 Home 键时

程序运行界面如图 1.97 所示。此时按下 Home 键，查看 Logcat 日志，如图 1.98 所示。

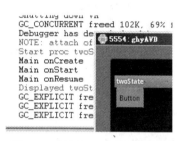

图 1.97　按下 Home 键前的效果　　　　图 1.98　按下 Home 键的日志信息

再重新进入 twoState 项目后查看一下日志，如图 1.99 所示。

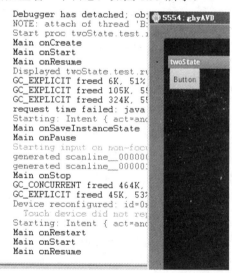

图 1.99　按下 Home 键后再重新进入项目

通过日志可以发现 Main.java 并没有调用 onRestoreInstanceState()方法。

（2）长按 Home 键运行其他程序时

重新运行项目，启动 AVD 中的浏览器，然后按下 Home 键，再启动 twoState 项目，在界面中长按 Home 键重新运行浏览器，打印日志如图 1.100 所示。

图 1.100　长按 Home 的日志效果

这种情况下当恢复显示 twoState 的 Activity 时，方法 onRestoreInstanceState()不被调用。

（3）按下电源按键，关闭屏幕时被调用，当重新按下电源显示屏幕时方法 onRestoreInstanceState()也不被调用。

（4）从 ActivityA 中启动一个新的 Activity 时，ActivityA 的 onSaveInstanceState()方法被调用，当按下 Back 按钮重新回到 ActivityA 时，ActivityA 的 onRestoreInstanceState()方法不被调用。

（5）屏幕方向切换时 onSaveInstanceState()方法和 onRestoreInstanceState()方法均被调用。

总而言之，onSaveInstanceState()的调用遵循一个重要原则，即当系统"未经你许可"时销毁了你的 Activity 时，则 onSaveInstanceState()会被系统调用，它必须提供一个机会让你保存数据，onRestoreInstanceState 被调用的前提是 Activity 确实是被系统销毁了，而不是人为的因素。另外，onRestoreInstanceState 的 Bundle 参数也会传递到 onCreate()方法中，也可以选择在 onCreate()方法中做数据还原。

1.11　LinearLayout 布局对齐方式和 Dialog 提示的使用

在 Android 中对话框的使用占据了很大的技术内容，与在 C#中使用对话框的技术相同，但如果创建复杂布局界面的对话框，则程序员还是必须要用手工写代码的方式来重新定义对话框中的内容与对话框中控件间的位置，这时在学习对话框之前就有必要先掌握一下布局对齐的基本使用知识。

Android 中的对齐功能主要由 android:gravity 属性来进行定义，它主要的值有 right、center、center_horizontal、right|center_vertical、bottom|center_horizontal 等。

（1）右对齐（right）效果

下面就来实现一个 right 右对齐的效果，main.xml 布局文件代码如下：

```
<?xml version="1.0" encoding="utf-8"?>
<LinearLayout xmlns:android="http://schemas.android.com/apk/res/android"
    android:orientation="vertical" android:layout_width="fill_parent"
    android:layout_height="fill_parent" android:gravity="right">
    <Button android:text="Button" android:id="@+id/button1"
```

```
            android:layout_width="wrap_content" android:layout_height="wrap_content"></Button>
</LinearLayout>
```

程序运行效果如图 1.101 所示。

图 1.101 right 右对齐效果

（2）水平和垂直居中（center）的效果

再实现一个水平和垂直居中 center 的效果，需要说明的是 Android 中的布局标签 LinearLayout 是可以嵌套的，main.xml 布局文件的代码如下：

```
<?xml version="1.0" encoding="utf-8"?>
<LinearLayout xmlns:android="http://schemas.android.com/apk/res/android"
    android:orientation="vertical" android:layout_width="fill_parent"
    android:layout_height="fill_parent" android:gravity="center">
    <LinearLayout android:orientation="vertical"
        android:layout_width="100px" android:layout_height="100px"
        android:background="#FFFFFF">
    </LinearLayout>
</LinearLayout>
```

程序运行效果如图 1.102 所示。

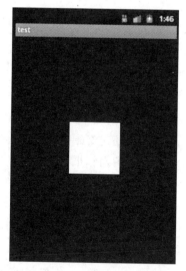

图 1.102 水平和垂直居中 center 的效果

（3）水平居中（center_horizontal）的效果

再实现一个水平居中 center_horizontal 的效果，main.xml 代码如下：

```
<?xml version="1.0" encoding="utf-8"?>
<LinearLayout xmlns:android="http://schemas.android.com/apk/res/android"
    android:orientation="vertical" android:layout_width="fill_parent"
```

```
        android:layout_height="fill_parent" android:gravity="center_horizontal">
    <LinearLayout android:orientation="vertical"
        android:layout_width="100px" android:layout_height="100px"
        android:background="#FFFFFF">
    </LinearLayout>
</LinearLayout>
```

程序运行效果如图 1.103 所示。

图 1.103　水平居中 center_horizontal 的效果

（4）右对齐垂直居中（right|center_vertical）的效果

再实现一个右对齐垂直居中（right|center_vertical）的效果，main.xml 代码如下：

```
<?xml version="1.0" encoding="utf-8"?>
<LinearLayout xmlns:android="http://schemas.android.com/apk/res/android"
    android:orientation="vertical" android:layout_width="fill_parent"
    android:layout_height="fill_parent" android:gravity="right|center_vertical">
    <LinearLayout android:orientation="vertical"
        android:layout_width="100px" android:layout_height="100px"
        android:background="#FFFFFF">
    </LinearLayout>
</LinearLayout>
```

程序运行效果如图 1.104 所示。

图 1.104　右对齐垂直居中（right|center_vertical）的效果

右对齐垂直居中（right|center_vertical）的效果和左对齐垂直居中（left|center_vertical）的效果正好相反。

（5）底部居中（bottom|center_horizontal）对齐的效果

再实现一个底部居中（bottom|center_horizontal）对齐的效果，main.xml 代码如下：

```
<?xml version="1.0" encoding="utf-8"?>
<LinearLayout xmlns:android="http://schemas.android.com/apk/res/android"
    android:orientation="vertical" android:layout_width="fill_parent"
    android:layout_height="fill_parent" android:gravity="bottom|center_horizontal">
    <LinearLayout android:orientation="vertical"
        android:layout_width="100px" android:layout_height="100px"
        android:background="#FFFFFF">
    </LinearLayout>
</LinearLayout>
```

程序运行效果如图 1.105 所示。

图 1.105　底部居中（bottom|center_horizontal）对齐的效果

（6）左对齐高度添充的效果

本示例要实现一个左对齐高度填充的效果，但这个效果并没有使用到 android:gravity 属性，而且 LinearLayout 标签的高度和宽度的属性已经被更改，完整的 main.xml 代码如下：

```
<?xml version="1.0" encoding="utf-8"?>
<LinearLayout xmlns:android="http://schemas.android.com/apk/res/android"
    android:orientation="vertical" android:layout_width="fill_parent"
    android:layout_height="fill_parent">
    <LinearLayout android:orientation="vertical"
        android:layout_width="100px" android:layout_height="fill_parent"
        android:background="#FFFFFF">
    </LinearLayout>
</LinearLayout>
```

程序运行效果如图 1.106 所示。

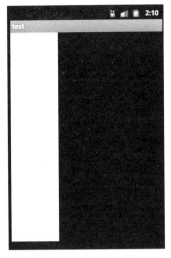

图 1.106　左对齐高度添充的效果

到此常用的布局对齐属性已经介绍完毕,掌握这些对齐属性的使用是设计复杂界面布局的基础。下面再继续学习 Android 中与对话框有关的技术点。

移动通信设备的屏幕尺寸和显示器比起来比较小,所以要在有限的空间里提供更多的信息供客户查看,这时对话框的使用就成为学习 Android 必须要掌握的技术之一,Android 对对话框的支持提供了很多自带的工具类,使用这些工具类可以非常方便地创建默认风格或自定义风格的对话框样式。

对话框父类是 Dialog,它的继承关系结构如图 1.107 所示。

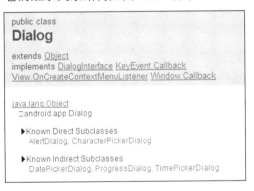

图 1.107　Dialog 类的继承关系结构

从 Dialog 类的继承关系中可以看到它的父类是 Object 类,而它具有两个非常重要的子类 AlertDialog 和 CharacterPickerDialog,其中在平时的项目开发中 AlertDialog 类最为常用,这也是本节学习的重点所在。

1.11.1　使用自定义对话框实现登录功能(对话框与 Activity 通信)

在开发 Web 项目时经常使用到浮动的 DIV 来实现一些模式的对话框,这些对话框上面有时是登录用的表单,有时是增加数据用的表单,有时则是最普通的消息提示文本,在 Android 中这种技术也经常用到,当然是使用对话框实现的,本小节就在 Android 中实现一个经典的自定义对话框样式并结合登录功能的示例。

学习本示例时要着重掌握以下几点：

- 广播的使用。
- 使用 IntentFilter 对象及动态创建广播的方法。
- 从广播中取得 Intent 对象中的数据值。

创建名称为 loginDialog 的 Android 项目，更改其中的文件 Main.java 代码如下：

```java
public class Main extends Activity {

    private TextView textView1;

    private BroadcastReceiver broadcastReceiverRef = new BroadcastReceiver() {

        @Override
        public void onReceive(Context arg0, Intent arg1) {
            Log.v("---------", "" + arg1.getAction());
            //也可以不写下方的 if 语句，但写 if 语句是为了多个
//IntentFilter 对象共用 1 个广播 Broadcast 的情况
            if (arg1.getAction().equals("getLoginUsername")) {
                String getLoginUsername = arg1.getStringExtra("loginUsername")
                        .toString();
                textView1.setText(getLoginUsername);
            }
        }
    };

    @Override
    protected void onResume() {

        super.onResume();
       //下面代码也可以不写在 onResume()方法而写在 onCreate()方法中
        IntentFilter idRef = new IntentFilter();
        idRef.addAction("getLoginUsername");

        this.registerReceiver(broadcastReceiverRef, idRef);
    }

    @Override
    public void onCreate(Bundle savedInstanceState) {
        super.onCreate(savedInstanceState);
        setContentView(R.layout.main);

        textView1 = (TextView) this.findViewById(R.id.mainTextView);

        LoginDialog ldRef = new LoginDialog(this);
        ldRef.show();

    }
```

}

其中代码：

```
private BroadcastReceiver broadcastReceiverRef = new BroadcastReceiver() {
    @Override
    public void onReceive(Context arg0, Intent arg1) {
        Log.v("---------", "" + arg1.getAction());
        if (arg1.getAction().equals("getLoginUsername")) {
            String getLoginUsername = arg1.getStringExtra("loginUsername")
                    .toString();
            textView1.setText(getLoginUsername);
        }
    }
};
```

上述代码的作用是创建一个广播接收者，专门用来接收传递过来的消息，然后取出其中的数据进而进一步的处理。而代码：

```
IntentFilter idRef = new IntentFilter();
idRef.addAction("getLoginUsername");

this.registerReceiver(broadcastReceiverRef, idRef);
```

上述代码的作用是在系统中注册这个广播接收者，并且关联一个 IntentFilter 对象，用来识别动作 action 名称为 getLoginUsername 的 Intent 传递过来的数据。

文件 main.xml 的代码比较简单，只有一个 TextView，代码如下：

```xml
<?xml version="1.0" encoding="utf-8"?>
<LinearLayout xmlns:android="http://schemas.android.com/apk/res/android"
    android:orientation="vertical" android:layout_width="fill_parent"
    android:layout_height="fill_parent">
    <TextView android:id="@+id/mainTextView" android:layout_width="fill_parent"
        android:layout_height="wrap_content" android:text="" />
</LinearLayout>
```

下面继续创建一个自定义对话框布局文件 logindialog.xml，这个布局文件是专门为登录界面的要求而设计，所以必须把代码中的每一个标签的属性进行上机实验、有效地测试并掌握这种界面布局的代码使用，布局文件程序如下：

```xml
<?xml version="1.0" encoding="utf-8"?>
<LinearLayout xmlns:android="http://schemas.android.com/apk/res/android"
    android:orientation="vertical" android:layout_width="300dip"
    android:layout_height="fill_parent" android:layout_marginLeft="20dip"
    android:layout_marginRight="20dip">
    <TextView android:text="账号：" android:id="@+id/logindialog_textView1"
        android:layout_width="fill_parent" android:layout_height="wrap_content"
        android:textSize="15dip" android:layout_marginLeft="20dip"
        android:layout_marginRight="20dip"></TextView>
    <EditText android:layout_height="wrap_content"
```

```xml
            android:id="@+id/logindialog_usernameEditText" android:text="a"
            android:layout_width="match_parent" android:textSize="15dip"
            android:layout_marginLeft="20dip" android:layout_marginRight="20dip"></EditText>
        <TextView android:text="密码： " android:id="@+id/logindialog_textView1"
            android:layout_width="fill_parent" android:layout_height="wrap_content"
            android:textSize="15dip" android:layout_marginLeft="20dip"
            android:layout_marginRight="20dip"></TextView>
        <EditText android:layout_height="wrap_content"
            android:id="@+id/logindialog_passwordEditText" android:text=""
            android:layout_width="match_parent" android:textSize="15dip"
            android:layout_marginLeft="20dip" android:layout_marginRight="20dip"></EditText>
        <LinearLayout android:layout_height="wrap_content"
            android:layout_width="fill_parent" android:orientation="horizontal"
            android:id="@+id/linearLayout1" android:gravity="center">
            <Button android:layout_height="wrap_content" android:id="@+id/logindialog_loginButton"
                android:layout_width="100dip" android:text="登录"></Button>
            <Button android:layout_height="wrap_content" android:id="@+id/logindialog_exitAppButton"
                android:layout_width="100dip" android:text="退出"></Button>
        </LinearLayout>
    </LinearLayout>
```

与布局文件 logindialog.xml 对应的还有一个 Dialog 对象，这个 Dialog 对象的 java 文件名为 LoginDialog.java，代码如下：

```java
public class LoginDialog extends Dialog {
    private Button logindialog_loginButton;
    private Button logindialog_exitAppButton;
    private EditText logindialog_usernameEditText;

    public LoginDialog(Context context) {
        super(context);
    }

    @Override
    protected void onCreate(Bundle savedInstanceState) {
        super.onCreate(savedInstanceState);
//下面的代码关联1个logindialog.xml布局文件
        this.setContentView(R.layout.logindialog);
        //取得布局文件中的控件对象
        logindialog_loginButton = (Button) this
                .findViewById(R.id.logindialog_loginButton);
        logindialog_exitAppButton = (Button) this
                .findViewById(R.id.logindialog_exitAppButton);
        logindialog_usernameEditText = (EditText) this
                .findViewById(R.id.logindialog_usernameEditText);

        this.setTitle("登录界面");
        //设置登录按钮的单击监听事件
        logindialog_loginButton.setOnClickListener(new View.OnClickListener() {

            public void onClick(View arg0) {
```

```
                    Intent sendLoginUsername = new Intent();
                //设置动作名称
                    sendLoginUsername.setAction("getLoginUsername");

                    sendLoginUsername.putExtra("loginUsername",
                            logindialog_usernameEditText.getText().toString());

                    LoginDialog.this.getContext().sendBroadcast(sendLoginUsername);
                    dismiss();

            }
        });

    }
}
```

自定义的 Dialog 类 LoginDialog.java 继承自 Dialog 类，可以在 LoginDialog.java 文件中验证用户名和密码的正确性，但为了练习广播的使用，所以不在此文件中进行登录数据的有效性验证。

上面程序中的代码段：

```
                    Intent sendLoginUsername = new Intent();
                //设置动作名称
                    sendLoginUsername.setAction("getLoginUsername");

                    sendLoginUsername.putExtra("loginUsername",
                            logindialog_usernameEditText.getText().toString());

                    LoginDialog.this.getContext().sendBroadcast(sendLoginUsername);
                    dismiss();
```

上述代码的功能是生成一个 Intent 对象，然后把这个 Intent 作为广播发送出去，关键的代码如下：

sendLoginUsername.setAction("getLoginUsername");

上面代码中的参数 getLoginUsername 一定要和 Main.java 文件中的代码一样，这样才可以正确取到指定 Intent 的广播数据，Main.java 文件中的代码片段如下：

```
            if (arg1.getAction().equals("getLoginUsername")) {
                String getLoginUsername = arg1.getStringExtra("loginUsername")
                        .toString();
                textView1.setText(getLoginUsername);
            }
```

文件 AndroidManifest.xml 的代码如下：

```
<?xml version="1.0" encoding="utf-8"?>
<manifest xmlns:android="http://schemas.android.com/apk/res/android"
    package="loginDialog.test.run" android:versionCode="1"
    android:versionName="1.0">
    <application android:icon="@drawable/icon" android:label="@string/app_name">
        <activity android:name=".Main" android:label="@string/app_name">
```

```
                <intent-filter>
                    <action android:name="android.intent.action.MAIN" />
                    <category android:name="android.intent.category.LAUNCHER" />
                </intent-filter>
            </activity>
        </application>
</manifest>
```

初始运行效果如图 1.108 所示。

在界面中输入 username 为 a，然后单击登录按钮出现登录的结果，如图 1.109 所示。

图 1.108　程序初始运行效果　　　　图 1.109　登录结果

到此本示例就正式结束。在这个示例中主要要掌握 3 个技术点：

- 广播的收发。
- 自定义对话框布局 XML 文件和创建对话框对应的 Dialog 类。
- 两个 Activity 的数据交互用 Intent 对象传递。

1.11.2　AlertDialog 对话框的使用

本示例继续学习 AlertDialog 类，其实 AlertDialog 类有很多的使用方法，本节就把常见的对话框使用情况一一列举出来，尽快地掌握 AlertDialog 是创建可操作性强的软件的必备条件。

使用 AlertDialog 类时不能 new 实例化，因为 AlertDialog 类的构造方法为 protected 保护的，如图 1.110 所示。

图 1.110　AlertDialog 的构造方法修饰

那如何取得 AlertDialog 类的实例呢？使用 AlertDialog 类的内置类 Builder 的 create()方法来取得 AlertDialog 类的对象，方法声明如图 1.111 所示。

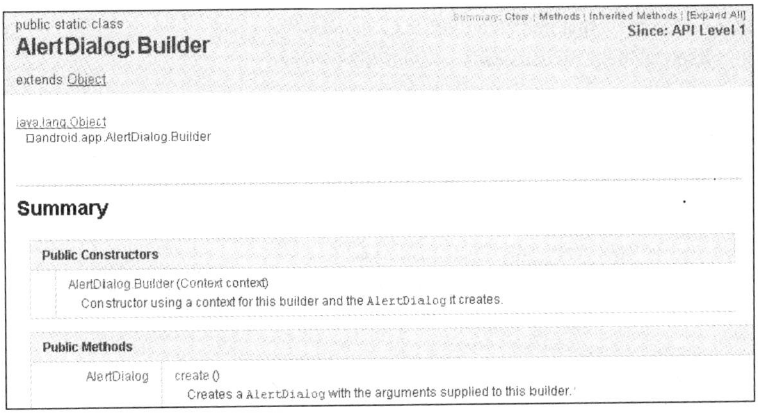
Summary: Ctors | Methods | Inherited Methods | [Expand All]
Since: API Level 1

public static class
AlertDialog.Builder
extends Object

java.lang.Object
 android.app.AlertDialog.Builder

Summary

Public Constructors

AlertDialog.Builder(Context context)
Constructor using a context for this builder and the AlertDialog it creates.

Public Methods

AlertDialog create()
Creates a AlertDialog with the arguments supplied to this builder.

图 1.111　内置类 Builder 和 create()方法的声明

创建名称为 AlertDialog 的 Android 项目，更改文件 main.xml 的代码如下：

```
<?xml version="1.0" encoding="utf-8"?>
<LinearLayout xmlns:android="http://schemas.android.com/apk/res/android"
    android:orientation="vertical" android:layout_width="fill_parent"
    android:layout_height="fill_parent">
    <Button android:text="1 个确定按钮" android:id="@+id/button1"
        android:layout_width="wrap_content" android:layout_height="wrap_content"></Button>
    <Button android:text="2 个按钮(确定和取消)" android:id="@+id/button2"
        android:layout_width="wrap_content" android:layout_height="wrap_content"></Button>
    <Button android:text="动态创建对话框中的内容" android:id="@+id/button3"
        android:layout_width="wrap_content" android:layout_height="wrap_content"></Button>
    <Button android:text="对话框中是单选列表" android:id="@+id/button4"
        android:layout_width="wrap_content" android:layout_height="wrap_content"></Button>
    <Button android:text="对话框中是复选列表" android:id="@+id/button5"
        android:layout_width="wrap_content" android:layout_height="wrap_content"></Button>
    <Button android:text="对话框中是普通列表" android:id="@+id/button6"
        android:layout_width="wrap_content" android:layout_height="wrap_content"></Button>
</LinearLayout>
```

布局文件 main.xml 里面都是 Button 控件，单击不同的 Button 显示出不同的对话框样式，而 main.xml 文件对应的 Activity 类 Main.java 代码如下：

```
public class Main extends Activity {
    private Button button1;
    private Button button2;
    private Button button3;
    private Button button4;
    private Button button5;
    private Button button6;

    @Override
    public void onCreate(Bundle savedInstanceState) {
        super.onCreate(savedInstanceState);
        setContentView(R.layout.main);
```

```java
button1 = (Button) this.findViewById(R.id.button1);
button1.setOnClickListener(new View.OnClickListener() {
    public void onClick(View arg0) {
        Log.v("----------", "单击了");
        AlertDialog adRef = new AlertDialog.Builder(Main.this).create();
        adRef.setIcon(android.R.drawable.btn_star);
        adRef.setTitle("标题");
        adRef.setMessage("我是消息内容");
        adRef.setButton("确定", new DialogInterface.OnClickListener() {
            public void onClick(DialogInterface arg0, int arg1) {
                // 无功能
            }
        });
        adRef.show();
    }
});

// ////////////////////

button2 = (Button) this.findViewById(R.id.button2);
button2.setOnClickListener(new View.OnClickListener() {
    public void onClick(View arg0) {

        AlertDialog adRef = new AlertDialog.Builder(Main.this).create();
        adRef.setMessage("我是消息内容");
        adRef.setIcon(android.R.drawable.btn_star);
        adRef.setTitle("标题");
        adRef.setButton("确定", new DialogInterface.OnClickListener() {
            public void onClick(DialogInterface arg0, int arg1) {
                Log.v("----------", "单击了2个按钮对话框中的确定");
            }
        });
        adRef.setButton2("取消", new DialogInterface.OnClickListener() {
            public void onClick(DialogInterface arg0, int arg1) {
                Log.v("----------", "单击了2个按钮对话框中的取消");
            }
        });
        adRef.show();
    }
});

// ////////////////////

button3 = (Button) this.findViewById(R.id.button3);
button3.setOnClickListener(new View.OnClickListener() {
    public void onClick(View arg0) {

        LayoutInflater inflater = Main.this.getLayoutInflater();
        View twoEditTextLayoutRef = inflater.inflate(
                R.layout.dialogtwoedittext, null);
```

```java
final EditText editText1 = (EditText) twoEditTextLayoutRef
        .findViewById(R.id.editText1);
final EditText editText2 = (EditText) twoEditTextLayoutRef
        .findViewById(R.id.editText2);
editText1.setText("username1");
editText2.setText("username2");

AlertDialog adRef = new AlertDialog.Builder(Main.this).create();
adRef.setView(twoEditTextLayoutRef);
adRef.setTitle("标题");
adRef.setIcon(android.R.drawable.btn_star);

adRef.setButton("取值", new DialogInterface.OnClickListener() {

    public void onClick(DialogInterface arg0, int arg1) {
        Log.v("***********", editText1.getText().toString()
                + " " + editText2.getText().toString());
    }

});

adRef.show();
    }
});

// ////////////////////

button4 = (Button) this.findViewById(R.id.button4);
button4.setOnClickListener(new View.OnClickListener() {
    public void onClick(View arg0) {

        final String[] userInfoArray = new String[] { "我是 A", "我是 B",
                "我是 C", "我是 D" };

        AlertDialog adRef = new AlertDialog.Builder(Main.this)
                .setSingleChoiceItems(userInfoArray, 1,
                        new DialogInterface.OnClickListener() {
                            public void onClick(DialogInterface arg0,
                                    int arg1) {
                                Log.v("您选中了:", userInfoArray[arg1]);
                                arg0.dismiss();
                            }
                        }).create();
        adRef.show();
    }
});

// ////////////////////
// 全选的思路是用 ListView 的 setItemChecked 方法
```

```java
// 而反选使用的是 SparseBooleanArray 布尔数组法
// 此数组来自于 ListView 的 getCheckedItemPositions()方法
// 另外需要留意的是,setMultiChoiceItems()方法的第 2 个参数
// 是设置 CheckBox 控件打勾与否的默认值,在这里传入 null 值
// 那如果有默认值怎么办呢?很简单!通过代码来设置默认值就可以了
// 不使用第 2 个参数
button5 = (Button) this.findViewById(R.id.button5);
button5.setOnClickListener(new View.OnClickListener() {
    public void onClick(View arg0) {

        final String[] userInfoArray = new String[] { "我是 1", "我是 2",
                "我是 3", "我是 4", "我是 5", "我是 6", "我是 7", "我是 8", "我是 9", "我是 10" };

        final AlertDialog adRef = new AlertDialog.Builder(Main.this)
                .setMultiChoiceItems(
                        userInfoArray,
                        null,
                        new OnMultiChoiceClickListener() {
                            public void onClick(DialogInterface arg0,
                                    int arg1, boolean arg2) {
                                Log.v("zzzzzzzzzzzz", "" + arg1 + "    "
                                        + arg2);
                            }
                        }).create();

        boolean[] defaultValueBooleanArray = new boolean[] { true,
                false, true, false, true, false, true, false, true,
                false };

        adRef.setButton("全选", new DialogInterface.OnClickListener() {
            public void onClick(DialogInterface arg0, int arg1) {
                ListView lv = adRef.getListView();
                for (int i = 0; i < lv.getCount(); i++) {
                    lv.setItemChecked(i, true);
                }

                // setAccessible()方法的解释:
                // 它提供了将反射的对象标记为在使用时
                // 取消默认 Java 语言访问控制检查的能力
                // 值为 true 则指示反射的对象在使用时应该取消 Java 语言访问检查。
                // 按着反射代码的功能来进行实现
                // 值为 false 则指示反射的对象应该实施 Java 语言访问检查
                // field.set(adRef, false)的功能是将 field 字段在 adRef 对象上
                // 设置最新的值,这个值为 false
                try {
                    Field field = adRef.getClass().getSuperclass()
                            .getDeclaredField("mShowing");
                    field.setAccessible(true);
                    // 将 mShowing 变量设为 false,表示对话框已关闭
                    field.set(adRef, false);
```

```java
            } catch (Exception e) {

            }
        }
    });

    adRef.setButton3("反选", new DialogInterface.OnClickListener() {
        public void onClick(DialogInterface arg0, int arg1) {

            SparseBooleanArray sbaRef = adRef.getListView()
                    .getCheckedItemPositions();

            ListView lv = adRef.getListView();

            for (int i = 0; i < lv.getCount(); i++) {
                lv.setItemChecked(i, !sbaRef.get(i));
            }

            try {
                Field field = adRef.getClass().getSuperclass()
                        .getDeclaredField("mShowing");
                field.setAccessible(true);
                // 将 mShowing 变量设为 false,表示对话框已关闭
                field.set(adRef, false);
            } catch (Exception e) {

            }

        }
    });
    adRef.setButton2("确定", new DialogInterface.OnClickListener() {
        public void onClick(DialogInterface arg0, int arg1) {

            ListView lv = adRef.getListView();

            for (int i = 0; i < lv.getCount(); i++) {
                if (lv.getCheckedItemPositions().get(i)) {
                    Log.v("多选选中了: ", ""
                            + lv.getAdapter().getItemId(i) + "   "
                            + lv.getAdapter().getItem(i));
                }
            }

            try {
                Field field = adRef.getClass().getSuperclass()
                        .getDeclaredField("mShowing");
                field.setAccessible(true);
                // 将 mShowing 变量设为 true,表示对话框未关闭
                field.set(adRef, true);
                adRef.dismiss();
```

```java
                    } catch (Exception e) {

                    }
                }
            });
            adRef.show();
            // 一定要在这里设置初始的默认值!
            // 注意，此循环一定要写在 show()方法的下面才可以进初始化
            for (int i = 0; i < defaultValueBooleanArray.length; i++) {
                adRef.getListView().setItemChecked(i,
                        defaultValueBooleanArray[i]);
            }
        }
    });

    // ////////////////////

    button6 = (Button) this.findViewById(R.id.button6);
    button6.setOnClickListener(new View.OnClickListener() {
        public void onClick(View arg0) {

            final String[] userInfoArray = new String[] { "我是 1", "我是 2",
                    "我是 3", "我是 4" };
            AlertDialog adRef = new AlertDialog.Builder(Main.this)
                    .setItems(userInfoArray,
                            new DialogInterface.OnClickListener() {
                                public void onClick(DialogInterface arg0,
                                        int arg1) {
                                    Log.v("您选中了:", userInfoArray[arg1]);
                                    arg0.dismiss();
                                }
                            }).create();
            adRef.show();
        }
    });

    }
}
```

名称为"确定"的 button5 按钮中的单击事件还可以这样写，也能实现取出打勾 checkbox 控件的索引值：

```java
dialog.setButton2("确定", new DialogInterface.OnClickListener() {
    public void onClick(DialogInterface arg0, int arg1) {
        try {
            ListView listView = dialog.getListView();
            long[] checkedArray = listView.getCheckItemIds();
            for (int i = 0; i < checkedArray.length; i++) {
                Log.v("!", "" + checkedArray[i]);
            }
```

```
            Field field = dialog.getClass().getSuperclass()
                           .getDeclaredField("mShowing");
            field.setAccessible(true);
            // 将 mShowing 变量设为 false，表示对话框已关闭
            field.set(dialog, true);
            arg0.dismiss();
        } catch (Exception e) {
        }
    }
});
```

由于在本示例中使用到了自定义对话框，那么自定义对话框的布局文件 dialogtwoedittext.xml 的代码如下：

```
<?xml version="1.0" encoding="utf-8"?>
<LinearLayout xmlns:android="http://schemas.android.com/apk/res/android"
    android:orientation="vertical" android:layout_width="fill_parent"
    android:layout_height="fill_parent">
    <EditText android:layout_height="wrap_content" android:id="@+id/editText1"
        android:layout_width="match_parent"></EditText>
    <EditText android:layout_height="wrap_content" android:id="@+id/editText2"
        android:layout_width="match_parent"></EditText>
</LinearLayout>
```

程序初始运行效果如图 1.112 所示。

单击"1 个确定按钮"出现如图 1.113 所示。

图 1.112　程序初始运行

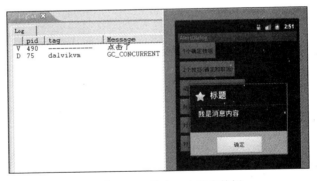

图 1.113　单击 1 个确定按钮

单击"2 个按钮（确定和取消）"按钮出现如图 1.114 所示。

在图 1.114 中单击"确定"按钮，LogCat 出现日志信息，如图 1.115 所示。

图 1.114　单击 2 个按钮（确定和取消）

图 1.115　单击确定后 LogCat 的消息

单击"动态创建对话框中的内容"按钮出现，如图 1.116 所示的效果。

图 1.116　单击动态创建对话框中的内容

在图 1.116 中单击"取值"按钮 LogCat 出现日志信息如图 1.117 所示。

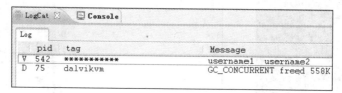
图 1.117　取出来的值

继续单击"对话框中是单选列表"按钮出现如图 1.118 所示。在图 1.118 中单击"我是 C"，然后在 LogCat 面板中出现日志信息，如图 1.119 所示。

图 1.118　单击对话框中是单选列表

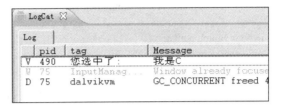
图 1.119　单击我是 C

单击"对话框中是复选列表"按钮后出现 checkbox 列表，可以在界面中单击"全选"或"反选"按钮，最终操作出来的界面如图 1.120 所示。在图 1.120 中单击"确定"按钮，查看选中的值，然后在 LogCat 面板中出现日志信息，如图 1.121 所示。

图 1.120　单击对话框中是复选列表

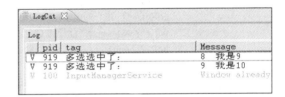
图 1.121　取出来的 checkbox 的值

最后单击"对话框中是普通列表"按钮出现如图 1.122 所示。在图 1.122 中单击"我是 4"，然后在 LogCat 面板中出现日志信息，如图 1.123 所示。

图 1.122　单击对话框中是普通列表

图 1.123　单击我是 4

1.11.3 ProgressDialog 对话框的使用

Android 系统也支持进度条对话框，本节来演示其实现方法，新建名称为 ProgressDialog_1_11_3 的 Android 项目，设计文件 Main.java 的代码如下：

```java
public class Main extends Activity implements Runnable {

    private Button button;
    private Button button2;

    private ProgressDialog pdRef;

    int progressValue = 0;

    @Override
    public void onCreate(Bundle savedInstanceState) {
        super.onCreate(savedInstanceState);
        setContentView(R.layout.main);
        button = (Button) this.findViewById(R.id.button1);
        button2 = (Button) this.findViewById(R.id.button2);

        button.setOnClickListener(new OnClickListener() {
            public void onClick(View arg0) {
                progressValue = 0;
                pdRef = new ProgressDialog(Main.this);
                pdRef.setTitle("进度");
                pdRef.setMessage("百分比 0%");
                pdRef.show();

                Thread thread = new Thread(Main.this);
                thread.start();
            }

        });

        button2.setOnClickListener(new OnClickListener() {
            public void onClick(View arg0) {
                progressValue = 0;
                pdRef = new ProgressDialog(Main.this);
                pdRef.setTitle("进度");
                pdRef.setMessage("百分比 0%");
                pdRef.setProgressStyle(ProgressDialog.STYLE_HORIZONTAL);
                pdRef.show();

                Thread thread = new Thread(Main.this);
                thread.start();
            }

        });
```

```java
    }
    
    private Handler handler = new Handler() {
    
        @Override
        public void handleMessage(Message msg) {
            super.handleMessage(msg);
            if (Integer.parseInt(msg.getData().getString("progressValue")
                    .toString()) > 100) {
                pdRef.dismiss();
            } else {
                pdRef.setProgress(Integer.parseInt(msg.getData().getString(
                        "progressValue").toString()));
                pdRef.setMessage(msg.getData().getString("progressValue")
                        .toString());
            }
        }
        
    };
    
    public void run() {
        try {
            while (progressValue < 101) {
                progressValue = progressValue + 20;
                
                Bundle BundleRef = new Bundle();
                BundleRef.putString("progressValue", "" + progressValue);
                
                Message sendMessage = new Message();
                sendMessage.setData(BundleRef);
                
                Thread.sleep(1000);
                handler.sendMessage(sendMessage);
            }
        } catch (InterruptedException e) {
            e.printStackTrace();
        }
        
    }
}
```

程序初始运行效果如图 1.124 所示。单击上面的 "Button" 出现进度对话框，如图 1.125 所示。

图 1.124 初始显示界面

图 1.125 圆型进度运行起来了

当进度到达 100%时进度对话框自动关闭。当单击下面的"Button"时出现进度对话框,如图 1.126 所示。

图 1.126 条型进度运行起来了

当进度到达 100%时进度对话框自动关闭。

1.11.4 对话框中的内容是列表条目的情况并取消后退按钮

其实在 AlertDialog 对话框中还可以是列表条目的信息,以列表的方式来进行消息的展示,本示例就来实现这样的效果。

新建 Android 项目,名称为 Dialog_ListItems,更改文件 main.xml 的代码,加入两个 Button 按钮,程序代码如下:

```xml
<?xml version="1.0" encoding="utf-8"?>
<LinearLayout xmlns:android="http://schemas.android.com/apk/res/android"
    android:orientation="vertical" android:layout_width="fill_parent"
    android:layout_height="fill_parent">
    <Button android:text="无按钮对话框" android:id="@+id/button1"
        android:layout_width="wrap_content" android:layout_height="wrap_content"></Button>
    <Button android:text="有按钮对话框" android:id="@+id/button2"
        android:layout_width="wrap_content" android:layout_height="wrap_content"></Button>
</LinearLayout>
```

更改文件 Main.java 的程序代码如下：

```java
public class Main extends Activity {

    private String[] itemsList = new String[] { "aaa", "bbb", "ccc", "ddd" };

    private Button button1;
    private Button button2;

    @Override
    public void onCreate(Bundle savedInstanceState) {
        super.onCreate(savedInstanceState);
        setContentView(R.layout.main);

        button1 = (Button) this.findViewById(R.id.button1);
        button2 = (Button) this.findViewById(R.id.button2);

        button1.setOnClickListener(new View.OnClickListener() {

            public void onClick(View arg0) {
                AlertDialog.Builder builderNoButton = new AlertDialog.Builder(
                        Main.this);
                builderNoButton.setCancelable(false);
                builderNoButton.setItems(itemsList, new OnClickListener() {
                    public void onClick(DialogInterface arg0, int arg1) {
                        Log.v("选中了：", itemsList[arg1]);
                    }
                });
                builderNoButton.show();

            }
        });

        button2.setOnClickListener(new View.OnClickListener() {

            public void onClick(View arg0) {
                AlertDialog.Builder builderHasButton = new AlertDialog.Builder(
                        Main.this);
                builderHasButton.setPositiveButton("确定", new OnClickListener() {
                    public void onClick(DialogInterface arg0, int arg1) {
                        Log.v("选中了：", "确定按钮");
                    }
                });

                builderHasButton.setCancelable(false);
                builderHasButton.setItems(itemsList, null);
                builderHasButton.show();

            }
        });
```

```
        }
    }
```

代码 builderNoButton.setCancelable(**false**);的作用是防止用户按钮回退/取消按钮 关闭对话框。
从程序中可以看到，在 AlertDialog 对话框中显示消息列表的主要代码是：

```
private String[] itemsList = new String[] { "aaa", "bbb", "ccc", "ddd" };
builderHasButton.setItems(itemsList, null);
```

程序运行初始效果如图 1.127 所示。单击"无按钮对话框"，弹出对话框，如图 1.128 所示。

图 1.127 初始运行效果　　　　　　　图 1.128 弹出无按钮对话框

单击"ccc"后在 LogCat 中出现相关信息，如图 1.129 所示。再次单击"有按钮对话框"，结果如图 1.130 所示。

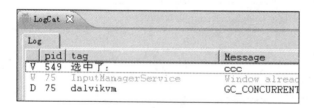

图 1.129 显示 ccc 信息　　　　　　　图 1.130 弹出有按钮对话框

单击"确定"按钮，LogCat 打印出来的信息如图 1.131 所示。

第1章 初识 Android

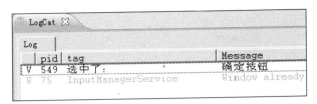

图 1.131 单击确定后的信息

1.11.5 使用自定义 XML 布局文件填充 AlertDialog 对话框的另外一种方法

前面已经实现了使用一个自定义的 XML 布局文件来作为对话框中的内容，其实还有另外一种代码也可以实现。

新建名称为 simpleXmlAlertDialog 的 Android 项目，创建自定义的布局文件 mydialog.xml 文件，代码如下：

```xml
<?xml version="1.0" encoding="utf-8"?>
<LinearLayout xmlns:android="http://schemas.android.com/apk/res/android"
    android:orientation="vertical" android:layout_width="300px"
    android:layout_height="fill_parent" android:padding="30px"
    android:gravity="center">
    <EditText android:layout_height="wrap_content"
        android:layout_width="match_parent" android:id="@+id/editText1"
        android:text="EditText1"></EditText>
    <EditText android:layout_height="wrap_content"
        android:layout_width="match_parent" android:id="@+id/editText2"
        android:text="EditText2"></EditText>
    <Button android:text="确定" android:id="@+id/button1"
        android:layout_width="100px" android:layout_height="wrap_content"></Button>
</LinearLayout>
```

文件 main.xml 代码还是原始状态，并没有改变，为了增加程序的完整性和可读性，代码还是有必要列出来，程序如下：

```xml
<?xml version="1.0" encoding="utf-8"?>
<LinearLayout xmlns:android="http://schemas.android.com/apk/res/android"
    android:orientation="vertical" android:layout_width="fill_parent"
    android:layout_height="fill_parent">
    <TextView android:layout_width="fill_parent"
        android:layout_height="wrap_content" android:text="@string/hello" />
</LinearLayout>
```

文件 Main.java 的代码如下：

```java
public class Main extends Activity {
    private Button button1;
    private EditText ev1;
    private EditText ev2;

    @Override
    public void onCreate(Bundle savedInstanceState) {
```

```
        super.onCreate(savedInstanceState);
        setContentView(R.layout.main);

        final Dialog dialog = new Dialog(this);
        dialog.setTitle("我是标题");
        dialog.setCancelable(false);
        dialog.setContentView(R.layout.mydialog);
        dialog.show();

        button1 = (Button) dialog.findViewById(R.id.button1);
        ev1 = (EditText) dialog.findViewById(R.id.editText1);
        ev2 = (EditText) dialog.findViewById(R.id.editText2);

        button1.setOnClickListener(new OnClickListener() {
            public void onClick(View arg0) {
                Log.v("信息是: ", ev1.getText().toString() + "    "
                        + ev2.getText().toString());
                dialog.dismiss();
            }
        });
    }
}
```

软件初始运行效果如图 1.132 所示。

单击界面中的"确定"按钮后，在 LogCat 中打印出信息，如图 1.133 所示。

图 1.132　初始运行效果

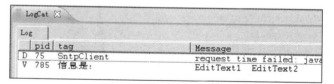

图 1.133　LogCat 中的打印信息

1.11.6　实现自动关闭对话框

自动关闭对话框的功能主要使用 Handler 对象来实现，该对象的 postDelayed 方法用来实现延时多少秒去执行某个任务。

先来实现 Handler 对象延时执行的简单示例，创建名称为 handlerTest 的 Android 项目，Main.java 的代码如下：

```
public class Main extends Activity {
    @Override
```

```
public void onCreate(Bundle savedInstanceState) {
    super.onCreate(savedInstanceState);
    setContentView(R.layout.main);

    Handler handler = new Handler();
    handler.postDelayed(new Runnable() {
        public void run() {
            Log.v("!","5 秒后打印出我！");
        }
    }, 5000);

    }
}
```

程序执行后在 LogCat 中打印相关的信息，如图 1.134 所示。

```
V  529   !                          5秒后打印出我！
```

图 1.134　打印结果

结合 Handler 对象的使用，再创建一个名称为 autoCloseAlertDialog 的 Android 项目，来实现一个自动关闭 AlertDialog 对话框的效果，Main.java 代码如下：

```
public class Main extends Activity {
    /** Called when the activity is first created. */
    @Override
    public void onCreate(Bundle savedInstanceState) {
        super.onCreate(savedInstanceState);
        setContentView(R.layout.main);

        final AlertDialog adRef = new AlertDialog.Builder(this).create();
        adRef.setTitle("我是标题");
        adRef.setMessage("3 秒后我被自动关闭");
        adRef.show();

        Handler handler = new Handler();
        handler.postDelayed(new Runnable() {

            public void run() {
                adRef.dismiss();
            }
        }, 3000);

    }
}
```

程序运行后弹出对话框，如图 1.135 所示。

图 1.135　弹出自动关闭的对话框

1.11.7　toast 提示的使用

Android 中可以使用 Dialog 来显示数据，还有一种方式也能用来显示一些提示类的信息，它就是 toast，使用起来非常的简单。

创建名称为 toastBegin 的 Android 项目，Main.java 的代码如下：

```java
public class Main extends Activity {
    private Handler handler = new Handler();

    private Button button1;
    private Button button2;
    private Button button3;
    private Button button4;
    private Button button5;
    private Button button6;

    public void showToast() {
        handler.post(new Runnable() {
            public void run() {
                Toast.makeText(getApplicationContext(), "我来自其他线程！",
                        Toast.LENGTH_SHORT).show();

            }
        });
    }

    @Override
    public void onCreate(Bundle savedInstanceState) {
        super.onCreate(savedInstanceState);
        setContentView(R.layout.main);

        button1 = (Button) this.findViewById(R.id.button1);
        button2 = (Button) this.findViewById(R.id.button2);
        button3 = (Button) this.findViewById(R.id.button3);
        button4 = (Button) this.findViewById(R.id.button4);
        button5 = (Button) this.findViewById(R.id.button5);
```

```java
button6 = (Button) this.findViewById(R.id.button6);

button1.setOnClickListener(new OnClickListener() {
    public void onClick(View arg0) {
        Toast.makeText(Main.this, "长时间显示的文本", Toast.LENGTH_LONG).show();
    }
});

button2.setOnClickListener(new OnClickListener() {
    public void onClick(View arg0) {
        Toast.makeText(Main.this, "短时间显示的文本", Toast.LENGTH_SHORT)
                .show();
    }
});

button3.setOnClickListener(new OnClickListener() {
    public void onClick(View arg0) {

        View viewRef = Main.this.getLayoutInflater().inflate(
                R.layout.toastview, null);
        Toast toastRef = new Toast(Main.this);
        toastRef.setView(viewRef);
        toastRef.show();

    }
});
button4.setOnClickListener(new OnClickListener() {
    public void onClick(View arg0) {
        Toast toastRef = Toast.makeText(getApplicationContext(),
                "自定义显示位置", Toast.LENGTH_LONG);
        toastRef.setGravity(Gravity.CENTER, 0, 0);
        toastRef.show();
    }
});

button5.setOnClickListener(new OnClickListener() {
    public void onClick(View arg0) {
        Toast toastRef = Toast.makeText(getApplicationContext(), "带图片",
                Toast.LENGTH_LONG);
        toastRef.setGravity(Gravity.CENTER, 0, 0);
        LinearLayout toastView = (LinearLayout) toastRef.getView();
        ImageView imageCodeProject = new ImageView(
                getApplicationContext());
        imageCodeProject.setImageResource(R.drawable.icon);
        toastView.addView(imageCodeProject, 0);
        toastRef.show();
    }
});
button6.setOnClickListener(new OnClickListener() {
    public void onClick(View arg0) {
```

```
                    new Thread(new Runnable() {
                        public void run() {
                            showToast();
                        }
                    }).start();
                }
            });

        }
    }
```

创建自定义的布局文件 toastview.xml，代码如下：

```
<?xml version="1.0" encoding="utf-8"?>
<LinearLayout xmlns:android="http://schemas.android.com/apk/res/android"
    android:orientation="vertical" android:layout_width="fill_parent"
    android:layout_height="fill_parent">
    <AnalogClock android:id="@+id/analogClock1"
        android:layout_width="wrap_content" android:layout_height="wrap_content"></AnalogClock>
</LinearLayout>
```

程序运行后单击"默认长时间显示"和"默认短时间显示"按钮出现提示信息，如图 1.136 所示。

图 1.136　默认长时间显示和默认短时间显示

程序运行后单击"自定义显示 View"按钮出现提示信息，如图 1.137 所示。
单击"自定义显示位置"按钮出现提示信息，如图 1.138 所示。

图 1.137　自定义显示 View 提示效果　　　　图 1.138　"自定义显示位置"按钮提示信息

单击"带图片"按钮出现提示信息，如图 1.139 所示。
单击"线程中启动"按钮出现提示信息，如图 1.140 所示。

图 1.139　"带图片"按钮提示效果　　　　图 1.140　"线程中启动"按钮提示效果

1.11.8　设置 Dialog 对话框的尺寸

在前面小节中介绍了 Dialog 对象使用继承 Dialog 类结合自定义 XML 的布局法，如果对话框中的内容比较简单，则可以直接对 Dialog 对象进行操作而达到弹出对话框显示提示信息的作用。

新建名称为 DialogSimple 的 Android 项目，Main.java 的代码如下：

```java
public class Main extends Activity {
    private Button newButton;

    @Override
    public void onCreate(Bundle savedInstanceState) {
        super.onCreate(savedInstanceState);
        setContentView(R.layout.main);

        newButton = new Button(this);
        newButton.setText("点我关闭对话框");

        final Dialog dialogRef = new Dialog(this);
```

```
            dialogRef.setTitle("我是标题");
            dialogRef.setContentView(newButton);
            dialogRef.getWindow().setLayout(200, 150);
            dialogRef.show();

            newButton.setOnClickListener(new OnClickListener() {
                public void onClick(View arg0) {
                    dialogRef.dismiss();
                }
            });

        }
    }
```

显示效果如图 1.141 所示。

图 1.141　显示效果

1.11.9　PopupWindow 对话框

对象 PopupWindow 可以弹出对话框并且可以设置弹出的位置，创建名称为 popupDialog 的 Android 项目，Main.java 的代码如下：

```
public class Main extends Activity {
    private Button newButton;
    private Button button1;

    @Override
    public void onCreate(Bundle savedInstanceState) {
        super.onCreate(savedInstanceState);
        setContentView(R.layout.main);

        button1 = (Button) this.findViewById(R.id.button1);

        button1.setOnClickListener(new OnClickListener() {
            public void onClick(View arg0) {
                newButton = new Button(Main.this);
                newButton.setText("关闭");

                final PopupWindow popupWindow = new PopupWindow(Main.this);
                popupWindow.setContentView(newButton);
                popupWindow.setWidth(200);
                popupWindow.setHeight(100);
                popupWindow
```

```
                .showAtLocation(Main.this
                        .findViewById(R.id.linearLayout1),
                        Gravity.CENTER, 0, 0);

                newButton.setOnClickListener(new OnClickListener() {
                    public void onClick(View arg0) {
                        popupWindow.dismiss();
                    }
                });
            }
        });
    }
}
```

程序运行效果如图 1.142 所示。

图 1.142 运行效果

1.12 抽象类 Window 与布局分析工具 Hierarchy View

　　前面介绍的 Activity 其实就是用户的界面，但在真正的 Android 界面组成部件中，Activity 中的控件是在抽象类 Window 中进行组织的，每一个 Activity 都有一个 Window 对象，Window 抽象类是界面的容器，在 Window 中可以包含 View 或 ViewGroup 控件，例如列表、下拉控件、表格控件等。

　　下面来创建一个 Android 项目来查看一下 Window 与界面布局和控件之间的关系。创建名称为 Window_Test 的 Android 项目，一切都是默认的，文件 Main.java 的代码如下：

```
public class Main extends Activity {
    /** Called when the activity is first created. */
    @Override
    public void onCreate(Bundle savedInstanceState) {
        super.onCreate(savedInstanceState);
        setContentView(R.layout.main);
    }
}
```

文件 main.xml 的代码如下：

```xml
<?xml version="1.0" encoding="utf-8"?>
<LinearLayout xmlns:android="http://schemas.android.com/apk/res/android"
    android:orientation="vertical" android:layout_width="fill_parent"
    android:layout_height="fill_parent">
    <TextView android:layout_width="fill_parent"
        android:layout_height="wrap_content" android:text="@string/hello" />
</LinearLayout>
```

文件 strings.xml 代码如下：

```xml
<?xml version="1.0" encoding="utf-8"?>
<resources>
    <string name="hello">Hello World, Main!</string>
    <string name="app_name">Window_Test</string>
</resources>
```

程序运行后就是最普通不过的界面，如图 1.143 所示。

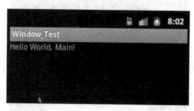

图 1.143　程序初始运行效果

在图 1.143 中并不能查看到整个 AVD 中界面的布局，所以得使用 ADT 中的工具 Hierarchy View，该工具在菜单中的位置如图 1.144 所示。

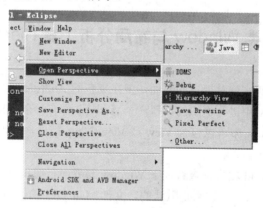

图 1.144　打开 Hierarchy View 菜单

单击 Hierarchy View 命令后切换到 "Hierarchy View" 透视图，在 Window 面板中出现 AVD 设备中存在的 Window 列表如图 1.145 所示。

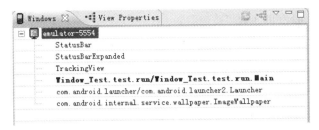

图 1.145　界面 Window 布局列表

其中在图 1.145 中就有刚才创建项目中的 Window，名称为 test.run.Main。单击 test.run.Main 条目在面板 Tree View 中显示出了当前 Window 的界面布局结构，如图 1.146 所示。

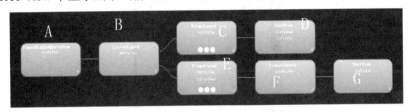

图 1.146　Window 窗口布局结构图

从图 1.146 中可以看到，界面布局的父组件是名称为"PhoneWindow$DecorView"的对象，在对象中组成了整个用户能看到的视觉效果，A 就是 Window 类的实现类"PhoneWindow$DecorView"，组件 B 是整个界面的总布局对象，在 B 中存在 C 和 E 组件，C 组件是当前应用程序的标题布局，可以看到 C 组件的详细信息如图 1.147 所示。

而 D 对象则是标题布局中的文本<TextView>标签，详细结构如图 1.148 所示。

图 1.147　C 组件的结构图　　　　图 1.148　D 组件的结构图

依此类推，E 组件就是当前 Activity 的"body 体"布局，组件 F 就是"body 体"中的布局方式，其实也就是 main.xml 文件中的代码：

```
<LinearLayout xmlns:android="http://schemas.android.com/apk/res/android"
    android:orientation="vertical" android:layout_width="fill_parent"
```

```
android:layout_height="fill_parent">
```

G 组件就是 main.xml 文件中的代码：

```
<TextView android:layout_width="fill_parent"
    android:layout_height="wrap_content" android:text="@string/hello" />
```

进一步确认这个 TextView，如图 1.149 所示。

图 1.149　G 组件的结构图

当然也可以通过 Window 类来设置一些基本的界面外观样式，比如无标题全屏幕显示，请将文件 Main.java 代码更改如下：

```java
public class Main extends Activity {
    /** Called when the activity is first created. */
    @Override
    public void onCreate(Bundle savedInstanceState) {
        super.onCreate(savedInstanceState);

        requestWindowFeature(Window.FEATURE_NO_TITLE);
        this.getWindow().setFlags(WindowManager.LayoutParams.FLAG_FULLSCREEN,
                WindowManager.LayoutParams.FLAG_FULLSCREEN);

        setContentView(R.layout.main);
    }
}
```

一定要在 setContentView 方法的前面来进行 Window 对象的初始值的设置。程序运行效果如图 1.150 所示。

图 1.150 无标题全屏的效果

1.13 控制控件位置和大小的常用属性

在 Android 中控制控件位置和大小的属性，如图 1.151 所示。

```
android:layout_gravity
android:layout_width
android:layout_height
android:layout_margin
android:layout_marginLeft
android:layout_marginTop
android:layout_marginRight
android:layout_marginBottom
```

```
android:padding
android:paddingLeft
android:paddingTop
android:paddingRight
android:paddingBottom
```

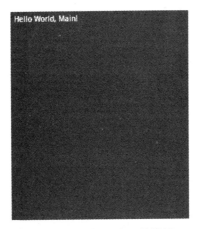

图 1.151 定位控件位置的属性

属性 android:layout_gravity 是对齐的方式，在前面的章节中已经介绍过，还有其他的常用属性值，比如 top 向上对齐，bottom 向下对齐，left 靠左对齐，right 靠右对齐，center_vertical 垂直居中，center_horizontal 水平居中等。

为了演示控件位置属性的使用，新建名称为 positionAndSizeSimple 的 Android 项目，更改文件 main.xml 的代码如下：

```
<?xml version="1.0" encoding="utf-8"?>
<LinearLayout xmlns:android="http://schemas.android.com/apk/res/android"
    android:orientation="vertical" android:layout_width="fill_parent"
    android:layout_height="fill_parent">
    <TextView android:layout_width="fill_parent"
        android:layout_height="wrap_content" android:text="@string/hello"
        android:background="#00ffff" android:layout_marginLeft="30px" />
    <LinearLayout android:orientation="vertical"
        android:layout_width="fill_parent" android:layout_height="fill_parent"
        android:background="#cccccc" android:layout_marginTop="10px"
        android:gravity="center">
        <TextView android:layout_width="fill_parent"
```

```
                android:layout_height="wrap_content" android:text="@string/hello"
                android:background="#00ffff" android:layout_marginLeft="40px"
                android:gravity="center" />
    </LinearLayout>
</LinearLayout>
```

程序运行效果如图 1.152 所示。

图 1.152　程序运行效果

1.14　设置应用程序背景图片

新建名称为 setBackground 的 Android 项目，更改 Main.java 的代码如下：

```
public class Main extends Activity {
    @Override
    public void onCreate(Bundle savedInstanceState) {
        super.onCreate(savedInstanceState);
        setContentView(R.layout.main);
        this.getWindow().setBackgroundDrawableResource(R.drawable.background);
    }
}
```

在项目 res 目录中添加图片资源，如图 1.153 所示。

运行效果如图 1.154 所示。

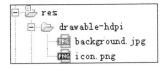

图 1.153　添加图片资源

图 1.154　运行效果

第 2 章 View 与 ViewGroup 类和控件事件

在第 2 章将和大家一起讨论的是如何在 Android 中做界面的布局，也就是像 Web 技术中使用 DIV 做页面的界面布局一样。

本章应该着重掌握的知识点：

- 动态创建控件的不同方式，这个知识点相对来讲是相当的重要，有些情况下在 Android 中是需要根据不同的情况来创建 View 对象，本教程主要以 3 个方法来介绍
- 5 大常见布局对象的使用
- 控件事件的回调函数
- 动态创建控件添加 margin 属性

2.1 View 和 ViewGroup 类的概述

类 View 是以矩形的方式显示在屏幕上，是用户界面控件的基础。它可以处理重绘功能和事件。

控件是类，控件也是对象，所以在 Android 中所有的界面控件都是 View 的子类，View 类的继承结构关系如图 2.1 所示。

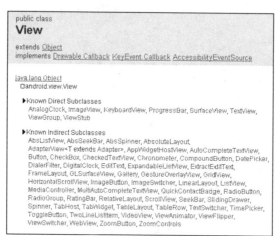

图 2.1 View 类的继承关系结构

从图 2.1 图中可以看到，View 的子类有很多，大多数都是程序员使用的 UI 界面控件，但这些控件并没有进行有效的管理，例如分组、排列等，所以在这里不得不提及 View 类的子类 ViewGroup 类，ViewGroup 类是把其他的 View 控件打包起来的容器，用 ViewGroup 容器去管理、组织这些控件，当然也支持控件的嵌套，ViewGroup 类的继承结构如图 2.2 所示。

图 2.2　ViewGroup 类的继承结构关系

ViewGroup 类的确可以以组或嵌套的方式管理容器中的组件,但怎么在容器中去管理这些控件呢,比如每种布局方式的不同,在其中的控件也显示不同的位置,这就需要用 ViewGroup 的子类"布局类 Layout"来进行实现了,ViewGroup 类的子类例如 FrameLayout、LinearLayout、RelativeLayout 等这些布局模型类是用不同的方式来管理容器中的 View 控件的摆放位置及显示方式等功能。但具体到被摆放控件的具体高度和宽度时,就需要每个布局类中的内置类 LayoutParams 类来进行处理,LayoutParams 类的结构如图 2.3 所示。

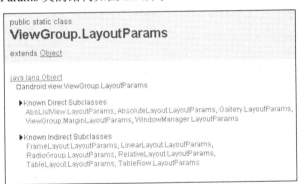

图 2.3　LayoutParams 类的结构

类 LayoutParams 是 ViewGroup 的内置类,LayoutParams 也有子类的具体实现。上面介绍类的继承关系总结构如图 2.4 所示。

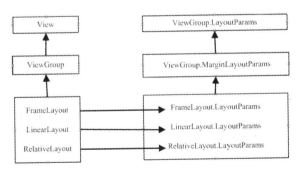

图 2.4　主要类的结构图

2.2 View 类的构造函数

View 类有 3 个构造函数，从官方的 API DOC 中就可以查询到，如图 2.5 所示。

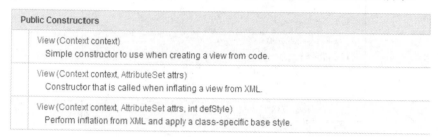

图 2.5　View 类的 3 个构造函数

其中最为常用的是前两种，本书主要介绍这两种构造方法的使用。

Android 中的 Button 类也是 View 类的子类，所以 View 类持有的特性 Button 类也持有，Button 的继承关系如图 2.6 所示。

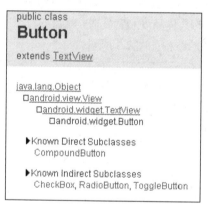

图 2.6　按钮 Button 的继承关系图

在下面的示例中，将通过动态实例化按钮并添加到布局中的方法来体会一下 View 构造方法的使用。

2.2.1　View(Context context)构造方法的使用

为了实验 View(Context context)构造方法的使用，新建名称为 view_1 的项目，来动态创建一个 Button 按钮控件到布局文件中，文件 main.xml 的代码只添加了 TextView 控件的 id，代码如下：

```
<?xml version="1.0" encoding="utf-8"?>
<LinearLayout xmlns:android="http://schemas.android.com/apk/res/android"
    android:orientation="vertical" android:layout_width="fill_parent"
    android:layout_height="fill_parent">
    <TextView android:layout_width="fill_parent"
        android:layout_height="wrap_content" android:text="@string/hello"
        android:id="@+id/textView1" />
</LinearLayout>
```

名称为 Main.java 的 Activity 代码如下：

```java
public class Main extends Activity {
    /** Called when the activity is first created. */
    @Override
    public void onCreate(Bundle savedInstanceState) {
        super.onCreate(savedInstanceState);
        setContentView(R.layout.main);

        Button button1 = new Button(this);
        button1.setText("确定 1");

        Button button2 = new Button(this);
        button2.setText("确定 2");

        ViewParent vpRef = this.findViewById(R.id.textView1).getParent();
        LinearLayout llRef = (LinearLayout) vpRef;
        llRef.addView(button1, 100, 100);
        llRef.addView(button2, 100, 100);
    }
}
```

其中对象 ViewParent 是 1 个接口，被 ViewGroup 对象所实现，实现结构如图 2.7 所示。程序运行后的效果如图 2.8 所示。

图 2.7　ViewGroup 实现 ViewParent 接口　　　图 2.8　程序运行效果

在界面中可以看到动态创建按钮并且成功添加到布局中了。

在上面的示例中为了找到 LinearLayout 对象，必须通过 TextView 控件的 getParent()方法来实现，其实还有另外一种方法，就是将<LinearLayout>标签添加 id 属性，更改后的 main.xml 的代码如下：

```xml
<?xml version="1.0" encoding="utf-8"?>
<LinearLayout xmlns:android="http://schemas.android.com/apk/res/android"
    android:orientation="vertical" android:layout_width="fill_parent"
    android:layout_height="fill_parent" android:id="@+id/linearlayout1">
    <TextView android:layout_width="fill_parent"
```

```
            android:layout_height="wrap_content" android:text="@string/hello" />
</LinearLayout>
```

还要更改 Main.java 的代码如下：

```java
public class Main extends Activity {
    @Override
    public void onCreate(Bundle savedInstanceState) {
        super.onCreate(savedInstanceState);
        setContentView(R.layout.main);

        // Context(this);
        Button button1 = new Button(this);
        button1.setText("我是按钮 1");

        Button button2 = new Button(this);
        button2.setText("我是按钮 2");

        Button button3 = new Button(this);
        button3.setText("我是按钮 3");

        LinearLayout llRef = (LinearLayout) this
                .findViewById(R.id.linearlayout1);
        llRef.addView(button1, 100, 100);
        llRef.addView(button2, 100, 100);
        llRef.addView(button3, 100, 100);
    }
}
```

程序运行后的效果如图 2.9 所示。

图 2.9　另外一种创建 Button 并添加到布局的办法

2.2.2　View(Context context, AttributeSet attrs)构造方法的使用

此构造方法的主要使用场合是在创建自定义控件时使用，并可以创建自定义的属性，本小节分两步来实现这个构造方法的使用。

第 1 步：先创建 1 个自定义的控件。

新建名称为 view_2 的 Android 项目，然后新建 1 个继承自 Button 类的自定义 Button 对象类 MyButton.java 文件，代码如下：

```java
public class MyButton extends Button {

    public MyButton(Context context, AttributeSet attrs) {
        super(context, attrs);
    }

    @Override
    protected void onDraw(Canvas canvas) {
        super.onDraw(canvas);

        Paint paint = new Paint();
        paint.setColor(Color.BLACK);
        paint.setTextSize(15);

        this.setText("我是按钮");

        canvas.drawText("我是画上去的", 66, 66, paint);

    }
}
```

并且还要更改 main.xml 文件

```xml
<?xml version="1.0" encoding="utf-8"?>
<LinearLayout xmlns:android="http://schemas.android.com/apk/res/android"
    android:orientation="vertical" android:layout_width="fill_parent"
    android:layout_height="fill_parent">
    <TextView android:layout_width="fill_parent"
        android:layout_height="wrap_content" android:text="@string/hello" />
    <mypackage.MyButton android:text="Button"
        android:id="@+id/button1" android:layout_width="200px"
        android:layout_height="80px"></mypackage.MyButton>
</LinearLayout>
```

系统的主启动 Activity 文件 Main.java 代码无须改动，默认即可。程序运行结果如图 2.10 所示。

图 2.10 程序第 1 次运行效果

第 2 步：上面的示例虽然自定义了 1 个 Button 按钮控件并且显示在布局界面上，但自定义的按钮控件并没有扩展系统自带 Button 对象的属性，有时候自定义的控件是需要自定义的属性来作为功能上的扩展，所以创建 1 个名称为 attrs.xml 来配置这个自定义属性的信息。

在 res 中的 values 目录下创建 attrs.xml 文件，代码如下：

```xml
<?xml version="1.0" encoding="utf-8"?>
<resources>
    <declare-styleable name="MyButtonProperties">
        <attr name="tag" format="string" />
    </declare-styleable>
</resources>
```

代码<declare-styleable name="MyButtonProperties">的主要的功能是定义一个自定义的属性集名称，通过这个名称就可以找到这个属性集中的自定义属性信息，包括自定义属性的名称和属性的类型，在这里自定义了一个属性名是 tag，这个属性 tag 对应的数据类型是 string 字符串类型。

上面的步骤创建了 attrs.xml 文件，这个文件也是资源的一种，所以在 R.java 文件中也自动生成了 attrs.xml 文件的资源索引，代码如下：

```java
package test.run;

public final class R {
    public static final class attr {
        public static final int tag = 0x7f010000;
    }

    public static final class drawable {
        public static final int icon = 0x7f020000;
    }

    public static final class id {
        public static final int button1 = 0x7f050000;
    }

    public static final class layout {
        public static final int main = 0x7f030000;
    }

    public static final class string {
        public static final int app_name = 0x7f040001;
        public static final int hello = 0x7f040000;
    }

    public static final class styleable {
        public static final int[] MyButtonProperties = { 0x7f010000 };
        public static final int MyButtonProperties_tag = 0;
    };
}
```

在 R.java 文件中 styleable 内置类中生成了这个属性集的字段：

```java
public static final int[] MyButtonProperties = { 0x7f010000 };
```

如何用程序代码找到这个属性呢？当然是如下的字段了：

```java
public static final int MyButtonProperties_tag = 0;
```

那么main.xml布局文件中的自定义控件使用哪个名称来作为属性名呢？代码如下：

```java
public static final int tag = 0x7f010000;
```

继续更改MyButton.java的代码如下：

```java
public class MyButton extends Button {

    private String ghyDeclareTagValue = "";

    public MyButton(Context context, AttributeSet attrs) {
        super(context, attrs);
        TypedArray taRef = context.obtainStyledAttributes(attrs,
                R.styleable.MyButtonProperties);

        String ghyDeclareTagValue = taRef
                .getString(R.styleable.MyButtonProperties_tag);

        if (ghyDeclareTagValue == null || "".equals(ghyDeclareTagValue)) {
            this.ghyDeclareTagValue = "没有值";
        } else {
            this.ghyDeclareTagValue = ghyDeclareTagValue;
        }

    }

    @Override
    protected void onDraw(Canvas canvas) {
        super.onDraw(canvas);

        Paint paint = new Paint();
        paint.setColor(Color.BLACK);
        paint.setTextSize(15);

        this.setText("我是按钮");

        canvas.drawText(ghyDeclareTagValue, 66, 66, paint);

    }
}
```

把文件main.xml的代码更改如下：

```xml
<?xml version="1.0" encoding="utf-8"?>
<LinearLayout xmlns:android="http://schemas.android.com/apk/res/android"
    xmlns:mystyle="http://schemas.android.com/apk/res/test.run"
    android:orientation="vertical" android:layout_width="fill_parent"
    android:layout_height="fill_parent">
```

```xml
<TextView android:layout_width="fill_parent"
    android:layout_height="wrap_content" android:text="@string/hello" />
<mypackage.MyButton android:text="Button"
    android:id="@+id/button1" android:layout_width="200px"
    android:layout_height="80px"></mypackage.MyButton>
<mypackage.MyButton android:text="Button"
    android:id="@+id/button1" android:layout_width="200px"
    android:layout_height="80px" mystyle:tag="我是高洪岩"></mypackage.MyButton>
<mypackage.MyButton android:text="Button"
    android:id="@+id/button1" android:layout_width="200px"
    android:layout_height="80px" mystyle:tag="我是高洪岩"></mypackage.MyButton>
</LinearLayout>
```

文件 main.xml 在原来代码的基础上加入了命名空间的限制代码：

```
xmlns:mystyle="http://schemas.android.com/apk/res/test.run"
```

那么这个命名空间的主要作用就是定义这个自定义属性 tag 的使用范围在 test.run 包下，而 mystyle 是属性 tag 的前缀，test.run 这个包不是随便起的，是根据文件 AndroidManifest.xml 中的 package 的值来进行定义，代码如下：

```
<manifest xmlns:android="http://schemas.android.com/apk/res/android"
    package="test.run" android:versionCode="1" android:versionName="1.0">
```

代码设计完毕后运行程序，效果如图 2.11 所示。

图 2.11　程序运行结果

2.3　View 单线程模型特性与在非 UI 线程中更新界面异常的实验

在 Android 中是不允许在用户的线程中更新 UI 界面的，也不建议用主线程来更新 UI 界面，因为在主线程中更新 UI 容易造成 UI 更新的延迟，比如从网络上取得数据再显示在 UI 中，根据网络环境的不同，将数据显示在 UI 中的速度也不一样，也容易造成后面 UI 控件更新的阻塞，所以这不是一个优秀的设计手段。在 Android 中的解决办法就是使用 Handler 对象来实现，Handler 对象的使用已经在第 1 章中有所涉及。

如果使用用户线程更新 UI 界面将会出现异常，我们来看看这种异常情况。创建实验用的 Android 项目，名称为 UserThreadUpdateUI_ExceptionTest。

新建实现 Runnable 接口的 java 文件 UpdateText.java，代码如下：

```java
public class UpdateText implements Runnable {
    private TextView textView1;

    public void setTextView1(TextView textView1) {
        this.textView1 = textView1;
    }

    public void run() {
        textView1.setText("我是最新版的 texView 文本");
    }
}
```

文件 Main.java 代码如下：

```java
public class Main extends Activity {
    private TextView textView1;
    private Button button1;

    @Override
    public void onCreate(Bundle savedInstanceState) {
        super.onCreate(savedInstanceState);
        setContentView(R.layout.main);

        textView1 = (TextView) this.findViewById(R.id.textView1);
        button1 = (Button) this.findViewById(R.id.button1);
        button1.setOnClickListener(new OnClickListener() {

            public void onClick(View arg0) {
                UpdateText utRef = new UpdateText();
                utRef.setTextView1(textView1);
                Thread thread = new Thread(utRef);
                thread.start();
            }
        });
    }
}
```

程序运行后单击按钮出现如下异常信息：

android.view.ViewRoot$CalledFromWrongThreadException: Only the original thread that created a view hierarchy can touch its views.

出错的原因就是在用户线程中更新 View 界面，这是错误的做法，所以要使用 Handler 对象发送 Message 消息的方式或发送广播的方式来解决这样的问题。

2.4 动态创建 View 和 ViewGroup 控件

在开发 Android 项目时，有时避免不了的是要动态创建控件并添加到布局界面中，本小节将介绍常用的动态创建控件的方式，使读者能掌握动态创建控件的基本技术。

2.4.1 第一种创建控件的办法

本章的开始已经介绍过动态创建 Button 控件并显示在布局中，本小节将在前面知识的基础上再实现一些相对复杂的创建控件的方法。

本示例中的 main.xml 布局文件的代码存在 LinearLayout 嵌套的情况，所以将动态创建的控件分别添加到外部 LinearLayout 和内部 LinearLayout 中，还要实现一个动态创建 LinearLayout 对象的功能，将动态创建的控件添加进这个新建的 LinearLayout 布局中。

新建名称为 addView1 的 Android 项目，文件 Main.java 的代码如下：

```java
public class Main extends Activity {
    @Override
    public void onCreate(Bundle savedInstanceState) {
        super.onCreate(savedInstanceState);
        setContentView(R.layout.main);

        LinearLayout outllRef = (LinearLayout) this
                .findViewById(R.id.outLayout);
        LinearLayout innerllRef = (LinearLayout) this
                .findViewById(R.id.innerLayout);

        Button outLayoutButton1 = new Button(this);
        outLayoutButton1.setHeight(50);
        outLayoutButton1.setWidth(50);
        outLayoutButton1.setText("我是外部按钮 1");

        Button outLayoutButton2 = new Button(this);
        outLayoutButton2.setHeight(50);
        outLayoutButton2.setWidth(50);
        outLayoutButton2.setText("我是外部按钮 2");

        TextView outLayoutTextView1 = new TextView(this);
        outLayoutTextView1.setText("我是外部 textView1");

        outllRef.addView(outLayoutButton1);// 添加进外部 layout 中
        outllRef.addView(outLayoutButton2);// 添加进外部 layout 中
        outllRef.addView(outLayoutTextView1);// 添加进外部 layout 中

        TextView innerLayoutTextView1 = new TextView(this);
        innerLayoutTextView1.setText("我是内部创建的 textView1");
        innerllRef.addView(innerLayoutTextView1);// 添加进内部 layout 中
        // 动态创建 layout 布局 不需要设置 LayoutParams 参数
```

```
        LinearLayout newllRef = new LinearLayout(this);
        newllRef.setOrientation(LinearLayout.VERTICAL);

        TextView newLayoutTextView1 = new TextView(this);
        newLayoutTextView1.setText("zzzzzzzzzzzzzzzzzzzz");

        TextView newLayoutTextView2 = new TextView(this);
        newLayoutTextView2.setText("动态布局中的 textView1");

        // 添加进新建的 layout 布局中，并设置宽和高
        newllRef.addView(newLayoutTextView1, LayoutParams.FILL_PARENT,
            LayoutParams.WRAP_CONTENT);
        newllRef.addView(newLayoutTextView2, LayoutParams.FILL_PARENT,
            LayoutParams.WRAP_CONTENT);

        outllRef.addView(newllRef);

    }
}
```

在这里需要注意，在 Android 中动态添加控件时，控件的宽和高尽量在 addView 方法中进行定义，使用的是 addView 方法的第 2 和第 3 个参数。

将布局文件 main.xml 的代码更改如下：

```
<?xml version="1.0" encoding="utf-8"?>
<LinearLayout xmlns:android="http://schemas.android.com/apk/res/android"
    android:orientation="vertical" android:layout_width="fill_parent"
    android:layout_height="fill_parent" android:id="@+id/outLayout">
    <TextView android:layout_width="wrap_content"
        android:layout_height="wrap_content" android:text="我是默认 1" />
    <LinearLayout android:orientation="vertical"
        android:layout_width="fill_parent" android:layout_height="wrap_content"
        android:id="@+id/innerLayout">
    </LinearLayout>
    <TextView android:layout_width="wrap_content"
        android:layout_height="wrap_content" android:text="我是默认 2" />
</LinearLayout>
```

程序运行效果如图 2.12 所示。

图 2.12　程序运行效果

2.4.2 第二种创建控件的办法

第一种动态创建控件的思路是每创建一个 UI 控件就添加到布局中，这种方式有一个缺点，就是如果动态创建的控件比较多时略显麻烦，还要用代码的方式更改其外观样式及其他的属性和事件，Android 中动态创建的控件也可以来自 XML 布局文件，采用 inflate 的方式来进行填充，本小节就来实现这个效果。

创建名称为 addView2 的 Android 项目，将文件 Main.java 的代码更改如下：

```java
public class Main extends Activity {
    /** Called when the activity is first created. */
    @Override
    public void onCreate(Bundle savedInstanceState) {
        super.onCreate(savedInstanceState);
        setContentView(R.layout.main);

        LayoutInflater layoutInflaterRef = this.getLayoutInflater();
        layoutInflaterRef.inflate(R.layout.buttonlayout, (ViewGroup) this
                .findViewById(R.id.innerLayOut));
    }
}
```

在文件 Main.java 代码中关联了 1 个布局文件 buttonlayout.xml，使用 inflate 方法将这个布局文件装载到内存后转化成 View 对象，再将这个 View 添加到 innerLayout 布局的内部。

LayoutInflater 对象的 inflate 方法有两个参数，在这里第 2 个参数传值或传 null 是有区别的：如果 inflate 第 2 个参数为 null，则 inflate 方法返回的对象就是第 1 个参数的 XML 布局文件转化成的 view 对象，并以返回值的方式进行返回。

如果 inflate 第 2 个参数不是 null，则将第 1 个参数转成的 View 对象的父结点设置为第 2 个参数，第 2 个参数对象的类型为 ViewGroup，返回值 View 是填充后的整体布局。

主布局文件 main.xml 的代码如下：

```xml
<?xml version="1.0" encoding="utf-8"?>
<LinearLayout android:id="@+id/outLayOut"
    xmlns:android="http://schemas.android.com/apk/res/android"
    android:orientation="vertical" android:layout_width="fill_parent"
    android:layout_height="fill_parent">
    <TextView android:layout_width="fill_parent"
        android:layout_height="wrap_content" android:text="我是最上面 text" />
    <LinearLayout android:id="@+id/innerLayOut"
        android:orientation="vertical" android:layout_width="wrap_content"
        android:layout_height="wrap_content">
    </LinearLayout>
    <TextView android:layout_width="fill_parent"
        android:layout_height="wrap_content" android:text="我是垫底的 text" />
</LinearLayout>
```

布局文件 main.xml 也采用双层 LinearLayout 布局嵌套的方式来组织界面。

自定义填充的布局文件 buttonlayout.xml 的代码如下：

```xml
<?xml version="1.0" encoding="utf-8"?>
<LinearLayout xmlns:android="http://schemas.android.com/apk/res/android"
    android:orientation="vertical" android:layout_width="fill_parent"
    android:layout_height="fill_parent">
    <Button android:text="Button" android:id="@+id/button1"
        android:layout_width="wrap_content" android:layout_height="wrap_content"></Button>
    <Button android:text="Button" android:id="@+id/button2"
        android:layout_width="wrap_content" android:layout_height="wrap_content"></Button>
</LinearLayout>
```

程序运行效果如图 2.13 所示。

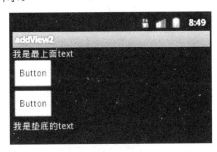

图 2.13　程序运行效果

在前面知识点中介绍了 LayoutInflater 对象的 inflate 方法有两个参数，并用文字的方式分别介绍了参数的功能作用，下面就用一个项目来详细介绍一下第 2 个参数在使用上的区别。

新建名称为 inflate_nullTest 的 Android 项目，main.xml 的代码如下：

```xml
<?xml version="1.0" encoding="utf-8"?>
<LinearLayout xmlns:android="http://schemas.android.com/apk/res/android"
    android:orientation="vertical" android:layout_width="fill_parent"
    android:layout_height="fill_parent" android:id="@+id/outLinearLayout">
    <AnalogClock android:id="@+id/analogClock1"
        android:layout_width="wrap_content" android:layout_height="wrap_content"></AnalogClock>
</LinearLayout>
```

文件 testlayout.xml 的代码如下：

```xml
<?xml version="1.0" encoding="utf-8"?>
<LinearLayout xmlns:android="http://schemas.android.com/apk/res/android"
    android:orientation="vertical" android:layout_width="fill_parent"
    android:layout_height="fill_parent">
    <Button android:text="Button" android:id="@+id/button1"
        android:layout_width="wrap_content" android:layout_height="wrap_content"></Button>
</LinearLayout>
```

文件 Main.java 的核心代码如下：

```java
public class Main extends Activity {
    private LinearLayout outLinearLayout;

    @Override
    public void onCreate(Bundle savedInstanceState) {
        super.onCreate(savedInstanceState);
```

```
            setContentView(R.layout.main);
            outLinearLayout = (LinearLayout) this
                    .findViewById(R.id.outLinearLayout);

            View view = this.getLayoutInflater().inflate(R.layout.testlayout, null);
            outLinearLayout.addView(view);
        }
    }
```

程序运行效果如图 2.14 所示。

使用 Hierarchy View 工具查看一下布局的结构如图 2.15 所示。

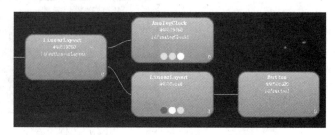

图 2.14　第 2 个参数为 null 的程序运行效果　　图 2.15　第 2 个参数为 null 的程序布局结构

通过这个实验可以发现，使用代码：

View view = **this**.getLayoutInflater().inflate(R.layout.*testlayout*, **null**);

上述代码的功能只是单纯的将 testlayout.xml 布局文件转成 View 对象，这个 View 对象在内存中并没有加入到布局界面中，如果想加入必须使用下述代码：

outLinearLayout.addView(view);

如果第 2 个参数不为 null 会是什么结果呢？更改 Main.java 的核心代码如下：

```
public class Main extends Activity {
    private LinearLayout outLinearLayout;

    @Override
    public void onCreate(Bundle savedInstanceState) {
        super.onCreate(savedInstanceState);
        setContentView(R.layout.main);
        outLinearLayout = (LinearLayout) this
                .findViewById(R.id.outLinearLayout);

        View view = this.getLayoutInflater().inflate(R.layout.testlayout,
                outLinearLayout);
        outLinearLayout.addView(view);
    }
}
```

程序运行后报出异常：

Caused by: java.lang.IllegalStateException: The specified child already has a parent. You must call removeView() on the child's parent first.

异常信息提示子元素已经有 1 个父元素了，不可以重复指定父元素，为什么会出现这种情况呢？因为代码：

```
View view = this.getLayoutInflater().inflate(R.layout.testlayout,
            outLinearLayout);
```

上述代码不仅将 testlayout.xml 转成 View 对象了，还把这个 View 对象自动添加到 outLinearLayout 对象的内部，所以就出现这样的错误，如何更改呢？既然是自动添加的，就不需要手动进行添加了，所以更改代码如下：

```
public class Main extends Activity {
    private LinearLayout outLinearLayout;

    @Override
    public void onCreate(Bundle savedInstanceState) {
        super.onCreate(savedInstanceState);
        setContentView(R.layout.main);
        outLinearLayout = (LinearLayout) this
                .findViewById(R.id.outLinearLayout);

        this.getLayoutInflater().inflate(R.layout.testlayout, outLinearLayout);
    }
}
```

程序运行结果如图 2.16 所示。

使用 Hierarchy View 工具查看一下布局的结构如图 2.17 所示。

 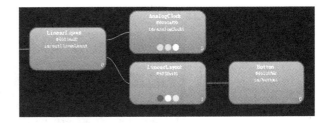

图 2.16　第 2 个参数不为空的运行结果　　　图 2.17　第 2 个参数不为 null 的程序布局结构

2.4.3　第三种创建控件的办法

前面两种动态创建控件的办法是：

（1）创建每个动态控件后再添加到布局中。

（2）从一个布局 XML 文件中批量创建控件，再添加到布局中。

本示例要实现一个和第 2 种方法比较接近的方式去创建控件并更改布局，相近的方式都是使用 inflate 填充法，不同的是本示例把界面中的全部旧控件替换成新插入的控件。

创建名称为 addView3 的 Android 项目。新建要填充的布局文件 newlayout.xml，代码如下：

```xml
<?xml version="1.0" encoding="utf-8"?>
<LinearLayout xmlns:android="http://schemas.android.com/apk/res/android"
    android:orientation="vertical" android:layout_width="fill_parent"
    android:layout_height="fill_parent">
    <TextView android:text="TextView" android:id="@+id/textView1"
        android:layout_width="wrap_content" android:layout_height="wrap_content"></TextView>
    <Button android:text="Button" android:id="@+id/button1"
        android:layout_height="wrap_content" android:layout_width="wrap_content"></Button>
</LinearLayout>
```

更改文件 Main.java 的代码如下：

```java
public class Main extends Activity {
    @Override
    public void onCreate(Bundle savedInstanceState) {
        super.onCreate(savedInstanceState);

        View newlayoutViewRef = this.getLayoutInflater().inflate(
                R.layout.newlayout, null);
        TextView textView1 = (TextView) newlayoutViewRef
                .findViewById(R.id.textView1);
        textView1.setText("我是 textView1");
        Button button1 = (Button) newlayoutViewRef.findViewById(R.id.button1);
        button1.setText("我是 button1");

        setContentView(newlayoutViewRef);//本示例的核心代码
    }
}
```

程序运行结果如图 2.18 所示。

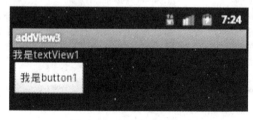

图 2.18　程序运行结果

2.5　界面布局的空间分配与权重

在 DIV+CSS 技术中，可以设置 DIV 标签的宽度为 100%，这样宽度就和剩余空间的宽度一样，在 android 中也存在这样的技术，可以达到同样的效果，使用的属性是 android:layout_weight。

新建名称为 layout_weight 的 Android 项目，将文件 main.xml 代码更改如下：

```xml
<?xml version="1.0" encoding="utf-8"?>
<LinearLayout xmlns:android="http://schemas.android.com/apk/res/android"
    android:orientation="vertical" android:layout_width="fill_parent"
    android:layout_height="fill_parent">
    <Button android:text="Button" android:id="@+id/button1"
        android:layout_width="fill_parent" android:layout_height="wrap_content"></Button>
    <Button android:text="Button" android:id="@+id/button2"
        android:layout_width="fill_parent" android:layout_height="wrap_content"></Button>
</LinearLayout>
```

从代码来看，并没有什么特别之处，布局中有两个按钮以垂直的方式来进行排列。程序运行效果如图 2.19 所示。

图 2.19　默认效果

在图 2.19 中的界面下方有大量的空白，有时候在做界面时不希望留有大面积的空白，希望某个控件将剩余的空白空间进行自适应的填充，属性 android:layout_weight 的作用就体现出来了，这个属性有个比喻，就是"分红"，可以通过这个属性把剩余的空间留给指定的 View 或 ViewGroup 控件，来达到填充空白空间的效果，再继续下面的实验。

将文件 main.xml 的代码更改如下：

```xml
<?xml version="1.0" encoding="utf-8"?>
<LinearLayout xmlns:android="http://schemas.android.com/apk/res/android"
    android:orientation="vertical" android:layout_width="fill_parent"
    android:layout_height="fill_parent">
    <Button android:text="Button" android:id="@+id/button1"
        android:layout_width="fill_parent" android:layout_height="wrap_content"></Button>
    <Button android:text="Button" android:id="@+id/button2"
        android:layout_width="fill_parent" android:layout_height="wrap_content"
        android:layout_weight="1"></Button>
</LinearLayout>
```

在上面的代码中将 id 为 button2 的按钮的 android:layout_weight 属性设置为 1，代表这个按钮占据所有的剩余空间，也就是将剩余的空间全部留给 button2，程序运行效果如图 2.20 所示。

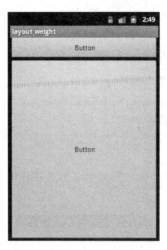

图 2.20 button2 占据所有剩余空间

再改动 main.xml 的代码如下：

```xml
<?xml version="1.0" encoding="utf-8"?>
<LinearLayout xmlns:android="http://schemas.android.com/apk/res/android"
    android:orientation="vertical" android:layout_width="fill_parent"
    android:layout_height="fill_parent">
    <Button android:text="Button" android:id="@+id/button1"
        android:layout_width="fill_parent" android:layout_height="wrap_content"
        android:layout_weight="1"></Button>
    <Button android:text="Button" android:id="@+id/button2"
        android:layout_width="fill_parent" android:layout_height="wrap_content"
        android:layout_weight="1"></Button>
</LinearLayout>
```

在上面的代码中可以发现，按钮 button1 和 button2 的属性 android:layout_weight 都设置了值 1，也就代表着 button1 和 button2 平分布局空间，程序运行效果如图 2.21 所示。

上面的步骤是使 button1 和 button2 平分布局空间，如果不想平分空间怎么办？再做一个实验，更改 main.xml 的代码如下：

```xml
<?xml version="1.0" encoding="utf-8"?>
<LinearLayout xmlns:android="http://schemas.android.com/apk/res/android"
    android:orientation="vertical" android:layout_width="fill_parent"
    android:layout_height="fill_parent">
    <Button android:text="Button" android:id="@+id/button1"
        android:layout_width="fill_parent" android:layout_height="wrap_content"
        android:layout_weight="2"></Button>
    <Button android:text="Button" android:id="@+id/button2"
        android:layout_width="fill_parent" android:layout_height="wrap_content"
        android:layout_weight="1"></Button>
</LinearLayout>
```

代码将 button1 的 android:layout_weight 属性设置为 2，将 button2 的 android:layout_weight 属性设置为 1，也就是说，剩下的整个布局界面空间被分成 3 等分，button1 占据 2，button2 占据 1，

程序运行效果如图 2.22 所示。

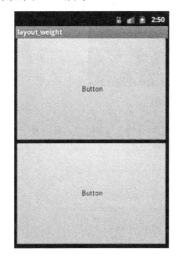
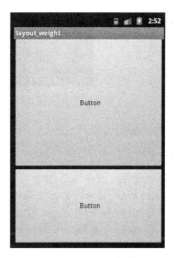

图 2.21 button1 和 button2 平分布局空间　　　图 2.22 不平分的效果

2.6 常用布局

在前面的学习中读者已经大体了解了 View 和 ViewGroup 组件的关系，也知道它们常用的直接或间接子类。下面将为大家介绍一下开发 android 应用程序比较重要的技术：UI 布局。

想学习 UI 布局就得掌握 ViewGroup 主要的布局子类，不同的布局方式控件也有不同的布局定位。

2.6.1 RelativeLayout 相对布局实验

布局 RelativeLayout 这种技术非常类似于 Web 开发中的 DIV+CSS 布局，使用这种布局可以使界面中的控件非常灵活化的摆放处理，而且还可以自适应不同的屏幕大小，但由于其具有非常大的灵活性，所以在学习时还要细心了解 RelativeLayout 布局主要的属性效果。

1. RelativeLayout 相对布局实验

下面开始学习 RelativeLayout 布局的使用，新建名称为 relativeLayout_test 的 Android 项目，将 main.xml 文件的代码更改如下：

```xml
<?xml version="1.0" encoding="utf-8"?>
<RelativeLayout xmlns:android="http://schemas.android.com/apk/res/android"
    android:layout_width="fill_parent" android:layout_height="fill_parent">
    <TextView android:id="@+id/textView1" android:layout_width="wrap_content"
        android:layout_height="wrap_content" android:text="I am text1"
        android:layout_marginLeft="10px" android:layout_marginTop="10px" />
</RelativeLayout>
```

上面的代码设置 TextView 有 left 和 top 外边距 margin 效果，程序运行效果如图 2.23 所示。

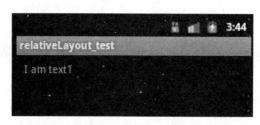

图 2.23　程序运行效果

图 2.23 运行的效果没有什么奇特之处，在平时的使用中也经常遇到，继续更改 main.xml 文件的代码如下：

```xml
<?xml version="1.0" encoding="utf-8"?>
<RelativeLayout xmlns:android="http://schemas.android.com/apk/res/android"
    android:layout_width="fill_parent" android:layout_height="fill_parent">
    <TextView android:id="@+id/textView1" android:layout_width="wrap_content"
        android:layout_height="wrap_content" android:text="I am text1"
        android:layout_marginLeft="10px" android:layout_marginTop="10px" />
    <TextView android:id="@+id/textView2" android:layout_width="wrap_content"
        android:layout_height="wrap_content" android:text="I am text2"
        android:layout_toRightOf="@+id/textView1" />
</RelativeLayout>
```

上面的代码在布局文件 main.xml 文件中添加了两个 TextView 控件，textView2 控件使用 android:layout_toRightOf 属性设置 textView2 控件在 textView1 的右边，程序运行效果如图 2.24 所示。

图 2.24　textView2 在 textView1 的右边

再继续更改 main.xml 文件的代码如下：

```xml
<?xml version="1.0" encoding="utf-8"?>
<RelativeLayout xmlns:android="http://schemas.android.com/apk/res/android"
    android:layout_width="fill_parent" android:layout_height="fill_parent">
    <TextView android:id="@+id/textView1" android:layout_width="wrap_content"
        android:layout_height="wrap_content" android:text="I am text1"
        android:layout_marginLeft="10px" android:layout_marginTop="10px" />
    <TextView android:id="@+id/textView2" android:layout_width="wrap_content"
        android:layout_height="wrap_content" android:text="I am text2"
        android:layout_below="@+id/textView1" />
</RelativeLayout>
```

从代码中可以看到，textView2 控件使用 android:layout_below 属性设置 textView2 控件在 textView1 的下方，程序运行效果如图 2.25 所示。

图 2.25　textView2 在 textView1 的下方

继续更改 main.xml 文件的代码如下：

```xml
<?xml version="1.0" encoding="utf-8"?>
<RelativeLayout xmlns:android="http://schemas.android.com/apk/res/android"
    android:layout_width="fill_parent" android:layout_height="fill_parent">
    <TextView android:id="@+id/textView1" android:layout_width="wrap_content"
        android:layout_height="wrap_content" android:text="I am text1"
        android:layout_marginLeft="10px" android:layout_marginTop="10px" />
    <TextView android:id="@+id/textView2" android:layout_width="wrap_content"
        android:layout_height="wrap_content" android:text="I am text2"
        android:layout_below="@+id/textView1" android:layout_marginTop="30px" />
</RelativeLayout>
```

上面的代码通过属性 android:layout_marginTop 设置 textView2 与上方的控件距离为 30px 单位，程序运行效果如图 2.26 所示。

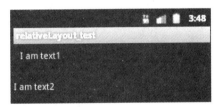

图 2.26　textView2 距离上方控件 30px

还要继续实验，继续更改 main.xml 文件的代码如下：

```xml
<?xml version="1.0" encoding="utf-8"?>
<RelativeLayout xmlns:android="http://schemas.android.com/apk/res/android"
    android:layout_width="fill_parent" android:layout_height="fill_parent">
    <TextView android:id="@+id/textView1" android:layout_width="wrap_content"
        android:layout_height="wrap_content" android:text="I am text1"
        android:layout_marginLeft="10px" android:layout_marginTop="10px" />
    <TextView android:id="@+id/textView2" android:layout_width="wrap_content"
        android:layout_height="wrap_content" android:text="I am text2"
        android:layout_below="@+id/textView1" android:layout_marginTop="30px"
        android:layout_alignParentRight="true" />
</RelativeLayout>
```

从上面的程序代码中可以发现，控件 textView2 使用属性 android:layout_alignParentRight 设置 textView2 当前控件在父控件的右边，也就是右侧对齐，但在这里一定要注意 textView2 控件的宽度设置值为 wrap_content，因为如果设置为 fill_parent，则 textView2 右边没有剩余的空间进行右对齐，程序运行效果如图 2.27 所示。

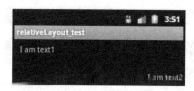

图 2.27 textView2 右对齐

最后一次更改了，将文件 main.xml 的代码更改如下：

```xml
<?xml version="1.0" encoding="utf-8"?>
<RelativeLayout xmlns:android="http://schemas.android.com/apk/res/android"
    android:layout_width="fill_parent" android:layout_height="fill_parent">
    <TextView android:id="@+id/textView1" android:layout_width="wrap_content"
        android:layout_height="wrap_content" android:text="I am text1"
        android:layout_marginLeft="10px" android:layout_marginTop="10px" />
    <TextView android:id="@+id/textView2" android:layout_width="wrap_content"
        android:layout_height="wrap_content" android:text="I am text2"
        android:layout_marginLeft="100px" android:layout_alignTop="@+id/textView1" />
</RelativeLayout>
```

上面的代码设置控件 textView2 与 textView1 的 top 值一样，这样的功能是通过属性 android:layout_alignTop 完成的，程序运行效果如图 2.28 所示。

在上面的示例中已经介绍了常用的对齐属性，其实 android 还有其他的对齐属性供程序员使用，列表如图 2.29 所示。

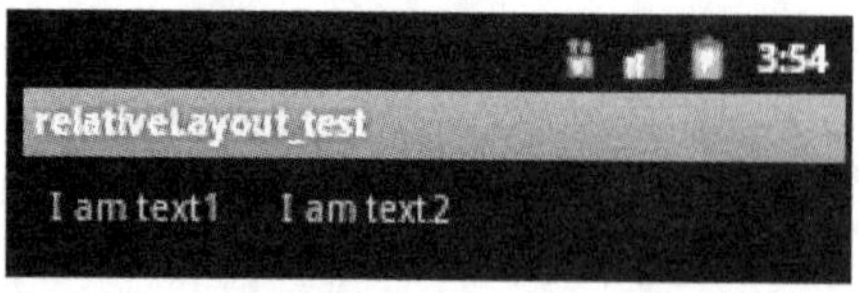

图 2.28 textView1 和 textView2 的 top 值一样

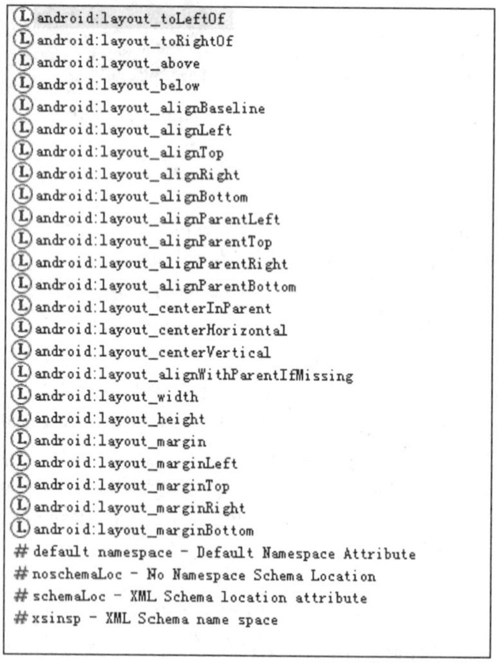

图 2.29 全部对齐属性集合

以控件高度的一半画一个水平线，这个水平线即是基线，属性 android:layout_alignBaseline 的作用是将控件定位到某一个控件文本 text 属性的基线上，也就是位置以 text 文本位置为对齐方式。

属性 android:layout_alignWithParentIfMissing 的含义是当对齐的控件不存在或不可见时参照父控件的位置。

从图 2.29 中已经看到对齐属性的不同仅仅在于上下左右等这些因素，在使用上没有非常大的区别。

2. RelativeLayout 相对布局动态创建控件及定位实验

上面的示例是将静态控件进行相对布局的定位，有时候动态创建控件时也需要相对布局的定位。新建名称为 relativeLayoutCreateView 的 Android 项目，将布局文件 main.xml 的代码更改如下：

```xml
<?xml version="1.0" encoding="utf-8"?>
<RelativeLayout xmlns:android="http://schemas.android.com/apk/res/android"
    android:layout_width="fill_parent" android:layout_height="fill_parent"
    android:id="@+id/relativeLayout">
    <TextView android:id="@+id/textView1" android:layout_width="wrap_content"
        android:layout_height="wrap_content" android:text="@string/hello"
        android:layout_marginLeft="30px" android:layout_marginTop="30px" />
</RelativeLayout>
```

程序代码并没有什么不同之处，仅仅 textView1 通过属性 android:layout_marginTop 和 android:layout_marginLeft 设置 top 和 left 距离各为 30px 像素。

改动 Activity 文件 Main.java 的代码如下：

```java
public class Main extends Activity {
    /** Called when the activity is first created. */
    @Override
    public void onCreate(Bundle savedInstanceState) {
        super.onCreate(savedInstanceState);
        setContentView(R.layout.main);

        ViewGroup vg = (ViewGroup) this.findViewById(R.id.relativeLayout);

        TextView textView1 = (TextView) this.findViewById(R.id.textView1);

        RelativeLayout.LayoutParams lp = new RelativeLayout.LayoutParams(
                LayoutParams.WRAP_CONTENT, LayoutParams.WRAP_CONTENT);
        lp.topMargin = 30;
        lp.leftMargin = 30;
        lp.addRule(RelativeLayout.RIGHT_OF, R.id.textView1);

        TextView textView2 = new TextView(this);
        textView2.setId(100);
        textView2.setText("zzzzzzzzzzzzz");
        textView2.setLayoutParams(lp);

        vg.addView(textView2);

        RelativeLayout.LayoutParams lpRef2 = new RelativeLayout.LayoutParams(
                LayoutParams.WRAP_CONTENT, LayoutParams.WRAP_CONTENT);
        lpRef2.addRule(RelativeLayout.BELOW, 100);
```

```
            lpRef2.addRule(RelativeLayout.ALIGN_LEFT, 100);
            TextView textView3 = new TextView(this);
            textView3.setText("我是 TextView3");
            textView3.setBackgroundColor(Color.RED);
            textView3.setLayoutParams(lpRef2);

            vg.addView(textView3);

    }
}
```

程序运行效果如图 2.30 所示。

图 2.30　程序运行效果

3. RelativeLayout 相对布局小实验——创建登录界面

到此常用的相对布局属性已经介绍完毕，下面就使用相对布局来创建一个登录界面。

创建名称为 RelativeLayout_login 的 Android 项目，更改 main.xml 布局文件的代码如下：

```
<?xml version="1.0" encoding="utf-8"?>
<RelativeLayout xmlns:android="http://schemas.android.com/apk/res/android"
    android:orientation="vertical" android:layout_width="fill_parent"
    android:layout_height="fill_parent" android:padding="10px">
    <TextView android:id="@+id/textView1" android:text="用户名："
        android:layout_width="fill_parent" android:layout_alignParentLeft="true"
        android:layout_height="wrap_content"></TextView>
    <EditText android:layout_width="fill_parent" android:id="@+id/editText1"
        android:layout_height="wrap_content" android:text=""
        android:layout_below="@+id/textView1"></EditText>
    <TextView android:id="@+id/textView2" android:text="用户名："
        android:layout_width="fill_parent" android:layout_alignParentLeft="true"
        android:layout_height="wrap_content" android:layout_below="@+id/editText1"></TextView>
    <EditText android:layout_width="fill_parent" android:id="@+id/editText2"
        android:layout_height="wrap_content" android:text=""
        android:layout_below="@+id/textView2"></EditText>
    <Button android:text="关闭" android:layout_width="100px"
        android:id="@+id/button2" android:layout_height="wrap_content"
        android:layout_alignParentRight="true" android:layout_below="@+id/editText2"></Button>
    <Button android:text="登录" android:layout_width="100px"
        android:id="@+id/button1" android:layout_height="wrap_content"
        android:layout_below="@+id/editText2" android:layout_toLeftOf="@+id/button2"></Button>
</RelativeLayout>
```

程序运行效果如图 2.31 所示。

第2章 View 与 ViewGroup 类和控件事件

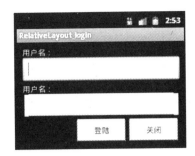

图 2.31 使用相对布局实现登录界面

在这里留意一个小知识点：

- android: layout_gravity 指本元素相对于父元素的对齐方向。
- android: gravity 指定本元素的所有子元素的重力方向。

2.6.2 TableLayout 布局的使用

在 Android 中也支持用表格进行界面的布局，使用的标签是<TableLayout>，还有 TableLayout 的特性是垂直排列，TableRow 的特性是水平排列。

1. TableLayout 布局的使用

为了在 Android 中实验表格布局的效果，新建名称为 TableLayoutTest 的 Android 项目，将布局文件 main.xml 的代码更改如下：

```xml
<?xml version="1.0" encoding="utf-8"?>
<LinearLayout xmlns:android="http://schemas.android.com/apk/res/android"
    android:orientation="vertical" android:layout_width="fill_parent"
    android:layout_height="fill_parent">
    <TableLayout android:layout_width="fill_parent"
        android:layout_height="wrap_content" android:id="@+id/tableLayout1">
        <TableRow android:gravity="left">
            <TextView android:id="@+id/usernameTextView" android:text="账号："
                android:layout_width="wrap_content" android:layout_height="wrap_content"></TextView>
            <EditText android:id="@+id/usernameEditText"
                android:layout_weight="1" android:layout_height="wrap_content"></EditText>
        </TableRow>
        <TableRow android:gravity="left">
            <TextView android:id="@+id/passwordTextView" android:text="密码："
                android:layout_width="wrap_content" android:layout_height="wrap_content"></TextView>
            <EditText android:id="@+id/passwordEditText"
                android:layout_weight="1" android:layout_height="wrap_content"></EditText>
        </TableRow>
        <LinearLayout android:gravity="center"
            android:orientation="horizontal" android:layout_width="fill_parent"
            android:layout_height="fill_parent">
            <Button android:id="@+id/button1" android:layout_width="100px"
                android:layout_height="wrap_content" android:text="登录"></Button>
            <Button android:id="@+id/button2" android:layout_width="100px"
```

```
                android:layout_height="wrap_content" android:text="取消"></Button>
            </LinearLayout>
        </TableLayout>
</LinearLayout>
```

从上面的布局代码中可以看到<TableLayout>与<LinearLayout>是可以嵌套的，使用<LinearLayout>布局的主要目的是将两个 button 按钮进行水平排列。程序初始运行效果如图 2.32 所示。

图 2.32　程序运行效果

当然，<TableLayout>表格布局也可以互相嵌套，将文件 main.xml 的代码更改如下：

```xml
<?xml version="1.0" encoding="utf-8"?>
<LinearLayout xmlns:android="http://schemas.android.com/apk/res/android"
    android:orientation="vertical" android:layout_width="fill_parent"
    android:layout_height="fill_parent">
    <TableLayout android:layout_width="fill_parent"
        android:layout_height="wrap_content" android:id="@+id/tableLayout1">
        <TableRow android:gravity="left">
            <TextView android:id="@+id/usernameTextView" android:text="账号："
                android:layout_width="wrap_content" android:layout_height="wrap_content"></TextView>
            <EditText android:id="@+id/usernameEditText"
                android:layout_weight="1" android:layout_height="wrap_content"></EditText>
        </TableRow>
        <TableRow android:gravity="left">
            <TextView android:id="@+id/passwordTextView" android:text="密码："
                android:layout_width="wrap_content" android:layout_height="wrap_content"></TextView>
            <EditText android:id="@+id/passwordEditText"
                android:layout_weight="1" android:layout_height="wrap_content"></EditText>
        </TableRow>
        <TableLayout android:layout_width="fill_parent"
            android:layout_height="wrap_content" android:id="@+id/tableLayout1">
            <TableRow android:gravity="center">
                <Button android:id="@+id/button1" android:layout_width="100px"
                    android:layout_height="wrap_content" android:text="登录"></Button>
                <Button android:id="@+id/button2" android:layout_width="100px"
                    android:layout_height="wrap_content" android:text="取消"></Button>
            </TableRow>
        </TableLayout>
    </TableLayout>
</LinearLayout>
```

程序运行后，也能实现相同的布局效果，运行结果如图 2.33 所示。

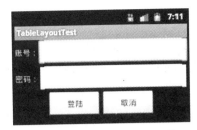

图 2.33　另一种布局显示效果

2. TableLayout 布局的进一步使用

关于标签<TableLayout>也有一些比较独特的属性，本示例就来演示一下。

新建名称为 tableLayoutMore 的 Android 项目，将文件 main.xml 代码更改如下：

```xml
<?xml version="1.0" encoding="utf-8"?>
<LinearLayout xmlns:android="http://schemas.android.com/apk/res/android"
    android:orientation="vertical" android:layout_width="fill_parent"
    android:layout_height="fill_parent">

    <TableLayout android:layout_width="fill_parent"
        android:layout_height="wrap_content">
        <TableRow>
            <TextView android:text="aaaaa"></TextView>
            <TextView android:text="bbbbb"></TextView>
            <TextView android:text="ccccc"></TextView>
        </TableRow>
    </TableLayout>
    <TextView android:layout_width="fill_parent"
        android:layout_height="wrap_content"   android:text="----华丽的分割线--------------------"></TextView>

    <TableLayout android:layout_width="fill_parent"
        android:layout_height="wrap_content" android:stretchColumns="0,1">
        <TableRow>
            <TextView android:text="aaaaaaaaa 我是小霸王我要占位置"></TextView>
            <TextView android:text="--被挤了我也是小霸王中的霸王"></TextView>
            <TextView android:text="ccccc"></TextView>
        </TableRow>
    </TableLayout>

    <TextView android:layout_width="fill_parent"
        android:layout_height="wrap_content"   android:text="----华丽的分割线--------------------"></TextView>
    <TableLayout android:layout_width="fill_parent"
        android:layout_height="wrap_content" android:shrinkColumns="0,1">
        <TableRow>
            <TextView android:text="aaaaaaaaa 我是小霸王我要占位置"></TextView>
            <TextView android:text="bbbbbbbbb 我也是小霸王中的霸王"></TextView>
            <TextView android:text="--我有我的位置哈哈哈哈哈哈--"></TextView>
        </TableRow>
```

```xml
    </TableLayout>

    <TextView android:layout_width="fill_parent"
        android:layout_height="wrap_content"    android:text="----华丽的分割线-----------------------"></TextView>

    <TableLayout android:layout_width="fill_parent"
        android:layout_height="wrap_content">
        <TableRow>
            <TextView android:text="11111"></TextView>
            <TextView android:text="22222"></TextView>
            <TextView android:text="33333"></TextView>
        </TableRow>
        <TableRow>
            <TextView android:text=""></TextView>
            <TextView android:text="22222"></TextView>
            <TextView android:text="33333"></TextView>
        </TableRow>
    </TableLayout>

    <TextView android:layout_width="fill_parent"
        android:layout_height="wrap_content"    android:text="----华丽的分割线-----------------------"></TextView>

    <TableLayout android:layout_width="fill_parent"
        android:layout_height="fill_parent" android:collapseColumns="1,2">
        <TableRow>
            <TextView android:text="xxxxx"></TextView>
            <TextView android:text="yyyyy"></TextView>
            <TextView android:text="zzzzz"></TextView>
        </TableRow>
    </TableLayout>

</LinearLayout>
```

程序运行效果如图 2.34 所示。

图 2.34　程序运行效果

在上面的布局文件代码中使用了 3 个陌生的属性，下面分别介绍一下。

- 属性 android:collapseColumns="1,2"的作用是隐藏表里的第 2 和第 3 列，因为索引从 0 开始。
- 属性 android:stretchColumns="0,1"的作用是设置第 1 和第 2 列为可拉伸的列，右边列的数据一直往后排，有可能排到屏幕的外面呈不显示的状态，主要用于列中的文字比较少的情况下，列具有伸缩特点。
- 属性 android:shrinkColumns="0,1"的作用是设置第 1 和第 2 列为可伸缩的列，如果列里面的数据过多导致后面的数据无法显示出来时，则自动换行进行显示。

如果属性值为*星号则代表全部的列。

下面的代码实现 1 行 3 列的效果，这里再次实现 android:stretchColumns 属性的应用效果：

```xml
<?xml version="1.0" encoding="utf-8"?>
<LinearLayout xmlns:android="http://schemas.android.com/apk/res/android"
    android:orientation="vertical" android:layout_width="fill_parent"
    android:layout_height="fill_parent">
    <TableLayout android:layout_height="wrap_content"
        android:id="@+id/tableLayout1" android:layout_width="match_parent"
        android:stretchColumns="0,1">
        <TableRow android:id="@+id/tableRow1" android:layout_width="wrap_content"
            android:layout_height="wrap_content">
            <Button android:text="Button1" android:id="@+id/button1"
                android:layout_width="wrap_content" android:layout_height="wrap_content"></Button>
            <Button android:text="Button2" android:id="@+id/button1"
                android:layout_width="wrap_content" android:layout_height="wrap_content"></Button>
            <Button android:text="Button3" android:id="@+id/button1"
                android:layout_width="wrap_content" android:layout_height="wrap_content"></Button>
        </TableRow>
    </TableLayout>
</LinearLayout>
```

运行效果如图 2.35 所示。

图 2.35　程序运行效果

3. 在 TableLayout 布局中动态创建行

可以在 TableLayout 布局中动态创建 TableRow 行对象，还可以动态添加 View，新建名称为 createTableLayoutRow 的 Android 项目，main.xml 布局文件代码如下：

```xml
<?xml version="1.0" encoding="utf-8"?>
```

```xml
<TableLayout android:id="@+id/tableLayout1"
    android:layout_width="fill_parent" android:layout_height="fill_parent"
    xmlns:android="http://schemas.android.com/apk/res/android">
</TableLayout>
```

文件 Main.java 代码如下：

```java
public class Main extends Activity {
    private TableLayout tableLayout;

    @Override
    public void onCreate(Bundle savedInstanceState) {
        super.onCreate(savedInstanceState);
        setContentView(R.layout.main);

        tableLayout = (TableLayout) this.findViewById(R.id.tableLayout1);

        Button button1 = new Button(this);
        button1.setText("button1");
        Button button2 = new Button(this);
        button2.setText("button2");
        Button button3 = new Button(this);
        button3.setText("button3");
        //第 3 个参数为 weight 的值
        TableRow.LayoutParams params = new TableRow.LayoutParams(
                LayoutParams.WRAP_CONTENT, LayoutParams.WRAP_CONTENT, 1);
        button3.setLayoutParams(params);

        TableRow newRow1 = new TableRow(this);
        newRow1.addView(button1);

        TableRow newRow2 = new TableRow(this);
        newRow2.addView(button2);
        newRow2.addView(button3);

        tableLayout.addView(newRow1);
        tableLayout.addView(newRow2);

    }
}
```

程序运行结果如图 2.36 所示。

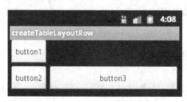

图 2.36　程序运行结果

2.6.3 FrameLayout 布局的使用

在 Android 布局方式中还有一种框架布局,这种布局的特点是所有的界面控件默认都显示在屏幕的左上角,这种布局非常适合于"展示"类的效果,比如图片的展示、产品的展示等环境中。

新建名称为 FrameLayout 的 Android 项目,将文件 main.xml 的代码更改如下:

```xml
<?xml version="1.0" encoding="utf-8"?>
<FrameLayout xmlns:android="http://schemas.android.com/apk/res/android"
    android:orientation="vertical" android:layout_width="fill_parent"
    android:layout_height="fill_parent">

    <Button android:text="Button" android:id="@+id/button1"
        android:layout_width="300px" android:layout_height="300px"
        android:background="#ff0000"></Button>
    <Button android:text="Button" android:id="@+id/button2"
        android:layout_width="200px" android:layout_height="200px"
        android:background="#0000ff"></Button>
    <Button android:text="Button" android:id="@+id/button3"
        android:layout_width="100px" android:layout_height="100px"
        android:background="#00ff00"></Button>

</FrameLayout>
```

程序运行效果如图 2.37 所示。

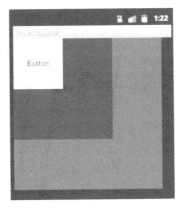

图 2.37 FrameLayout 运行效果

2.6.4 AbsoluteLayout 布局的实验

在 Android 中还支持绝对坐标的布局,新建名称为 absLayoutTest 的 Android 项目,更改布局文件 main.xml 的代码如下:

```xml
<?xml version="1.0" encoding="utf-8"?>
<AbsoluteLayout android:id="@+id/absoluteLayout1"
    android:layout_width="fill_parent" android:layout_height="fill_parent"
    xmlns:android="http://schemas.android.com/apk/res/android">
    <Button android:layout_width="wrap_content" android:text="Button"
        android:layout_height="wrap_content" android:id="@+id/button1"
```

```
            android:layout_x="62dip" android:layout_y="64dip"></Button>
        <Button android:layout_width="wrap_content" android:layout_x="194dip"
            android:text="Button" android:layout_height="wrap_content" android:id="@+id/button3"
            android:layout_y="224dip"></Button>
        <Button android:layout_width="wrap_content" android:text="Button"
            android:layout_height="wrap_content" android:id="@+id/button2"
            android:layout_x="120dip" android:layout_y="142dip"></Button>
</AbsoluteLayout>
```

通过设置 android:layout_x 和 android:layout_y 属性就可以确定控件的摆放位置。程序运行结果如图 2.38 所示。

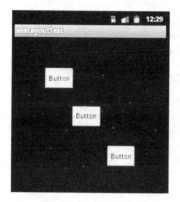

图 2.38 绝对坐标的显示效果

2.6.5 用程序来实现 margin 的实验

前面介绍了在 LinearLayout 布局中将动态创建的 EditText 控件的宽度设置为和屏幕一样宽，有些时候还需要设置动态创建控件的 margin 属性，但可惜的是，View 和 ViewGroup 对象并没有 setMargin 类似的方法，所以必须还要依赖于 LinearLayout.LayoutParams 布局参数类来实现。

新建名称为 marginTest 的 Android 项目，布局文件 main.xml 的代码很简单，如下所示：

```
<?xml version="1.0" encoding="utf-8"?>
<LinearLayout xmlns:android="http://schemas.android.com/apk/res/android"
    android:orientation="vertical" android:layout_width="fill_parent"
    android:layout_height="fill_parent" android:id="@+id/linearLayout1">
</LinearLayout>
```

将文件 Main.java 的代码更改如下：

```
public class Main extends Activity {
    @Override
    public void onCreate(Bundle savedInstanceState) {
        super.onCreate(savedInstanceState);
        setContentView(R.layout.main);

        Button button = new Button(this);
        button.setText("this is my button");
        button.setWidth(200);
        button.setHeight(100);
```

```
        LinearLayout.LayoutParams lpRef = new LinearLayout.LayoutParams(
                LayoutParams.WRAP_CONTENT, LayoutParams.WRAP_CONTENT);
        lpRef.setMargins(50, 0, 0, 0);

        button.setLayoutParams(lpRef);

        LinearLayout llRef = (LinearLayout) this
                .findViewById(R.id.linearLayout1);

        llRef.addView(button);

    }
}
```

程序运行后，margin 的效果出现了，如图 2.39 所示。

图 2.39　margin 效果

由于对象 android.widget.LinearLayout.LayoutParams 类的父类是：android.view.ViewGroup.MarginLayoutParams，如图 2.40 所示。

而 MarginLayoutParams 对象的子类有 FrameLayout.LayoutParams、LinearLayout.LayoutParams、RelativeLayout.LayoutParams、RadioGroup.LayoutParams、TableLayout.LayoutParams、TableRow.LayoutParams，如图 2.41 所示。

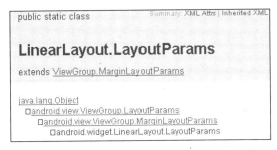

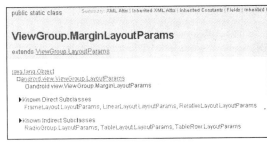

图 2.40　LinearLayout.LayoutParams 类的继承关系　　图 2.41　MarginLayoutParams 对象的子类

所以，MarginLayoutParams 对象的子类也有设置 margin 属性的方法，继承下来了。

2.7　控件事件

在 Android 中使用控件的事件主要有两种方式：回调法及监听法。由于接口监听法已经在第 1

章中有所涉及，因此这里不会重复讲述。

所谓回调事件，大多数是指重写 Activity 对象和 View 对象中的回调事件方法。

1. onKeyDown 和 onKeyUp 方法

回调事件 onKeyDown 方法声明在接口 KeyEvent.Callback 中，如图 2.42 所示。

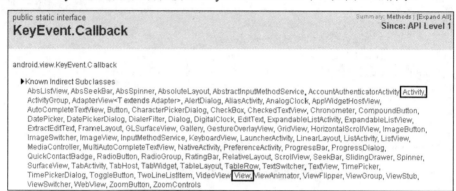

图 2.42　KeyEvent.Callback 接口的信息

从图 2.42 中可以看到，Activity 和 View 对象实现了这个接口，那么这两个对象的子类也同样持有这个方法。

接口 KeyEvent.Callback 中的方法列表如图 2.43 所示。

Public Methods	
abstract boolean	onKeyDown (int keyCode, KeyEvent event) Called when a key down event has occurred.
abstract boolean	onKeyLongPress (int keyCode, KeyEvent event) Called when a long press has occurred.
abstract boolean	onKeyMultiple (int keyCode, int count, KeyEvent event) Called when multiple down/up pairs of the same key have occurred in a row.
abstract boolean	onKeyUp (int keyCode, KeyEvent event) Called when a key up event has occurred.

图 2.43　接口 KeyEvent.Callback 中的方法

其中方法 onKeyDown(int keyCode, KeyEvent event)的参数解释为：

- 参数 keyCode：代表按下了哪个键。
- 参数 event：按下键盘后的相关按键信息的封装实体类，也就是事件信息对象。
- 返回值类型为 boolean 布尔：如果返回为 true，则代表终止这个事件的传播；如果返回为 false，则代表传播这个事件。

用实验来验证一下 onKeyDown 回调方法的使用，新建名称为 event_1 的 Android 项目，新建自定义按钮 MyButton.java 文件，代码如下：

```
public class MyButton extends Button {
```

```java
    public MyButton(Context context) {
        super(context);
    }

    @Override
    public boolean onKeyDown(int keyCode, KeyEvent event) {
        // super.onKeyDown(keyCode, event);
        Log.v("!MyButton event.hashCode()=", "" + event.hashCode());
        Log.v("!", "进入了 MyButton 的 onKeyDown 方法 keyCode=" + keyCode
                + " event.isLongPress()=" + event.isLongPress());
        return false;
    }
}
```

Activity 文件 Main.java 的代码如下：

```java
public class Main extends Activity {
    private LinearLayout ll;

    @Override
    public void onCreate(Bundle savedInstanceState) {
        super.onCreate(savedInstanceState);
        setContentView(R.layout.main);
        ll = (LinearLayout) this.findViewById(R.id.linearLayout1);

        MyButton mybutton = new MyButton(this);
        mybutton.setText("我的文本");
        ll.addView(mybutton);
    }

    @Override
    public boolean onKeyDown(int keyCode, KeyEvent event) {
        Log.v("!", "进入了 Main 的 onKeyDown 方法");
        Log.v("!Main event.hashCode()=", "" + event.hashCode());
        return true;
    }
}
```

运行这个项目，使 Button 控件获得焦点，然后按下键盘的 1 数字键，出现的效果如图 2.44 所示。

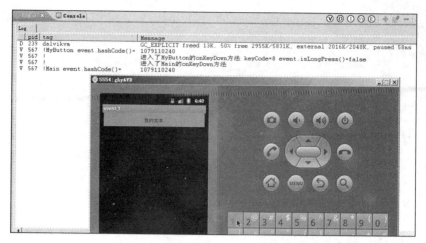

图 2.44 Button 获得焦点按下数字 1 键

从图 2.44 中可以看到，不仅 Button 控件执行了 onKeyDown 回调，而且 Activity 对象 Main.java 也执行了 onKeyDown 方法，原因是在 MyButton 控件的 onKeyDown 方法中返回了 false，代表这个事件并没有彻底结束，而是进行事件的传播，如果将 MyButton.java 的 onKeyDown 方法由 return false; 改回 return true; 后再运行这个项目，再一次使 MyButton 获得焦点并按下数字 1 键，如图 2.45 所示。

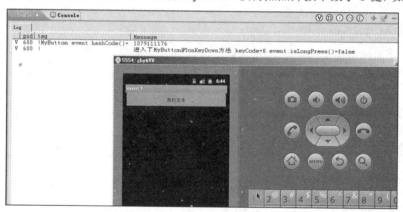

图 2.45 返回 true 的效果

从图 2.45 中可以看到，由于 MyButton 控件中的 onKeyDown 返回为 true，事件并没有被传播，所以只执行了 MyButton 中的 onKeyDown 方法。

onKeyUp 方法和 onKeyDown 方法使用上完全一样。

2. onTouchEvent 方法

方法 public boolean onTouchEvent(MotionEvent event) 的功能是获得用户触摸或滑动屏幕的动作，参数 MotionEvent 的作用是封装当前触摸屏幕的信息的实体，包括位置、时间等属性，本示例就用一个实验来实现一下触摸屏幕时动态创建一个 Button 控件，当滑动时 Button 也进行跟随的效果。

新建名称为 event_2 的 Android 项目，文件 Main.java 的代码如下：

```java
public class Main extends Activity {
    private Button button;
    private AbsoluteLayout absoluteLayout1;

    @Override
    public void onCreate(Bundle savedInstanceState) {
        super.onCreate(savedInstanceState);
        setContentView(R.layout.main);
        absoluteLayout1 = (AbsoluteLayout) this
                .findViewById(R.id.absoluteLayout1);
    }

    OnClickListener listener = new OnClickListener() {
        public void onClick(View arg0) {
            if (button != null) {
                absoluteLayout1.removeView(button);
            }
        }
    };

    @Override
    public boolean onTouchEvent(MotionEvent event) {
        AbsoluteLayout.LayoutParams params = new AbsoluteLayout.LayoutParams(
                LayoutParams.WRAP_CONTENT, LayoutParams.WRAP_CONTENT, Integer
                        .parseInt("" + (int) event.getX()), Integer.parseInt(""
                        + (int) (event.getY() - 50)));
        Log.v("!", "x=" + event.getX() + " y=" + event.getY());
        if (event.getAction() == MotionEvent.ACTION_DOWN) {
            button = new Button(this);
            button.setOnClickListener(listener);
            button.setText("按下了！");
            button.setLayoutParams(params);
            absoluteLayout1.addView(button);
        }
        if (event.getAction() == MotionEvent.ACTION_MOVE) {
            button.setText("移动了！");
            button.setLayoutParams(params);
        }
        if (event.getAction() == MotionEvent.ACTION_UP) {
            button.setText("抬起了！");
            button.setLayoutParams(params);
        }
        return true;
    }
}
```

当用鼠标按下屏幕时，动态创建出 1 个 Button 控件，如图 2.46 所示。当滑动鼠标时，Button 的 text 文本属性发生改变，并且按钮的位置也跟着光标移动，如图 2.47 所示。

图 2.46 动态创建出的 Button 按钮

图 2.47 Button 按钮随着光标移动

当抬起鼠标时，Button 的 text 属性也被改变，效果如图 2.48 所示。

图 2.48 抬起鼠标

onTouchEvent 方法返回值的作用和 onKeyDown 的返回值作用一样。

提 示

方法 getRawX()和 getRawY()获得的是相对屏幕的位置，而方法 getX()和 getY()获得的是相对 view 控件的触摸位置坐标。

3. onTrackballEvent 方法

方法 public boolean onTrackballEvent(MotionEvent event)的作用是感应轨迹球事件，新建名称为 event_3 的 Android 项目，文件 Main.java 的代码如下：

```
public class Main extends Activity {
    @Override
    public void onCreate(Bundle savedInstanceState) {
        super.onCreate(savedInstanceState);
        setContentView(R.layout.main);
    }

    @Override
    public boolean onTrackballEvent(MotionEvent event) {
        if (event.getAction() == MotionEvent.ACTION_DOWN) {
            Log.v("!", "按下轨迹球了");
        }
        if (event.getAction() == MotionEvent.ACTION_UP) {
```

```
            Log.v("!", "抬起轨迹球了");
        }
        return super.onTrackballEvent(event);
    }
}
```

程序运行后在 AVD 中按下 F6 键激活轨迹球，单击鼠标再抬起，LogCat 打印出来的信息如图 2.49 所示。

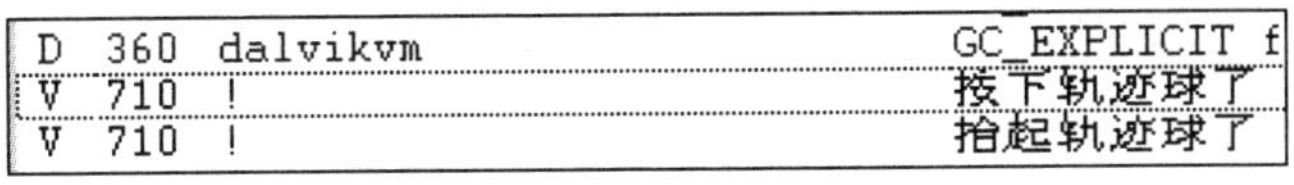

```
D  360  dalvikvm              GC_EXPLICIT f
V  710  !                     按下轨迹球了
V  710  !                     抬起轨迹球了
```

图 2.49　单击轨迹球的日志

第 3 章 Android 的 UI 控件

在第 3 章将和大家一起讨论的是 Anroid 中的控件，Android 中的控件较多，但 Android 是一个客户端的技术，所以 UI 界面的展示是每一个 Anddroid 程序员必须要面对的问题，好在 google 公司提供了很多功能强大的控件可供使用，了解并熟悉这些控件的使用是每个 Android 程序员应该面对的问题，而并不仅仅只是接触一些 TextView 或 EditText 这样的控件，从高级应用上来讲，有些控件的使用原理或使用方式，或控件自带的基本功能都对以后的开发提供了非常好的思路，所以掌握 Android 控件非常重要。

本章应该着重掌握以下知识点：

- Android 中针对控件的 Adapter 适配器技术
- ListView 控件的熟悉使用
- 对话框 Dialog 的使用
- GridView 控件的使用
- TabHost 控件的使用
- 9Patch 工具的使用

3.1 UI 控件与 Adapter 和 ListView 对象

在软件开发行业，如果做一些与 UI 界面有关的技术，就避免不了与许多种类的界面控件打交道。Android 同.Net 平台的 WinForm 技术一样，软件项目也都是由许多控件组合而成，所以学习 Android 必须要对其内部的常用控件达到熟练使用的程度。

Android 常用控件的列表如图 3.1 所示。

从图 3.1 中的英语控件名称可以大致了解到，Android 控件的数量和功能比较繁多，而且大多数都是非常常用的控件，但在使用这些控件时有一个非常重要的知识点不得不提，那就是 Adapter 接口和 ListView 对象，在介绍控件之前先来介绍一下它们之间的关系。

从以下代码：

`package test.test.run;`

图 3.1 Android 常用控件列表

```
import android.app.Activity;
import android.os.Bundle;
import android.widget.Adapter;
import android.widget.ListView;

public class Main extends Activity {
    @Override
    public void onCreate(Bundle savedInstanceState) {
        super.onCreate(savedInstanceState);
        setContentView(R.layout.main);

        Adapter adapter1;
        ListView listView1;

    }
}
```

从上述代码可以看到，Adapter 和 ListView 控件都是在 android.widget 包中，查看一下 doc，就可以了解到这两个对象的详细信息。

3.2 Adapter 接口

从图 3.2 中可以看到，Adapter 是一个接口，并且这个接口有很多的实现类或子接口，比如常用的有 ArrayAdapter、ListAdapter 和 SimpleAdapter 等。

那么这个 Adapter 接口到底有什么作用呢？其实 Adapter 接口的主要作用是在具有 Adapter 特性的 View 控件和数据源之间架起一个桥梁，通过 Adapter 接口就可以实现将数据源中的数据显示到 View 控件中，Adapter 也支持对数据源的访问，访问数据源的方法列表如图 3.3 所示。

图 3.2 Adapter 接口信息　　　　　图 3.3 Adapter 访问数据源的方法

接口 Adapter 还可以把数据源中的每一个数据条目变成一个 View 控件显示到布局界面上供用户查看和使用。

3.3 ListAdapter 接口

Adapter 接口提供了对数据源访问的基本形式，比如取得数据 id 的方法 getItemId(int position)，取得数据内容的方法 getItem(int position)等，Adapter 接口有一个最重要的子接口即 ListAdapter 接口，这个接口的作用是使 ListView 控件与数据源之间建立起一个桥梁，通过这个桥梁，ListAdapter 接口就可以把数据源中的数据显示到 ListView 控件中，接口 ListAdapter 声明如图 3.4 所示。

图 3.4　ListAdapter 接口的声明与详细信息

从图 3.4 中可以看到，ArrayAdapter 是 ListAdapter 的直接子类，而 ArrayAdapter 是泛型类，那么比较好地使用自定义 Adapter 对象的方式是继承自 ArrayAdapter 类。

新建 extendsArrayAdapter 项目，新建 Userinfo.java 类，有 3 个属性，即 id、username 和 password。
新建 Adapter 适配器类文件 GhyArrayAdapter.java，代码如下：

```
package extadapter;

import java.util.List;

import android.content.Context;
import android.widget.ArrayAdapter;
import entity.Userinfo;

public class GhyArrayAdapter extends ArrayAdapter<Userinfo> {
    public GhyArrayAdapter(Context context, int textViewResourceId,
            List<Userinfo> objects) {
        super(context, textViewResourceId, objects);
    }
}
```

Activity 对象 Main.java 核心代码如下：

```
public class Main extends Activity {
    @Override
    public void onCreate(Bundle savedInstanceState) {
        super.onCreate(savedInstanceState);
        setContentView(R.layout.main);

        List<Userinfo> userinfoList = new ArrayList<Userinfo>();
        for (int i = 0; i < 10; i++) {
```

```
            Userinfo userinfo = new Userinfo("" + (i + 1),
                    "username" + (i + 1), "password" + (i + 1));
            userinfoList.add(userinfo);
        }

        GhyArrayAdapter adapter = new GhyArrayAdapter(this,
                android.R.layout.simple_list_item_1, userinfoList);
        System.out.println(adapter.getItem(2).getUsername());

    }
}
```

程序运行后正确打印出指定索引位置的 username 属性值，如图 3.5 所示。

通过这个示例可以看到继承自 ArrayAdapter 类可以非常方便地操作集合中的数据，而这样的功能在 BaseAdapter 中却不存在，因为 BaseAdapter 提供的方法功能非常有限，如图 3.6 所示。

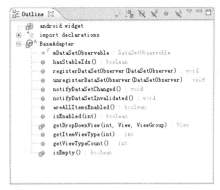

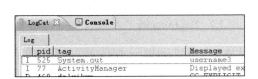

图 3.5　正确打印 username 值　　　　图 3.6　BaseAdapter 方法列表

那么，在上文提到的 ListView 又是什么呢？下面我们来介绍 ListView 对象的使用。

3.4　ListView 对象

对象 ListView 是实现一个垂直滚动的列表，列表中的数据来自于 ListAdapter 对象，它的类声明信息如图 3.7 所示。

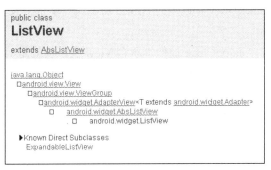

图 3.7　ListView 对象的声明

所以 ListView 对象结合 ListAdapter 的实现类就能实现将 ListAdapter 实现类中的数据显示到垂直列表中。而 ListAdapter 接口的实现类有许多，比较常用的有 ArrayAdapter 对象。

3.5　ArrayAdapter 对象

ArrayAdapter 对象是 ListAdapter 的实现类，类声明如图 3.8 所示。

但在图 3.8 中并没有看到 ListAdapter 字样，因为 ArrayAdapter 类的父类 BaseAdapter 实现了 ListAdapter 接口，BaseAdapter 类声明如图 3.9 所示。

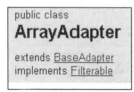
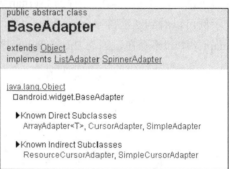

图 3.8　ArrayAdapter 类声明　　　　图 3.9　BaseAdapter 类声明

默认情况下，ArrayAdapter 对象将其中的数据显示到 1 个 TextView 控件中，所以在布局文件中一定要有这个 TextView 控件，但 Android 本身已经提供了许多默认的布局文件可供程序员使用，不需要再重复创建它们就可以将 Adapter 中的数据显示到 View 中。

上面介绍的就是适配器对象 Adapter 及实现类的大体结构，后面将介绍 Android 中的控件，而大多数控件是需要 Adapter 适配器对象的，请拭目以待。

3.6　AnalogClock 和 DigitalClock 控件

在 Android 的控件列表中，有两种时钟控件：AnalogClock 和 DigitalClock 控件，这两种控件的用法非常简单，这两个对象中都没有过多的 API 方法，属"拿来主义"。其中 AnalogClock 控件的类继承结构如图 3.10 所示。

从图 3.10 中可以看到，AnalogClock 控件继承自 View 类，也就是实现了基本的事件处理及自绘的功能。而 DigitalClock 控件的类继承结构如图 3.11 所示。

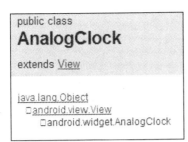

图 3.10　AnalogClock 控件的类继承结构　　图 3.11　DigitalClock 控件的类继承结构

从图 3.11 中可以发现，DigitalClock 控件继承自 TextView 对象，所以确定 DigitalClock 控件可以显示一些文本及设置文本的一些特性，例如文字颜色、字体大小和文本背景颜色等。

进入 Eclipse，创建名称为 ui_1 的 Android 项目，在 main.xml 文件中加入 AnalogClock 和 DigitalClock 控件，加入后的效果如图 3.12 所示。

图 3.12　Main.xml 文件中的 AnalogClock 和 DigitalClock 控件

加入控件后的 main.xml 文件的代码如下：

```xml
<?xml version="1.0" encoding="utf-8"?>
<LinearLayout xmlns:android="http://schemas.android.com/apk/res/android"
    android:orientation="vertical" android:layout_width="fill_parent"
    android:layout_height="fill_parent">
    <TextView android:layout_width="fill_parent"
        android:layout_height="wrap_content" android:text="@string/hello" />
    <AnalogClock android:id="@+id/analogClock1"
        android:layout_width="wrap_content" android:layout_height="wrap_content"></AnalogClock>
    <DigitalClock android:text="DigitalClock" android:id="@+id/digitalClock1"
        android:layout_width="wrap_content" android:layout_height="wrap_content"></DigitalClock>
</LinearLayout>
```

与布局文件 main.xml 对应的 Activity 类 Main.java 代码更改如下：

```java
public class Main extends Activity {

    private static final String TAG = "MainActivity";

    @Override
    public void onCreate(Bundle savedInstanceState) {
        super.onCreate(savedInstanceState);
```

```
            setContentView(R.layout.main);

            DigitalClock digitalClock1 = (DigitalClock) this
                    .findViewById(R.id.digitalClock1);
            // 文字大小 30
            digitalClock1.setTextSize(30);
            // 文字背景颜色为红色
            digitalClock1.setBackgroundColor(Color.RED);
            // 文字颜色为黑色
            digitalClock1.setTextColor(Color.BLACK);//

        }
}
```

AnalogClock 和 DigitalClock 控件使用很简单,放到布局中就可以应用了。运行 AVD 设备就可以在虚拟机中显示出当前的 AVD 设备时间,如图 3.13 所示。

图 3.13 在 AVD 设备中运行的 AnalogClock 和 DigitalClock 控件

3.7 AutoCompleteTextView 控件的使用与 XML 数据源

AutoCompleteTextView 控件的功能和 http://www.baidu.com 主页搜索辅助的下拉菜单相似,都是用来提高软件使用人员的录入方便性,以及提高软件可操作性的一种手段。

这个控件有一个比较重要的知识点,AutoCompleteTextView 控件中的方法 public void setAdapter (T adapter)是用来设置下拉辅助输入框中的数据来源。

创建名称为 ui_2 的项目,在主布局文件 main.xml 中加入如下代码:

```xml
<?xml version="1.0" encoding="utf-8"?>
<LinearLayout xmlns:android="http://schemas.android.com/apk/res/android"
    android:orientation="vertical" android:layout_width="fill_parent"
    android:layout_height="fill_parent">
    <AutoCompleteTextView android:text=""
        android:id="@+id/autoCompleteTextView1" android:layout_width="fill_parent"
        android:layout_height="wrap_content"></AutoCompleteTextView>
</LinearLayout>
```

将类 Main.java 文件的代码更改如下：

```java
public class Main extends Activity {

    @Override
    public void onCreate(Bundle savedInstanceState) {
        super.onCreate(savedInstanceState);
        setContentView(R.layout.main);

        List names = new ArrayList();
        names.add("高洪岩 1");
        names.add("高洪岩 2");
        names.add("高洪岩 3");

        AutoCompleteTextView textView = (AutoCompleteTextView) findViewById(R.id.autoCompleteTextView1);
        textView.setThreshold(1);

        ArrayAdapter adapter = new ArrayAdapter(this,
                android.R.layout.simple_dropdown_item_1line, names);
        textView.setAdapter(adapter);
    }
}
```

其中代码 textView.setThreshold(1);的功能是设置输入几个文字后进行自动提示，这里设置为 1，也就是输入 1 个字母时就进行下拉辅助提示。

本示例中控件 AutoCompleteTextView 使用方法 setAdapter()关联 1 个 ArrayAdapter 对象。而 ArrayAdapter 对象将类型为 List 的对象 names 中的每一个条目放入 id 为 android.R.layout.simple_dropdown_item_1line 的布局中，这个布局 id 对应的 XML 文件名是 simple_dropdown_item_1line.xml，文件中的布局配置代码如下：

```xml
<?xml version="1.0" encoding="utf-8"?>
<TextView xmlns:android="http://schemas.android.com/apk/res/android"
    android:id="@android:id/text1" style="?android:attr/dropDownItemStyle"
    android:textAppearance="?android:attr/textAppearanceLargeInverse"
    android:singleLine="true" android:layout_width="match_parent"
    android:layout_height="?android:attr/listPreferredItemHeight"
    android:ellipsize="marquee" />
```

文件 simple_dropdown_item_1line.xml 中只有 1 个 TextView 控件，用来显示 Adapter 中数据集的每一个数据条目。

程序运行效果如图 3.14 所示。

图 3.14 运行效果

上面的示例是在 AutoCompleteTextView 控件中加载自定义 List 对象中的数据，AutoCompleteTextView 控件还可以从 XML 文件中加载数据。新建名称为 AutoCompleteTextView_dataFromXML 的 Android 项目，创建任意名称的 XML 文件 arrays.xml 文件，将这个文件放入 res/values 目录下，代码的写法与格式如下：

```xml
<?xml version="1.0" encoding="utf-8"?>
<resources>
    <string-array name="city">
        <item>北京</item>
        <item>上海</item>
        <item>广州</item>
        <item>深圳 1</item>
        <item>深圳 2</item>
        <item>深圳 3</item>
        <item>深圳 4</item>
        <item>深圳 5</item>
        <item>深圳 6</item>
    </string-array>
</resources>
```

文件 Main.java 代码如下：

```java
public class Main extends Activity {
    private AutoCompleteTextView actvRef;

    @Override
    public void onCreate(Bundle savedInstanceState) {
        super.onCreate(savedInstanceState);
        setContentView(R.layout.main);

        actvRef = (AutoCompleteTextView) this
                .findViewById(R.id.autoCompleteTextView1);
        actvRef.setThreshold(1);

        ArrayAdapter adapter = ArrayAdapter.createFromResource(this,
                R.array.city, android.R.layout.simple_dropdown_item_1line);
        actvRef.setAdapter(adapter);
```

        }
    }

程序运行结果如图 3.15 所示。

图 3.15 从 XML 取得的数据

3.8 Button 控件

Button 按钮控件在以前的章节已经广泛使用过,它的使用还是比较简单的。创建名称为 ui_3 的 Android 项目,并且更改布局文件 main.xml 的代码如下:

```xml
<?xml version="1.0" encoding="utf-8"?>
<LinearLayout xmlns:android="http://schemas.android.com/apk/res/android"
    android:orientation="vertical" android:layout_width="fill_parent"
    android:layout_height="fill_parent">
    <TextView android:layout_width="fill_parent"
        android:layout_height="wrap_content" android:text="@string/hello" />
    <Button android:text="Button" android:id="@+id/button1"
        android:layout_width="wrap_content" android:layout_height="wrap_content"></Button>
    <Button android:text="Button" android:id="@+id/button2"
        android:layout_width="wrap_content" android:layout_height="wrap_content"
        android:drawableTop="@drawable/icon" android:drawablePadding="5px"></Button>
    <Button android:text="Button" android:id="@+id/button3"
        android:layout_width="wrap_content" android:layout_height="wrap_content"
        android:drawableBottom="@drawable/icon" android:drawablePadding="5px"></Button>
    <Button android:text="Button" android:id="@+id/button4"
        android:layout_width="wrap_content" android:layout_height="wrap_content"
        android:drawableLeft="@drawable/icon" android:drawablePadding="5px"></Button>
    <Button android:text="Button" android:id="@+id/button5"
        android:layout_width="wrap_content" android:layout_height="wrap_content"
```

```
                    android:drawableRight="@drawable/icon" android:drawablePadding="5px"></Button>
</LinearLayout>
```

属性 android:drawableTop 的功能是定义图像资源放在文字的哪个方向，而属性 android:drawablePadding 是定义图像与文字的间距是多少。

在类 Main.java 文件加入如下代码：

```
public class Main extends Activity {

    private static final String TAG = "MainActivity";

    @Override
    public void onCreate(Bundle savedInstanceState) {
        super.onCreate(savedInstanceState);
        setContentView(R.layout.main);

        Button button1 = (Button) this.findViewById(R.id.button1);

        button1.setOnClickListener(new OnClickListener() {

            public void onClick(View arg0) {

                Log.v(TAG, "单击了按钮！");

            }
        });

    }
}
```

运行项目后单击界面中的 Button 按钮，在 LogCat 中输出字符串，如图 3.16 所示。

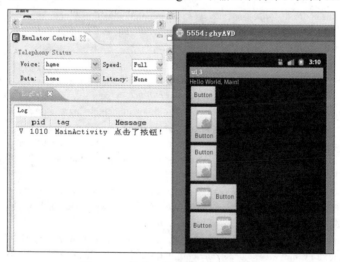

图 3.16　单击 Button 按钮打印日志

3.9 CheckBox 控件

Android 中 CheckBox 控件的功能和 HTML 语言中的<input type="checkbox">功能一样，外观都是以"打勾"的方式来进行选项的选中和取消。

创建名称为 android1 的 Android 项目，更改布局文件 main.xml 的代码如下：

```xml
<?xml version="1.0" encoding="utf-8"?>
<LinearLayout xmlns:android="http://schemas.android.com/apk/res/android"
    android:orientation="vertical" android:layout_width="fill_parent"
    android:layout_height="fill_parent">
    <CheckBox android:tag="a" android:text="  爱好 a" android:id="@+id/checkBox1"
        android:layout_width="wrap_content" android:layout_height="wrap_content"></CheckBox>
    <CheckBox android:tag="b" android:text="  爱好 b" android:id="@+id/checkBox2"
        android:layout_width="wrap_content" android:layout_height="wrap_content"></CheckBox>
    <CheckBox android:tag="c" android:text="  爱好 c" android:id="@+id/checkBox3"
        android:layout_width="wrap_content" android:layout_height="wrap_content"></CheckBox>
    <CheckBox android:tag="d" android:text="  爱好 d" android:id="@+id/checkBox4"
        android:layout_width="wrap_content" android:layout_height="wrap_content"></CheckBox>
    <CheckBox android:tag="e" android:text="  爱好 e" android:id="@+id/checkBox5"
        android:layout_width="wrap_content" android:layout_height="wrap_content"></CheckBox>
    <Button android:text="Button" android:id="@+id/button1"
        android:layout_width="wrap_content" android:layout_height="wrap_content"></Button>
</LinearLayout>
```

在类 Main.java 文件中加入如下代码：

```java
public class Main extends Activity {

    private CheckBox checkbox1;
    private CheckBox checkbox2;
    private CheckBox checkbox3;
    private CheckBox checkbox4;
    private CheckBox checkbox5;

    private ArrayList<Integer> checkBoxIdList = new ArrayList();

    private Button button1;

    @Override
    public void onCreate(Bundle savedInstanceState) {
        super.onCreate(savedInstanceState);
        setContentView(R.layout.main);
        checkbox1 = (CheckBox) this.findViewById(R.id.checkBox1);
        checkbox2 = (CheckBox) this.findViewById(R.id.checkBox2);
        checkbox3 = (CheckBox) this.findViewById(R.id.checkBox3);
        checkbox4 = (CheckBox) this.findViewById(R.id.checkBox4);
        checkbox5 = (CheckBox) this.findViewById(R.id.checkBox5);

        checkBoxIdList.add(checkbox1.getId());
```

```
checkBoxIdList.add(checkbox2.getId());
checkBoxIdList.add(checkbox3.getId());
checkBoxIdList.add(checkbox4.getId());
checkBoxIdList.add(checkbox5.getId());

button1 = (Button) this.findViewById(R.id.button1);

checkbox1.setChecked(true);
checkbox3.setChecked(true);
checkbox5.setChecked(true);

button1.setOnClickListener(new OnClickListener() {
    public void onClick(View arg0) {
        CheckBox eachCheckBoxRef = null;
        for (int i = 0; i < checkBoxIdList.size(); i++) {
            eachCheckBoxRef = (CheckBox) Main.this
                    .findViewById(checkBoxIdList.get(i));
            if (eachCheckBoxRef.isChecked()) {
                Log.v("选中了: ", "" + eachCheckBoxRef.getTag());
            }
        }
    }
});
    }
}
```

运行项目后选中"爱好 b"和"爱好 d"选项，呈打勾状态，如图 3.17 所示。

单击按钮后在 LogCat 界面中打印选中的信息，如图 3.18 所示。

图 3.17　呈两个打勾的界面状态

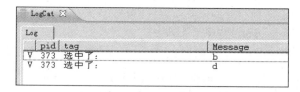

图 3.18　打印结果

3.10　CheckedTextView 控件

在 Android 技术中实现打勾的 checked 效果其实还有另外一个控件也可以实现，它就是

CheckedTextView 控件。

新建名称为 android2 的 Android 项目，更改布局文件 main.xml 的代码如下：

```xml
<?xml version="1.0" encoding="utf-8"?>
<ScrollView xmlns:android="http://schemas.android.com/apk/res/android"
    android:id="@+id/scrollView1" android:layout_width="match_parent"
    android:layout_height="wrap_content">
    <LinearLayout android:padding="10px" android:orientation="vertical"
        android:layout_width="fill_parent" android:layout_height="fill_parent">
        <CheckedTextView android:tag="a1" android:id="@+id/checkedTextView1"
            android:layout_width="fill_parent" android:layout_height="wrap_content"
            android:checkMark="?android:attr/listChoiceIndicatorMultiple"
            android:text="checkedTextView1" />
        <CheckedTextView android:tag="a2" android:id="@+id/checkedTextView2"
            android:layout_width="fill_parent" android:layout_height="wrap_content"
            android:checkMark="?android:attr/listChoiceIndicatorMultiple"
            android:text="checkedTextView2" />
        <CheckedTextView android:tag="a3" android:id="@+id/checkedTextView3"
            android:layout_width="fill_parent" android:layout_height="wrap_content"
            android:checkMark="?android:attr/listChoiceIndicatorMultiple"
            android:text="checkedTextView4" />
        <CheckedTextView android:tag="a4" android:id="@+id/checkedTextView4"
            android:layout_width="fill_parent" android:layout_height="wrap_content"
            android:checkMark="?android:attr/listChoiceIndicatorMultiple"
            android:text="checkedTextView5" />
        <CheckedTextView android:tag="a5" android:id="@+id/checkedTextView5"
            android:layout_width="fill_parent" android:layout_height="wrap_content"
            android:checkMark="?android:attr/listChoiceIndicatorMultiple"
            android:text="checkedTextView6" />
        <Button android:text="Button" android:id="@+id/button1"
            android:layout_width="wrap_content" android:layout_height="wrap_content"></Button>
        <CheckedTextView android:tag="A" android:id="@+id/checkedTextViewa"
            android:layout_width="fill_parent" android:layout_height="wrap_content"
            android:checkMark="?android:attr/listChoiceIndicatorSingle"
            android:text="checkedTextViewa" />
        <CheckedTextView android:tag="B" android:id="@+id/checkedTextViewb"
            android:layout_width="fill_parent" android:layout_height="wrap_content"
            android:checkMark="?android:attr/listChoiceIndicatorSingle"
            android:text="checkedTextViewb" />
        <CheckedTextView android:tag="C" android:id="@+id/checkedTextViewc"
            android:layout_width="fill_parent" android:layout_height="wrap_content"
            android:checkMark="?android:attr/listChoiceIndicatorSingle"
            android:text="checkedTextViewc" />
        <CheckedTextView android:tag="D" android:id="@+id/checkedTextViewd"
            android:layout_width="fill_parent" android:layout_height="wrap_content"
            android:checkMark="?android:attr/listChoiceIndicatorSingle"
            android:text="checkedTextViewd" />
        <Button android:text="Button" android:id="@+id/button2"
            android:layout_width="wrap_content" android:layout_height="wrap_content"></Button>
    </LinearLayout>
```

```
</ScrollView>
```

由于本示例的控件比较多,为了能全部显示它们,所以用了带滚动条效果的<ScrollView>控件。

文件 Main.java 的代码如下:

```java
public class Main extends Activity {
    private CheckedTextView checkedTextViewMul1;
    private CheckedTextView checkedTextViewMul2;
    private CheckedTextView checkedTextViewMul3;
    private CheckedTextView checkedTextViewMul4;
    private CheckedTextView checkedTextViewMul5;

    private CheckedTextView checkedTextViewSinglea;
    private CheckedTextView checkedTextViewSingleb;
    private CheckedTextView checkedTextViewSinglec;
    private CheckedTextView checkedTextViewSingled;

    private Button getMulCheckedTextValue;
    private Button getSingleCheckedTextValue;
    private ArrayList<Integer> mulCheckedTextViewIdArray = new ArrayList();
    private ArrayList<Integer> singleCheckedTextViewIdArray = new ArrayList();

    @Override
    public void onCreate(Bundle savedInstanceState) {
        super.onCreate(savedInstanceState);
        setContentView(R.layout.main);

        getMulCheckedTextValue = (Button) this.findViewById(R.id.button1);
        getSingleCheckedTextValue = (Button) this.findViewById(R.id.button2);

        checkedTextViewMul1 = (CheckedTextView) this
                .findViewById(R.id.checkedTextView1);
        checkedTextViewMul1.setChecked(true);
        checkedTextViewMul2 = (CheckedTextView) this
                .findViewById(R.id.checkedTextView2);
        checkedTextViewMul3 = (CheckedTextView) this
                .findViewById(R.id.checkedTextView3);
        checkedTextViewMul3.setChecked(true);
        checkedTextViewMul4 = (CheckedTextView) this
                .findViewById(R.id.checkedTextView4);
        checkedTextViewMul5 = (CheckedTextView) this
                .findViewById(R.id.checkedTextView5);
        checkedTextViewMul5.setChecked(true);

        mulCheckedTextViewIdArray.add(checkedTextViewMul1.getId());
        mulCheckedTextViewIdArray.add(checkedTextViewMul2.getId());
        mulCheckedTextViewIdArray.add(checkedTextViewMul3.getId());
        mulCheckedTextViewIdArray.add(checkedTextViewMul4.getId());
        mulCheckedTextViewIdArray.add(checkedTextViewMul5.getId());

        OnClickListener checkedTextViewMulListenerRef = new OnClickListener() {
```

```java
            public void onClick(View arg0) {
                ((CheckedTextView) arg0).toggle();
            }
        };

        checkedTextViewMul1.setOnClickListener(checkedTextViewMulListenerRef);
        checkedTextViewMul2.setOnClickListener(checkedTextViewMulListenerRef);
        checkedTextViewMul3.setOnClickListener(checkedTextViewMulListenerRef);
        checkedTextViewMul4.setOnClickListener(checkedTextViewMulListenerRef);
        checkedTextViewMul5.setOnClickListener(checkedTextViewMulListenerRef);

        getMulCheckedTextValue.setOnClickListener(new OnClickListener() {
            public void onClick(View arg0) {
                for (int i = 0; i < mulCheckedTextViewIdArray.size(); i++) {
                    CheckedTextView findCheckedTextViewRef = (CheckedTextView) Main.this
                            .findViewById(mulCheckedTextViewIdArray.get(i));
                    if (findCheckedTextViewRef.isChecked() == true) {
                        Log.v("打勾的checkbox值是", ""
                                + findCheckedTextViewRef.getTag());
                    }
                }
            }
        });

        checkedTextViewSinglea = (CheckedTextView) this
                .findViewById(R.id.checkedTextViewa);
        checkedTextViewSingleb = (CheckedTextView) this
                .findViewById(R.id.checkedTextViewb);
        checkedTextViewSinglec = (CheckedTextView) this
                .findViewById(R.id.checkedTextViewc);
        checkedTextViewSingled = (CheckedTextView) this
                .findViewById(R.id.checkedTextViewd);

        singleCheckedTextViewIdArray.add(checkedTextViewSinglea.getId());
        singleCheckedTextViewIdArray.add(checkedTextViewSingleb.getId());
        singleCheckedTextViewIdArray.add(checkedTextViewSinglec.getId());
        singleCheckedTextViewIdArray.add(checkedTextViewSingled.getId());

        OnClickListener checkedTextViewSinglelListenerRef = new OnClickListener() {
            public void onClick(View arg0) {
                for (int i = 0; i < singleCheckedTextViewIdArray.size(); i++) {
                    if (singleCheckedTextViewIdArray.get(i).intValue() != ((CheckedTextView) arg0)
                            .getId()) {
                        ((CheckedTextView) Main.this
                                .findViewById(singleCheckedTextViewIdArray
                                        .get(i))).setChecked(false);
                    } else {
```

```java
                            ((CheckedTextView) Main.this
                                    .findViewById(singleCheckedTextViewIdArray
                                            .get(i))).setChecked(true);
                        }
                    }
                }
            };

            checkedTextViewSinglea
                    .setOnClickListener(checkedTextViewSinglelListenerRef);
            checkedTextViewSingleb
                    .setOnClickListener(checkedTextViewSinglelListenerRef);
            checkedTextViewSinglec
                    .setOnClickListener(checkedTextViewSinglelListenerRef);
            checkedTextViewSingled
                    .setOnClickListener(checkedTextViewSinglelListenerRef);

            getSingleCheckedTextValue.setOnClickListener(new OnClickListener() {
                public void onClick(View arg0) {
                    for (int i = 0; i < singleCheckedTextViewIdArray.size(); i++) {
                        CheckedTextView eachCheckedTextViewRef = ((CheckedTextView) Main.this
                                .findViewById(singleCheckedTextViewIdArray.get(i)));
                        if (eachCheckedTextViewRef.isChecked() == true) {
                            Log.v("单选选中了:", ""
                                    + eachCheckedTextViewRef.getTag().toString());
                        }
                    }
                }
            });

        }
    }
```

程序初始运行效果如图 3.19 所示。

将多选控件的状态改成如图 3.20 所示。

图 3.19　初始运行效果　　　　　图 3.20　更变 checked 状态后界面

单击上方的 Button 按钮取出多选 CheckedTextView 控件的状态值,如图 3.21 所示。再把单选的状态改成如图 3.22 所示。

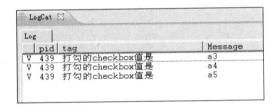
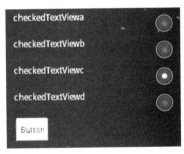

图 3.21　第 1 个按钮按下的效果　　　　　图 3.22　答案 c 被选中

单击下方的 Button 在 LogCat 打印结果信息,如图 3.23 所示。

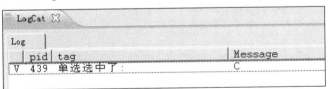

图 3.23　单选结果打印

3.11　Chronometer 控件

Chronometer 控件是记录当前的已过时间,使用的情况是手机正在接听电话时计录已过时间。新建名称为 ui_6 的 Android 项目,将文件 Main.java 的代码更改如下:

```java
public class Main extends Activity {

    private TextView showTime;
    private TextView showEnd;
    private Button startButton;
    private Button stopButton;
    private Chronometer chronometerRef;

    private int runNum = -2;

    @Override
    public void onCreate(Bundle savedInstanceState) {
        super.onCreate(savedInstanceState);
        setContentView(R.layout.main);

        showTime = (TextView) this.findViewById(R.id.textView1);
        showEnd = (TextView) this.findViewById(R.id.textView2);
        startButton = (Button) this.findViewById(R.id.start);
        stopButton = (Button) this.findViewById(R.id.stop);

        chronometerRef = (Chronometer) this.findViewById(R.id.chronometer1);

        chronometerRef
                .setOnChronometerTickListener(new OnChronometerTickListener() {

                    public void onChronometerTick(Chronometer arg0) {
                        runNum++;
                        showTime.setText("" + runNum);
                    }
                });

        startButton.setOnClickListener(new OnClickListener() {
            public void onClick(View arg0) {
                showEnd.setText("");
                showTime.setText("");

                chronometerRef.setBase(SystemClock.elapsedRealtime());
                chronometerRef.start();
                startButton.setEnabled(false);
                stopButton.setEnabled(true);

            }
        });

        stopButton.setOnClickListener(new OnClickListener() {

            public void onClick(View arg0) {
                chronometerRef.stop();
                startButton.setEnabled(true);
                stopButton.setEnabled(false);
```

```
            showEnd.setText("用了："+ runNum + "秒");
            showEnd.setText("用了："+ runNum + "秒");
            runNum = -2;
        }
    });

    }
}
```

更改布局文件 main.xml 的代码如下：

```xml
<?xml version="1.0" encoding="utf-8"?>
<LinearLayout xmlns:android="http://schemas.android.com/apk/res/android"
    android:orientation="vertical" android:layout_width="fill_parent"
    android:layout_height="fill_parent">
    <Chronometer android:text="Chronometer" android:id="@+id/chronometer1"
        android:layout_width="wrap_content" android:layout_height="wrap_content"></Chronometer>
    <TextView android:layout_width="fill_parent" android:id="@+id/textView1"
        android:layout_height="wrap_content" android:text="--------------" />
    <TextView android:layout_width="fill_parent" android:id="@+id/textView2"
        android:layout_height="wrap_content" android:text="--------------" />
    <Button android:text="开始" android:id="@+id/start"
        android:layout_width="wrap_content" android:layout_height="wrap_content"></Button>
    <Button android:text="结束" android:id="@+id/stop"
        android:layout_width="wrap_content" android:layout_height="wrap_content"></Button>
</LinearLayout>
```

程序初始运行效果如图 3.24 所示。

单击"开始"按钮，时间开始前进，再单击"结束"按钮计时停止，停止时的效果如图 3.25 所示。

图 3.24　程序初始运行效果

图 3.25　停止时的运行效果

3.12　DatePicker 和 TimePicker 控件

控件 DatePicker 和 TimePicker 的功能是选择指定的日期和时间，使用比较简单。

新建名称为 ui_7 的 Android 项目，将布局文件 main.xml 代码更改如下：

```xml
<?xml version="1.0" encoding="utf-8"?>
<LinearLayout xmlns:android="http://schemas.android.com/apk/res/android"
```

```
android:orientation="vertical" android:layout_width="fill_parent"
android:layout_height="fill_parent">
<DatePicker android:id="@+id/datePicker1"
    android:layout_width="wrap_content" android:layout_height="wrap_content"></DatePicker>
<DatePicker android:id="@+id/datePicker2"
    android:layout_width="wrap_content" android:layout_height="wrap_content"></DatePicker>
</LinearLayout>
```

程序运行的结果如图 3.26 所示。在图 3.26 中看到的日期格式并不是中文的"年月日"格式，那如何才可以显示出"年月日"的格式呢，改系统时区即可。

在 AVD 界面中单击 Home 按钮返回 Android 系统的主界面，再单击"menu"菜单弹出如图 3.27 所示的菜单界面。

图 3.26 非中文格式的日期显示

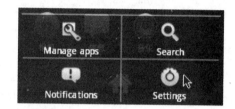

图 3.27 单击 menu 后弹出菜单

在图 3.27 中单击"Settings"菜单项后选中图 3.28 所示菜单选项。再单击图 3.29 所示的菜单。

图 3.28 设置语言和键盘

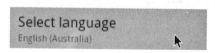

图 3.29 单击选择语言的菜单项

在图 3.29 中选择最下方的"中文(简体)"语言，如图 3.30 所示。

再次运行刚才的 Android 项目，正确显示出了中文格式，如图 3.31 所示。

图 3.30 选择中文简体

图 3.31 正确显示中文格式

再次更改布局文件 main.xml 的代码如下：

```xml
<?xml version="1.0" encoding="utf-8"?>
<LinearLayout xmlns:android="http://schemas.android.com/apk/res/android"
    android:orientation="vertical" android:layout_width="fill_parent"
    android:layout_height="fill_parent">
    <DatePicker android:id="@+id/datePicker1"
        android:layout_width="wrap_content" android:layout_height="wrap_content"></DatePicker>
    <DatePicker android:id="@+id/datePicker2"
        android:layout_width="wrap_content" android:layout_height="wrap_content"></DatePicker>
    <Button android:text="Button" android:id="@+id/button1"
        android:layout_width="wrap_content" android:layout_height="wrap_content"></Button>
</LinearLayout>
```

将 Activity 对象文件 Main.java 的代码更改如下：

```java
public class Main extends Activity {
    private DatePicker dp1;
    private DatePicker dp2;
    private Button button1;

    @Override
    public void onCreate(Bundle savedInstanceState) {
        super.onCreate(savedInstanceState);
        setContentView(R.layout.main);
        dp1 = (DatePicker) this.findViewById(R.id.datePicker1);
        dp2 = (DatePicker) this.findViewById(R.id.datePicker2);
        button1 = (Button) this.findViewById(R.id.button1);

        dp1.init(2000, 0, 1, new OnDateChangedListener() {
            public void onDateChanged(DatePicker arg0, int arg1, int arg2,
                    int arg3) {
                Log.v("dp1 日期改变了！", "设置的日期为" + arg1 + "年" + (arg2 + 1) + "月"
                        + arg3 + "日");
            }
        });
        dp2.init(2000, 1, 1, new OnDateChangedListener() {
            public void onDateChanged(DatePicker arg0, int arg1, int arg2,
                    int arg3) {
                Log.v("dp1 日期改变了！", "设置的日期为" + arg1 + "年" + (arg2 + 1) + "月"
                        + arg3 + "日");
            }
        });

        button1.setOnClickListener(new OnClickListener() {
            public void onClick(View arg0) {
                Log.v("获取的时间 dp1：", "" + dp1.getYear() + " "
                        + (dp1.getMonth() + 1) + " " + dp1.getDayOfMonth());
                Log.v("获取的时间 dp2：", "" + dp2.getYear() + " "
                        + (dp2.getMonth() + 1) + " " + dp2.getDayOfMonth());

            }
        });
```

            }
        }

程序运行后，更改 dp1 和 dp2 控件的时间，在 LogCat 上打印出相关的日志信息，再单击下面的 Button 按钮打印出最终的日期，运行效果如图 3.32 所示。

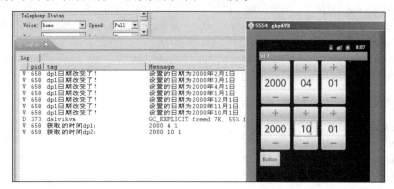

图 3.32　更改 dp1 和 dp2 的时间及单击 Button 按钮打印日志

当然也可以用代码取得当前的日期和时间，然后再初始化 DatePicker 控件的默认日期，代码如下：

```
public class test {

    public static void main(String[] args) {
        Calendar calendar = Calendar.getInstance();
        int year = calendar.get(Calendar.YEAR);
        int monthOfYear = calendar.get(Calendar.MONTH);
        int dayOfMonth = calendar.get(Calendar.DAY_OF_MONTH);
        System.out.println(year + " " + (monthOfYear + 1) + " " + dayOfMonth);

    }

}
```

下面是另外一种情况：

```
public static void main(String[] args) {
    long time = System.currentTimeMillis();
    Calendar calendarRef = Calendar.getInstance();
    calendarRef.setTimeInMillis(time);
}
```

前面已经将控件 DatePicker 介绍完毕，还有 1 个 TimePicker 控件需要进一步的学习。

新建名称为 ui_7_1 的 Android 项目，更改布局文件 main.xml 的代码如下：

```
<?xml version="1.0" encoding="utf-8"?>
<LinearLayout xmlns:android="http://schemas.android.com/apk/res/android"
    android:orientation="vertical" android:layout_width="fill_parent"
    android:layout_height="fill_parent">
    <TimePicker android:id="@+id/timePicker1"
```

android:layout_width=*"wrap_content"* android:layout_height=*"wrap_content"*></TimePicker>
</LinearLayout>

继续更改 Main.java 的代码如下：

```java
public class Main extends Activity {
    private TimePicker tp;

    @Override
    public void onCreate(Bundle savedInstanceState) {
        super.onCreate(savedInstanceState);
        setContentView(R.layout.main);
        tp = (TimePicker) this.findViewById(R.id.timePicker1);
        tp.setIs24HourView(true);
        tp.setOnTimeChangedListener(new OnTimeChangedListener() {
            public void onTimeChanged(TimePicker arg0, int arg1, int arg2) {
                Log.v("取得的时间为：", "" + arg1 + " " + arg2);
            }
        });
    }
}
```

更改时间时在 LogCat 面板中显示出选择的时间，运行效果如图 3.33 所示。

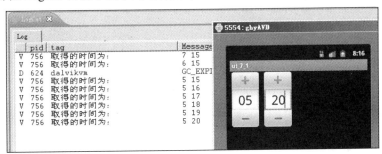

图 3.33　TimePicker 更改时间时打印日志

前面的 DatePicker 和 TimePicker 控件都是提供用户选择日期的功能，有时需要在界面上显示出当前的 "年月日时分秒" 的信息，这时这两个控件就无能为力了，而且前面学习过的 AnalogClock 和 DigitalClock 控件也都不能满足这样的需求，那只有自己动手，丰衣足食。

下面自己手动写代码显示当前的日期和时间。创建名称为 getDateAndTimeShow 的 Android 项目，新建线程工具类 GetDateAndTimeTools.java，代码如下：

```java
public class GetDateAndTimeTools extends Thread {

    private Context context;
    private Handler handler;

    public GetDateAndTimeTools(Context context, Handler handler) {
        super();
        this.context = context;
        this.handler = handler;
    }
```

```java
@Override
public void run() {

    try {
        super.run();
        while (1 == 1) {
            Calendar calendar = Calendar.getInstance();
            String year = "" + calendar.get(Calendar.YEAR);
            String month = "" + calendar.get(Calendar.MONTH);
            String day = "" + calendar.get(Calendar.DAY_OF_MONTH);
            String hour = "" + calendar.get(Calendar.HOUR);
            String minute = "" + calendar.get(Calendar.MINUTE);
            String second = "" + calendar.get(Calendar.SECOND);

            String returnDateAndTimeString = year + "-" + month + "-" + day
                    + " " + hour + ":" + minute + ":" + second;

            Bundle bundleRef = new Bundle();
            bundleRef.putString("dateAndTime", returnDateAndTimeString);

            Message message = new Message();
            message.what = 100;// 消息的 id 值
            message.setData(bundleRef);

            handler.sendMessage(message);

            Thread.sleep(1000);

        }
    } catch (InterruptedException e) {
        // TODO Auto-generated catch block
        e.printStackTrace();
    }

}
```

文件 Main.java 的代码如下：

```java
public class Main extends Activity {
    private TextView textView1;

    @Override
    public void onCreate(Bundle savedInstanceState) {
        super.onCreate(savedInstanceState);
        setContentView(R.layout.main);
        textView1 = (TextView) this.findViewById(R.id.textView1);

        Handler handler = new Handler() {
            @Override
```

```
            public void handleMessage(Message msg) {
                textView1.setText("" + msg.getData().get("dateAndTime"));
                super.handleMessage(msg);
            }
        };

        GetDateAndTimeTools getRef = new GetDateAndTimeTools(this, handler);
        getRef.start();

    }
}
```

文件 main.xml 的代码如下:

```
<?xml version="1.0" encoding="utf-8"?>
<LinearLayout xmlns:android="http://schemas.android.com/apk/res/android"
    android:orientation="vertical" android:layout_width="fill_parent"
    android:layout_height="fill_parent">
    <TextView android:id="@+id/textView1" android:layout_width="fill_parent"
        android:layout_height="wrap_content" android:text="" />
</LinearLayout>
```

程序运行后每隔 1 秒显示当前的日期和时间，如图 3.34 所示。

图 3.34　时时显示日期和时间

3.13　EditText 控件

EditText 是一个单行文本域的控件，使用起来比较简单。

创建名称为 ui_8 的 Android 项目，更改 Main.java 文件的代码如下:

```
public class Main extends Activity {

    private EditText editText1;

    @Override
    public void onCreate(Bundle savedInstanceState) {
        super.onCreate(savedInstanceState);
        setContentView(R.layout.main);

        editText1 = (EditText) this.findViewById(R.id.editText1);

        editText1.setOnFocusChangeListener(new OnFocusChangeListener() {
            public void onFocusChange(View arg0, boolean arg1) {
```

```
                Log.v("onFocusChange", "onFocusChange=" + arg1);
            }
        });
        // arg1:start 代表发生改变位置的索引
        // arg2:count 文本内容发生减少时减少的字符个数
        // arg3:after 文本内容发生增加时增加的字符个数
        //设置 editText1 控件中的内容改变时的监听
        editText1.addTextChangedListener(new TextWatcher() {

            public void onTextChanged(CharSequence arg0, int arg1, int arg2,
                    int arg3) {
                Log.v("onTextChanged", "onTextChanged");
            }

            public void beforeTextChanged(CharSequence arg0, int arg1,
                    int arg2, int arg3) {
                Log.v("beforeTextChanged", "beforeTextChanged");
            }

            public void afterTextChanged(Editable arg0) {
                Log.v("afterTextChanged", "afterTextChanged");
            }
        });

    }
}
```

更改布局文件 main.xml 的代码如下：

```
<?xml version="1.0" encoding="utf-8"?>
<LinearLayout xmlns:android="http://schemas.android.com/apk/res/android"
    android:orientation="vertical" android:layout_width="fill_parent"
    android:layout_height="fill_parent">
    <EditText android:layout_width="match_parent"
        android:layout_height="wrap_content" android:text="EditText"
        android:id="@+id/editText1"></EditText>
    <EditText android:layout_width="match_parent"
        android:layout_height="wrap_content" android:text="EditText"
        android:id="@+id/editText2"></EditText>
</LinearLayout>
```

程序运行效果如图 3.35 所示。

图 3.35　初始运行效果

从图 3.35 中可以看到，默认情况下 EditText1 获得焦点后打印日志信息的布尔值为 true，当把焦点用鼠标从 EditText1 转到 EditText2 时，LogCat 打印效果如图 3.36 所示。

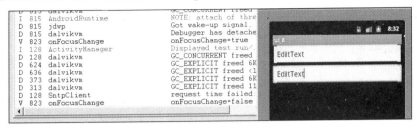

图 3.36　EditText1 失焦时打印的值是 false

把 EditText1 中的文本删除两个字母后的 LogCat 日志结果如图 3.37 所示。

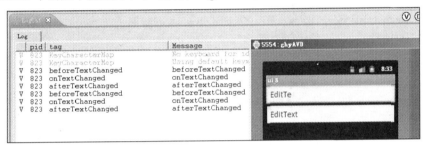

图 3.37　删除两个字母后的日志信息

另外，EditText 控件还可以限制输入的字符，本示例在名称为 editText_inputType 的 Android 项目中，main.xml 布局文件的代码如下：

```xml
<?xml version="1.0" encoding="utf-8"?>
<LinearLayout xmlns:android="http://schemas.android.com/apk/res/android"
    android:orientation="vertical" android:layout_width="fill_parent"
    android:layout_height="fill_parent">
<TextView android:text="只能输入小数" android:id="@+id/textView1"
        android:layout_width="wrap_content" android:layout_height="wrap_content"></TextView>
<EditText android:id="@+id/editText1" android:layout_height="wrap_content"
        android:layout_width="match_parent"                                    android:text=""
android:inputType="numberDecimal"></EditText>
<TextView android:text="只能输入整数" android:id="@+id/textView2"
        android:layout_width="wrap_content" android:layout_height="wrap_content"></TextView>
<EditText android:id="@+id/editText2" android:layout_height="wrap_content"
        android:layout_width="match_parent" android:text="" android:inputType="numberSigned"></EditText>
<TextView android:text="输入密码" android:id="@+id/textView3"
        android:layout_width="wrap_content" android:layout_height="wrap_content"></TextView>
<EditText android:id="@+id/editText3" android:layout_height="wrap_content"
        android:layout_width="match_parent" android:text="" android:inputType="textPassword"></EditText>
</LinearLayout>
```

所谓的限制输入的字符就是 Android 根据 android:inputType 属性值的不同而显示出不同的输入法界面。

由于在 Android 中并没有"多行文本域",所以 EditText 控件通过设置属性完全可以实现多行文本域的效果,示例布局代码在 editTextMuliLine 项目中,main.xml 布局文件的代码如下:

```xml
<?xml version="1.0" encoding="utf-8"?>
<LinearLayout xmlns:android="http://schemas.android.com/apk/res/android"
    android:orientation="vertical" android:layout_width="fill_parent"
    android:layout_height="fill_parent">
    <EditText android:id="@+id/editText1" android:layout_height="wrap_content"
        android:text="" android:layout_width="match_parent" android:hint="请在这里输入用户名"></EditText>
    <EditText android:id="@+id/editText2" android:layout_height="wrap_content"
        android:text="" android:layout_width="match_parent" android:gravity="top"
        android:lines="5"></EditText>
    <EditText android:id="@+id/editText3" android:layout_height="wrap_content"
        android:text="请在这里输入用户名" android:layout_width="match_parent"></EditText>
    <EditText android:id="@+id/editText4" android:layout_height="wrap_content"
        android:text="请在这里输入用户名" android:layout_width="match_parent"></EditText>
    <EditText android:id="@+id/editText5" android:layout_height="wrap_content"
        android:text="请在这里输入用户名" android:layout_width="match_parent"
        android:editable="false"></EditText>
</LinearLayout>
```

文件 Main.java 的代码如下:

```java
public class Main extends Activity {
    private EditText editText2;
    private EditText editText3;
    private EditText editText4;

    @Override
    public void onCreate(Bundle savedInstanceState) {
        super.onCreate(savedInstanceState);
        setContentView(R.layout.main);

        editText2 = (EditText) this.findViewById(R.id.editText2);
        for (int i = 0; i < 20; i++) {
            editText2.append((i + 1) + "我是多行文本,我很宽,我很高");
        }

        editText3 = (EditText) this.findViewById(R.id.editText3);
        editText3.setSelection(2, 4);

        editText4 = (EditText) this.findViewById(R.id.editText4);
        editText4.selectAll();

    }
}
```

程序运行效果如图 3.38 所示。属性 android:hint 代表提示信息,当在 EditText 输入文字后,hint 字符就消失,如图 3.39 所示。

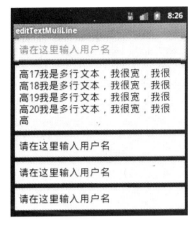
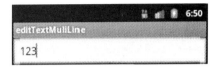

图 3.38　多行运行效果　　　　　　图 3.39　hint 消失

当单击 EditText3 时部分文本自动被选中，如图 3.40 所示。单击 EditText4 时文本全部被选中，如图 3.41 所示。

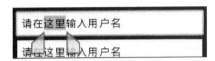

图 3.40　部分选中的文本　　　　　　图 3.41　文本全部被选中

而 EditText5 中的文本是不允许进行编辑的，但可以用长按事件来复制文本，如图 3.42 所示。

图 3.42　可以复制不可以编辑的 EditText

其实在图 3.42 中可以看到，多行文本域 EditText2 并没有显示出完整的文本内容，而且也没有由于文本过多显示出垂直滚动条，其实可以用 ScrollView 和 EditText 结合的方式来做出垂直滚动条的效果。

新建名称为 ScrollView_EditText 的 Android 项目，布局文件 main.xml 的代码如下：

```
<?xml version="1.0" encoding="utf-8"?>
<LinearLayout xmlns:android="http://schemas.android.com/apk/res/android"
    android:orientation="vertical" android:layout_width="fill_parent"
    android:layout_height="fill_parent">
    <ScrollView android:id="@+id/scrollView1"
```

```
                android:layout_width="match_parent" android:layout_height="wrap_content">
                <LinearLayout android:layout_width="match_parent"
                    android:id="@+id/linearLayout1" android:layout_height="match_parent">
                    <EditText android:text="EditText" android:id="@+id/editText1"
                        android:layout_width="fill_parent" android:layout_height="fill_parent"></EditText>
                </LinearLayout>
            </ScrollView>
</LinearLayout>
```

文件 Main.java 的代码如下:

```
public class Main extends Activity {
    private EditText editText1;

    @Override
    public void onCreate(Bundle savedInstanceState) {
        super.onCreate(savedInstanceState);
        setContentView(R.layout.main);

        editText1 = (EditText) this.findViewById(R.id.editText1);
        for (int i = 0; i < 20; i++) {
            editText1.append((i + 1) + "我是多行文本,我很宽,我很高");
        }
    }
}
```

程序运行效果如图 3.43 所示。

图 3.43 具有垂直滚动条的 EditText 控件

3.14 Gallery 控件和 ImageSwitcher 控件

Gallery 控件的功能是以画廊的方式展示数据，比较常用的是展示图片。
创建名称为 ui_9 的 Android 项目，在 res/drawable-ldpi 目录中加入若干图片，如图 3.44 所示。

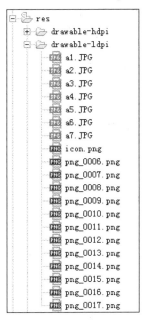

图 3.44 加入图片

由于控件 Gallery 显示的数据是来自于 Adapter 适配器，所以还需要新建一个名称为 ImageAdapter.java 适配器类文件，类代码如下：

```java
public class ImageAdapter extends BaseAdapter {

    // 关联 Context 上下文
    private Context context;
    // 往 Integer[]数组中加入图片的 id
    private Integer[] mImageIds = { R.drawable.png_0006, R.drawable.png_0007,
            R.drawable.png_0008, R.drawable.a1, R.drawable.a2, R.drawable.a3,
            R.drawable.a4, R.drawable.a5, R.drawable.a6, R.drawable.a7 };

    // 构造方法，需要传入 Context 对象，因为要创建 ImageView 控件
    public ImageAdapter(Context c) {
        context = c;
    }

    // 这个属性非常重要，决定 Gallery 控件显示多少张图片
    public int getCount() {
        return mImageIds.length;
    }
```

```java
// 默认代码，如果想取指定位置的对象，则必须要重定义这个方法中的代码
public Object getItem(int position) {
    return position;
}

// 默认代码，如果想取指定位置的 id 对象，则必须要重定义这个方法中的代码
public long getItemId(int position) {
    return position;
}

// 返回值 View 代表每 1 个显示在 Gallery 控件中的图片
public View getView(int position, View convertView, ViewGroup parent) {
    ImageView i = new ImageView(context);
    i.setPadding(10, 10, 10, 10);
    i.setImageResource(mImageIds[position]);
    i.setLayoutParams(new Gallery.LayoutParams(200, 300));
    i.setScaleType(ImageView.ScaleType.CENTER_INSIDE);
    return i;
}
}
```

还需要更改 Main.java 文件的代码如下：

```java
public class Main extends Activity {
    @Override
    public void onCreate(Bundle savedInstanceState) {
        super.onCreate(savedInstanceState);
        setContentView(R.layout.main);

        Gallery g = (Gallery) findViewById(R.id.gallery1);
        g.setAdapter(new ImageAdapter(this));

        g.setOnItemClickListener(new OnItemClickListener() {
            public void onItemClick(AdapterView parent, View v, int position,
                    long id) {
                Toast.makeText(Main.this, "" + position, Toast.LENGTH_SHORT)
                        .show();
            }
        });

    }
}
```

程序运行效果如图 3.45 所示。

图 3.45　Gallery 运行效果

控件 Gallery 的运行效果就是在水平方向进行图片的滚动，但如果想接近于画廊的效果，那么大多数的情况下都使用 Gallery 控件当缩略图，然后再使用显示图片的控件显示大尺寸的图片。

再创建一个名称为 switchImage 的 Android 项目，在这个项目中要演示一个具有缩略图和显示完整图片功能的示例。

新建自定义图片适配器类 ImageAdapter.java，代码如下：

```java
public class ImageAdapter extends BaseAdapter {

    public List<Integer> imageList = new ArrayList();

    private Context context;

    public ImageAdapter(Context context) {
        super();
        this.context = context;
        // 往 imageList 中存图片的 ID
        imageList.add(R.drawable.a1);
        imageList.add(R.drawable.a2);
        imageList.add(R.drawable.a3);
        imageList.add(R.drawable.a4);
        imageList.add(R.drawable.a5);
        imageList.add(R.drawable.a6);
        imageList.add(R.drawable.a7);
        imageList.add(R.drawable.a8);
        imageList.add(R.drawable.a9);
        imageList.add(R.drawable.a10);
        imageList.add(R.drawable.a11);
        imageList.add(R.drawable.a12);
        imageList.add(R.drawable.a13);
        imageList.add(R.drawable.a14);
        imageList.add(R.drawable.a15);
        imageList.add(R.drawable.a16);
        imageList.add(R.drawable.a17);
        imageList.add(R.drawable.a18);
```

```java
    }

    public int getCount() {
        return imageList.size();
    }

    public Object getItem(int arg0) {
        // TODO Auto-generated method stub
        return null;
    }

    public long getItemId(int arg0) {
        // TODO Auto-generated method stub
        return 0;
    }

    public View getView(int arg0, View arg1, ViewGroup arg2) {
        ImageView returnImage = new ImageView(context);
        returnImage.setImageResource(imageList.get(arg0));
        returnImage.setLayoutParams(new Gallery.LayoutParams(100, 100));
        returnImage.setScaleType(ImageView.ScaleType.CENTER_INSIDE);
        return returnImage;
    }
}
```

相关的图片资源列表如图 3.46 所示。

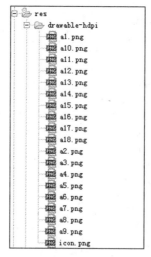

图 3.46 添加的图片资源列表信息

文件 main.xml 的代码如下：

```xml
<?xml version="1.0" encoding="utf-8"?>
<LinearLayout xmlns:android="http://schemas.android.com/apk/res/android"
    android:orientation="vertical" android:layout_width="fill_parent"
```

```xml
            android:layout_height="fill_parent">
    <ImageView android:id="@+id/imageView1" android:scaleType="centerInside"
            android:layout_weight="1" android:layout_height="fill_parent"
            android:layout_width="fill_parent" android:src="@drawable/icon"></ImageView>
    <Gallery android:id="@+id/gallery1" android:layout_width="fill_parent"
            android:layout_height="wrap_content"></Gallery>
</LinearLayout>
```

Activity 对象 Main.java 文件的代码如下：

```java
public class Main extends Activity {
    private ImageView imageView1;
    private ImageAdapter iaRef;

    @Override
    public void onCreate(Bundle savedInstanceState) {
        super.onCreate(savedInstanceState);
        setContentView(R.layout.main);

        iaRef = new ImageAdapter(this);

        imageView1 = (ImageView) this.findViewById(R.id.imageView1);
        imageView1.setImageResource(iaRef.imageList.get(0));

        Gallery GalleryRef = (Gallery) this.findViewById(R.id.gallery1);
        GalleryRef.setAdapter(iaRef);

        GalleryRef.setOnItemClickListener(new OnItemClickListener() {
            public void onItemClick(AdapterView<?> arg0, View arg1, int arg2,
                    long arg3) {
                Animation ani1 = AnimationUtils.loadAnimation(Main.this,
                        android.R.anim.fade_in);

                imageView1.setImageResource(iaRef.imageList.get(arg2));
                imageView1.startAnimation(ani1);
            }
        });

    }
}
```

程序运行效果如图 3.47 所示。

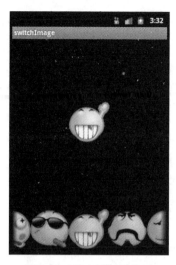

图 3.47 缩略图效果显示

3.15 TextView 控件

下面的示例将演示 TextView 控件的基本使用，实现的功能是更改 TextView 控件的字体颜色和基本的样式，还有添加超级链接 A 标签的效果。

创建名称为 textViewColor 的 Android 项目，将文件 main.xml 的代码更改如下：

```xml
<?xml version="1.0" encoding="utf-8"?>
<LinearLayout xmlns:android="http://schemas.android.com/apk/res/android"
    android:orientation="vertical" android:layout_width="fill_parent"
    android:layout_height="fill_parent">
    <TextView android:text="TextView" android:id="@+id/textView1"
        android:layout_width="wrap_content" android:layout_height="wrap_content"></TextView>
    <TextView android:text="TextView" android:id="@+id/textView2"
        android:layout_width="wrap_content" android:layout_height="wrap_content"
        android:textColorLink="#0F0"></TextView>
</LinearLayout>
```

属性 android:textColorLink="#0F0" 的功能是定义当前超链接文本的字体颜色。

继续更改文件 Main.java 的代码如下：

```java
public class Main extends Activity {

    private TextView tv1;
    private TextView tv2;

    @Override
    public void onCreate(Bundle savedInstanceState) {
        super.onCreate(savedInstanceState);
        setContentView(R.layout.main);
```

```
        tv1 = (TextView) this.findViewById(R.id.textView1);
        tv2 = (TextView) this.findViewById(R.id.textView2);

        tv1.setText("去往");
        tv1.setTextColor(Color.RED);

        // 有顺序问题
        tv2.setAutoLinkMask(Linkify.ALL);
        tv2.setText("http://www.baidu.com");
    }
}
```

程序运行初始效果如图3.48所示。

图3.48 显示红色普通文本和带下划线的绿色超级链接文本

下面的示例将继续演示设置控件 TextView 字体的样式，在 Android 中已经预置了一些基本的字体样式，文档声明如图3.49所示。

图3.49 预置的字体样式

创建名称为 textViewFontStyle 的 Android 项目，更改文件 main.xml 的代码如下：

```xml
<?xml version="1.0" encoding="utf-8"?>
<LinearLayout xmlns:android="http://schemas.android.com/apk/res/android"
    android:orientation="vertical" android:layout_width="fill_parent"
    android:layout_height="fill_parent">
    <TextView android:text="TextView" android:id="@+id/textView1"
        android:layout_width="wrap_content" android:layout_height="wrap_content"></TextView>
    <TextView android:text="TextView" android:id="@+id/textView2"
        android:layout_width="wrap_content" android:layout_height="wrap_content"></TextView>
    <TextView android:text="TextView" android:id="@+id/textView3"
        android:layout_width="wrap_content" android:layout_height="wrap_content"></TextView>
    <TextView android:text="TextView" android:id="@+id/textView4"
```

```xml
        android:layout_width="wrap_content" android:layout_height="wrap_content"></TextView>
    <TextView android:text="TextView" android:id="@+id/textView5"
        android:layout_width="wrap_content" android:layout_height="wrap_content"></TextView>
    <TextView android:text="TextView" android:id="@+id/textView6"
        android:layout_width="wrap_content" android:layout_height="wrap_content"></TextView>
    <TextView android:text="TextView" android:id="@+id/textView7"
        android:layout_width="wrap_content" android:layout_height="wrap_content"></TextView>
    <TextView android:text="TextView" android:id="@+id/textView8"
        android:layout_width="wrap_content" android:layout_height="wrap_content"></TextView>
</LinearLayout>
```

更改 Main.java 的代码如下：

```java
public class Main extends Activity {

    private TextView tv1;
    private TextView tv2;
    private TextView tv3;
    private TextView tv4;
    private TextView tv5;
    private TextView tv6;
    private TextView tv7;
    private TextView tv8;

    @Override
    public void onCreate(Bundle savedInstanceState) {
        super.onCreate(savedInstanceState);
        setContentView(R.layout.main);

        tv1 = (TextView) this.findViewById(R.id.textView1);
        tv2 = (TextView) this.findViewById(R.id.textView2);
        tv3 = (TextView) this.findViewById(R.id.textView3);
        tv4 = (TextView) this.findViewById(R.id.textView4);
        tv5 = (TextView) this.findViewById(R.id.textView5);
        tv6 = (TextView) this.findViewById(R.id.textView6);
        tv7 = (TextView) this.findViewById(R.id.textView7);
        tv8 = (TextView) this.findViewById(R.id.textView8);

        tv1.setText("Typeface.DEFAULT");
        tv1.setTypeface(Typeface.DEFAULT);

        tv2.setText("Typeface.DEFAULT_BOLD");
        tv2.setTypeface(Typeface.DEFAULT_BOLD);

        tv3.setText("Typeface.MONOSPACE");
        tv3.setTypeface(Typeface.MONOSPACE);

        tv4.setText("Typeface.SANS_SERIF");
        tv4.setTypeface(Typeface.SANS_SERIF);

        tv5.setText("Typeface.SERIF");
```

```
            tv5.setTypeface(Typeface.SERIF);

            tv6.setText("Typeface.BOLD");
            tv6.setTypeface(null, Typeface.BOLD);

            tv7.setText("Typeface.BOLD_ITALIC");
            tv7.setTypeface(null, Typeface.BOLD_ITALIC);

            tv8.setText("Typeface.ITALIC");
            tv8.setTypeface(null, Typeface.ITALIC);

        }
    }
```

程序运行后显示不同样式的字体，如图 3.50 所示。

图 3.50　显示不同样式的字体

还可以对控件 TextView 设置行数及最大和最小行数，示例代码在名称为 testView_test 的项目中，布局文件的代码如下：

```xml
<?xml version="1.0" encoding="utf-8"?>
<LinearLayout xmlns:android="http://schemas.android.com/apk/res/android"
    android:orientation="vertical" android:layout_width="fill_parent"
    android:layout_height="fill_parent">
    <TextView android:layout_width="fill_parent"
        android:layout_height="wrap_content" android:lines="3"
        android:text="---------元素需要指定 href 或 name 属性。文本和图像都可包含在锚内。作为锚的图像有一个边框表明该链接是否访问过。要避免显示此边框，你可以设置 IMG 元素的 BORDER 标签属性为 0 或者省略 BORDER 标签属性。你还可以使用样式表 CSS 来覆盖 A 和 IMG 元素的默认外观。注意，TABLE 对象当包含在 A 标签内时可能工作不正常。如果对 A 元素应用 time2 行为的话，那么该元素仅当在时间线上激活时才会变成链接。此元素在 Microsoft? Internet Explorer 3.0 的 HTML 和脚本中可用。" />
    <TextView android:layout_width="fill_parent"
        android:background="#ff0000"
        android:layout_height="wrap_content" android:maxLines="20"
        android:minLines="5" android:text="---------元素需要指定 href 或 name 属性。" />
</LinearLayout>
```

程序运行效果如图 3.51 所示。

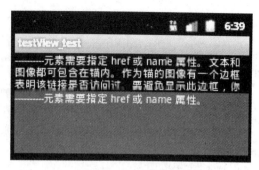

图 3.51 行数的测试

3.16 ImageView 和 ImageButton 控件

控件 ImageView 的功能是显示图片，而控件 ImageButton 的功能是带有图片的按钮。

创建名称为 ui_11 的 Android 项目，并且在项目中的 res/drawable-ldpi 目录中添加几张图片资源，如图 3.52 所示。

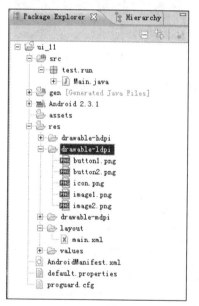

图 3.52 添加图片

布局文件 main.xml 的代码更改如下：

<?xml version="*1.0*" encoding="*utf-8*"?>
<LinearLayout xmlns:android="*http://schemas.android.com/apk/res/android*"
    android:orientation="*vertical*" android:layout_width="*fill_parent*"
    android:layout_height="*fill_parent*" android:background="*#ff0000*">
    <TextView android:layout_width="*fill_parent*"
        android:layout_height="*wrap_content*" android:text="*@string/hello*" />

```xml
<ImageButton android:layout_width="wrap_content"
    android:id="@+id/imageButton1" android:src="@drawable/button1"
    android:layout_height="wrap_content" android:background="#FF000000"></ImageButton>
<ImageButton android:layout_width="wrap_content"
    android:id="@+id/imageButton2" android:src="@drawable/button2"
    android:layout_height="wrap_content" android:background="#00000000"></ImageButton>
<ImageView android:src="@drawable/icon" android:id="@+id/imageView1"
    android:layout_width="wrap_content" android:layout_height="wrap_content"></ImageView>
</LinearLayout>
```

文件 Main.java 的代码如下：

```java
public class Main extends Activity {
    private ImageButton button1;
    private ImageButton button2;
    private ImageView image1;

    @Override
    public void onCreate(Bundle savedInstanceState) {
        super.onCreate(savedInstanceState);
        setContentView(R.layout.main);

        button1 = (ImageButton) this.findViewById(R.id.imageButton1);
        button2 = (ImageButton) this.findViewById(R.id.imageButton2);

        image1 = (ImageView) this.findViewById(R.id.imageView1);

        button1.setOnClickListener(new OnClickListener() {
            public void onClick(View arg0) {
                image1.setImageResource(R.drawable.image1);
            }
        });

        button2.setOnClickListener(new OnClickListener() {
            public void onClick(View arg0) {
                image1.setImageResource(R.drawable.image2);
            }
        });

    }
}
```

程序初始运行效果如图 3.53 所示。

单击第 1 个按钮出现效果如图 3.54 所示。

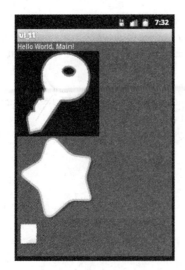
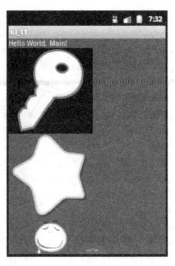

图 3.53　初始运行效果　　　　图 3.54　ImageView 换了图片

3.17　MultiAutoCompleteTextView 控件

在控件 AutoCompleteTextView 中 1 次只能选择 1 个选项并且支持模糊查询功能，而在 MultiAutoCompleteTextView 控件中是可以选择多个选项并且也支持模糊查询的功能。

新建名称为 ui_12 的 Android 项目，更改布局文件 main.xml 的代码如下：

```
<?xml version="1.0" encoding="utf-8"?>
<LinearLayout xmlns:android="http://schemas.android.com/apk/res/android"
    android:orientation="vertical" android:layout_width="fill_parent"
    android:layout_height="fill_parent">
    <MultiAutoCompleteTextView android:text=""
        android:id="@+id/multiAutoCompleteTextView1" android:layout_width="fill_parent"
        android:layout_height="wrap_content"></MultiAutoCompleteTextView>
</LinearLayout>
```

更改 Main.java 的代码如下：

```
public class Main extends Activity {
    private static final String[] COUNTRIES = new String[] { "高洪岩 1", "高洪岩 2",
            "高洪岩 3", "高洪岩 4", "高洪岩 5" };

    @Override
    public void onCreate(Bundle savedInstanceState) {
        super.onCreate(savedInstanceState);
        setContentView(R.layout.main);
        ArrayAdapter<String> adapter = new ArrayAdapter<String>(this,
                android.R.layout.simple_dropdown_item_1line, COUNTRIES);
        MultiAutoCompleteTextView textView = (MultiAutoCompleteTextView) findViewById(R.id.multiAutoCompleteTextView1);
        textView.setAdapter(adapter);
```

```
        textView.setThreshold(1);
        textView.setTokenizer(new MultiAutoCompleteTextView.CommaTokenizer());

        textView.setOnItemClickListener(new OnItemClickListener() {
            public void onItemClick(AdapterView<?> arg0, View arg1, int arg2,
                    long arg3) {
                Log.v("textView.setOnItemClickListener", ""
                        + ((TextView) arg1).getText());
            }
        });

    }
}
```

控件 MultiAutoCompleteTextView 的运行效果和 AutoCompleteTextView 控件运行效果基本一致，如图 3.55 所示。

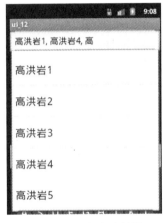

图 3.55 具有多选效果

3.18 ProgressBar 控件

控件 ProgressBar 专门实现进度条效果，使用起来非常简单。

创建名称为 progressBarTest 的 Android 项目，将文件 Main.java 的代码更改如下：

```
public class Main extends Activity {
    private ProgressBar pb;
    private Button button;

    @Override
    public void onCreate(Bundle savedInstanceState) {
        super.onCreate(savedInstanceState);
        setContentView(R.layout.main);
        pb = (ProgressBar) this.findViewById(R.id.progressBar1);
        button = (Button) this.findViewById(R.id.button1);
```

```
                button.setOnClickListener(new View.OnClickListener() {
                    public void onClick(View arg0) {
                        pb.setProgress(50);
                        pb.setSecondaryProgress(100);
                    }
                });
        }
}
```

更改布局文件 main.xm 的代码如下：

```
<?xml version="1.0" encoding="utf-8"?>
<LinearLayout xmlns:android="http://schemas.android.com/apk/res/android"
    android:orientation="vertical" android:layout_width="fill_parent"
    android:layout_height="fill_parent">
    <ProgressBar android:id="@+id/progressBar1"
        android:layout_width="fill_parent" android:layout_height="wrap_content"
        style="?android:attr/progressBarStyleHorizontal" android:max="100"
        android:progress="25" android:secondaryProgress="75"></ProgressBar>
    <Button android:text="Button" android:id="@+id/button1"
        android:layout_width="wrap_content" android:layout_height="wrap_content"></Button>
</LinearLayout>
```

值 "progressBarStyleHorizontal" 为 theme 皮肤的名称，使用@android:attr/引用这个皮肤属性样式。

程序初始运行效果如图 3.56 所示。单击 "Button" 按钮后运行效果如图 3.57 所示。

图 3.56　初始运行效果

图 3.57　改变进度值了

3.19　RadioGroup 与 RadioButton 控件

在 HTML 语言中实现 radio 控件单选效果时，必须设置标签的 name 属性值为相同，目的是将这些 radio 设置为一组，彼此之间互斥，但在 Android 中想要实现相同的效果必须用到 RadioGroup 控件。

新建名称为 ui_14 的 Android 项目，更改 main.xml 配置文件的代码如下：

```
<?xml version="1.0" encoding="utf-8"?>
<LinearLayout xmlns:android="http://schemas.android.com/apk/res/android"
    android:orientation="vertical" android:layout_width="fill_parent"
```

```xml
android:layout_height="fill_parent">
<RadioGroup android:id="@+id/radioGroup1"
    android:layout_height="wrap_content" android:layout_width="match_parent"
    android:orientation="horizontal">
        <RadioButton android:id="@+id/radio1" android:tag="bj"
            android:layout_height="wrap_content" android:text="北京"
            android:layout_width="wrap_content"></RadioButton>
        <RadioButton android:id="@+id/radio2" android:tag="sh"
            android:layout_height="wrap_content" android:text="上海"
            android:layout_width="wrap_content"></RadioButton>
        <RadioButton android:id="@+id/radio3" android:tag="sz"
            android:layout_height="wrap_content" android:text="深圳"
            android:layout_width="wrap_content" android:checked="true"></RadioButton>
        <RadioButton android:id="@+id/radio4" android:tag="gz"
            android:layout_height="wrap_content" android:text="广州"
            android:layout_width="wrap_content"></RadioButton>
</RadioGroup>
<TextView android:text="TextView" android:id="@+id/textView1"
    android:layout_width="wrap_content" android:layout_height="wrap_content"></TextView>
<Button android:text="Button" android:id="@+id/button2"
    android:layout_width="wrap_content" android:layout_height="wrap_content"></Button>
</LinearLayout>
```

更改 Main.java 文件的代码如下：

```java
public class Main extends Activity {
    private Button button2;
    private TextView textView1;
    private RadioGroup radioGroup1;

    @Override
    public void onCreate(Bundle savedInstanceState) {
        super.onCreate(savedInstanceState);
        setContentView(R.layout.main);
        radioGroup1 = (RadioGroup) this.findViewById(R.id.radioGroup1);
        button2 = (Button) this.findViewById(R.id.button2);
        textView1 = (TextView) this.findViewById(R.id.textView1);

        radioGroup1.setOnCheckedChangeListener(new OnCheckedChangeListener() {
            public void onCheckedChanged(RadioGroup arg0, int arg1) {
                textView1.setText(((RadioButton) Main.this.findViewById(arg1))
                        .getText().toString());
            }
        });

        button2.setOnClickListener(new OnClickListener() {
            public void onClick(View arg0) {
                Log.v("!!", ((RadioButton) Main.this.findViewById(radioGroup1
                        .getCheckedRadioButtonId())).getTag().toString());
            }
        });
```

        }
}

程序运行初始效果如 3.58 所示。

用鼠标单击"上海"单选按钮后，出现如图 3.59 所示。

图 3.58　初始界面

图 3.59　在 TextView 中显示选中了上海

单击"Button"按钮在 LogCat 中打印 tag 值，如图 3.60 所示。

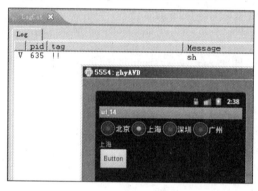
图 3.60　显示 tag 的值

在 XML 布局文件中设置某个 radio 控件为默认选中状态的代码也可以按下述方式编写：

```
<RadioGroup android:checkedButton="@+id/radio3"
    android:layout_height="wrap_content" android:layout_width="match_parent"
    android:id="@+id/radioGroup1">
```

使用<RadioGroup>标签的 android:checkedButton 属性可以设置默认选中的 radio 控件。

3.20　RatingBar 控件

控件 RatingBar 的功能是实现选择等级。

新建名称为 ui_15 的 Android 项目，更改 main.xml 文件的代码如下：

```
<?xml version="1.0" encoding="utf-8"?>
<LinearLayout xmlns:android="http://schemas.android.com/apk/res/android"
    android:orientation="vertical" android:layout_width="fill_parent"
    android:layout_height="fill_parent">
    <RatingBar android:id="@+id/ratingBar1" android:layout_width="wrap_content"
```

```xml
        android:layout_height="wrap_content"></RatingBar>
    <RatingBar android:id="@+id/ratingBar2" android:layout_width="wrap_content"
        android:layout_height="wrap_content"></RatingBar>
    <Button android:text="Button" android:id="@+id/button1"
        android:layout_width="wrap_content" android:layout_height="wrap_content"></Button>
    <RatingBar android:id="@+id/ratingBar3" android:layout_width="wrap_content"
        android:layout_height="wrap_content" style="?android:attr/ratingBarStyleSmall"
        android:numStars="5" android:max="5" android:progress="3"></RatingBar>
    <RatingBar android:id="@+id/ratingBar4" android:layout_width="wrap_content"
        android:layout_height="wrap_content" style="?android:attr/ratingBarStyleIndicator"
        android:numStars="5" android:max="5" android:progress="3"></RatingBar>
</LinearLayout>
```

更改 Main.java 文件的代码如下：

```java
public class Main extends Activity {
    private RatingBar ratingBar1Ref;
    private RatingBar ratingBar2Ref;
    private Button button1;

    @Override
    public void onCreate(Bundle savedInstanceState) {
        super.onCreate(savedInstanceState);
        setContentView(R.layout.main);

        ratingBar1Ref = (RatingBar) this.findViewById(R.id.ratingBar1);
        ratingBar2Ref = (RatingBar) this.findViewById(R.id.ratingBar2);
        button1 = (Button) this.findViewById(R.id.button1);

        ratingBar1Ref.setNumStars(5);
        ratingBar1Ref.setStepSize(0.5F);

        ratingBar2Ref.setNumStars(5);
        ratingBar2Ref.setStepSize(0.5F);

        button1.setOnClickListener(new OnClickListener() {

            public void onClick(View arg0) {
                Log.v("rb1 value=", "" + ratingBar1Ref.getProgress());
                Log.v("rb2 value=", "" + ratingBar2Ref.getProgress());
            }

        });
    }
}
```

本项目演示的控件 RatingBar 运行效果是出现 5 颗星，操作者可以用单击或鼠标拖曳的方式来进行星级的选择，如图 3.61 所示。

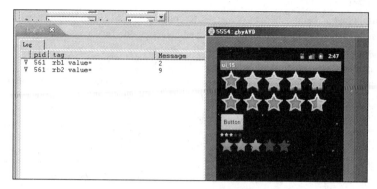

图 3.61 运行结果

3.21 SeekBar 控件

控件 SeekBar 的功能是提供一个可以用鼠标拖曳的进条度,当然也可以用程序代码来控制显示的进度值。

创建名称为 ui_16 的 Android 项目,更改 main.xml 配置文件的代码如下:

```
<?xml version="1.0" encoding="utf-8"?>
<LinearLayout xmlns:android="http://schemas.android.com/apk/res/android"
    android:orientation="vertical" android:layout_width="fill_parent"
    android:layout_height="fill_parent">
    <SeekBar android:layout_height="wrap_content" android:id="@+id/seekBar1"
        android:layout_width="match_parent"></SeekBar>
    <TextView android:text="TextView" android:id="@+id/textView1"
        android:layout_width="wrap_content" android:layout_height="wrap_content"></TextView>
</LinearLayout>
```

更改 Main.java 文件的代码如下:

```
public class Main extends Activity {
    private SeekBar seekBarRef;
    private TextView textView1;

    @Override
    public void onCreate(Bundle savedInstanceState) {
        super.onCreate(savedInstanceState);
        setContentView(R.layout.main);
        seekBarRef = (SeekBar) this.findViewById(R.id.seekBar1);
        textView1 = (TextView) this.findViewById(R.id.textView1);
        seekBarRef.setMax(100);
        seekBarRef.setProgress(80);

        seekBarRef.setOnSeekBarChangeListener(new OnSeekBarChangeListener() {

            public void onStopTrackingTouch(SeekBar arg0) {
                Log.v("onStopTrackingTouch", "onStopTrackingTouch");
```

```
                textView1.setText("" + (arg0.getProgress() + 1));
            }

            public void onStartTrackingTouch(SeekBar arg0) {
                Log.v("onStartTrackingTouch", "onStartTrackingTouch");
                textView1.setText("" + (arg0.getProgress() + 1));
            }

            public void onProgressChanged(SeekBar arg0, int arg1, boolean arg2) {
                Log.v("onProgressChanged", "onProgressChanged");
                textView1.setText("" + (arg0.getProgress() + 1));
            }
        });

    }
}
```

程序初始运行效果如图 3.62 所示。

用鼠标单击并且拉动 SeekBar 控件的进度条后,在 LogCat 面板中打印此操作过程中的进度信息,如图 3.63 所示。

图 3.62　程序初始运行

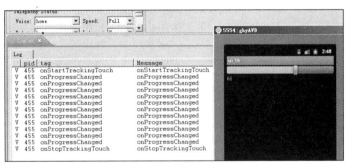

图 3.63　LogCat 打印的进度信息

3.22　ListView 对象和 Spinner 控件

控件 Spinner 的作用是提供一个用户选择的列表,在这个列表中用户可以选择指定的单选选项,下面进入 Spinner 控件的学习。

3.22.1　Spinner 控件初步使用

创建名称为 ui_17 的 Android 项目,更改 main.xml 配置文件的代码如下:

```xml
<?xml version="1.0" encoding="utf-8"?>
<LinearLayout xmlns:android="http://schemas.android.com/apk/res/android"
    android:orientation="vertical" android:layout_width="fill_parent"
    android:layout_height="fill_parent">
    <Spinner android:layout_width="match_parent"
```

```xml
        android:layout_height="wrap_content" android:id="@+id/spinner1"></Spinner>
    <TextView android:text="TextView" android:id="@+id/textView1"
        android:layout_width="wrap_content" android:layout_height="wrap_content"></TextView>
    <TextView android:text="单击按钮时我改变文本" android:id="@+id/textView2"
        android:layout_width="wrap_content" android:layout_height="wrap_content"></TextView>
    <Button android:text="Button" android:id="@+id/button1"
        android:layout_width="wrap_content" android:layout_height="wrap_content"></Button>
</LinearLayout>
```

更改 Main.java 的代码如下：

```java
public class Main extends Activity {
    private Button button1;
    private TextView textView1;
    private TextView textView2;
    private Spinner spinner;

    @Override
    public void onCreate(Bundle savedInstanceState) {
        super.onCreate(savedInstanceState);
        setContentView(R.layout.main);
        button1 = (Button) this.findViewById(R.id.button1);
        textView1 = (TextView) this.findViewById(R.id.textView1);
        textView2 = (TextView) this.findViewById(R.id.textView2);
        spinner = (Spinner) this.findViewById(R.id.spinner1);

        List listData = new ArrayList();
        listData.add("accp1");
        listData.add("accp2");
        listData.add("accp3");
        listData.add("accp4");
        listData.add("accp5");
        listData.add("accp6");
        listData.add("accp7");
        listData.add("accp8");

        ArrayAdapter arrayAdapterRef = new ArrayAdapter(this,
                android.R.layout.simple_spinner_item, listData);

        arrayAdapterRef
                .setDropDownViewResource(android.R.layout.simple_spinner_dropdown_item);

        spinner.setAdapter(arrayAdapterRef);

        spinner.setOnItemSelectedListener(new OnItemSelectedListener() {

            public void onItemSelected(AdapterView<?> arg0, View arg1,
                    int arg2, long arg3) {
                textView1.setText(((TextView) arg1).getText());
            }
```

```
            public void onNothingSelected(AdapterView<?> arg0) {
                // TODO Auto-generated method stub

            }
        });

        button1.setOnClickListener(new OnClickListener() {

            public void onClick(View arg0) {
                textView2.setText("" + spinner.getSelectedItem());

            }
        });

    }
}
```

程序初始运行效果如图 3.64 所示。

单击 Spinner 控件弹出列表选项界面，如图 3.65 所示。

图 3.64　Spinner 控件初始效果　　　　图 3.65　弹出选项列表

选中 "accp4" 选项后在 TextView1 控件显示出当前的值，如图 3.66 所示。

如果单击 Button 按钮则在 TextView2 中显示当前 Spinner 控件选中的文本，如图 3.67 所示。

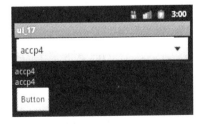

图 3.66　在 TextView1 控件显示出当前的值　　图 3.67　单击 Button 按钮在 TextView2 控件显示文本

另外，控件 Spinner 中的数据还可以来自字符串数组资源，在名称为 Spinner_testtest 项目中创建 arrays.xml 资源文件，内容如下：

```xml
<?xml version="1.0" encoding="utf-8"?>
<resources>
    <string-array name="ghyStringArray">
        <item>AA1</item>
        <item>AA2</item>
        <item>AA3</item>
        <item>AA4</item>
        <item>AA5</item>
    </string-array>
</resources>
```

文件 main.xml 的代码如下：

```xml
<?xml version="1.0" encoding="utf-8"?>
<LinearLayout xmlns:android="http://schemas.android.com/apk/res/android"
    android:orientation="vertical" android:layout_width="fill_parent"
    android:layout_height="fill_parent">
    <Spinner android:id="@+id/spinner1" android:layout_width="match_parent"
        android:layout_height="wrap_content"></Spinner>
    <Spinner android:id="@+id/spinner2" android:layout_width="match_parent"
        android:layout_height="wrap_content" android:entries="@array/ghyStringArray"
        android:prompt="@string/spinner2Title"></Spinner>
</LinearLayout>
```

在 strings.xml 资源文件中添加节点：
`<string name="spinner2Title">请选择选项在 Spinner2 中</string>`

文件 Main.java 的代码如下：

```java
public class Main extends Activity {
    private Spinner s1;
    private Spinner s2;

    @Override
    public void onCreate(Bundle savedInstanceState) {
        requestWindowFeature(Window.FEATURE_INDETERMINATE_PROGRESS);
        requestWindowFeature(Window.FEATURE_PROGRESS);
        super.onCreate(savedInstanceState);
        setContentView(R.layout.main);
        setProgressBarIndeterminateVisibility(true);
        setProgressBarVisibility(true);
        setProgress(5000);

        s1 = (Spinner) this.findViewById(R.id.spinner1);
        s2 = (Spinner) this.findViewById(R.id.spinner2);

        String[] stringArray = this.getResources().getStringArray(
                R.array.ghyStringArray);
        ArrayAdapter adapter = new ArrayAdapter(this,
                android.R.layout.simple_spinner_item, stringArray);
        adapter
                .setDropDownViewResource(android.R.layout.simple_spinner_dropdown_item);
```

```
            s1.setPrompt("请选择选项在 Spinner1 中");
            s1.setAdapter(adapter);

    }
}
```

程序运行结果如图 3.68 所示。

图 3.68 来自于字符串数组的运行结果

并且在 Activity 标题栏中显示进度条的效果,如图 3.69 所示。

图 3.69 标题中有进度条显示的功能

3.22.2 在 ListView 控件中显示文本列表功能

上面演示的是使用 Spinner 控件来进行弹出列表框选择其中条目的示例,其实在 Android 中还有一个重量级的列表对象,它就是 ListView。在 Android 用到列表的地方,ListView 的使用率基本达到了 90%,所以掌握 ListView 对象的使用是掌握 Android 处理数据和展示数据的基本技能。

单独使用 ListView 控件可以不弹出对话框来进行列表选项的选择,因为整个界面只存在 1 个 ListView 控件。

创建名称为 ui_17_1_testList 的 Android 项目,布局文件 main.xml 的代码如下:

```
<?xml version="1.0" encoding="utf-8"?>
<LinearLayout xmlns:android="http://schemas.android.com/apk/res/android"
    android:orientation="vertical" android:layout_width="fill_parent"
    android:layout_height="fill_parent">
    <ListView android:layout_height="wrap_content" android:id="@+id/listView1"
        android:layout_width="match_parent"></ListView>
</LinearLayout>
```

文件 Main.java 的代码如下：

```java
public class Main extends Activity {
    private ListView listView1;
    private ArrayList cityList = new ArrayList();

    @Override
    public void onCreate(Bundle savedInstanceState) {
        super.onCreate(savedInstanceState);
        setContentView(R.layout.main);

        listView1 = (ListView) this.findViewById(R.id.listView1);

        cityList.add("北京 1");
        cityList.add("北京 2");
        cityList.add("北京 3");
        cityList.add("北京 4");

        ArrayAdapter arrayAdapterRef = new ArrayAdapter(this,
                android.R.layout.simple_list_item_1, cityList);

        listView1.setAdapter(arrayAdapterRef);

        listView1.setOnItemClickListener(new OnItemClickListener() {

            public void onItemClick(AdapterView<?> arg0, View arg1, int arg2,
                    long arg3) {
                Log.v("listView1.setOnItemClickListener", ""
                        + cityList.get(arg2));
            }
        });

    }
}
```

程序运行后单击列表中的条目，在 LogCat 中打印出当前选中条目的文本内容，如图 3.70 所示。

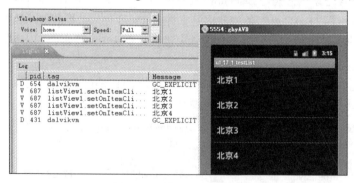

图 3.70　显示出文本内容

3.22.3 在 ListView 控件中使用多选 checkedbox 控件

创建名称为 ui_17_1_checkbox 的 Android 项目,本示例要演示一个在 ListView 中显示 checkbox 控件及有全选、反选和取值的功能。

将布局文件 main.xml 的代码更改如下:

```xml
<?xml version="1.0" encoding="utf-8"?>
<LinearLayout xmlns:android="http://schemas.android.com/apk/res/android"
    android:orientation="vertical" android:layout_width="fill_parent"
    android:layout_height="fill_parent">
    <ListView android:layout_weight="1" android:id="@+id/listView1"
        android:layout_height="wrap_content" android:layout_width="match_parent"></ListView>
    <LinearLayout android:gravity="center"
        android:orientation="horizontal" android:layout_width="fill_parent"
        android:layout_height="wrap_content">
        <Button android:text="全选" android:id="@+id/button1"
            android:layout_width="wrap_content" android:layout_height="wrap_content"></Button>
        <Button android:text="反选" android:id="@+id/button2"
            android:layout_width="wrap_content" android:layout_height="wrap_content"></Button>
        <Button android:text="取值" android:id="@+id/button3"
            android:layout_width="wrap_content" android:layout_height="wrap_content"></Button>
    </LinearLayout>
</LinearLayout>
```

更改文件 Main.java 的代码如下:

```java
public class Main extends Activity {
    private List cityList = new ArrayList();
    private ListView listView;

    private Button button1;// 全选
    private Button button2;// 反选
    private Button button3;// 取值

    @Override
    public void onCreate(Bundle savedInstanceState) {
        super.onCreate(savedInstanceState);
        setContentView(R.layout.main);

        listView = (ListView) this.findViewById(R.id.listView1);
        button1 = (Button) this.findViewById(R.id.button1);
        button2 = (Button) this.findViewById(R.id.button2);
        button3 = (Button) this.findViewById(R.id.button3);

        final boolean[] isCheckedArray = new boolean[8];
        isCheckedArray[0] = false;
        isCheckedArray[1] = true;// 默认值为 true
        isCheckedArray[2] = false;
        isCheckedArray[3] = true;// 默认值为 true
        isCheckedArray[4] = false;
```

```java
isCheckedArray[5] = true;// 默认值为 true
isCheckedArray[6] = false;
isCheckedArray[7] = true;// 默认值为 true

cityList.add("accp1");
cityList.add("accp2");
cityList.add("accp3");
cityList.add("accp4");
cityList.add("accp5");
cityList.add("accp6");
cityList.add("accp7");
cityList.add("accp8");

ArrayAdapter adapter = new ArrayAdapter(this,
        android.R.layout.simple_list_item_multiple_choice, cityList);

listView.setChoiceMode(ListView.CHOICE_MODE_MULTIPLE);
listView.setAdapter(adapter);

listView.setOnItemClickListener(new OnItemClickListener() {
    public void onItemClick(AdapterView<?> arg0, View arg1, int arg2,
            long arg3) {
        Log.v("-----------", "" + ((TextView) arg1).getText());
    }
});

// 赋初始值
for (int i = 0; i < isCheckedArray.length; i++) {
    listView.setItemChecked(i, isCheckedArray[i]);
}

// 全选
button1.setOnClickListener(new OnClickListener() {
    public void onClick(View arg0) {
        Log.v("单击了全选", "单击了全选");

        for (int i = 0; i < isCheckedArray.length; i++) {
            listView.setItemChecked(i, true);
        }
    }
});

// 反选
button2.setOnClickListener(new OnClickListener() {
    public void onClick(View arg0) {
        Log.v("单击了反选", "单击了反选");

        SparseBooleanArray sparseBooleanArrayRef = listView
                .getCheckedItemPositions();
        for (int i = 0; i < sparseBooleanArrayRef.size(); i++) {
```

```
                    if (sparseBooleanArrayRef.get(i) == true) {
                        listView.setItemChecked(i, false);
                    } else {
                        listView.setItemChecked(i, true);
                    }
                }
            }
        });

        // 取值
        button3.setOnClickListener(new OnClickListener() {
            public void onClick(View arg0) {
                Log.v("单击了取值", "单击了取值");

                SparseBooleanArray sparseBooleanArrayRef = listView
                        .getCheckedItemPositions();
                for (int i = 0; i < sparseBooleanArrayRef.size(); i++) {
                    if (sparseBooleanArrayRef.get(i) == true) {
                        Log.v("值为：", "" + listView.getAdapter().getItemId(i)
                                + " " + listView.getAdapter().getItem(i));
                    }
                }
            }
        });

    }
}
```

程序初始运行效果如图 3.71 所示。

从图 3.71 中可以看到，运行时有默认选中的值，这时单击"全选"按钮后界面效果如图 3.72 所示。

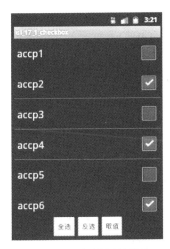 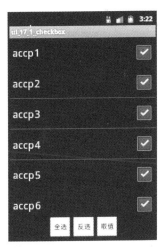

图 3.71　初始运行效果　　　　　图 3.72　单击全选后呈全部选中状态

然后再单击"取值"按钮，看看能否正确取出全部选中的值，LogCat 面板效果如图 3.73 所示。

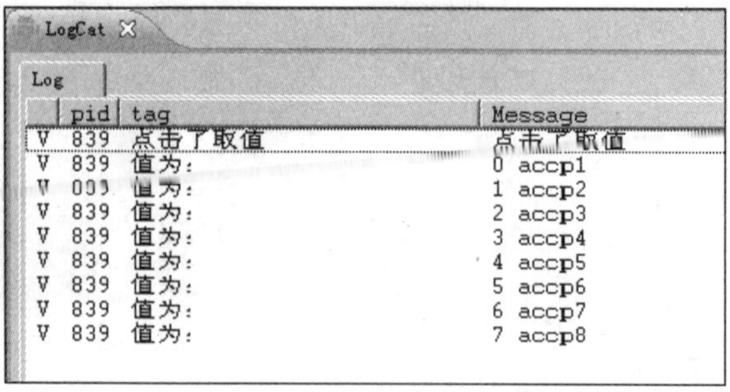

图 3.73 打印出全部选中的值

然后再单击"反选",取消全部的选中状态,这时用鼠标单击"accp1"、"accp3"和"accp5"选项,在LogCat中打印出了当前单击条目的文本信息,如图 3.74 所示。

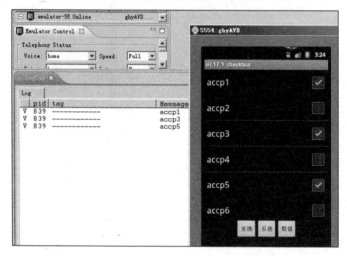

图 3.74 打印 accp1 和 accp3 和 accp5 单击的文本

这时再单击"取值"看看能否把这3个选项的信息打印出来,在LogCat中打印出的信息如图 3.75 所示。

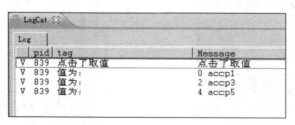

图 3.75 选中 3 项后的取值信息

3.22.4 在 ListView 控件中使用单选 radioButton 控件

前面的演示是在 ListView 控件中显示多选 checkedbox 控件来进行全选、反选和取值的示例,其实在 ListView 中的每一个条目还可以是 radioButton 类型的控件。

创建名称为 ui_17_3_radio 的 Android 项目,更改布局文件 main.xml 的代码如下:

```
<?xml version="1.0" encoding="utf-8"?>
```

```xml
<LinearLayout xmlns:android="http://schemas.android.com/apk/res/android"
    android:orientation="vertical" android:layout_width="fill_parent"
    android:layout_height="fill_parent">
    <ListView android:layout_weight="1" android:id="@+id/listView1"
        android:layout_height="wrap_content" android:layout_width="match_parent"></ListView>
</LinearLayout>
```

更改文件 Main.java 的代码如下：

```java
public class Main extends Activity {
    private List cityList = new ArrayList();
    private ListView listView;

    @Override
    public void onCreate(Bundle savedInstanceState) {
        super.onCreate(savedInstanceState);
        setContentView(R.layout.main);

        // ***************************
        // List Entity username isShow
        listView = (ListView) this.findViewById(R.id.listView1);

        final boolean[] isCheckedArray = new boolean[8];
        isCheckedArray[0] = false;
        isCheckedArray[1] = true;
        isCheckedArray[2] = false;
        isCheckedArray[3] = false;
        isCheckedArray[4] = false;
        isCheckedArray[5] = false;
        isCheckedArray[6] = false;
        isCheckedArray[7] = false;

        cityList.add("accp1");
        cityList.add("accp2");
        cityList.add("accp3");
        cityList.add("accp4");
        cityList.add("accp5");
        cityList.add("accp6");
        cityList.add("accp7");
        cityList.add("accp8");

        ArrayAdapter adapter = new ArrayAdapter(this,
                android.R.layout.simple_list_item_single_choice, cityList);

        listView.setChoiceMode(ListView.CHOICE_MODE_SINGLE);
        listView.setAdapter(adapter);

        listView.setOnItemClickListener(new OnItemClickListener() {
            public void onItemClick(AdapterView<?> arg0, View arg1, int arg2,
                    long arg3) {
                Log.v("------------", "" + ((TextView) arg1).getText());
```

```
            }
        });

        // 设默认选中状态
        for (int i = 0; i < isCheckedArray.length; i++) {
            listView.setItemChecked(i, isCheckedArray[i]);
        }

    }
}
```

程序初始运行效果如图 3.76 所示。

图 3.76　有 1 个 radioButton 为默认选中

当用鼠标单击"accp5"时在 LogCat 控制台打印出相关的日志信息，如图 3.77 所示。

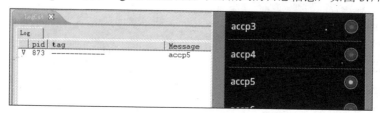

图 3.77　选中的单选选项文本内容

3.22.5　在 ListView 中自定义布局内容

强大的 ListView 控件也支持自定义条目布局，本示例就来实现一个自定义 ListView 中条目布局的实例。

创建名称为 ui_17_4_extLayout 的 Android 项目，更改 main.xml 配置文件的代码如下：

```
<?xml version="1.0" encoding="utf-8"?>
<LinearLayout xmlns:android="http://schemas.android.com/apk/res/android"
    android:orientation="vertical" android:layout_width="fill_parent"
    android:layout_height="fill_parent">
```

```xml
    <ListView android:layout_height="wrap_content" android:id="@+id/listView1"
        android:layout_width="match_parent"></ListView>
</LinearLayout>
```

文件 Main.java 的代码如下:

```java
public class Main extends Activity {
    /** Called when the activity is first created. */
    @Override
    public void onCreate(Bundle savedInstanceState) {
        super.onCreate(savedInstanceState);
        setContentView(R.layout.main);

        ListView lv1 = (ListView) this.findViewById(R.id.listView1);

        PhoneItemAdapter phoneItemAdapterRef = new PhoneItemAdapter(this);
        lv1.setAdapter(phoneItemAdapterRef);

        lv1.setOnItemClickListener(new OnItemClickListener() {

            public void onItemClick(AdapterView<?> arg0, View arg1, int arg2,
                    long arg3) {
                Log.v("单击了=====", ((TextView) arg1
                        .findViewById(R.id.phoneNumTextView1)).getText()
                        .toString());

            }
        });

    }
}
```

还要创建自定义的 ListView 条目布局文件 extlayout.xml 代码如下:

```xml
<?xml version="1.0" encoding="utf-8"?>
<LinearLayout xmlns:android="http://schemas.android.com/apk/res/android"
    android:orientation="horizontal" android:layout_width="fill_parent"
    android:layout_height="50px" android:padding="0px">
    <LinearLayout android:orientation="horizontal"
        android:layout_width="180px" android:layout_height="50px"
        android:gravity="center|left" android:layout_weight="1">
        <ImageView android:id="@+id/imageView1"
            android:layout_height="wrap_content" android:layout_width="40px"
            android:src="@drawable/myphone"></ImageView>
        <TextView android:text="i am phoneNum" android:id="@+id/phoneNumTextView1"
            android:layout_width="wrap_content" android:layout_height="wrap_content"></TextView>
    </LinearLayout>
    <LinearLayout android:orientation="vertical"
        android:layout_width="wrap_content" android:layout_height="50px"
        android:gravity="center">
        <ImageView android:id="@+id/imageView2"
            android:layout_height="wrap_content" android:layout_width="wrap_content"
```

```
                android:src="@drawable/no"></ImageView>
            <TextView android:text="thisTime" android:id="@+id/phoneTimeTextView1"
                android:layout_width="wrap_content" android:layout_height="wrap_content"></TextView>
        </LinearLayout>
    </LinearLayout>
```

预览效果如图 3.78 所示。

图 3.78　布局效果

创建自定义布局的适配器 PhoneItemAdapter.java 代码如下：

```java
public class PhoneItemAdapter extends BaseAdapter {

    private List<PhoneInfoEntity> phoneList = new ArrayList();

    private Context context;

    public PhoneItemAdapter(Context context) {
        super();
        this.context = context;

        PhoneInfoEntity pie1 = new PhoneInfoEntity("13811111111",
                "2000-1-1 20-20-20");
        PhoneInfoEntity pie2 = new PhoneInfoEntity("13822222222",
                "2000-1-2 20-20-20");
        PhoneInfoEntity pie3 = new PhoneInfoEntity("13833333333",
                "2000-1-3 20-20-20");
        PhoneInfoEntity pie4 = new PhoneInfoEntity("13844444444",
                "2000-1-4 20-20-20");
        PhoneInfoEntity pie5 = new PhoneInfoEntity("13855555555",
                "2000-1-5 20-20-20");
        PhoneInfoEntity pie6 = new PhoneInfoEntity("13866666666",
                "2000-1-6 20-20-20");
        PhoneInfoEntity pie7 = new PhoneInfoEntity("13877777777",
                "2000-1-7 20-20-20");
        PhoneInfoEntity pie8 = new PhoneInfoEntity("13888888888",
                "2000-1-8 20-20-20");

        phoneList.add(pie1);
        phoneList.add(pie2);
        phoneList.add(pie3);
        phoneList.add(pie4);
        phoneList.add(pie5);
        phoneList.add(pie6);
        phoneList.add(pie7);
        phoneList.add(pie8);
```

```java
    }

    public int getCount() {
        return phoneList.size();
    }

    public Object getItem(int arg0) {
        // TODO Auto-generated method stub
        return null;
    }

    public long getItemId(int arg0) {
        // TODO Auto-generated method stub
        return 0;
    }

    public View getView(int arg0, View arg1, ViewGroup arg2) {

        LayoutInflater inflater = ((Activity) context).getLayoutInflater();
        View twoEditTextLayoutRef = inflater.inflate(R.layout.extlayout, null);

        TextView tv1 = (TextView) twoEditTextLayoutRef
                .findViewById(R.id.phoneNumTextView1);
        TextView tv2 = (TextView) twoEditTextLayoutRef
                .findViewById(R.id.phoneTimeTextView1);

        tv1.setText(phoneList.get(arg0).getPhoneNum());
        tv2.setText(phoneList.get(arg0).getTime());

        return twoEditTextLayoutRef;
    }
}
```

创建实体类 PhoneInfoEntity.java 的代码如下：

```java
package entity;

public class PhoneInfoEntity {

    private String phoneNum;

    public PhoneInfoEntity(String phoneNum, String time) {
        super();
        this.phoneNum = phoneNum;
        this.time = time;
    }

    private String time;

    public String getPhoneNum() {
```

```
            return phoneNum;
    }

    public void setPhoneNum(String phoneNum) {
        this.phoneNum = phoneNum;
    }

    public String getTime() {
        return time;
    }

    public void setTime(String time) {
        this.time = time;
    }
}
```

还要添加两个图标元素，如图 3.79 所示。

程序初始运行效果如图 3.80 所示。

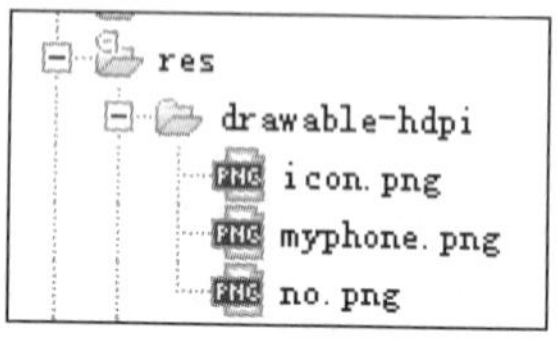

图 3.79　添加图标资源　　　　　　　图 3.80　运行效果

在条目中进行单击时，LogCat 打印的日志信息如图 3.81 所示。

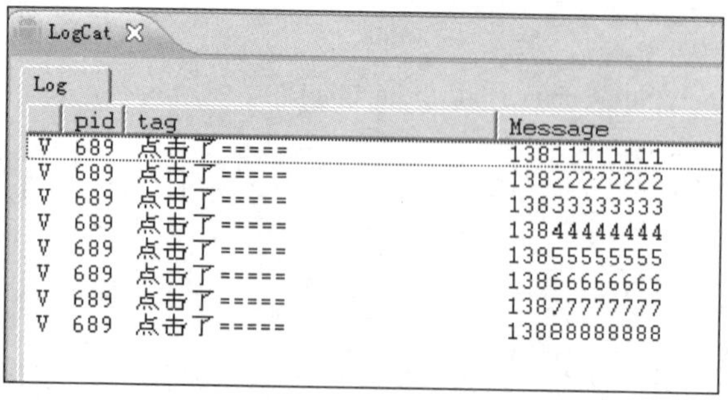

图 3.81　打印日志

3.22.6 在 ListView 中添加及删除条目

新建名称为 listview_add_remove 的 Android 项目，更改 main.xml 的代码如下：

```xml
<?xml version="1.0" encoding="utf-8"?>
<LinearLayout xmlns:android="http://schemas.android.com/apk/res/android"
    android:orientation="vertical" android:layout_width="fill_parent"
    android:layout_height="fill_parent">
    <LinearLayout android:orientation="vertical"
        android:layout_width="fill_parent" android:layout_height="wrap_content">
        <Button android:text="Button" android:id="@+id/button1"
            android:layout_width="wrap_content" android:layout_height="wrap_content"></Button>
        <Button android:text="Button" android:id="@+id/button2"
            android:layout_width="wrap_content" android:layout_height="wrap_content"></Button>
    </LinearLayout>
    <LinearLayout android:orientation="vertical"
        android:layout_width="fill_parent" android:layout_height="fill_parent">
        <ListView android:id="@+id/listView1" android:layout_height="wrap_content"
            android:layout_width="match_parent"></ListView>
    </LinearLayout>
</LinearLayout>
```

文件 Main.java 的代码如下：

```java
public class Main extends Activity {
    private Button button1;
    private Button button2;
    private ListView listView1;
    final private List dataList = new ArrayList();

    @Override
    public void onCreate(Bundle savedInstanceState) {
        super.onCreate(savedInstanceState);
        setContentView(R.layout.main);

        listView1 = (ListView) this.findViewById(R.id.listView1);
        button1 = (Button) this.findViewById(R.id.button1);
        button2 = (Button) this.findViewById(R.id.button2);

        dataList.add("我是条目啊，序号为：1");
        dataList.add("我是条目啊，序号为：2");
        dataList.add("我是条目啊，序号为：3");
        dataList.add("我是条目啊，序号为：4");
        dataList.add("我是条目啊，序号为：5");
        dataList.add("我是条目啊，序号为：6");
        dataList.add("我是条目啊，序号为：7");
        dataList.add("我是条目啊，序号为：8");
        dataList.add("我是条目啊，序号为：9");
        dataList.add("我是条目啊，序号为：10");

        final ArrayAdapter adapter = new ArrayAdapter(this,
```

```
                android.R.layout.simple_list_item_1, dataList);
        listView1.setAdapter(adapter);

        button1.setOnClickListener(new OnClickListener() {
            public void onClick(View arg0) {
                dataList.add(0, "zzz");
                adapter.notifyDataSetChanged();
                Log.v("!", "dataList size()=" + dataList.size());
            }
        });

        button2.setOnClickListener(new OnClickListener() {
            public void onClick(View arg0) {
                dataList.remove(0);
                adapter.notifyDataSetChanged();
                Log.v("!", "dataList size()=" + dataList.size());
            }
        });

    }
}
```

程序运行后单击上面的 Button 按钮 1 次，再单击下面的 Button 按钮 2 次，LogCat 打印日志及 AVD 界面效果如图 3.82 所示。

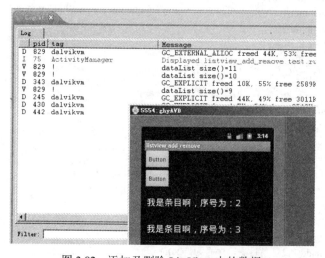

图 3.82　添加及删除 ListView 中的数据

3.22.7　在 ListView 中使用带图标的自定义布局

前面虽然用 ListView 实现全选及反选等效果，但 ListView 中每一个条目都是字符串类型，但在项目开发中，为了程序界面的美观，通常都需要有图标的美化，这时前面所学习到的知识就要得到一个更新。

新建名称为 duduli_multiCheckboxTest 的 Android 项目，新建实体类 Userinfo.java，代码如下：

```java
package entity;

public class Userinfo {
    private String username;
    private int imageId;
    private boolean isChecked;
    //……get 和 set 方法省略
}
```

创建自定义 Adapter 适配器文件 DuduliAdapter.java，代码如下：

```java
public class DuduliAdapter extends BaseAdapter {

    private List<Userinfo> userinfoList = new ArrayList<Userinfo>();
    private Context context;
    final private ArrayList<Boolean> checkedStateArray=new ArrayList<Boolean>();

    public ArrayList getCheckedStateArray() {
        return checkedStateArray;
    }

    public DuduliAdapter(Context context, List<Userinfo> userinfoList) {
        super();
        this.context = context;
        this.userinfoList = userinfoList;
        for (int i = 0; i < userinfoList.size(); i++) {
            checkedStateArray.add(userinfoList.get(i).isChecked());
        }
    }

    public int getCount() {
        return userinfoList.size();
    }

    public Object getItem(int arg0) {
        return null;
    }

    public long getItemId(int arg0) {
        return 0;
    }

    public void setItemCheck(int index, boolean setCheckedValue) {
        checkedStateArray.set(index, setCheckedValue);
        this.notifyDataSetChanged();
    }

    public void add(Userinfo userinfo){
        userinfoList.add(userinfo);
        checkedStateArray.add(userinfo.isChecked());
        this.notifyDataSetChanged();
```

```java
    }

    public void remove(int index){
        userinfoList.remove(index);
        checkedStateArray.remove(index);
        this.notifyDataSetChanged();
    }

    public void setAllChecked() {
        for (int i = 0; i < userinfoList.size(); i++) {
            checkedStateArray.set(i,true);
        }
        this.notifyDataSetChanged();
    }

    public View getView(int arg0, View arg1, ViewGroup arg2) {

        Log.v("!", "arg1=" + arg1 + " arg2=" + arg2);

        View layoutView = null;

        Userinfo userinfo = userinfoList.get(arg0);
        final int currentIndex = arg0;
        if (arg1 == null) {
            layoutView = ((Activity) context).getLayoutInflater().inflate(
                    R.layout.dudulilayout, null);
        } else {
            layoutView = arg1;
        }

        ImageView imageView = (ImageView) layoutView
                .findViewById(R.id.imageView1);
        TextView textView = (TextView) layoutView.findViewById(R.id.textView1);
        CheckBox checkBox = (CheckBox) layoutView.findViewById(R.id.checkBox1);
        checkBox.setOnCheckedChangeListener(new OnCheckedChangeListener() {
            public void onCheckedChanged(CompoundButton arg0, boolean arg1) {
                Log.v("!","set value="+arg1);
                checkedStateArray.set(currentIndex, arg1);
            }
        });

        textView.setText("" + userinfo.getUsername());
        imageView.setImageResource(userinfo.getImageId());
        checkBox.setChecked(checkedStateArray.get(arg0));

        return layoutView;
    }
}
```

主 Activity 对象文件 Main.java 的代码如下：

```java
public class Main extends Activity {
    private ListView listView;

    private Button button1;
    private Button button2;
    private Button button3;
    private Button button4;
    private Button button5;
    private List<Userinfo> userinfoList;

    private DuduliAdapter adapter;

    @Override
    public void onCreate(Bundle savedInstanceState) {
        super.onCreate(savedInstanceState);
        setContentView(R.layout.main);

        button1 = (Button) this.findViewById(R.id.button1);
        button2 = (Button) this.findViewById(R.id.button2);
        button3 = (Button) this.findViewById(R.id.button3);
        button4 = (Button) this.findViewById(R.id.button4);
        button5 = (Button) this.findViewById(R.id.button5);

        userinfoList = new ArrayList<Userinfo>();
        for (int i = 0; i < 100; i++) {
            Userinfo userinfo = new Userinfo();
            userinfo.setUsername("duduli-" + (i + 1));
            userinfo.setImageId(R.drawable.duduli);
            if (i % 2 != 0) {
                userinfo.setChecked(true);
            }
            userinfoList.add(userinfo);
        }

        listView = (ListView) this.findViewById(R.id.listView1);
        adapter = new DuduliAdapter(this, userinfoList);
        listView.setAdapter(adapter);

        // 全选
        button1.setOnClickListener(new OnClickListener() {
            public void onClick(View arg0) {
                adapter.setAllChecked();
            }
        });

        // 反选
        button2.setOnClickListener(new OnClickListener() {
            public void onClick(View arg0) {
```

```java
                    ArrayList<Boolean> checkedStateArray = adapter
                            .getCheckedStateArray();
                    for (int i = 0; i < checkedStateArray.size(); i++) {
                        if (checkedStateArray.get(i) == true) {
                            adapter.setItemCheck(i, false);
                        } else {
                            adapter.setItemCheck(i, true);
                        }
                    }
                }
            });

            // 取值
            button3.setOnClickListener(new OnClickListener() {
                public void onClick(View arg0) {
                    ArrayList<Boolean> checkedStateArray = adapter
                            .getCheckedStateArray();
                    for (int i = 0; i < checkedStateArray.size(); i++) {
                        Log.v("!", "" + (i + 1) + " " + checkedStateArray.get(i));
                    }
                }
            });

            // 增加
            button4.setOnClickListener(new OnClickListener() {
                public void onClick(View arg0) {
                    double randomDouble = Math.random();
                    Userinfo userinfo = new Userinfo();
                    userinfo.setUsername("duduli-" + randomDouble);
                    userinfo.setImageId(R.drawable.duduli);
                    userinfo.setChecked(true);
                    userinfo.setChecked(false);
                    adapter.add(userinfo);
                    listView.setSelection(userinfoList.size());
                }
            });

            // 删除
            button5.setOnClickListener(new OnClickListener() {
                public void onClick(View arg0) {
                    ArrayList<Boolean> checkedStateArray = adapter
                            .getCheckedStateArray();
                    Log.v("!", "size=" + checkedStateArray.size());
                    for (int i = checkedStateArray.size() - 1; i >= 0; i--) {
                        if (checkedStateArray.get(i) == true)
                            adapter.remove(i);
                    }
                    listView.setSelection(0);
                }
            });
```

       }
}

程序运行后的效果如图 3.83 所示。

图 3.83　运行效果

3.23　VideoView 控件

虽然在 Android 平台上有很多优秀的媒体播放软件，但如果想自己实现一个最简单的媒体播放器却是再简单不过的事情，因为 Android 平台已经自带了一个名称叫 VideoView 的控件，这个控件专门负责媒体的播放，包含音频和视频文件的播放。

新建名称为 videoviewTest 的 Android 项目，在布局文件 main.xml 中加入 VideoView 控件代码：

```
<?xml version="1.0" encoding="utf-8"?>
<LinearLayout xmlns:android="http://schemas.android.com/apk/res/android"
    android:orientation="vertical" android:layout_width="fill_parent"
    android:layout_height="fill_parent" android:gravity="center">
    <VideoView android:id="@+id/videoView1" android:layout_width="wrap_content"
        android:layout_height="wrap_content"></VideoView>
</LinearLayout>
```

更改 Activity 对象 Main.java 的代码如下：

```
public class Main extends Activity {
    public static final String TAG = "MainActivity";
    private VideoView videoView1;
    private Uri UriRef;
    private MediaController mediaControllerRef;

    @Override
    public void onCreate(Bundle savedInstanceState) {
        super.onCreate(savedInstanceState);
```

```
// 水平方向
this.setRequestedOrientation(ActivityInfo.SCREEN_ORIENTATION_LANDSCAPE);

setContentView(R.layout.main);

// 找到 videoView 控件
videoView1 = (VideoView) findViewById(R.id.videoView1);
// 取得 Uri 对象
UriRef = Uri.parse("sdcard/qtbhl.mp4");
// 创建媒体播放控制器
mediaControllerRef = new MediaController(this);
// 关联媒体播放控制器
videoView1.setMediaController(mediaControllerRef);

videoView1.setVideoURI(UriRef);
videoView1.start();
    }

}
```

然后在 AVD 的 sdcard 中加入这个文件，如图 3.84 所示。

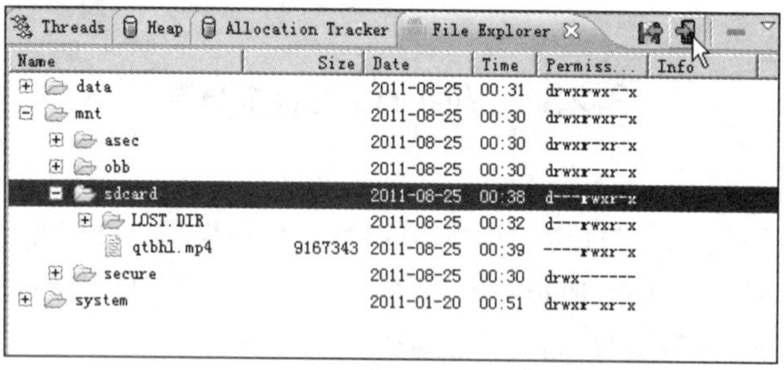

图 3.84　添加 1 个 mp4 文件到 sdcard 中

程序运行时横屏显示这个视频文件，如图 3.85 所示。

图 3.85　横屏显示视频文件

3.24 SimpleAdapter 对象

SimpleAdapter 对象是适配器的 1 种，它的功能是将界面中控件的 id 与 Map 中的同名 key 进行绑定，然后将当前 key 对应的值显示到这个控件里。

新建名称为 SimpleAdapter 的 Android 项目，文件 Main.java 的代码如下：

```java
public class Main extends Activity {
    @Override
    public void onCreate(Bundle savedInstanceState) {
        super.onCreate(savedInstanceState);
        setContentView(R.layout.main);

        List listMap = new ArrayList();
        for (int i = 0; i < 15; i++) {
            Map rowMap = new HashMap();
            rowMap.put("username", "gaohongyan" + i);
            rowMap.put("password", "password" + i);
            rowMap.put("address", "address" + i);
            listMap.add(rowMap);
        }

        SimpleAdapter simpleAdapterRef = new SimpleAdapter(this, listMap,
                R.layout.eachrowlayout, new String[] { "username", "password",
                        "address" }, new int[] { R.id.textView1,
                        R.id.textView2, R.id.textView3 });

        ListView lvRef = (ListView) this.findViewById(R.id.listView1);
        lvRef.setAdapter(simpleAdapterRef);

    }
}
```

布局文件 main.xml 的代码如下：

```xml
<?xml version="1.0" encoding="utf-8"?>
<LinearLayout xmlns:android="http://schemas.android.com/apk/res/android"
    android:orientation="vertical" android:layout_width="fill_parent"
    android:layout_height="fill_parent">
    <ListView android:id="@+id/listView1" android:layout_height="fill_parent"
        android:layout_width="fill_parent"></ListView>
</LinearLayout>
```

新建布局文件 eachrowlayout.xml 的代码如下：

```xml
<?xml version="1.0" encoding="utf-8"?>
<LinearLayout xmlns:android="http://schemas.android.com/apk/res/android"
    android:orientation="vertical" android:layout_width="fill_parent"
    android:layout_height="40px">
    <LinearLayout android:orientation="horizontal"
```

```xml
        android:layout_width="fill_parent" android:layout_height="25px"
        android:gravity="left|center_vertical">
    <TextView android:text="TextView" android:id="@+id/textView1"
        android:layout_width="wrap_content" android:layout_height="wrap_content"></TextView>
    <TextView android:text="TextView" android:id="@+id/textView2"
        android:layout_width="wrap_content" android:layout_height="wrap_content"
        android:paddingLeft="20dip"></TextView>
    <TextView android:text="TextView" android:id="@+id/textView3"
        android:layout_width="wrap_content" android:layout_height="wrap_content"
        android:paddingLeft="20dip"></TextView>
    </LinearLayout>
    <LinearLayout android:orientation="vertical" android:gravity="bottom"
        android:layout_width="fill_parent" android:layout_height="fill_parent"
        android:layout_marginTop="5px">
    <TextView android:text="" android:id="@+id/textView4"
        android:layout_width="fill_parent" android:layout_height="2px"
        android:background="#C0F"></TextView>
    </LinearLayout>
</LinearLayout>
```

程序运行效果如图 3.86 所示。

图 3.86　SimpleAdapter 运行效果

虽然 SimpleAdapter 对象能非常方便地将 Map 的 key 和控件的 id 值进行绑定来显示 List 中 Map 对象中的数据，但还是有一些细节上的限制，就是 SimpleAdapter 支持显示的对象类型仅仅是 TextView 的子类和 ImageView 的子类。当然如果想实现复杂的 UI 界面，SimpleAdapter 对象很明显不能适应这样的情况，还得使用自定义的 XML 布局文件结合自定义 Adapter 适配器对象来进行界面的生成。

3.25 WebView 对象

为了演示用 AVD 访问外部的 URL 网址，用 MyEclipse 新建一个 Web 项目，然后布属到 tomcat 中，运行效果如图 3.87 所示。

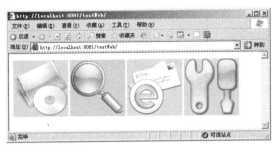

图 3.87 普通的 web 项目在运行

新建名称为 Webview 的 Android 项目，布局文件 main.xml 的代码如下：

```
<?xml version="1.0" encoding="utf-8"?>
<LinearLayout xmlns:android="http://schemas.android.com/apk/res/android"
    android:orientation="vertical" android:layout_width="fill_parent"
    android:layout_height="fill_parent">
    <WebView android:id="@+id/webView1" android:layout_width="match_parent"
        android:layout_height="match_parent"></WebView>
</LinearLayout>
```

在实现本项目时，需要在 AndroidManifest.xml 文件中添加访问 Internet 的权限配置代码：

```
<uses-permission android:name="android.permission.INTERNET" />
```

（1）使用 URL 方式打开网页

更改 Main.java 文件的代码如下：

```
public class Main extends Activity {
    @Override
    public void onCreate(Bundle savedInstanceState) {
        super.onCreate(savedInstanceState);
        setContentView(R.layout.main);

        WebView wvRef = (WebView) this.findViewById(R.id.webView1);
        wvRef.getSettings().setJavaScriptEnabled(true);
        wvRef.loadUrl("http://10.0.2.2:8081/testWeb/");
    }
}
```

程序运行结果如图 3.88 所示。

图 3.88　URL 直接访问

（2）使用 HTML 字符串方式

更改文件 Main.java 的代码如下：

```java
public class Main extends Activity {
    @Override
    public void onCreate(Bundle savedInstanceState) {
        super.onCreate(savedInstanceState);
        setContentView(R.layout.main);
        WebView wv = (WebView) this.findViewById(R.id.webView1);

        // wv.loadUrl("http://10.0.2.2:8081/testWeb/");

        try {
            AssetManager amRef = this.getAssets();
            // 想取 char
            InputStream isRef = amRef.open("index.html");
            InputStreamReader isrRef = new InputStreamReader(isRef);
            char[] charArray = new char[2];
            StringBuffer sbRef = new StringBuffer();
            int readLength = isrRef.read(charArray);
            while (readLength != -1) {
                sbRef.append(charArray, 0, readLength);
                readLength = isrRef.read(charArray);
            }
            Log.v("ccc", "" + sbRef.toString());

            wv.loadDataWithBaseURL("file:///android_asset/", sbRef.toString(),
                    "text/html", "utf-8", "");

        } catch (IOException e) {
            // TODO Auto-generated catch block
            e.printStackTrace();
        }

    }
}
```

还要在 assets 目录中存储自定义的 HTML 文件和图片资源，如图 3.89 所示。

图 3.89 自定义 HTML 文件和图片资源

其中 index.html 的代码如下:

```
<!DOCTYPE html PUBLIC "-//W3C//DTD XHTML 1.0 Transitional//EN" "http://www.w3.org/TR/xhtml1/DTD/xhtml1-transitional.dtd">
<html xmlns="http://www.w3.org/1999/xhtml">
<head>
<meta http-equiv="Content-Type" content="text/html; charset=utf-8" />
<title>无标题文档</title>
</head>
<body>
<table width="621" border="1">
  <tr>
    <td align="center"><img src="images/png-0001.png" /></td>
    <td align="center"><img src="images/png-0002.png" /></td>
    <td align="center"><img src="images/png-0003.png" /></td>
    <td align="center"><img src="images/png-0004.png" /></td>
  </tr>
</table>
<p>大中国</p>
</body>
</html>
```

程序运行效果如图 3.90 所示。

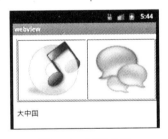

图 3.90 运行效果

3.26 控件的显示与隐藏

在 Android 中控件的显示与隐藏也与 Web 技术中的盒子模型相似,新建名称为 visible_gone_test 的 Android 项目,布局文件 main.xml 的代码如下:

```
<?xml version="1.0" encoding="utf-8"?>
<LinearLayout xmlns:android="http://schemas.android.com/apk/res/android"
    android:orientation="vertical" android:layout_width="fill_parent"
    android:layout_height="fill_parent">
```

```xml
    <Button android:text="Button1" android:id="@+id/button1"
        android:layout_width="wrap_content" android:layout_height="wrap_content"
        android:visibility="visible"></Button>
    <Button android:text="你看不到我但我占着位置" android:id="@+id/button2"
        android:layout_width="wrap_content" android:layout_height="wrap_content"
        android:visibility="invisible"></Button>
    <EditText android:id="@+id/editText1" android:layout_height="wrap_content"
        android:layout_width="match_parent" android:text="EditText"></EditText>
    <Button android:text="你看不到我我也不占位置" android:id="@+id/button3"
        android:visibility="gone" android:layout_width="wrap_content"
        android:layout_height="wrap_content"></Button>
    <EditText android:id="@+id/editText1" android:layout_height="wrap_content"
        android:layout_width="match_parent" android:text="EditText"></EditText>
</LinearLayout>
```

程序运行结果如图 3.91 所示。

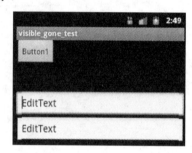

图 3.91 控件的显示与隐藏运行效果

3.27 GridView 对象

对象 GridView 是一个表格控件，可以在每个单元格中显示自定义的 View 或字符串，在项目中使用 GridView 的情况非常多。

3.27.1 GridView 中放置文字

新建名称为 gridview1 的 Android 项目，更改文件 main.xml 的代码如下：

```xml
<?xml version="1.0" encoding="utf-8"?>
<LinearLayout xmlns:android="http://schemas.android.com/apk/res/android"
    android:orientation="vertical" android:layout_width="fill_parent"
    android:layout_height="fill_parent">
    <GridView android:id="@+id/gridView1" android:layout_height="wrap_content"
        android:layout_width="match_parent" android:numColumns="5"></GridView>
</LinearLayout>
```

文件 Main.java 的代码如下：

```java
public class Main extends Activity {
    private GridView gv;
```

```java
@Override
public void onCreate(Bundle savedInstanceState) {
    super.onCreate(savedInstanceState);
    setContentView(R.layout.main);

    final ArrayList itemTextList = new ArrayList();
    itemTextList.add("a1");
    itemTextList.add("a2");
    itemTextList.add("a3");
    itemTextList.add("a4");
    itemTextList.add("a5");
    itemTextList.add("a6");
    itemTextList.add("a7");
    itemTextList.add("a8");

    ListAdapter listAdapter = new ArrayAdapter(this,
            android.R.layout.simple_list_item_1, itemTextList);

    gv = (GridView) this.findViewById(R.id.gridView1);
    gv.setAdapter(listAdapter);

    gv.setOnItemClickListener(new OnItemClickListener() {
        public void onItemClick(AdapterView<?> arg0, View arg1, int arg2,
                long arg3) {
            Log.v("选中了==", itemTextList.get(arg2).toString());
        }
    });
}
```

初始运行效果如图 3.92 所示。单击其中的条目后在 LogCat 中打印出相关的信息，如图 3.93 所示。

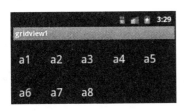

图 3.92　表格显示文字

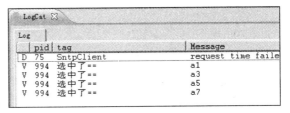

图 3.93　单击文字后的效果

3.27.2　在 GridView 中放置图片

新建名称为 gridview2 的 Android 项目，更改文件 main.xml 的代码如下：

```xml
<?xml version="1.0" encoding="utf-8"?>
<LinearLayout xmlns:android="http://schemas.android.com/apk/res/android"
    android:orientation="vertical" android:layout_width="fill_parent"
    android:layout_height="fill_parent">
```

```xml
<GridView android:id="@+id/gridView1" android:layout_height="fill_parent"
    android:layout_width="fill_parent" android:numColumns="3"
    android:horizontalSpacing="5px" android:verticalSpacing="5px"></GridView>
</LinearLayout>
```

更改文件 Main.java 的代码如下:

```java
public class Main extends Activity {
    private GridView gv;

    @Override
    public void onCreate(Bundle savedInstanceState) {
        super.onCreate(savedInstanceState);
        setContentView(R.layout.main);

        ListAdapter listAdapter = new ImageAdapter(this);

        gv = (GridView) this.findViewById(R.id.gridView1);
        gv.setAdapter(listAdapter);

    }
}
```

添加图标资源，如图 3.94 所示。

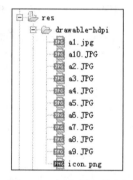

图 3.94 添加图标资源

为了显示图片，必须创建自定义的 Adapter 适配器类 ImageAdapter.java，代码如下：

```java
public class ImageAdapter extends BaseAdapter {

    private List<Integer> imageList = new ArrayList();

    private Context context;

    public ImageAdapter(Context context) {
        super();
        this.context = context;

        imageList.add(R.drawable.a1);
        imageList.add(R.drawable.a2);
        imageList.add(R.drawable.a3);
```

```java
        imageList.add(R.drawable.a4);
        imageList.add(R.drawable.a5);
        imageList.add(R.drawable.a6);
        imageList.add(R.drawable.a7);
        imageList.add(R.drawable.a8);
        imageList.add(R.drawable.a9);
        imageList.add(R.drawable.a10);
    }

    public int getCount() {
        return imageList.size();
    }

    public Object getItem(int arg0) {
        // TODO Auto-generated method stub
        return null;
    }

    public long getItemId(int arg0) {
        // TODO Auto-generated method stub
        return 0;
    }

    public View getView(int arg0, View arg1, ViewGroup arg2) {
        Log.v("run", "--------------------------");
        ImageView imageView = new ImageView(context);
        AbsListView.LayoutParams lpRef = new AbsListView.LayoutParams(80, 80);
        imageView.setLayoutParams(lpRef);
        imageView.setScaleType(ScaleType.CENTER_INSIDE);
        imageView.setImageResource(imageList.get(arg0));
        return imageView;
    }
}
```

提示 这里使用 AbsListView.LayoutParams 对象的原因是 setLayoutParams 设置的 LayoutParams 类型由父元素的类型决定。

程序运行效果如图 3.95 所示。

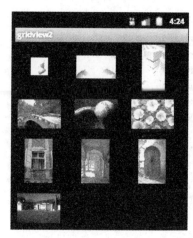

图 3.95 显示图片

3.27.3 在 GridView 中放置图片和文字

新建名称为 gridview3 的 Android 项目，但由于本示例要实现一个图标下方有文字的效果，所以自定义布局文件 imageandtext.xml 的代码如下：

```xml
<?xml version="1.0" encoding="utf-8"?>
<LinearLayout xmlns:android="http://schemas.android.com/apk/res/android"
    android:orientation="vertical" android:layout_width="fill_parent"
    android:layout_height="fill_parent" android:gravity="center">
    <ImageView android:layout_height="wrap_content" android:id="@+id/imageView1"
        android:layout_width="wrap_content" android:src="@drawable/icon"></ImageView>
    <TextView android:text="TextView" android:id="@+id/textView1"
        android:layout_width="wrap_content" android:layout_height="wrap_content"></TextView>
</LinearLayout>
```

还要自定义一个 Adapter，名称为 ImageAdapter.java，代码如下：

```java
public class ImageAdapter extends BaseAdapter {

    private List<EachIcon> eachIconList = new ArrayList();

    private Context context;

    public ImageAdapter(Context context) {
        super();
        this.context = context;

        EachIcon ei1 = new EachIcon(R.drawable.a1, "图标 1");
        EachIcon ei2 = new EachIcon(R.drawable.a2, "图标 2");
        EachIcon ei3 = new EachIcon(R.drawable.a3, "图标 3");
        EachIcon ei4 = new EachIcon(R.drawable.a4, "图标 4");
        EachIcon ei5 = new EachIcon(R.drawable.a5, "图标 5");
        EachIcon ei6 = new EachIcon(R.drawable.a6, "图标 6");
        EachIcon ei7 = new EachIcon(R.drawable.a7, "图标 7");
        EachIcon ei8 = new EachIcon(R.drawable.a8, "图标 8");
```

```java
        EachIcon ei9 = new EachIcon(R.drawable.a9, "图标 9");
        EachIcon ei10 = new EachIcon(R.drawable.a10, "图标 10");
        EachIcon ei11 = new EachIcon(R.drawable.a11, "图标 11");
        EachIcon ei12 = new EachIcon(R.drawable.a12, "图标 12");
        EachIcon ei13 = new EachIcon(R.drawable.a13, "图标 13");
        EachIcon ei14 = new EachIcon(R.drawable.a14, "图标 14");
        EachIcon ei15 = new EachIcon(R.drawable.a15, "图标 15");
        EachIcon ei16 = new EachIcon(R.drawable.a16, "图标 16");
        EachIcon ei17 = new EachIcon(R.drawable.a17, "图标 17");
        EachIcon ei18 = new EachIcon(R.drawable.a18, "图标 18");

        eachIconList.add(ei1);
        eachIconList.add(ei2);
        eachIconList.add(ei3);
        eachIconList.add(ei4);
        eachIconList.add(ei5);
        eachIconList.add(ei6);
        eachIconList.add(ei7);
        eachIconList.add(ei8);
        eachIconList.add(ei9);
        eachIconList.add(ei10);
        eachIconList.add(ei11);
        eachIconList.add(ei12);
        eachIconList.add(ei13);
        eachIconList.add(ei14);
        eachIconList.add(ei15);
        eachIconList.add(ei16);
        eachIconList.add(ei17);
        eachIconList.add(ei18);

    }

    public int getCount() {
        return eachIconList.size();
    }

    public Object getItem(int arg0) {
        // TODO Auto-generated method stub
        return null;
    }

    public long getItemId(int arg0) {
        // TODO Auto-generated method stub
        return 0;
    }

    public View getView(int arg0, View arg1, ViewGroup arg2) {
        Log.v("run", "---------------------------");

        LayoutInflater inflater = ((Activity) context).getLayoutInflater();
```

```java
                View twoEditTextLayoutRef = inflater.inflate(R.layout.imageandtext,
                        null);
                ImageView ivRef = (ImageView) twoEditTextLayoutRef
                        .findViewById(R.id.imageView1);
                TextView tvRef = (TextView) twoEditTextLayoutRef
                        .findViewById(R.id.textView1);

                ivRef.setImageResource(eachIconList.get(arg0).getImageSrcId());
                tvRef.setText(eachIconList.get(arg0).getIconString());
                return twoEditTextLayoutRef;
            }
}
```

实体类文件 EachIcon.java 的代码如下：

```java
package extlayout;

public class EachIcon {
    private int imageSrcId;
    private String iconString;

    public EachIcon(int imageSrcId, String iconString) {
        super();
        this.imageSrcId = imageSrcId;
        this.iconString = iconString;
    }

    public int getImageSrcId() {
        return imageSrcId;
    }

    public void setImageSrcId(int imageSrcId) {
        this.imageSrcId = imageSrcId;
    }

    public String getIconString() {
        return iconString;
    }

    public void setIconString(String iconString) {
        this.iconString = iconString;
    }
}
```

布局文件 main.xml 的代码如下：

```xml
<?xml version="1.0" encoding="utf-8"?>
<LinearLayout xmlns:android="http://schemas.android.com/apk/res/android"
    android:orientation="vertical" android:layout_width="fill_parent"
    android:layout_height="fill_parent">
```

```xml
<GridView android:id="@+id/gridView1" android:layout_height="wrap_content"
    android:layout_width="match_parent" android:numColumns="4"
    android:horizontalSpacing="5px" android:verticalSpacing="5px"></GridView>
</LinearLayout>
```

文件 Main.java 的代码如下：

```java
public class Main extends Activity {

    @Override
    public void onCreate(Bundle savedInstanceState) {
        super.onCreate(savedInstanceState);
        setContentView(R.layout.main);

        ListAdapter listAdapter = new ImageAdapter(this);

        GridView gv = (GridView) this.findViewById(R.id.gridView1);
        gv.setAdapter(listAdapter);

        gv.setOnItemClickListener(new OnItemClickListener() {
            public void onItemClick(AdapterView<?> arg0, View arg1, int arg2,
                    long arg3) {
                Log.v("-----------单击的索值是：", ""
                        + ((TextView) arg1.findViewById(R.id.textView1))
                            .getText().toString() + " 索引是:" + arg2);

            }
        });

    }
}
```

程序初始运行效果如图 3.96 所示。单击其中的几个图标后在 LogCat 上显示相关信息，如图 3.97 所示。

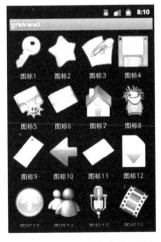

图 3.96 有图片有文字效果   图 3.97 LogCat 显示单击图片后的效果

3.28 菜单 Menu 控件之选项菜单

菜单在 Android 中的使用非常重要，几乎所有重要的选项和设置都在菜单中进行处理，所以掌握菜单的使用是开发操作方便软件的基础。

Android 系统支持 3 种形式的菜单：选项菜单、子菜单和上下文菜单，其中子菜单和上下文菜单项支持复选和单选按钮，但不支持图像，而选项菜单则刚好相反。

3.28.1 创建选项菜单

新建名称为 ui_23 的 Android 项目，修改 Main.java 文件的代码如下：

```java
public class Main extends Activity {

    private final static int MyMENU1_ID = 1;
    private final static int MyMENU2_ID = 2;
    private final static int MyMENU3_ID = 3;
    private final static int MyMENU4_ID = 4;
    private final static int MyMENU5_ID = 5;
    private final static int MyMENU6_ID = 6;
    private final static int MyMENU7_ID = 7;

    @Override
    public void onCreate(Bundle savedInstanceState) {
        super.onCreate(savedInstanceState);
        setContentView(R.layout.main);
    }

    @Override
    public boolean onCreateOptionsMenu(Menu menu) {
        Log.v("onCreateOptionsMenu", "执行了 onCreateOptionsMenu");

        menu.add(0, MyMENU1_ID, 1, "菜单 1").setIcon(R.drawable.menuico);
        menu.add(0, MyMENU2_ID, 2, "菜单 2").setIcon(R.drawable.menuico);
        menu.add(0, MyMENU3_ID, 3, "菜单 3").setIcon(R.drawable.menuico);
        menu.add(0, MyMENU4_ID, 4, "菜单 4").setIcon(R.drawable.menuico);
        menu.add(0, MyMENU5_ID, 5, "菜单 5").setIcon(R.drawable.menuico);
        menu.add(0, MyMENU6_ID, 6, "菜单 6").setIcon(R.drawable.menuico);
        menu.add(0, MyMENU7_ID, 7, "菜单 7").setIcon(R.drawable.menuico);

        return super.onCreateOptionsMenu(menu);
    }

    @Override
    public boolean onPrepareOptionsMenu(Menu menu) {
        Log.v("onPrepareOptionsMenu", "执行了 onPrepareOptionsMenu");

        menu.findItem(MyMENU1_ID).setVisible(false);
```

```
            return super.onPrepareOptionsMenu(menu);
    }

    @Override
    public boolean onOptionsItemSelected(MenuItem item) {
        switch (item.getItemId()) {
            case MyMENU1_ID:
                Log.v("菜单事件", "单击菜单 1");
                break;
            case MyMENU2_ID:
                Log.v("菜单事件", "单击菜单 2");
                break;
            case MyMENU3_ID:
                Log.v("菜单事件", "单击菜单 3");
                break;
            case MyMENU4_ID:
                Log.v("菜单事件", "单击菜单 4");
                break;
            case MyMENU5_ID:
                Log.v("菜单事件", "单击菜单 5");
                break;
            case MyMENU6_ID:
                Log.v("菜单事件", "单击菜单 6");
                break;
            case MyMENU7_ID:
                Log.v("菜单事件", "单击菜单 7");
                break;
        }
        return super.onOptionsItemSelected(item);
    }
}
```

类 Menu 的 add 方法有 4 个参数，按顺序分别是 groupId、itemId、order、title，第 1 个参数代表菜单的分组，第 2 个参数代表菜单条目的唯一标识，第 3 个参数代表菜单条目的顺序，而最后 1 个参数是菜单条目的文本内容。

还要在 res/ drawable-ldpi 目录添加 png 资源图片 menuico.png，如图 3.98 所示。

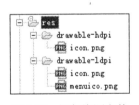

图 3.98　添加资源文件

程序运行后单击模拟器中的 按钮弹出菜单，并且在 LogCat 中看到打印出来的日志顺序，如图 3.99 所示。

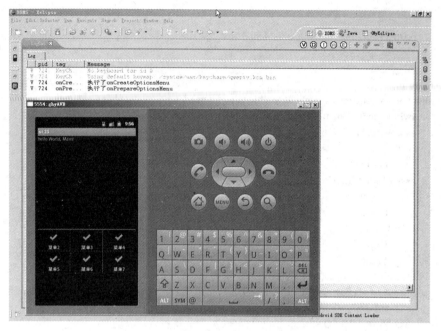

图 3.99 显示菜单并且查看回调函数运行顺序

在图 3.99 中可以看到，先执行的是 onCreateOptionsMenu 函数，后执行的是 onPrepareOptionsMenu 函数，并且在 onPrepareOptionsMenu 函数中隐藏了菜单 1，单击菜单 4 后在 LogCat 打印出相关的日志信息，如图 3.100 所示。

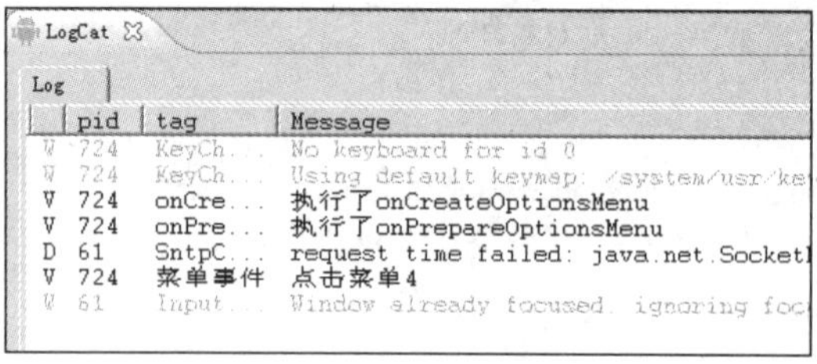

图 3.100 单击菜单 4

需要注意一个知识点，即 onCreateOptionsMenu 只执行一次，onPrepareOptionsMenu 多次运行。请自己上机尝试一下。

到此，创建菜单项、设置菜单项的单击事件并且关联处理方法的功能就演示完毕。从上面的截图中可以看到，Menu 默认只有 6 个菜单，如果多于 6 个菜单会出现什么样的效果呢？将 onPrepareOptionsMenu 方法中的隐藏菜单 1 的代码注释掉，变成如下代码：

```
@Override
public boolean onPrepareOptionsMenu(Menu menu) {
    Log.v("onPrepareOptionsMenu", "执行了 onPrepareOptionsMenu");

    // menu.findItem(MyMENU1_ID).setVisible(false);

    return super.onPrepareOptionsMenu(menu);
}
```

再次运行项目，出现的效果如图 3.101 所示。

菜单项数量一共 7 项，7 是大于 6 项的，所以自动出现 "More" 按钮，单击后出现的界面效果如图 3.102 所示。

图 3.101　出现 more 按钮

图 3.102　单击 more 后弹出的剩余菜单

3.28.2　为菜单加多选和单选功能

可以为菜单的子菜单加入多选和单选的功能，新建名称为 addmenu_checked_radio 的 Android 项目，Main.java 的代码如下：

```java
public class Main extends Activity {
    @Override
    public void onCreate(Bundle savedInstanceState) {
        super.onCreate(savedInstanceState);
        setContentView(R.layout.main);
    }

    @Override
    public boolean onPrepareOptionsMenu(Menu menu) {
        menu.findItem(11).setChecked(false);
        menu.findItem(12).setChecked(true);

        menu.findItem(21).setChecked(true);
        return super.onPrepareOptionsMenu(menu);
    }

    @Override
    public boolean onOptionsItemSelected(MenuItem item) {
        // 在这个方法中并不仅仅只是取得菜单的 checked 值
        // 通常情况下要把这个 checked 状态值保存进数据库
        // 然后在 onPrepareOptionsMenu 方法根据数据库中的数据
        // 进行菜单 checked 值的初始化
        Log.v("!", "进入了 onContextItemSelected menuId=" + item.getItemId());
```

```java
        if (item.getItemId() == 11) {
            Log.v("!", "id 为 11 的复选菜单的值为: " + !item.isChecked());
        }
        if (item.getItemId() == 12) {
            Log.v("!", "id 为 12 的复选菜单的值为: " + !item.isChecked());
        }
        if (item.getItemId() == 21) {
            Log.v("!", "id 为 21 的单选菜单的值为: " + !item.isChecked());
        }
        if (item.getItemId() == 22) {
            Log.v("!", "id 为 22 的单选菜单的值为: " + !item.isChecked());
        }
        return super.onContextItemSelected(item);
    }

    @Override
    public boolean onCreateOptionsMenu(Menu menu) {
        SubMenu menu1 = menu.addSubMenu(1, 1, 1, "我有 2 个复选菜单");
        menu1.setIcon(R.drawable.icon);
        menu1.setHeaderIcon(R.drawable.icon);

        menu1.add(1, 11, 1, "我是复选菜单 1").setCheckable(true);
        menu1.add(1, 12, 2, "我是复选菜单 2").setCheckable(true);

        // //////////////////////////////////////
        SubMenu menu2 = menu.addSubMenu(1, 2, 2, "我有 2 个单选菜单");
        menu2.add(100, 21, 1, "我是单选菜单 1").setCheckable(true);
        menu2.add(100, 22, 2, "我是单选菜单 2").setCheckable(true);
        // 方法第 3 个参数如果是 false 则设置菜单为多选菜单
        menu2.setGroupCheckable(100, true, true);
        return super.onCreateOptionsMenu(menu);
    }
}
```

程序运行后单击 AVD 设备的 menu 按钮,程序初始运行效果如图 3.103 所示。
单击"我有 2 个复选菜单"出现的效果如图 3.104 所示。

图 3.103 程序初始运行效果

图 3.104 显示 2 个复选菜单

单击"我是复选菜单 2"菜单项，出现的效果如图 3.105 所示。

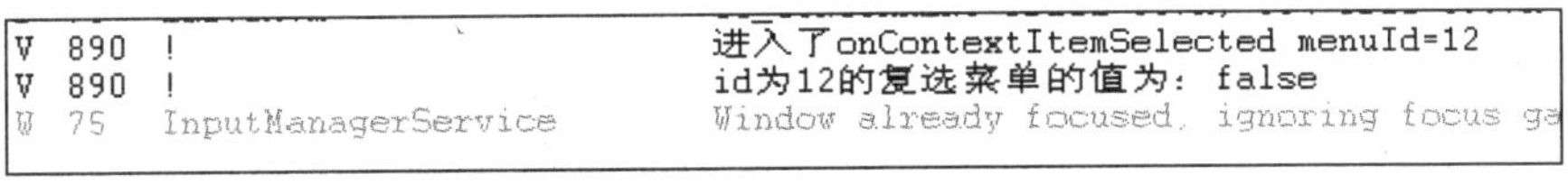

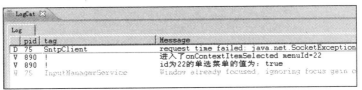

图 3.104 单击了我是复选菜单 2

回到主界面再次单击 menu 按钮，单击"我有 2 个单选菜单"选项，出现的界面如图 3.106 所示。

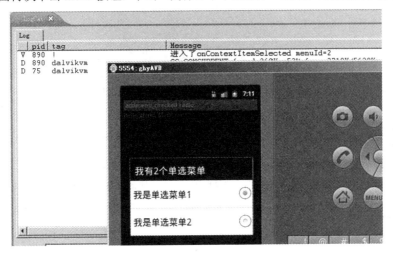

图 3.106 显示 2 个单选按钮

单击"我是单选菜单 2"选项出现打印结果如图 3.107 所示。

图 3.107 单击我是单选菜单 2

3.29 菜单 Menu 控件之子菜单

前面介绍的是选项菜单，其实在选项菜单中还支持子菜单的功能。

创建名称为 ui_24 的 Android 项目，更改 Main.java 文件的代码如下：

```java
public class Main extends Activity {

    private final static int MyMENU1_ID = 1;
    private final static int MyMENU2_ID = 2;

    private final static int MyMENU1_Sub1_ID = 11;
    private final static int MyMENU1_Sub2_ID = 12;

    @Override
    public void onCreate(Bundle savedInstanceState) {
        super.onCreate(savedInstanceState);
        setContentView(R.layout.main);
    }

    @Override
    public boolean onCreateOptionsMenu(Menu menu) {
        Log.v("onCreateOptionsMenu", "执行了 onCreateOptionsMenu");

        // 非子菜单功能"菜单 2"
        menu.add(0, MyMENU2_ID, 2, "菜单 2").setIcon(R.drawable.menuico);

        // 有子菜单功能"菜单 1"
        SubMenu sm1 = (SubMenu) menu.addSubMenu(0, MyMENU1_ID, 1, "菜单 1")
                .setIcon(R.drawable.menuico);
        sm1.setHeaderIcon(R.drawable.menuico);
        sm1.setHeaderTitle("菜单 1 的子菜单项");

        sm1.add(0, MyMENU1_Sub1_ID, 1, "菜单 1 子菜单 1").setIcon(R.drawable.menuico);
        sm1.add(0, MyMENU1_Sub2_ID, 2, "菜单 1 子菜单 2").setIcon(R.drawable.menuico);

        return super.onCreateOptionsMenu(menu);
    }

    @Override
    public boolean onPrepareOptionsMenu(Menu menu) {
        Log.v("onPrepareOptionsMenu", "执行了 onPrepareOptionsMenu");

        // menu.findItem(MyMENU1_ID).setVisible(false);

        return super.onPrepareOptionsMenu(menu);
    }

    @Override
```

```
public boolean onOptionsItemSelected(MenuItem item) {
    switch (item.getItemId()) {
    case MyMENU2_ID:
        Log.v("菜单事件","单击菜单 2");
        break;
    case MyMENU1_Sub1_ID:
        Log.v("菜单事件","单击菜单 1 中子菜单 1");
        break;
    case MyMENU1_Sub2_ID:
        Log.v("菜单事件","单击菜单 1 中子菜单 2");
        break;
    }
    return super.onOptionsItemSelected(item);
}
```

程序运行后，单击模拟器的 按钮运行效果如图 3.108 所示。

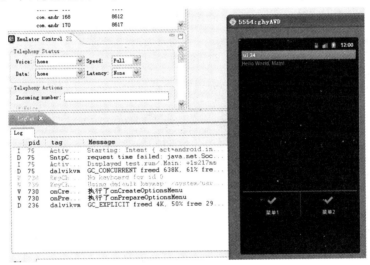

图 3.108 显示选项菜单

其中菜单 1 中有子菜单，而菜单 2 中没有子菜单，用鼠标单击"菜单 2"，则在 DDMS 中打印的日志内容如图 3.109 所示。

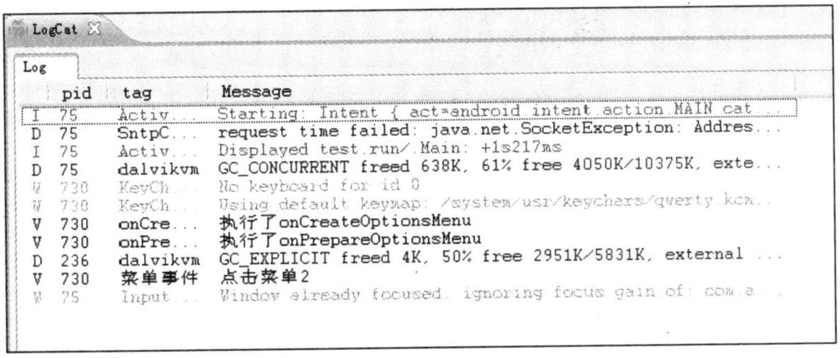

图 3.109 单击菜单 2 时的日志

这时再按下 MENU 按钮需要重新调出菜单，出现的效果如图 3.110 所示。

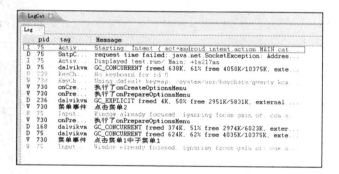

图 3.110　重复调用选项菜单

通过图 3.110 可以发现，选项菜单每一次弹出都要重新执行 1 次 onPrepareOptionsMenu 方法，这时用鼠标单击"菜单 1"则出现子菜单，如图 3.111 所示。

单击"菜单 1 子菜单 1"菜单项可以在 DDMS 中打印出日志信息，如图 3.112 所示。

图 3.111　单击菜单 1 出现子菜单　　　　　图 3.112　单击菜单 1 子菜单 1 的日志

如果单击"菜单 1"中的"子菜单 2"，则在控制台打印出来的日志结果如图 3.113 所示。

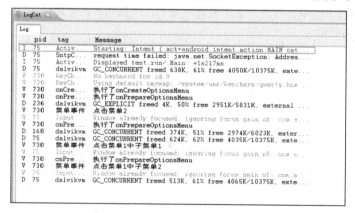

图 3.113　单击菜单 1 子菜单 2 的日志

3.30 菜单 Menu 控件之上下文菜单

前面已经将选项菜单介绍完毕，Android 还支持针对控件级的菜单，叫做"上下文菜单"。创建名称为 ui_25 的 Android 项目，更改文件 Main.java 的代码如下：

```java
public class Main extends Activity {

    private final static int MyMENU1_ID = 1;
    private final static int MyMENU2_ID = 2;
    private final static int MyMENU3_ID = 3;

    private final static int MyMENU1_Sub1_ID = 11;
    private final static int MyMENU1_Sub2_ID = 12;

    @Override
    public void onCreate(Bundle savedInstanceState) {
        super.onCreate(savedInstanceState);
        setContentView(R.layout.main);

        EditText et1 = (EditText) this.findViewById(R.id.editText1);
        EditText et2 = (EditText) this.findViewById(R.id.editText2);

        this.registerForContextMenu(et1);
        this.registerForContextMenu(et2);
    }

    @Override
    public void onCreateContextMenu(ContextMenu menu, View v,
            ContextMenuInfo menuInfo) {
        if (v.getId() == R.id.editText1) {

            Log.v("onCreateContextMenu", "执行了 onCreateContextMenu");

            // 非子菜单功能"菜单 2"
            menu.add(0, MyMENU2_ID, 2, "菜单 2").setIcon(R.drawable.menuico);

            // 有子菜单功能"菜单 1"
            SubMenu sm1 = (SubMenu) menu.addSubMenu(0, MyMENU1_ID, 1, "菜单 1")
                    .setIcon(R.drawable.menuico);
            sm1.setHeaderIcon(R.drawable.menuico);
            sm1.setHeaderTitle("菜单 1 的子菜单项");

            sm1.add(0, MyMENU1_Sub1_ID, 1, "菜单 1 子菜单 1").setIcon(
                    R.drawable.menuico);
            sm1.add(0, MyMENU1_Sub2_ID, 2, "菜单 1 子菜单 2").setIcon(
                    R.drawable.menuico);
        }
        if (v.getId() == R.id.editText2) {
```

```java
            Log.v("onCreateContextMenu", "执行了 onCreateContextMenu");

            // 非子菜单功能"菜单 3"
            menu.add(0, MyMENU3_ID, 1, "菜单 3").setIcon(R.drawable.menuico);

        }
    }

    @Override
    public boolean onContextItemSelected(MenuItem item) {
        switch (item.getItemId()) {
        case MyMENU1_ID:
            Log.v("菜单事件", "单击菜单 1");
            break;
        case MyMENU2_ID:
            Log.v("菜单事件", "单击菜单 2");
            break;
        case MyMENU3_ID:
            Log.v("菜单事件", "单击菜单 3");
            break;
        case MyMENU1_Sub1_ID:
            Log.v("菜单事件", "单击菜单 1 中子菜单 1");
            break;
        case MyMENU1_Sub2_ID:
            Log.v("菜单事件", "单击菜单 1 中子菜单 2");
            break;
        }
        return super.onOptionsItemSelected(item);
    }
}
```

布局文件 main.xml 的代码如下：

```xml
<?xml version="1.0" encoding="utf-8"?>
<LinearLayout xmlns:android="http://schemas.android.com/apk/res/android"
    android:orientation="vertical" android:layout_width="fill_parent"
    android:layout_height="fill_parent">
    <EditText android:text="EditText" android:id="@+id/editText1"
        android:layout_height="wrap_content" android:layout_width="match_parent"></EditText>
    <EditText android:text="EditText" android:id="@+id/editText2"
        android:layout_height="wrap_content" android:layout_width="match_parent"></EditText>
</LinearLayout>
```

程序运行后，长单击 EditText1 控件出现快捷菜单，如图 3.114 所示。

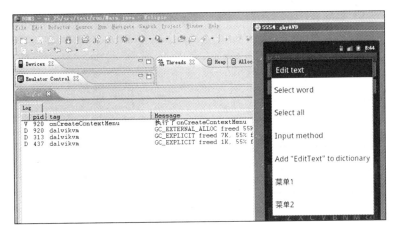

图 3.114　出现上下文菜单

单击"菜单 2"后在 LogCat 中打印的日志信息如图 3.115 所示。

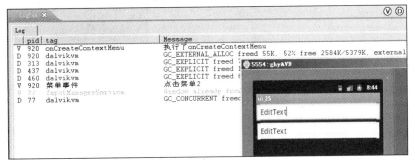

图 3.115　单击菜单 2

再次长单击 EditText1 控件,弹出上下文快捷菜单,如图 3.116 所示。

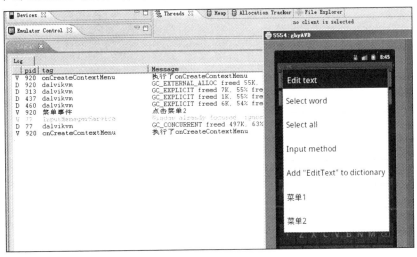

图 3.116　再次显示 editText1 的上下文菜单

单击"菜单 1"出现子菜单,效果如图 3.117 所示。

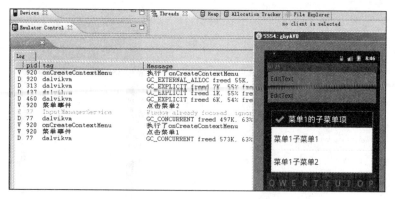

图 3.117　单击菜单 1 出现子菜单

单击"菜单 1 子菜单 1"选项，LogCat 出现日志如图 3.118 所示。

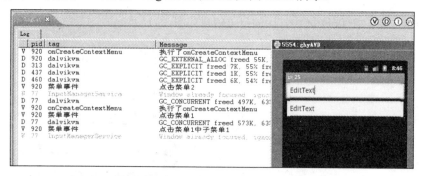

图 3.118　单击菜单 1 子菜单 1 的日志

需要注意的是，快捷菜单不能显示图标。

再长单击 EditText2 后弹出菜单，如图 3.119 所示。

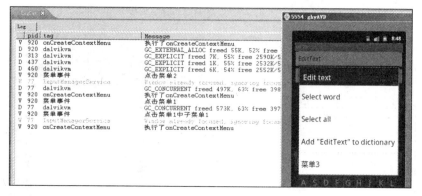

图 3.119　长按 editText2 显示上下文菜单

单击"菜单 3"后效果如图 3.120 所示。

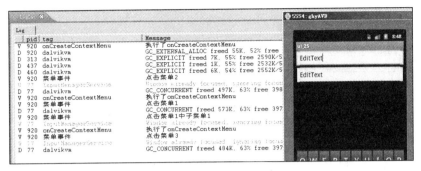

图 3.120 单击菜单 3 的效果

从上面日志的打印结果来看，onCreateContextMenu 是执行多次的。

上下文菜单还在 ListView 这个控件中应用比较多，比如长按 ListView 中的条目时弹出 1 个菜单以供选择是删除还是更改此条目，本示例再来继续实验一下 ListView 结合上下文菜单实现删除条目的功能，新建名称为 listViewAddMenu 的 Android 项目，文件 Main.java 的代码如下：

```java
public class Main extends Activity {
    private ListView listView1;
    private ArrayList bbslist = new ArrayList();
    private ArrayAdapter adapter;

    @Override
    public boolean onContextItemSelected(MenuItem item) {
        AdapterContextMenuInfo acmiRef = (AdapterContextMenuInfo) item
                .getMenuInfo();
        int removeIndex = acmiRef.position;
        bbslist.remove(removeIndex);
        adapter.notifyDataSetChanged();

        Log.v("!选中了：", "" + item.getItemId() + " 位置：" + acmiRef.position);

        return super.onContextItemSelected(item);
    }

    @Override
    public void onCreateContextMenu(ContextMenu menu, View v,
            ContextMenuInfo menuInfo) {
        super.onCreateContextMenu(menu, v, menuInfo);
        menu.add(0, 1, 1, "删除");
        menu.add(0, 2, 1, "不删除");
    }

    @Override
    public void onCreate(Bundle savedInstanceState) {
        super.onCreate(savedInstanceState);
        setContentView(R.layout.main);

        listView1 = (ListView) this.findViewById(R.id.listView1);
```

```
        bbslist.add("aaaaaaaaa1");
        bbslist.add("aaaaaaaaa2");
        bbslist.add("aaaaaaaaa3");
        bbslist.add("aaaaaaaaa4");
        bbslist.add("aaaaaaaaa5");
        bbslist.add("aaaaaaaaa6");
        bbslist.add("aaaaaaaaa7");
        bbslist.add("aaaaaaaaa8");
        bbslist.add("aaaaaaaaa9");
        bbslist.add("aaaaaaaaa10");

        adapter = new ArrayAdapter(this, android.R.layout.simple_list_item_1,
                        bbslist);
        listView1.setAdapter(adapter);

        this.registerForContextMenu(listView1);

    }
}
```

对如下代码进行一些解释：

```
AdapterContextMenuInfo acmiRef = (AdapterContextMenuInfo) item
                .getMenuInfo();
```

其实用程序 item.getMenuInfo()取出的对象是 android.view.ContextMenu.ContextMenuInfo，查看文档可以找到它的相关资料，如图 3.121 所示。

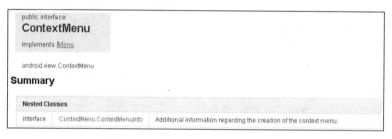

图 3.121　ContextMenuInfo 的信息

从图 3.121 中可以看到，在接口 ContextMenu 中有一个子接口 ContextMenuInfo，它在 android 的源代码中也有声明，代码如下：

```
public interface ContextMenu extends Menu {
    public interface ContextMenuInfo {
    }
}
```

那么，这个 ContextMenuInfo 接口也有其实现类，实现类如图 3.122 所示。

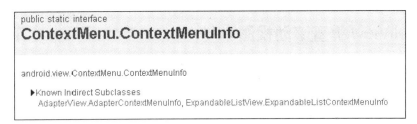

图 3.122　ContextMenuInfo 接口的实现类

从图 3.122 中可以看到，ContextMenuInfo 接口有 2 个实现类：AdapterView.AdapterContextMenuInfo 和 ExpandableListView.ExpandableListContextMenuInfo，其中在 AdapterView.AdapterContextMenuInfo 对象中就有我们想要得到当前单击 ListView 中的位置信息，AdapterContextMenuInfo 类的结构如图 3.123 所示。

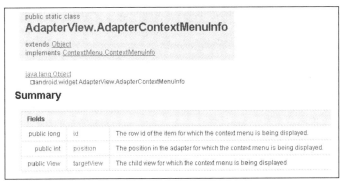

图 3.123　AdapterContextMenuInfo 类的结构

在图 3.123 可以看到，有一个名称为 position 的字段，这个字段就是取得 ListView 中单击的位置信息。

程序运行后单击第 2 个选项，弹出菜单的效果如图 3.124 所示。

单击"删除"菜单，条目"a2"的已被删除，如图 3.125 所示。

图 3.124　ListView 中弹出上下文菜单

图 3.125　删除第 2 项的列表内容

3.31 ScrollView 垂直滚动视图和 HorizontalScrollView 水平滚动视图

当屏幕中的控件太多，屏幕显示不全时，可以使用 ScrollView 控件来将显示的内容添加滚动条效果。

新建名称为 ScrollView 的 Android 项目，更改文件 main.xml 的代码如下：

```xml
<?xml version="1.0" encoding="utf-8"?>
<LinearLayout xmlns:android="http://schemas.android.com/apk/res/android"
    android:orientation="vertical" android:layout_width="fill_parent"
    android:layout_height="fill_parent">
    <ScrollView android:layout_weight="1"
        xmlns:android="http://schemas.android.com/apk/res/android"
        android:orientation="vertical" android:layout_width="fill_parent"
        android:layout_height="fill_parent">
        <LinearLayout xmlns:android="http://schemas.android.com/apk/res/android"
            android:orientation="vertical" android:layout_width="fill_parent"
            android:layout_height="fill_parent">
            <Button android:text="Button" android:id="@+id/button1"
                android:layout_width="wrap_content" android:layout_height="wrap_content"></Button>
            <Button android:text="Button" android:id="@+id/button2"
                android:layout_width="wrap_content" android:layout_height="wrap_content"></Button>
            <Button android:text="Button" android:id="@+id/button3"
                android:layout_width="wrap_content" android:layout_height="wrap_content"></Button>
            <Button android:text="Button" android:id="@+id/button4"
                android:layout_width="wrap_content" android:layout_height="wrap_content"></Button>
            <Button android:text="Button" android:id="@+id/button5"
                android:layout_width="wrap_content" android:layout_height="wrap_content"></Button>
            <Button android:text="Button" android:id="@+id/button6"
                android:layout_width="wrap_content" android:layout_height="wrap_content"></Button>
            <Button android:text="Button" android:id="@+id/button8"
                android:layout_width="wrap_content" android:layout_height="wrap_content"></Button>
            <Button android:text="Button" android:id="@+id/button10"
                android:layout_width="wrap_content" android:layout_height="wrap_content"></Button>
            <Button android:text="Button" android:id="@+id/button9"
                android:layout_width="wrap_content" android:layout_height="wrap_content"></Button>
            <Button android:text="Button" android:id="@+id/button7"
                android:layout_width="wrap_content" android:layout_height="wrap_content"></Button>
        </LinearLayout>
    </ScrollView>
    <LinearLayout xmlns:android="http://schemas.android.com/apk/res/android"
        android:orientation="horizontal" android:layout_width="fill_parent"
        android:layout_height="wrap_content">
        <EditText android:layout_height="wrap_content" android:text="EditText"
            android:id="@+id/editText1" android:layout_width="50px"></EditText>
        <EditText android:layout_height="wrap_content" android:text="EditText"
            android:id="@+id/editText2" android:layout_width="50px"></EditText>
    </LinearLayout>
</LinearLayout>
```

程序代码多但结构简单，多个 Button 按钮和两个 EditText 控件，这些控件在屏幕中是不能全部显示的，所以使用 ScrollView 控件来使这些控件具有滚动条，程序运行效果如图 3.126 所示。

图 3.126 添加 ScrollView 控件的效果

屏幕具有多个 button 控件，但屏幕显示不全，添加 ScrollView 控件后，右边有垂直滚动条。

水平滚动使用 HorizontalScrollView 标签来进行实现，示例代码是在名称为 horizontalscrollview_test 的 Android 项目中，main.xml 布局文件的代码如下：

```
<?xml version="1.0" encoding="utf-8"?>
<HorizontalScrollView xmlns:android="http://schemas.android.com/apk/res/android"
    android:id="@+id/horizontalScrollView1" android:layout_height="wrap_content"
    android:layout_width="match_parent">
    <LinearLayout android:id="@+id/linearLayout1"
        android:layout_width="match_parent" android:layout_height="match_parent"
        android:orientation="horizontal"><Button    android:id="@+id/button1"    android:text="Button"
android:layout_width="match_parent" android:layout_height="match_parent"></Button>
        <Button   android:id="@+id/button1"   android:text="Button"   android:layout_width="match_parent"
android:layout_height="match_parent"></Button>
        <Button   android:id="@+id/button1"   android:text="Button"   android:layout_width="match_parent"
android:layout_height="match_parent"></Button>
        ....里面有很多的 Button 按钮控件
    </LinearLayout>
</HorizontalScrollView>
```

程序运行后效果如图 3.127 所示。

图 3.127 水平滚动效果

3.32 DatePickerDialog 和 TimePickerDialog 对话框

在实现本节中的实验前,要先设置 AVD 设备的语言类别,不然日期的格式显示会有问题。
新建名称为 ui_27 的 Android 项目,更改 main.xml 文件的代码如下:

```xml
<?xml version="1.0" encoding="utf-8"?>
<LinearLayout xmlns:android="http://schemas.android.com/apk/res/android"
    android:orientation="vertical" android:layout_width="fill_parent"
    android:layout_height="fill_parent">
    <TextView android:id="@+id/dateText" android:layout_width="fill_parent"
        android:layout_height="wrap_content" android:text="" />
    <TextView android:id="@+id/timeText" android:layout_width="fill_parent"
        android:layout_height="wrap_content" android:text="" />
    <Button android:text="单击我显示日期" android:id="@+id/showDateButton"
        android:layout_width="wrap_content" android:layout_height="wrap_content"></Button>
    <Button android:text="单击我显示时间" android:id="@+id/showTimeButton"
        android:layout_width="wrap_content" android:layout_height="wrap_content"></Button>
</LinearLayout>
```

再更改 Main.java 文件的代码如下:

```java
public class Main extends Activity {

    private Button button1;
    private Button button2;
    private TextView textView1;
    private TextView textView2;

    DatePickerDialog.OnDateSetListener onDateSetListenerRef = new OnDateSetListener() {

        public void onDateSet(DatePicker arg0, int arg1, int arg2, int arg3) {

            textView2.setText(arg1 + "年" + (arg2 + 1) + "月" + arg3 + "日");
        }
    };

    TimePickerDialog.OnTimeSetListener onTimeSetListenerRef = new TimePickerDialog.OnTimeSetListener() {
        public void onTimeSet(TimePicker arg0, int arg1, int arg2) {
            textView1.setText(arg1 + "分" + arg2 + "秒");
        }
    };

    @Override
    public void onCreate(Bundle savedInstanceState) {
        super.onCreate(savedInstanceState);
        setContentView(R.layout.main);

        button1 = (Button) this.findViewById(R.id.showDateButton);
        button2 = (Button) this.findViewById(R.id.showTimeButton);
```

```
        textView1 = (TextView) this.findViewById(R.id.dateText);
        textView2 = (TextView) this.findViewById(R.id.timeText);

        button1.setOnClickListener(new OnClickListener() {
            public void onClick(View arg0) {
                new DatePickerDialog(Main.this, onDateSetListenerRef, 2000, 0,
                        1).show();
            }
        });

        button2.setOnClickListener(new OnClickListener() {
            public void onClick(View arg0) {
                new TimePickerDialog(Main.this, onTimeSetListenerRef, 14, 2,
                        true).show();
            }
        });

    }
}
```

项目运行时单击"单击我显示日期"按钮出现日期选择对话框,如图 3.128 所示。
设置好日期后单击"设置"按钮在 TextView 中显示中当前日期,如图 3.129 所示。

图 3.128 单击我显示日期

图 3.129 显示设置的日期值

项目运行时单击"单击我显示时间"按钮出现时间选择对话框如图 3.130 所示。
设置好时间后单击"设置"按钮在 TextView 中显示中当前时间,如图 3.131 所示。

图 3.130 单击我显示时间

图 3.131 显示设置后的时间值

3.33 TextView 控件小示例继续讨论

控件 TextView 还可以简单地实现滚动和...省略效果。

新建名称为 textViewProperties 的 Android 项目，更改 main.xml 的代码如下：

```xml
<?xml version="1.0" encoding="utf-8"?>
<LinearLayout xmlns:android="http://schemas.android.com/apk/res/android"
    android:orientation="vertical" android:layout_width="fill_parent"
    android:layout_height="fill_parent">
    <TextView android:layout_width="fill_parent"
        android:layout_height="wrap_content" android:text="我是中国人你也是中国人我们都是中国人你是地球人我也是地球人我们都是地球人"
        android:scrollHorizontally="true" android:ellipsize="marquee"
        android:singleLine="true" android:focusableInTouchMode="true"
        android:focusable="true" android:marqueeRepeatLimit="marquee_forever" />
    <TextView android:layout_width="fill_parent"
        android:layout_height="wrap_content" android:text="我是中国人你也是中国人我们都是中国人你是地球人我也是地球人我们都是地球人"
        android:ellipsize="start" android:singleLine="true" />
    <TextView android:layout_width="fill_parent"
        android:layout_height="wrap_content" android:text="我是中国人你也是中国人我们都是中国人你是地球人我也是地球人我们都是地球人"
        android:ellipsize="middle" android:singleLine="true" />
    <TextView android:layout_width="fill_parent"
        android:layout_height="wrap_content" android:text="我是中国人你也是中国人我们都是中国人你是地球人我也是地球人我们都是地球人"
        android:ellipsize="end" android:singleLine="true" />
</LinearLayout>
```

程序运行效果如图 3.132 所示。

Android 的 UI 控件 第 3 章

图 3.132 滚动和省略效果

在前面的示例中通过在 TextView 控件中使用 Linkify 对象来做匹配的链接，本示例将实现 1 个用户自定义的正则匹配 Linkify，新建名称为 textViewMoreTest 的 Android 项目，在 main.xml 布局中添加 4 个 TextView 控件，Main.java 核心代码如下：

```java
//TransformFilter 接口的作用为有成功的正则匹配后单击 Link 后转向的 Uri 地址
class MyTransformFilter implements TransformFilter {
    public String transformUrl(Matcher arg0, String arg1) {
        System.out.println(arg0 + " " + arg1);
        return "http://www.baidu.com?param=" + arg1.substring(3);
    }
}

// 接口 MatchFilter 的作用可以使正则匹配更加细化
// 可以在 acceptMatch 方法中进行更加详细的业务操作
class MyMatchFilter implements MatchFilter {
    public boolean acceptMatch(CharSequence arg0, int arg1, int arg2) {
        System.out.println(("" + arg0).substring(arg1, arg2));
        String tempString = ("" + arg0).substring(arg1, arg2);
        tempString = tempString.substring(3);
        tempString = tempString.substring(0, tempString.length() - 1);
        String[] numArray = tempString.split("\\_");
        if ((Integer.parseInt(numArray[0]) + Integer.parseInt(numArray[1])) == Integer
                .parseInt(numArray[2])) {
            return true;
        } else {
            return false;
        }
    }
}

public class Main extends Activity {
    private TextView textView1;
    private TextView textView2;
    private TextView textView3;
    private TextView textView4;

    @Override
    public void onCreate(Bundle savedInstanceState) {
        super.onCreate(savedInstanceState);
        setContentView(R.layout.main);

        // setAutoLinkMask 和 setText 方法调用有顺序问题
        textView1 = (TextView) this.findViewById(R.id.textView1);
```

```
            textView1.setAutoLinkMask(Linkify.ALL);
            textView1.setText("网址为: http://www.baidu.com 邮箱为: abc@sohu.com");

            // setText 和 addLinks 方法调用有顺序问题
            textView2 = (TextView) this.findViewById(R.id.textView2);
            textView2.setText("网址为: http://www.sohu.com 邮箱为: xyz@sohu.com");
            Linkify.addLinks(textView2, Linkify.ALL);

            textView3 = (TextView) this.findViewById(R.id.textView3);
            textView3.setText("欢迎高洪岩 123");
            int flags1 = Pattern.CASE_INSENSITIVE;
            Pattern p1 = Pattern.compile("高洪岩", flags1);
            Linkify.addLinks(textView3, p1, "http://www.gaohongyan.com");

            textView4 = (TextView) this.findViewById(R.id.textView4);
            textView4.setText("欢迎高洪岩 1_4_5!欢迎高洪岩 6_3_5!欢迎高洪岩 2_4_5!");
            int flags2 = Pattern.CASE_INSENSITIVE;
            Pattern p2 = Pattern.compile("高洪岩\\w*!", flags2);
            Linkify.addLinks(textView4, p2, "", new MyMatchFilter(),
                    new MyTransformFilter());

        }
    }
```

代码:

```
Linkify.addLinks(textView4, p2, "", new MyMatchFilter(),
        new MyTransformFilter());
```

第 3 个参数是设置返回 Uri 字符串的 scheme，如果没有设置，默认就是 http://。
程序运行后结果如图 3.133 所示。

图 3.133　LogCat 中的打印及运行效果

3.34　ToggleButton 对话框

控件 ToggleButton 的功能有些类似于"电池开关"，只有两种状态：打开和关闭。
新建名称为 ToggleButton 的 Android 项目，在布局文件 main.xml 加入两个 ToggleButton 控件，布局文件的代码如下：

```xml
<?xml version="1.0" encoding="utf-8"?>
<LinearLayout xmlns:android="http://schemas.android.com/apk/res/android"
    android:orientation="vertical" android:layout_width="fill_parent"
    android:layout_height="fill_parent">
    <ToggleButton android:textOn="打开1啦！" android:textOff="关闭1啦！"
        android:id="@+id/toggleButton1" android:layout_width="wrap_content"
        android:layout_height="wrap_content"></ToggleButton>
    <ToggleButton android:textOn="打开2啦！" android:textOff="关闭2啦！"
        android:id="@+id/toggleButton2" android:layout_width="wrap_content"
        android:layout_height="wrap_content"></ToggleButton>
</LinearLayout>
```

文件 Main.java 的代码如下：

```java
public class Main extends Activity {

    private ToggleButton tb1;
    private ToggleButton tb2;

    @Override
    public void onCreate(Bundle savedInstanceState) {
        super.onCreate(savedInstanceState);
        setContentView(R.layout.main);

        tb1 = (ToggleButton) this.findViewById(R.id.toggleButton1);
        tb1.setChecked(true);//默认为打开状态
        tb2 = (ToggleButton) this.findViewById(R.id.toggleButton2);

        tb1.setOnCheckedChangeListener(new OnCheckedChangeListener() {
            public void onCheckedChanged(CompoundButton arg0, boolean arg1) {
                Log.v("tb1 现在的值为：", "" + arg1);
            }
        });
        tb2.setOnCheckedChangeListener(new OnCheckedChangeListener() {
            public void onCheckedChanged(CompoundButton arg0, boolean arg1) {
                Log.v("tb2 现在的值为：", "" + arg1);
            }
        });

    }
}
```

程序运行后的效果如图 3.134 所示。

图 3.134 中第一个控件默认为选中状态，单击这两个开关控件，在 LogCat 中打印出相关的日志信息，效果如图 3.135 所示。

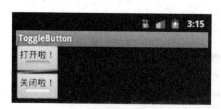
图 3.134 运行效果

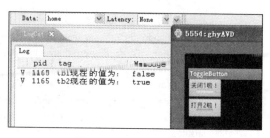
图 3.135 改变开关状态

3.35 ListActivity 对象

对象 ListActivity 可以让一个 Activity 变成一个 List 列表，在这个列表中可以显示相关的列表数据。

新建名称为 ListActivity_3_33 的 Android 项目，将文件 Main.java 的代码更改如下：

```java
public class Main extends ListActivity {
    private String[] cityArray = new String[] { "A", "B", "C", "D", "E" };

    @Override
    protected void onListItemClick(ListView l, View v, int position, long id) {
        super.onListItemClick(l, v, position, id);
        Log.v("单击了：", cityArray[position]);
    }

    @Override
    public void onCreate(Bundle savedInstanceState) {
        super.onCreate(savedInstanceState);
        // setContentView(R.layout.main);

        ArrayAdapter arrayAdapterRef = new ArrayAdapter(this,
                android.R.layout.simple_list_item_1, cityArray);

        this.setListAdapter(arrayAdapterRef);

    }
}
```

程序运行后并单击 List 中的条目 item，结果如图 3.136 所示。

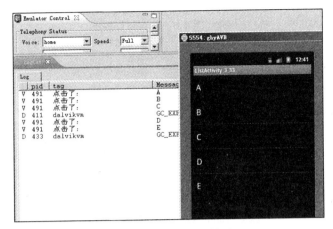

图 3.136　单击 Item 效果

3.36　TabHost 标签页控件

控件 TabHost 的功能和 Web 技术中的"选项卡"控件显示的外观一样，都是用最小的空间显示更多的数据，在 Android 中实现这种效果很简单。

新建名称为 ui_31 的 Android 项目，在 res/layout 目录下新建 TabHost 标签页控件中的 3 个子标签内容布局文件，存放在 layout 目录下，如图 3.137 所示。

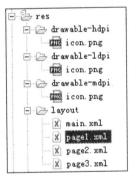

图 3.137　存放在 layout 目录下的 3 个布局文件

其中文件 page1.xml 的内容如下：

```xml
<?xml version="1.0" encoding="utf-8"?>
<LinearLayout xmlns:android="http://schemas.android.com/apk/res/android"
    android:orientation="vertical" android:layout_width="fill_parent"
    android:layout_height="fill_parent" android:id="@+id/page1">
    <Button android:text="我是page1" android:id="@+id/button1"
        android:layout_width="wrap_content" android:layout_height="wrap_content"></Button>
</LinearLayout>
```

文件 page2.xml 的文件内容如下：

```xml
<?xml version="1.0" encoding="utf-8"?>
```

```xml
<LinearLayout xmlns:android="http://schemas.android.com/apk/res/android"
    android:orientation="vertical" android:layout_width="fill_parent"
    android:layout_height="fill_parent" android:id="@+id/page2">
    <Button android:text="我是page2" android:id="@+id/button1"
        android:layout_width="wrap_content" android:layout_height="wrap_content"></Button>
</LinearLayout>
```

文件 page3.xml 的内容如下：

```xml
<?xml version="1.0" encoding="utf-8"?>
<LinearLayout xmlns:android="http://schemas.android.com/apk/res/android"
    android:orientation="vertical" android:layout_width="fill_parent"
    android:layout_height="fill_parent" android:id="@+id/page3">
    <Button android:text="我是page3" android:id="@+id/button1"
        android:layout_width="wrap_content" android:layout_height="wrap_content"></Button>
</LinearLayout>
```

文件 main.xml 的代码如下：

```xml
<?xml version="1.0" encoding="utf-8"?>

<LinearLayout xmlns:android="http://schemas.android.com/apk/res/android"
    android:orientation="vertical" android:layout_width="fill_parent"
    android:layout_height="fill_parent">
    <TabHost xmlns:android="http://schemas.android.com/apk/res/android"
        android:orientation="vertical" android:layout_width="fill_parent"
        android:layout_height="fill_parent" android:id="@android:id/tabhost">
        <LinearLayout android:orientation="vertical"
            android:layout_width="fill_parent" android:layout_height="fill_parent"
            android:padding="5dp">
            <TabWidget android:id="@android:id/tabs"
                android:layout_width="fill_parent" android:layout_height="wrap_content" />
            <FrameLayout android:id="@android:id/tabcontent"
                android:layout_width="fill_parent" android:layout_height="fill_parent"
                android:padding="5dp" />
        </LinearLayout>

    </TabHost>
</LinearLayout>
```

更改 Main.java 文件的代码如下：

```java
public class Main extends TabActivity {
    private TabHost tabHostRef;

    @Override
    public void onCreate(Bundle savedInstanceState) {
        super.onCreate(savedInstanceState);
        setContentView(R.layout.main);

        tabHostRef = this.getTabHost();
```

```
        LayoutInflater.from(this).inflate(R.layout.page1,
                tabHostRef.getTabContentView(), true);
        LayoutInflater.from(this).inflate(R.layout.page2,
                tabHostRef.getTabContentView(), true);
        LayoutInflater.from(this).inflate(R.layout.page3,
                tabHostRef.getTabContentView(), true);

        tabHostRef.addTab(tabHostRef.newTabSpec("第一页").setIndicator("第一页啊")
                .setContent(R.id.page1));
        tabHostRef.addTab(tabHostRef.newTabSpec("第二页").setIndicator("第二页啊")
                .setContent(R.id.page2));
        tabHostRef.addTab(tabHostRef.newTabSpec("第三页").setIndicator("第三页啊")
                .setContent(R.id.page3));

        tabHostRef.setCurrentTab(1);

        tabHostRef.setOnTabChangedListener(new OnTabChangeListener() {

            public void onTabChanged(String arg0) {
                Log.v("单击了======", "" + arg0);

            }
        });

    }
}
```

其中代码 newTabSpec("第二页")的参数值是当前标签页的唯一标识，并不是标题。程序初次运行效果如图 3.138 所示。

单击不同标签显示不同标签页中的内容，在 LogCat 面板中打印相关的信息，如图 3.139 所示。

图 3.138　第 2 页默认显示的初始效果

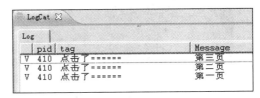

图 3.139　单击不同标签打印不同信息

3.37　控件显示内容的国际化 i18n

在 Android 中也支持控件显示内容的国际化，也就是平常所称的 i18n 技术，创建 Android 项目，名称为 i18nTest，布局文件 main.xml 的代码如下：

```
<?xml version="1.0" encoding="utf-8"?>
<LinearLayout xmlns:android="http://schemas.android.com/apk/res/android"
    android:orientation="vertical" android:layout_width="fill_parent"
```

```
        android:layout_height="fill_parent">
    <TextView android:layout_width="fill_parent"
        android:layout_height="wrap_content" android:text="@string/hello" />
    <ImageView android:src="@drawable/i18ntest"
        android:layout_height="wrap_content" android:id="@+id/imageView1"
        android:layout_width="wrap_content"></ImageView>
</LinearLayout>
```

将项目的资源 res 文件夹结构改成如图 3.140 所示。

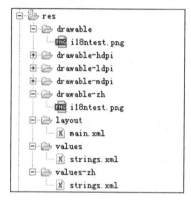

图 3.140 res 文件夹中的结构

从图 3.140 中可以看到，本项目支持两种语言的国际化，一个是默认的 English，另外一个就是 Chinese 中文，因为将文件夹命名为 values-zh 就代表当前目录中的资源是中文本地化的资源，项目运行后默认是 English 环境，所以显示英语的资源，显示效果如图 3.1341 所示。

将当前的 AVD 环境设置为简体中文，显示的界面内容如图 3.142 所示。

图 3.141 显示英语的资源

图 3.142 简体中文资源

3.38 Color 颜色的操作

在 Android 中美化 UI 界面的操作大多都与 Color 有关，所以本节将对 Color 的操作进行讲解。

新建名称为 colorTest 的 Android 项目，布局文件 main.xml 的代码如下：

```xml
<?xml version="1.0" encoding="utf-8"?>
<LinearLayout xmlns:android="http://schemas.android.com/apk/res/android"
    android:orientation="vertical" android:layout_width="fill_parent"
    android:layout_height="fill_parent">
```

```xml
<TextView android:id="@+id/textView1" android:layout_width="fill_parent"
    android:layout_height="wrap_content" android:text="@string/hello"
    android:background="#ffffff" />
<TextView android:id="@+id/textView2" android:layout_width="fill_parent"
    android:layout_height="wrap_content" android:text="@string/hello" />
<TextView android:id="@+id/textView3" android:layout_width="fill_parent"
    android:layout_height="wrap_content" android:text="@string/hello" />
<TextView android:id="@+id/textView4" android:layout_width="fill_parent"
    android:layout_height="wrap_content" android:text="@string/hello"
    android:background="@drawable/ghyColor1" />
<TextView android:id="@+id/textView5" android:layout_width="fill_parent"
    android:layout_height="wrap_content" android:text="@string/hello"
    android:background="@drawable/ghyColor2" />
<TextView android:id="@+id/textView6" android:layout_width="fill_parent"
    android:layout_height="wrap_content" android:text="@string/hello" />
<TextView android:id="@+id/textView7" android:layout_width="fill_parent"
    android:layout_height="wrap_content" android:text="@string/hello" />
</LinearLayout>
```

在 values 目录下创建一个名称为 color.xml 的颜色配置文件，代码如下：

```xml
<?xml version="1.0" encoding="utf-8"?>
<resources>
    <drawable name="ghyColor1">#50FFFFFF</drawable>
    <drawable name="ghyColor2">#80FF0000</drawable>
</resources>
```

文件 Main.java 的代码如下：

```java
public class Main extends Activity {
    private TextView textView2;
    private TextView textView3;
    private TextView textView6;
    private TextView textView7;

    @Override
    public void onCreate(Bundle savedInstanceState) {
        super.onCreate(savedInstanceState);
        setContentView(R.layout.main);

        textView2 = (TextView) this.findViewById(R.id.textView2);
        textView2.setBackgroundColor(Color.parseColor("#ff0000"));

        textView3 = (TextView) this.findViewById(R.id.textView3);
        // 设置背景透明度
        textView3.setBackgroundColor(Color.argb(100, 125, 125, 125));
        // 设置文字透明度
        textView3.setTextColor(Color.argb(100, 125, 125, 125));

        textView6 = (TextView) this.findViewById(R.id.textView6);
        Drawable drawableRef = this.getBaseContext().getResources()
```

```
            .getDrawable(R.drawable.ghyColor1);
    textView6.setBackgroundDrawable(drawableRef);
    // 取得屏幕高和宽 PX
    String width_heigth = this.getBaseContext().getResources()
            .getDisplayMetrics().widthPixels
            + " "
            + this.getBaseContext().getResources().getDisplayMetrics().heightPixels;
    textView6.setText(width_heigth);

    textView7 = (TextView) this.findViewById(R.id.textView7);
    textView7.setBackgroundResource(R.drawable.ghyColor2);
    textView7.setText(R.string.username);

    }
}
```

程序运行结果如图 3.143 所示。

图 3.143　程序运行结果

3.39　draw9Patch 工具的使用

在 Android 的 SDK 自带工具中还有一个比较重要的工具，它就是 draw9patch.bat，这个工具主要的作用是制作自适应拉伸的图片，因为在 Android 的移动终端设备中屏幕的大小是不固定的，所以就需要有种技术能使充当背景的图片具有自适应屏幕大小而又不失真的效果，这个工具就是用于制作这样的图片。

例如，有如图 3.144 所示的 PNG 的位图图片资源。位图的特性是具有存储高色彩且便于处理的图片资源，但它也有比较明显的缺点，就是放大失真，那如何做到将位图放入 Android 设备中随着屏幕的变化而图片又不失真呢？下面介绍其处理方法。

双击 android-sdk-windows\tools\draw9patch.bat 工具，打开工具后弹出如图 3.145 所示的界面。

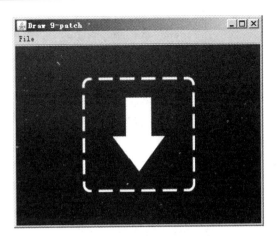

图 3.144 ghy.png 的位图资源　　　　图 3.145 打开工具后的界面

单击 File 菜单中的 Open 9-patch 菜单项，弹出选中图片的对话框，选中刚才介绍的 ghy.png 图片，出现如图 3.146 所示的界面。

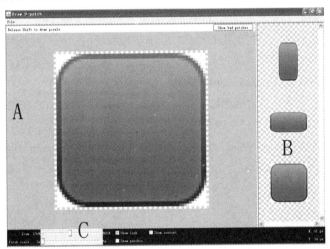

图 3.146 打开 ghy.png 后的软件界面

在图 3.146 中可以发现软件的界面分为 3 个主要部分，分别是 A 区、B 区和 C 区，其中 A 区是待处理图片资源的操作界面，而 B 区是对图片进行处理后的效果展示，分为 3 种情况展示，而 C 区则是一些控制选项。

那到底这个工具怎么用呢？在这里需要说明的一点是，不是任何图片资源都能做到自适应，那什么样的图片资源才能做到自适应呢？图片中有重复的区域就可以做到自适应，比如 ghy.png 图片中就有重复的区域，如图 3.147 所示。

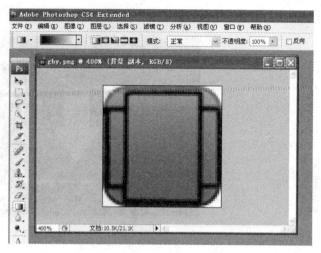

图 3.147　图片资源中有重复的区域

在图 3.147 中画上黑线的地方就可以做到有规律地进行重复，因为没有任何的文字和不规律的图形，完全是一种有规律的渐变，那在 draw9patch.bat 中如何做到呢？

首先一定要记住图 3.147 原始的图片效果，未来要和最终的效果进行比较。

在 draw9patch 工具中打开图片时，图片的四周都被自动扩充了 1 像素，这个 1 像素就是用来留给设计人员进行处理的，效果如图 3.148 所示。

然后在想得到图片自适应的边框边缘上画上黑线，如果想取消黑线时按住 Shift 键再单击鼠标就可以了，得到 9 格图片资源如图 3.149 所示。

图 3.148　自动扩充的边框

图 3.149　得到 9 格式图片资源

这时再比较一下原始图片和最终的图片自适应效果，如图 3.150 所示。

Android 的 UI 控件 第 3 章

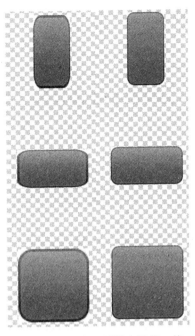

图 3.150　左边是原始右边是 9 格

从图 3.150 中可以发现，经过 draw9patch 工具处理后的 9 格图片在自适应时图片并没有失真，圆角还保留了原来的样式，因为没有被黑线画过的边缘呈原始显示状态，不参与图片的放大及缩小等情况。

在这里一定要留意的是，在 photoshop 中处理 PNG 图片的过程中，一定要设置当前图片的背景为透明，不能设置为白色。

3.40　以 9 格图片资源作为 Button 背景

将原始的 ghy.png 和经过保存后的 9 格图片放入名称为 drawableTest 的 Android 项目中，效果如图 3.151 所示。

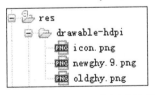

图 3.151　原始和 9 格图片资源

在 Main.java 代码中设计如下程序：

```
public class Main extends Activity {
    @Override
    public void onCreate(Bundle savedInstanceState) {
        super.onCreate(savedInstanceState);
```

```java
        setContentView(R.layout.main);
        
        Drawable drawableOld = this.getResources().getDrawable(
                R.drawable.oldghy);
        
        Drawable drawableNew = this.getResources().getDrawable(
                R.drawable.newghy);
        
        Log.v("!", "" + drawableOld.toString());
        Log.v("!", "" + drawableNew.toString());
        
    }
}
```

程序运行后在 LogCat 打印出来的效果如图 3.152 所示。

```
V  522  !                          android.graphics.drawable.BitmapDrawable@4051ab60
V  522  !                          android.graphics.drawable.NinePatchDrawable@4051af10
```

图 3.152 打印 drawable 对象类型

从图 3.152 中可以看到 9 格图片的对象资源是 NinePatchDrawable。如何在 Button 中应用这个 9 格图片资源呢？更改当前项目的 main.xml 代码如下：

```xml
<?xml version="1.0" encoding="utf-8"?>
<LinearLayout xmlns:android="http://schemas.android.com/apk/res/android"
    android:orientation="vertical" android:layout_width="fill_parent"
    android:layout_height="fill_parent">
    <Button android:text="Button" android:id="@+id/button1"
        android:layout_width="fill_parent" android:layout_height="wrap_content"></Button>
    <Button android:text="Button" android:id="@+id/button2"
        android:layout_width="wrap_content" android:layout_height="wrap_content"></Button>
</LinearLayout>
```

再次更改 Main.java 的代码如下：

```java
public class Main extends Activity {
    private Button button1;
    private Button button2;
    
    @Override
    public void onCreate(Bundle savedInstanceState) {
        super.onCreate(savedInstanceState);
        setContentView(R.layout.main);
        
        Drawable drawableOld = this.getResources().getDrawable(
                R.drawable.oldghy);
        
        Drawable drawableNew = this.getResources().getDrawable(
                R.drawable.newghy);
        
        Log.v("!", "" + drawableOld.toString());
        Log.v("!", "" + drawableNew.toString());
```

```
        button1 = (Button) this.findViewById(R.id.button1);
        button1.setBackgroundDrawable(drawableNew);

        button2 = (Button) this.findViewById(R.id.button2);
        button2.setBackgroundDrawable(drawableNew);

    }
}
```

程序运行后的效果如图 3.153 所示。

图 3.153　背景自适应的 Button

示例演示到这，Button 并没有图片状态的切换，也就是按下 Button 和不按 Button 的背景图片都是 1 个效果，应该如何改变呢？下面我们来看看如何改变按钮的状态。

3.41　使用 selector 改变按钮状态

新建名称为 buttonState 的 Android 项目，将图片资源放入项目中，如图 3.154 所示。

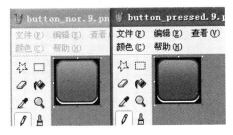

图 3.154　两张 Button 背景图片资源

在 res 目录中新建文件夹 drawable，在里面新建一个名称为 button.xml 的选择器文件，代码如下：

```xml
<?xml version="1.0" encoding="utf-8"?>
<selector xmlns:android="http://schemas.android.com/apk/res/android">
    <item android:state_pressed="true" android:drawable="@drawable/button_pressed" />
    <item android:state_focused="true" android:drawable="@drawable/button_pressed" />
    <item android:drawable="@drawable/button_nor" />
</selector>
```

项目的 res 资源如图 3.155 所示。

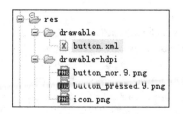

图 3.155 res 资源

更改 main.xml 布局文件的代码如下：

```xml
<?xml version="1.0" encoding="utf-8"?>
<LinearLayout xmlns:android="http://schemas.android.com/apk/res/android"
    android:orientation="vertical" android:layout_width="fill_parent"
    android:layout_height="fill_parent">
    <Button android:text="Button" android:id="@+id/button1"
        android:layout_width="fill_parent"          android:layout_height="wrap_content"
android:background="@drawable/button"></Button>
    <Button android:text="Button" android:id="@+id/button2"
        android:layout_width="wrap_content"         android:layout_height="wrap_content"
android:background="@drawable/button"></Button>
</LinearLayout>
```

程序运行后的效果如图 3.156 所示。

图 3.156 程序运行后的效果

按下 Button 时的状态自动切换背景，产生按下的效果。

第 4 章　Intent 对象

在第 4 章将和大家一起讨论的是与 Intent 有关的技术，虽然 Android 的 4 大核心组件并没有 Intent，但 Intent 在 Android 的开发中却提供了组件彼此之间交互的作用，可以说，没有 Intent 对象就没有 Android 的通信。

本章应该着重掌握如下知识点：

- 熟练掌握隐式调用与显式调用的区别
- 熟悉静/动态广播接收者 Broadcast Receiver 的使用
- 多个 Activity 之间的数据传递
- 通知的使用
- Intent 的 flag 标记的使用

4.1　Intent 对象必备技能

在 Android 中有 4 种核心技术，分别是 Activity、Service、Broadcast Receiver 和 Content Provider。这些组件除了 Content Provider 之外，彼此之间的通信都要使用 Intent 对象来进行，所以掌握 Intent 对象的使用是非常重要的技能点。

Intent 对象是一种可以在运行时动态绑定组件的关键技术，通过使用 Intent 对象，可以告诉系统你想要实现什么样的操作，也就是 Intent 对象里面包含请求的内容，请求再由 Android 操作系统接收到，然后到 IntentFilter 过滤器中找到已经注册的组件，再调用这个组件就完成了组件间通信的过程。

Intent 对象描述了要执行的结果是什么，但由于 Android 系统组件间的通信方式细节上都是不同的，所以 Intent 对象里面描述要执行的目的是模糊的、抽象的，但其描述的基本内容可以分为：componentName 组件名称、Action 动作名称、Data 数据、Category 类别、Extra 附加数据和 Flag 标志位 6 个部分。

下面将对主要的 Intent 描述内容进行示例演示。

4.1.1　指定 componentName 组件名称与显式调用

如果指定了 componentName 组件名称，那么这种调用就是显式调用，因为要执行的组件名称已经确定，只需要告诉 Android 系统调用指定的组件即可。

新建名称为 componentName 的 Android 项目，项目整理后共有两个 Activity 对象和两个 XML 布局文件，本示例要演示的内容就是指定 componentName 组件名称来达到显式调用组件的效果，是多个 Activity 对象之间的通信，也就是切换页面的效果。下面依次列举它们的代码。

文件 Main.java 的代码如下：

```java
public class Main extends Activity {
    @Override
    public void onCreate(Bundle savedInstanceState) {
        super.onCreate(savedInstanceState);
        setContentView(R.layout.main);

        Button button = (Button) this.findViewById(R.id.button1);
        button.setOnClickListener(new OnClickListener() {

            public void onClick(View arg0) {
                Intent gotoSecondActivity = new Intent();
                gotoSecondActivity.setClass(Main.this, Second.class);
                Main.this.startActivity(gotoSecondActivity);
            }
        });

    }
}
```

文件 Second.java 的代码如下：

```java
public class Second extends Activity {
    @Override
    public void onCreate(Bundle savedInstanceState) {
        super.onCreate(savedInstanceState);
        // 注意：关联自己的 xml 布局文件
        setContentView(R.layout.second);
    }
}
```

文件 main.xml 的代码如下：

```xml
<?xml version="1.0" encoding="utf-8"?>
<LinearLayout xmlns:android="http://schemas.android.com/apk/res/android"
    android:orientation="vertical" android:layout_width="fill_parent"
    android:layout_height="fill_parent">
    <Button android:text="Button" android:id="@+id/button1"
        android:layout_width="wrap_content" android:layout_height="wrap_content"></Button>
</LinearLayout>
```

文件 second.xml 的代码如下：

```xml
<?xml version="1.0" encoding="utf-8"?>
<LinearLayout xmlns:android="http://schemas.android.com/apk/res/android"
    android:orientation="vertical" android:layout_width="fill_parent"
    android:layout_height="fill_parent">
    <TextView android:layout_width="fill_parent"
        android:layout_height="wrap_content" android:text="this is second page!" />
</LinearLayout>
```

另外，还需要在文件 AndroidManifest.xml 中注册 Second.java 的 Activity 对象，代码如下：

```xml
<?xml version="1.0" encoding="utf-8"?>
<manifest xmlns:android="http://schemas.android.com/apk/res/android"
      package="componentName.test.run" android:versionCode="1"
      android:versionName="1.0">
    <application android:icon="@drawable/icon" android:label="@string/app_name">
        <activity android:name=".Main" android:label="@string/app_name">
            <intent-filter>
                <action android:name="android.intent.action.MAIN" />
                <category android:name="android.intent.category.LAUNCHER" />
            </intent-filter>
        </activity>

        <activity android:name=".Second" android:label="@string/app_name">
        </activity>
    </application>
</manifest>
```

程序运行时初始效果如图 4.1 所示。

单击 Button 按钮后转入 Second.java 文件并显示了 second.xml 布局，如图 4.2 所示。

图 4.1　程序初始运行效果

图 4.2　最终效果

以下代码：

```
gotoSecondActivity.setClass(Main.this, Second.class);
```

上述代码明确指定了要调用的组件名称，至此指定组件名称显式调用的情况就成功实现了。

4.1.2　指定 Action 动作名称与隐式调用

　　上面是明确指定组件名称的"显式调用"的情况，没有明确指出目标组件名称的情况则叫做"隐式调用"，Android 系统要使用 IntentFilter 过滤器来寻找与隐式 Intent 相匹配的组件对象，但匹配的成功与否与 3 个元素有关，它们是"action"、"category"和"data"。一个隐式的 Intent 调用必须通过这 3 个元素的匹配检查，如果检查成功则成功匹配，如果检查失败则不匹配。这个 IntentFiler 要在 AndroidManifest.xml 文件中进行注册，并且至少要有一个<action>元素，如果没有则任何的 Intent 都不匹配。

　　本节将要实现多个示例，这些示例的主要意图就是在 Android 中使用隐式调用。

1. 隐式调用并传递 Extra 附加数据与静态广播 BroadcastReceiver 的使用

　　本示例是一个隐式调用与传递 Extra 附加数据和广播 BroadcastReceiver 技术的使用示例，使用广播的好处是组件彼此间是以松耦合的方式来进行组织，有利于模块式的组件开发。

新建名称为 intent_4_3 的 Android 项目，将文件 Main.java 的代码更改如下：

```java
public class Main extends Activity {

    private Button button;

    @Override
    public void onCreate(Bundle savedInstanceState) {
        super.onCreate(savedInstanceState);
        setContentView(R.layout.main);

        button = (Button) this.findViewById(R.id.button1);

        button.setOnClickListener(new OnClickListener() {
            public void onClick(View arg0) {
                //创建 Intent 意图对象                添加动作名称为 intent_4_3
                Intent sendBroadcastIntent = new Intent("intent_4_3");
                sendBroadcastIntent.putExtra("sendText", "高洪岩发送广播");//在这个 Intent 意图对象中
                存储一些数据，通过广播携带到目的组件中
                //开始发送广播啦！
                Main.this.sendBroadcast(sendBroadcastIntent);
            }
        });

    }
}
```

广播是发送出去了，谁接收呢？

新建广播接收者，文件名称为 GhyBroadcastReceiver.java，程序代码如下：

```java
public class GhyBroadcastReceiver extends BroadcastReceiver {

    @Override
    public void onReceive(Context arg0, Intent arg1) {
        Log.v("GhyBroadcastReceiver", arg1.getStringExtra("sendText"));
        Toast
            .makeText(arg0, arg1.getStringExtra("sendText"),
                    Toast.LENGTH_LONG).show();

    }
}
```

一定要继承自 BroadcastReceiver 类，然后重写 onReceive 方法，在 onReceive 方法中从 Intent 对象中取出附加的数据值，并通过对话框显示出来。

上面的文件仅仅是创建了一个广播接收者，仅仅是一个普通的类，并没有集成到 Android 系统中，所以还需要在文件 AndroidManifest.xml 中注册这个广播接收者，程序代码如下：

```xml
<?xml version="1.0" encoding="utf-8"?>
<manifest xmlns:android="http://schemas.android.com/apk/res/android"
    package="test.run" android:versionCode="1" android:versionName="1.0">
    <application android:icon="@drawable/icon" android:label="@string/app_name">
        <activity android:name=".Main" android:label="@string/app_name">
            <intent-filter>
                <action android:name="android.intent.action.MAIN" />
                <category android:name="android.intent.category.LAUNCHER" />
            </intent-filter>
        </activity>
        <receiver android:name=".GhyBroadcastReceiver">
            <intent-filter>
                <action android:name="intent_4_3"></action>
            </intent-filter>
        </receiver>
    </application>
</manifest>
```

其中最为重要的是这个广播一定要响应名称为 intent_4_3 这个动作，关键代码如下：

```xml
<receiver android:name=".GhyBroadcastReceiver">
    <intent-filter>
        <action android:name="intent_4_3"></action>
    </intent-filter>
</receiver>
```

还有布局文件 main.xml 的代码如下：

```xml
<?xml version="1.0" encoding="utf-8"?>
<LinearLayout xmlns:android="http://schemas.android.com/apk/res/android"
    android:orientation="vertical" android:layout_width="fill_parent"
    android:layout_height="fill_parent">
    <Button android:text="Button" android:id="@+id/button1"
        android:layout_width="wrap_content" android:layout_height="wrap_content"></Button>
</LinearLayout>
```

程序初始运行效果如图 4.3 所示。

图 4.3 初始运行效果

整个界面只有一个按钮，单击这个按钮后将把动作 Action 名称为 intent_4_3 的广播发送出去，等待注册到 Android 系统中的广播接收者进行处理，本示例成功出现接收到的广播消息，如图 4.4 所示。

图 4.4 成功接收到广播

2. 隐式调用并传递 Extra 附加数据与动态广播 BroadcastReceiver 的使用

上面的示例仅仅是定义静态的广播，也就是在文件 AndroidManifest.xml 中定义广播接收者，而在 Android 中还支持动态创建的广播。

新建名称为 dyna_broadcast 的 Android 项目，继续创建自定义的广播接收类 MyBroadCast.java，程序代码如下：

```java
public class MyBroadCast extends BroadcastReceiver {

    @Override
    public void onReceive(Context arg0, Intent arg1) {
        Log.v("=====", arg1.getStringExtra("username"));

    }

}
```

发送广播，带 Button 按钮的布局文件 main.xml 的代码如下：

```xml
<?xml version="1.0" encoding="utf-8"?>
<LinearLayout xmlns:android="http://schemas.android.com/apk/res/android"
    android:orientation="vertical" android:layout_width="fill_parent"
    android:layout_height="fill_parent">
    <Button android:text="Button" android:id="@+id/button1"
        android:layout_width="wrap_content" android:layout_height="wrap_content"></Button>
</LinearLayout>
```

发送广播的 Activity 对象文件 Main.java 的代码如下：

```java
public class Main extends Activity {
    private IntentFilter myIntent = new IntentFilter();
    private MyBroadCast MyBroadCastRef = new MyBroadCast();
    private Button button1;

    @Override
```

```java
public void onCreate(Bundle savedInstanceState) {
    super.onCreate(savedInstanceState);
    setContentView(R.layout.main);

    myIntent.addAction("ghyBroadCast");
    registerReceiver(MyBroadCastRef, myIntent);

    button1 = (Button) this.findViewById(R.id.button1);

    button1.setOnClickListener(new OnClickListener() {
        public void onClick(View arg0) {
            Intent newIntent = new Intent();
            newIntent.putExtra("username", "gaohongyan");
            newIntent.setAction("ghyBroadCast");
            Main.this.sendBroadcast(newIntent);
        }
    });

}
@Override
protected void onStop() {
    unregisterReceiver(MyBroadCastRef);
}
}
```

程序运行后，单击界面中的 Button 按钮，在 LogCat 打印字符串 gaohongyan，说明动态创建广播、发送广播、接收广播的示例成功，程序运行效果如图 4.5 所示。

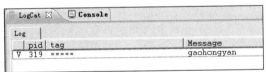

图 4.5　动态创建广播并成功接收广播信息

3. 不用广播的 Intent 隐式调用

使用广播的确可以触发 Intent 的 action 动作，当然使用 startActivity()方法也可以实现相同的功能，只是实现的方式不一样。

新建一个名称为 useStartActivityMethodAction 的 Android 项目，布局文件 main.xml 的代码如下：

```xml
<?xml version="1.0" encoding="utf-8"?>
<LinearLayout xmlns:android="http://schemas.android.com/apk/res/android"
    android:orientation="vertical" android:layout_width="fill_parent"
    android:layout_height="fill_parent">
    <Button android:text="Button" android:id="@+id/button1"
        android:layout_width="wrap_content" android:layout_height="wrap_content"></Button>
</LinearLayout>
```

文件 Main.java 的代码如下：

```java
public class Main extends Activity {
    private Button button1;

    @Override
    public void onCreate(Bundle savedInstanceState) {
        super.onCreate(savedInstanceState);
        setContentView(R.layout.main);
        button1 = (Button) this.findViewById(R.id.button1);
        button1.setOnClickListener(new OnClickListener() {
            public void onClick(View arg0) {
                Intent newIntent = new Intent("ghyActionName");
                Main.this.startActivity(newIntent);
            }
        });

    }
}
```

布局文件 second.xml 的代码如下：

```xml
<?xml version="1.0" encoding="utf-8"?>
<LinearLayout xmlns:android="http://schemas.android.com/apk/res/android"
    android:orientation="vertical" android:layout_width="fill_parent"
    android:layout_height="fill_parent">
    <TextView android:layout_width="fill_parent"
        android:layout_height="wrap_content" android:text="second page!" />
</LinearLayout>
```

文件 Second.java 的代码如下：

```java
public class Second extends Activity {
    @Override
    public void onCreate(Bundle savedInstanceState) {
        super.onCreate(savedInstanceState);
        setContentView(R.layout.second);

    }
}
```

为了使用 ActionName 动作名称，还需要在 AndroidManifest.xml 文件中注册这个 Activity 及 actionName 动作名称，代码如下：

```xml
<?xml version="1.0" encoding="utf-8"?>
<manifest xmlns:android="http://schemas.android.com/apk/res/android"
    package="useStartActivityMethodAction.test.run" android:versionCode="1"
    android:versionName="1.0">
    <application android:icon="@drawable/icon" android:label="@string/app_name">
        <activity android:name=".Main" android:label="@string/app_name">
            <intent-filter>
                <action android:name="android.intent.action.MAIN" />
                <category android:name="android.intent.category.LAUNCHER" />
            </intent-filter>
```

```xml
        </activity>
        <activity android:name=".Second" android:label="@string/app_name">
            <intent-filter>
                <action android:name="ghyActionName" />
                <category android:name="android.intent.category.DEFAULT" />
            </intent-filter>
        </activity>
    </application>
</manifest>
```

在这里一定要注意代码：

```
<category android:name="android.intent.category.DEFAULT" />
```

前面已经介绍过，通过 startActivity() 方法实现 Intent 隐式调用的情况下，匹配成功与否与 Action 和 Category 还有 Data 有关，本示例正是使用隐式调用，而且在 Intent 对象中偏偏又没有指定 category 匹配值，所以 Android 在进行匹配前，自动在 Intent 中加入了 android.intent.category.DEFAULT，如果不在 <intent-filter> 中加入 android.intent.category.DEFAULT，是没有办法匹配到 Activity 的。

程序初始运行显示 Button 按钮，如图 4.6 所示。

单击 Button 按钮后切换界面，如图 4.7 所示。

图 4.6　显示默认界面

图 4.7　切换后的界面

4. 使用系统自带的 ActionName 动作名称

前面是使用自定义的 ActionName 动作名称 ghyActionName 来实现隐式调用，在 Android 中也有一些系统自带的 ActionName 动作名称，只要在 Intent 意图对象中设置这些 Action 动作名称并将这个 Intent 意图启动，系统就能自动响应这个 Intent 对象并执行对应 Action 动作名称的处理。

新建名称为 useSystemAction 的 Android 项目，将文件 Main.java 的代码更改如下：

```java
public class Main extends Activity {
    @Override
    public void onCreate(Bundle savedInstanceState) {
        super.onCreate(savedInstanceState);
        setContentView(R.layout.main);

        Intent intent = new Intent(Intent.ACTION_DIAL);
        this.startActivity(intent);

    }
}
```

程序运行后直接把电话拨号界面显示出来，如图 4.8 所示。

图 4.8　使用系统自带的 Action 名称

4.1.3　指定 Action 的动作名称和 Data 数据

本节将要用两个示例来演示 Action 的动作名称和 Data 数据的使用，那 Intent 中的 Data 到底是什么呢？Data 是描述 Intent 要操作的数据 URI。

1. 使用系统的 Data 数据

在本小节中将要实现单击 Button 按钮时去往不同的网站。

创建名称为 openURL 的 Android 项目，文件 Main.java 的代码如下：

```java
public class Main extends Activity {
    @Override
    public void onCreate(Bundle savedInstanceState) {
        super.onCreate(savedInstanceState);
        setContentView(R.layout.main);

        new AlertDialog.Builder(this).setTitle("我是标题").setMessage("我是正文")
                .setPositiveButton("关闭", new OnClickListener() {
                    public void onClick(DialogInterface arg0, int arg1) {

                    }
                }).setNegativeButton("去往百度", new OnClickListener() {
                    public void onClick(DialogInterface arg0, int arg1) {
                        //URI 就是 Data
                        Uri uriRef = Uri.parse("http://www.baidu.com");

                        Intent gotoIntent = new Intent(Intent.ACTION_VIEW,
                                uriRef);
                        Main.this.startActivity(gotoIntent);

                    }
                }).show();
```

        }
}

程序初始运行效果如图4.9所示。

单击"去往百度"按钮,自动打开浏览器访问百度,如图4.10所示。

图4.9 初始运行效果　　　　　　图4.10 去往百度

再实现一个拨号的示例,使用的还是系统自带的 Data 数据。

新建名称为 callPhoneNum 的 Android 项目,文件 main.xml 的代码如下:

```xml
<?xml version="1.0" encoding="utf-8"?>
<LinearLayout xmlns:android="http://schemas.android.com/apk/res/android"
    android:orientation="vertical" android:layout_width="fill_parent"
    android:layout_height="fill_parent">
    <Button android:text="Button" android:id="@+id/button1"
        android:layout_width="wrap_content" android:layout_height="wrap_content"></Button>
</LinearLayout>
```

文件 Main.java 的代码如下:

```java
public class Main extends Activity {
    private Button button1;

    @Override
    public void onCreate(Bundle savedInstanceState) {
        super.onCreate(savedInstanceState);
        setContentView(R.layout.main);

        button1 = (Button) this.findViewById(R.id.button1);

        button1.setOnClickListener(new OnClickListener() {
            public void onClick(View arg0) {
                Intent newIntent = new Intent(Intent.ACTION_CALL, Uri
                        .parse("tel:13888888888"));
                Main.this.startActivity(newIntent);
            }
        });

    }
}
```

文件 AndroidManifest.xml 要加入拨号权限的代码，代码如下：

```xml
<?xml version="1.0" encoding="utf-8"?>
<manifest xmlns:android="http://schemas.android.com/apk/res/android"
    package="callPhoneNum.test.run" android:versionCode="1"
    android:versionName="1.0">


    <application android:icon="@drawable/icon" android:label="@string/app_name">
        <activity android:name=".Main" android:label="@string/app_name">
            <intent-filter>
                <action android:name="android.intent.action.MAIN" />
                <category android:name="android.intent.category.LAUNCHER" />
            </intent-filter>
        </activity>
    </application>
    <uses-permission android:name="android.permission.CALL_PHONE"></uses-permission>

</manifest>
```

初始运行效果如图 4.11 所示。

单击按钮后弹出指定拨号的界面如图 4.12 所示。

图 4.11　出现 Button 控件　　　　　图 4.12　进入拨号过程

4.1.4　两个 Activity 之间传递 Extra 字符串和 Extra 实体对象的实验

在使用 Intent 通信的过程中还可以附加一些其他的数据，这些数据在 Intent 中叫做 Extra。

1. 使用 startActivityForResult()方法回传数据

新建名称为 intent_4_2 的 Android 项目，更改文件 Main.java 的代码如下：

```java
public class Main extends Activity {
    private Button button1;
```

```java
    private Button button2;
    private EditText editText1;
    private EditText editText2;

    @Override
    protected void onActivityResult(int requestCode, int resultCode, Intent data) {
        if (requestCode == 100) {
            if (resultCode == Activity.RESULT_OK) {
                editText1.setText(data
                        .getStringExtra("giveMainStringFromSecond"));
            }

        }
        if (requestCode == 200) {
            if (resultCode == Activity.RESULT_OK) {
                Userinfo userinfo = (Userinfo) data
                        .getSerializableExtra("userinfo");
                editText2.setText("username=" + userinfo.getUsername()
                        + " password=" + userinfo.getPassword());
            }

        }
        super.onActivityResult(requestCode, resultCode, data);
    }

    @Override
    public void onCreate(Bundle savedInstanceState) {
        super.onCreate(savedInstanceState);
        setContentView(R.layout.main);

        button1 = (Button) this.findViewById(R.id.button1);
        button2 = (Button) this.findViewById(R.id.button2);
        editText1 = (EditText) this.findViewById(R.id.editText1);
        editText2 = (EditText) this.findViewById(R.id.editText2);

        button1.setOnClickListener(new OnClickListener() {
            public void onClick(View arg0) {
                Intent showPage1 = new Intent(Main.this, Second.class);
                Main.this.startActivityForResult(showPage1, 100);
            }
        });

        button2.setOnClickListener(new OnClickListener() {
            public void onClick(View arg0) {
                Intent showPage2 = new Intent(Main.this, Third.class);
                Main.this.startActivityForResult(showPage2, 200);
            }
        });
```

    }
}

 参数 requestCode 是当新打开的 Activity2 对象关闭时自动回传到 Activity1 中,而参数 resultCode 是代表操作的结果,系统默认自带的有两个常量值 RESULT_CANCELED 和 RESULT_OK,或者可以自定义这个值,在 RESULT_FIRST_USER 常量上进行累加整数。

由于本示例要实现多个 Activity 对象之间传递数据,所以在启动其他的 Activity 时要使用 startActivityForResult()方法,方法的第 2 个参数是启动 Activity 的唯一标识。

本示例要实现的界面操作为:A 启动 B,在 B 中处理一些数据,当关闭 B 时把 B 中处理的数据回传给 A。由于在 A 中还可以启动 C 组件、D 组件、E 组件,等等组件,所以要给这些 Activity 起一个唯一的标识。

文件 main.xml 的代码如下:

```xml
<?xml version="1.0" encoding="utf-8"?>
<LinearLayout xmlns:android="http://schemas.android.com/apk/res/android"
    android:orientation="vertical" android:layout_width="fill_parent"
    android:layout_height="fill_parent">
    <EditText android:text="" android:layout_width="match_parent"
        android:id="@+id/editText1" android:layout_height="wrap_content"></EditText>
    <Button android:text="Button" android:id="@+id/button1"
        android:layout_width="wrap_content" android:layout_height="wrap_content"></Button>
    <EditText android:text="" android:layout_width="match_parent"
        android:id="@+id/editText2" android:layout_height="wrap_content"></EditText>
    <Button android:text="Button" android:id="@+id/button2"
        android:layout_width="wrap_content" android:layout_height="wrap_content"></Button>
</LinearLayout>
```

在 main.xml 代码中配置了两个按钮和两个 EditText 控件。

文件 second.xml 的代码如下:

```xml
<?xml version="1.0" encoding="utf-8"?>
<LinearLayout xmlns:android="http://schemas.android.com/apk/res/android"
    android:orientation="vertical" android:layout_width="fill_parent"
    android:layout_height="fill_parent">
    <EditText android:text="" android:layout_width="match_parent"
        android:id="@+id/secondEditText" android:layout_height="wrap_content"></EditText>
    <Button android:text="Button" android:id="@+id/secondButton"
        android:layout_width="wrap_content" android:layout_height="wrap_content"></Button>

</LinearLayout>
```

文件 Second.java 的代码如下:

```java
public class Second extends Activity {
    private Button secondButton;
```

```java
    private EditText secondEditText;

    @Override
    public void onCreate(Bundle savedInstanceState) {
        super.onCreate(savedInstanceState);
        setContentView(R.layout.second);

        secondButton = (Button) this.findViewById(R.id.secondButton);
        secondEditText = (EditText) this.findViewById(R.id.secondEditText);

        secondButton.setOnClickListener(new OnClickListener() {
            public void onClick(View arg0) {

                String giveMainStringFromSecond = secondEditText.getText()
                        .toString();
                Intent intent = new Intent();
                intent.putExtra("giveMainStringFromSecond",
                        giveMainStringFromSecond);
                Second.this.setResult(Activity.RESULT_OK, intent);
                Second.this.finish();
            }
        });

    }
}
```

在上面的代码中使用到了 setResult()，这个方法的第 1 个参数是 Activity.RESULT_OK，证明在当前的 Activity 中操作业务的过程是正确的，并没有出现意外的情况，当关闭名称为 Second.java 的 Activity 界面时，Main.java 文件可以收到在 Second.java 文件中操作业务的状态值 Activity.RESULT_OK，通过这个值就可以判断 Second.java 文件中操作业务是成功还是失败。失败的常量值为 Activity.RESULT_CANCELED。

布局文件 third.xml 的代码如下：

```xml
<?xml version="1.0" encoding="utf-8"?>
<LinearLayout xmlns:android="http://schemas.android.com/apk/res/android"
    android:orientation="vertical" android:layout_width="fill_parent"
    android:layout_height="fill_parent">
    <Button android:text="Button" android:id="@+id/thirdButton"
        android:layout_width="wrap_content" android:layout_height="wrap_content"></Button>
</LinearLayout>
```

文件 Third.java 的代码如下：

```java
public class Third extends Activity {
    private Button thirdButton;

    @Override
    public void onCreate(Bundle savedInstanceState) {
        super.onCreate(savedInstanceState);
```

```java
        setContentView(R.layout.third);

        thirdButton = (Button) this.findViewById(R.id.thirdButton);

        thirdButton.setOnClickListener(new OnClickListener() {
            public void onClick(View arg0) {

                Userinfo userinfo = new Userinfo();
                userinfo.setUsername("A");
                userinfo.setPassword("B");

                Intent intent = new Intent();
                intent.putExtra("userinfo", userinfo);
                Third.this.setResult(Activity.RESULT_OK, intent);
                Third.this.finish();
            }
        });

    }
}
```

往 Main.java 中发送自定义数据 Userinfo 实体对象，实体类 Userinfo.java 的代码如下：

```java
package entity;

import java.io.Serializable;

public class Userinfo implements Serializable {

    public String getUsername() {
        return username;
    }

    public void setUsername(String username) {
        this.username = username;
    }

    public String getPassword() {
        return password;
    }

    public void setPassword(String password) {
        this.password = password;
    }

    private String username;
    private String password;

}
```

关键配置文件 AndroidManifest.xml 的代码如下：

```xml
<?xml version="1.0" encoding="utf-8"?>
<manifest xmlns:android="http://schemas.android.com/apk/res/android"
      package="intent_4_2.test.run" android:versionCode="1"
      android:versionName="1.0">
    <application android:icon="@drawable/icon" android:label="@string/app_name">
        <activity android:name=".Main" android:label="@string/app_name">
            <intent-filter>
                <action android:name="android.intent.action.MAIN" />
                <category android:name="android.intent.category.LAUNCHER" />
            </intent-filter>
        </activity>

        <activity android:name=".Second" android:label="@string/app_name">
        </activity>
        <activity android:name=".Third" android:label="@string/app_name">
        </activity>
    </application>
</manifest>
```

程序初始运行效果如图 4.13 所示。

单击上面的 Button 按钮弹出界面如图 4.14 所示，并且输入字母 ghy。

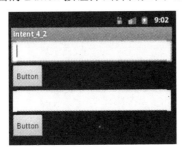

图 4.13　启动时的界面　　　　　　图 4.14　启动 Second.java

输入完成后单击 Button 按钮返回 main.xml 布局，如图 4.15 所示。正确取得返回的字符串数据。
再单击 main.xml 布局下面的按钮，弹出界面如图 4.16 所示，并且输入 a 和 b。

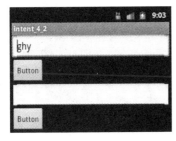

图 4.15　Main.java 接收到 Secon.java 传过来的数据　　　图 4.16　显示 Third.java 界面

直接单击 Button 按钮返回 main.xml 布局，在 main.java 中取得实体类 Userinfo 中的属性值，如图 4.17 所示。

如果在目标 Activity 直接按 Back 键，则在源 Activity 中可以通过如下代码进行判断是否取消返回值的操作：

图 4.17　Main.java 取得 Third.java 回传的 Userinfo 实体

```java
@Override
protected void onActivityResult(int requestCode, int resultCode, Intent data) {
    super.onActivityResult(requestCode, resultCode, data);
    if (requestCode == 100 && resultCode == Activity.RESULT_CANCELED) {
        Log.v("!", "取消了");
    }
}
```

2. 使用 Bundle 传递数据

除了 startActivityForResult()方法可以回传数据外，还有另外一种两个 Activity 之间可以传递数据的方式，即使用 Bundle 对象。

新建名称为 useBundleTwoActivity 的 Android 项目，将文件 Main.java 的代码更改如下：

```java
public class Main extends Activity {
    @Override
    public void onCreate(Bundle savedInstanceState) {
        super.onCreate(savedInstanceState);
        setContentView(R.layout.main);

        Bundle bundleRef = new Bundle();
        bundleRef.putString("username", "高洪岩");
        bundleRef.putInt("age", 100);

        Intent intentRef = new Intent();
        intentRef.setClass(Main.this, Two.class);
        intentRef.putExtras(bundleRef);

        this.startActivity(intentRef);
    }
}
```

接收数据的文件 Two.java 的代码如下：

```java
public class Two extends Activity {
    @Override
    public void onCreate(Bundle savedInstanceState) {
        super.onCreate(savedInstanceState);
```

```
        setContentView(R.layout.two);

        Bundle bunldeRef = this.getIntent().getExtras();
        String username = "username=" + bunldeRef.getString("username");
        int age = bunldeRef.getInt("age") + 1;

        Toast.makeText(this, username + " " + age, Toast.LENGTH_LONG).show();

    }
}
```

文件 AndroidManifest.xml 的代码如下：

```
<?xml version="1.0" encoding="utf-8"?>
<manifest xmlns:android="http://schemas.android.com/apk/res/android"
    package="useBundleTwoActivity.test.run" android:versionCode="1"
    android:versionName="1.0">

    <application android:icon="@drawable/icon" android:label="@string/app_name">
        <activity android:name=".Main" android:label="@string/app_name">
            <intent-filter>
                <action android:name="android.intent.action.MAIN" />
                <category android:name="android.intent.category.LAUNCHER" />
            </intent-filter>
        </activity>

        <activity android:name=".Two" android:label="@string/app_name">
        </activity>

    </application>
</manifest>
```

程序运行后马上切换界面，并且通过 Bundle 传递数据，如图 4.18 所示。

图 4.18　通过 Bundle 传递数据

4.1.5　category 类型的使用

Category 的作用是 Intent 的附加信息，通过 category 能使 Intent 对象的执行意图更加明确，为了演示 category 类型的使用，还要将隐式调用的技术重新复习一遍。

1. 复习隐式调用

新建一个 Android 项目，名称为 Intent_OtherType，将文件 main.xml 的代码更改如下：

```xml
<?xml version="1.0" encoding="utf-8"?>
<LinearLayout xmlns:android="http://schemas.android.com/apk/res/android"
    android:orientation="vertical" android:layout_width="fill_parent"
    android:layout_height="fill_parent">
    <Button android:text="Button" android:id="@+id/button1"
        android:layout_width="wrap_content" android:layout_height="wrap_content"></Button>
</LinearLayout>
```

文件 Main.java 的代码更改如下：

```java
public class Main extends Activity {
    private Button button1;

    @Override
    public void onCreate(Bundle savedInstanceState) {
        super.onCreate(savedInstanceState);
        setContentView(R.layout.main);

        button1 = (Button) this.findViewById(R.id.button1);

        button1.setOnClickListener(new OnClickListener() {
            public void onClick(View arg0) {

                Intent showMeIntent = new Intent();
                showMeIntent.setAction("showMe_ZZZZZ");
                Main.this.startActivity(showMeIntent);

            }
        });

    }
}
```

程序的意图很明显，即隐式调用 ActionName 动作名称为 showMe_ZZZZZ 的组件。那么，如果有多个组件同时满足 showMe_ZZZZZ 动作名称，该怎么办？

文件 ShowA.java 的代码如下：

```java
public class ShowA extends Activity {
    @Override
    public void onCreate(Bundle savedInstanceState) {
        super.onCreate(savedInstanceState);
        setContentView(R.layout.a);

    }
}
```

文件 ShowA.java 对应的布局 a.xml 代码如下：

```xml
<?xml version="1.0" encoding="utf-8"?>
<LinearLayout xmlns:android="http://schemas.android.com/apk/res/android"
    android:orientation="vertical" android:layout_width="fill_parent"
    android:layout_height="fill_parent">
    <TextView android:layout_width="fill_parent"
        android:layout_height="wrap_content" android:text="AAAAAAAAAAAAAAAA" />
</LinearLayout>
```

文件 ShowB.java 的代码如下：

```java
public class ShowB extends Activity {
    @Override
    public void onCreate(Bundle savedInstanceState) {
        super.onCreate(savedInstanceState);
        setContentView(R.layout.b);

    }
}
```

文件 ShowB.java 对应的布局 b.xml 代码如下：

```xml
<?xml version="1.0" encoding="utf-8"?>
<LinearLayout xmlns:android="http://schemas.android.com/apk/res/android"
    android:orientation="vertical" android:layout_width="fill_parent"
    android:layout_height="fill_parent">
    <TextView android:layout_width="fill_parent"
        android:layout_height="wrap_content" android:text="BBBBBBBBBBBBBBBB" />
</LinearLayout>
```

项目整体配置文件 AndroidManifest.xml 的代码如下：

```xml
<?xml version="1.0" encoding="utf-8"?>
<manifest xmlns:android="http://schemas.android.com/apk/res/android"
    package="Intent_OtherType.test.run" android:versionCode="1"
    android:versionName="1.0">

    <application android:icon="@drawable/icon" android:label="@string/app_name">
        <activity android:name=".Main" android:label="@string/app_name">
            <intent-filter>
                <action android:name="android.intent.action.MAIN" />
                <category android:name="android.intent.category.LAUNCHER" />
            </intent-filter>
        </activity>

        <activity android:name=".ShowA" android:label="ShowA_Label">
            <intent-filter>
                <action android:name="showMe_ZZZZZ" />
```

```xml
                <category android:name="android.intent.category.DEFAULT" />
            </intent-filter>
        </activity>

        <activity android:name=".ShowB" android:label="ShowB_Label">
            <intent-filter>
                <action android:name="showMe_ZZZZZ" />
                <category android:name="android.intent.category.DEFAULT" />
            </intent-filter>
        </activity>

    </application>
</manifest>
```

组件 ShowA 和 ShowB 共同捕获 ActionName 动作名称为 showMe_ZZZZZ，也就是说一旦发现有 Intent 包含 showMe_ZZZZZ 的动作名称，则 ShowA 和 ShowB 一起被激活，那结果是什么呢？运行这个程序就知道了。

程序初始运行效果如图 4.19 所示。

当单击 Button 按钮时，系统弹出选择组件界面，如图 4.20 所示。

图 4.19　初始运行效果　　　　　图 4.20　选择组件界面

在该界面中单击"ShowA_Label"显示界面如图 4.21 所示。

当在该界面中单击"ShowB_Label"时显示界面如图 4.22 所示。

图 4.21　单击 ShowA_Label 的效果　　　图 4.22　单击 ShowB_Label 的效果

通过这个示例不仅复习了 Intent 的隐式调用组件，也知道了有多个组件满足 ActionName 时的后期处理过程，但本节的主角 category 还没有出现！

2. category 使意图更加具体和明确

新建名称为 Intent_more_category 的 Android 项目，主要文件中的代码和前面的代码大体一样，但为了代码的完整性，还是一一列举出来吧！

文件 main.xml 的代码如下：

```xml
<?xml version="1.0" encoding="utf-8"?>
<LinearLayout xmlns:android="http://schemas.android.com/apk/res/android"
    android:orientation="vertical" android:layout_width="fill_parent"
    android:layout_height="fill_parent">
    <Button android:text="Button" android:id="@+id/button1"
        android:layout_width="wrap_content" android:layout_height="wrap_content"></Button>
</LinearLayout>
```

文件 Main.java 的代码如下：

```java
public class Main extends Activity {
    private Button button1;

    @Override
    public void onCreate(Bundle savedInstanceState) {
        super.onCreate(savedInstanceState);
        setContentView(R.layout.main);

        button1 = (Button) this.findViewById(R.id.button1);

        button1.setOnClickListener(new OnClickListener() {
            public void onClick(View arg0) {

                Intent showMeIntent = new Intent();
                showMeIntent.setAction("showMe_ZZZZZ");
                showMeIntent.addCategory("moreData");
                Main.this.startActivity(showMeIntent);

            }
        });

    }
}
```

布局文件 a.xml 的代码如下：

```xml
<?xml version="1.0" encoding="utf-8"?>
<LinearLayout xmlns:android="http://schemas.android.com/apk/res/android"
    android:orientation="vertical" android:layout_width="fill_parent"
    android:layout_height="fill_parent">
    <TextView android:layout_width="fill_parent"
        android:layout_height="wrap_content" android:text="AAAAAAAAAAAAAAAA" />
</LinearLayout>
```

文件 ShowA.java 的代码如下：

```java
public class ShowA extends Activity {
    @Override
    public void onCreate(Bundle savedInstanceState) {
        super.onCreate(savedInstanceState);
        setContentView(R.layout.a);

    }
}
```

布局文件 b.xml 的代码如下：

```xml
<?xml version="1.0" encoding="utf-8"?>
<LinearLayout xmlns:android="http://schemas.android.com/apk/res/android"
    android:orientation="vertical" android:layout_width="fill_parent"
    android:layout_height="fill_parent">
    <TextView android:layout_width="fill_parent"
        android:layout_height="wrap_content" android:text="BBBBBBBBBBBBBBBBB" />
</LinearLayout>
```

文件 ShowB.java 的代码如下：

```java
public class ShowB extends Activity {
    @Override
    public void onCreate(Bundle savedInstanceState) {
        super.onCreate(savedInstanceState);
        setContentView(R.layout.b);

    }
}
```

项目配置文件 AndroidManifest.xml 的代码如下：

```xml
<?xml version="1.0" encoding="utf-8"?>
<manifest xmlns:android="http://schemas.android.com/apk/res/android"
    package="Intent_more_category.test.run" android:versionCode="1"
    android:versionName="1.0">

    <application android:icon="@drawable/icon" android:label="@string/app_name">
        <activity android:name=".Main" android:label="@string/app_name">
            <intent-filter>
                <action android:name="android.intent.action.MAIN" />
                <category android:name="android.intent.category.LAUNCHER" />
            </intent-filter>
        </activity>

        <activity android:name=".ShowA" android:label="ShowA_Label">
            <intent-filter>
                <action android:name="showMe_ZZZZZ" />
                <category android:name="android.intent.category.DEFAULT" />
```

```xml
        </intent-filter>
    </activity>


    <activity android:name=".ShowB" android:label="ShowB_Label">
        <intent-filter>
            <action android:name="showMe_ZZZZZ" />
            <category android:name="android.intent.category.DEFAULT" />
            <category android:name="moreData" />
        </intent-filter>
    </activity>


</application>
</manifest>
```

程序初始运行效果如图 4.23 所示。

当单击 Button 按钮时并没有弹出一个选择组件的对话框，而是直接进入界面，如图 4.24 所示。

图 4.23 初始运行界面　　　　　　　图 4.24 直接进入 B 界面

通过这个示例可以发现，加入 category 使 Intent 的意图更加明确和具体。另外，当 Intent 请求中所有的 category 与组件中某一个 IntentFilter 的<category>标签完全匹配时，才会让 Intent 通过测试，而 IntentFilter 中其他多余的<category>声明并不会导致匹配失败。还可以这样去理解这个匹配原理：Intent 对象中的全部 Category 一定要在 IntentFilter 存在，但是允许 IntentFilter 中的 Category 比 Intent 对象中的 Category 多。

4.1.6 data 标签的使用

标签<data>主要的作用就是用来匹配 Intent 中的 Uri。

新建名称为 test_data1 的 Android 项目，核心 Activity 文件 Main.java 的代码如下：

```java
public class Main extends Activity {
    private Button button1;
    private Button button2;
    private Button button3;

    @Override
    public void onCreate(Bundle savedInstanceState) {
        super.onCreate(savedInstanceState);
        setContentView(R.layout.main);

        button1 = (Button) this.findViewById(R.id.button1);
        button1.setOnClickListener(new OnClickListener() {
            public void onClick(View arg0) {
```

```java
                    Intent intent = new Intent("gotoSecond", Uri
                            .parse("http://www.gaohongyan.com/lookupFile"));
                    Main.this.startActivity(intent);
                }
            });

            // 属性 android:path 用来匹配完整路径
            // 属性 android:pathPrefix="/bbs"匹配 path 中有/bbs 开头的即为匹配成功

            // 在这里需要注意的是：
            // 属性 android:pathPrefix 和 android:path 和 android:pathPattern
            // 有一个匹配成功则结果就是成功，也就是 or 关系！

            button2 = (Button) this.findViewById(R.id.button2);
            button2.setOnClickListener(new OnClickListener() {
                public void onClick(View arg0) {
                    Intent intent = new Intent("gotoThird", Uri
                            .parse("http://www.gaohongyan.com/bbs/userinfo.html"));
                    Main.this.startActivity(intent);
                }
            });

            button3 = (Button) this.findViewById(R.id.button3);
            button3.setOnClickListener(new OnClickListener() {
                public void onClick(View arg0) {
                    // Intent intent = new Intent("gotoFour", Uri
                    // .parse("http://www.gaohongyan.com/gaohongyan123.mp3"));
                    Intent intent = new Intent(
                            "gotoFour",
                            Uri
                                    .parse("http://www.gaohongyan.com/music/gaohongyan123.mp3"));
                    Main.this.startActivity(intent);
                }
            });

        }
    }
```

文件 AndroidManifest.xml 的代码如下：

```xml
<?xml version="1.0" encoding="utf-8"?>
<manifest xmlns:android="http://schemas.android.com/apk/res/android"
    package="test_data1.test.run" android:versionCode="1"
    android:versionName="1.0">
    <application android:icon="@drawable/icon" android:label="@string/app_name">
        <activity android:name=".Main" android:label="@string/app_name">
            <intent-filter>
                <action android:name="android.intent.action.MAIN" />
                <category android:name="android.intent.category.LAUNCHER" />
            </intent-filter>
```

```xml
    </activity>
    <activity android:name=".Test1" android:label="@string/app_name">
        <intent-filter>
            <action android:name="gotoSecond" />
            <category android:name="android.intent.category.DEFAULT" />
            <data android:scheme="http" android:host="www.gaohongyan.com"
                  android:path="/lookupFile"></data>
        </intent-filter>
    </activity>
    <activity android:name=".Test2" android:label="@string/app_name">
        <intent-filter>
            <action android:name="gotoThird" />
            <category android:name="android.intent.category.DEFAULT" />
            <data android:scheme="http" android:host="www.gaohongyan.com"
                  android:pathPrefix="/bbs"></data>
        </intent-filter>
    </activity>
    <activity android:name=".Test3" android:label="@string/app_name">
        <intent-filter>
            <action android:name="gotoFour" />
            <category android:name="android.intent.category.DEFAULT" />
            <data android:scheme="http" android:host="www.gaohongyan.com"
                  android:pathPattern=".*\\.mp3"></data>
        </intent-filter>
    </activity>
</application>
</manifest>
```

程序运行后单击 button1、button2 和 button3，出现界面如图 4.25 所示。

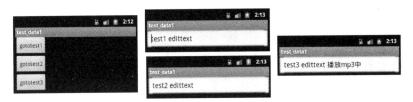

图 4.25　匹配 data 的界面

4.2　创建 Dialog 式的 Activity 登录实例

在开发项目时，登录界面有时候是通过对话框的形式来进行呈现，在对话框中输入数据然后再进入验证的阶段，本示例还是练习 Intent 对象和 startActivityForResult()方法的使用。

新建名称为 loginDialog2 的 Android 项目，将布局文件 main.xml 的代码更改如下：

```xml
<?xml version="1.0" encoding="utf-8"?>
```

```xml
<LinearLayout xmlns:android="http://schemas.android.com/apk/res/android"
    android:orientation="vertical" android:layout_width="fill_parent"
    android:layout_height="fill_parent">
    <Button android:text="登录" android:id="@+id/main_button1"
        android:layout_width="wrap_content" android:layout_height="wrap_content"></Button>
</LinearLayout>
```

文件 Main.java 的代码更改如下：

```java
public class Main extends Activity {
    private Button button1;

    @Override
    protected void onActivityResult(int requestCode, int resultCode, Intent data) {
        super.onActivityResult(requestCode, resultCode, data);
        if (requestCode == 100) {
            if (resultCode == Activity.RESULT_OK) {
                String username = data.getStringExtra("username");
                String password = data.getStringExtra("password");
                Log.v("登录的结果为：", "username=" + username + " password="
                        + password);
            }

        }
    }

    @Override
    public void onCreate(Bundle savedInstanceState) {
        super.onCreate(savedInstanceState);
        setContentView(R.layout.main);

        button1 = (Button) this.findViewById(R.id.main_button1);
        button1.setOnClickListener(new OnClickListener() {
            public void onClick(View arg0) {
                Intent loginIntent = new Intent();
                loginIntent.setClass(Main.this, LoginDialog.class);
                Main.this.startActivityForResult(loginIntent, 100);
            }
        });
    }
}
```

还要创建对话框样式的 Activity 对象和布局文件。对话框样式的 Activity 文件 LoginDialog.java 的代码如下：

```java
public class LoginDialog extends Activity {
    private Button logindialog_loginButton;
    private EditText logindialog_usernameEditText;
    private EditText logindialog_passwordEditText;

    @Override
```

```java
public void onCreate(Bundle savedInstanceState) {
    super.onCreate(savedInstanceState);
    setContentView(R.layout.logindialog);
    logindialog_loginButton = (Button) this
            .findViewById(R.id.logindialog_loginButton);

    logindialog_usernameEditText = (EditText) this
            .findViewById(R.id.logindialog_usernameEditText);

    logindialog_passwordEditText = (EditText) this
            .findViewById(R.id.logindialog_passwordEditText);

    logindialog_loginButton.setOnClickListener(new OnClickListener() {
        public void onClick(View arg0) {
            Intent returnIntent = new Intent();
            returnIntent.putExtra("username", logindialog_usernameEditText
                    .getText().toString());
            returnIntent.putExtra("password", logindialog_passwordEditText
                    .getText().toString());

            LoginDialog.this.setResult(Activity.RESULT_OK, returnIntent);
            LoginDialog.this.finish();
        }
    });
}
```

对话框界面的布局文件 logindialog.xml 的代码如下：

```xml
<?xml version="1.0" encoding="utf-8"?>
<LinearLayout xmlns:android="http://schemas.android.com/apk/res/android"
    android:orientation="vertical" android:layout_width="300dip"
    android:layout_height="fill_parent" android:layout_marginLeft="20dip"
    android:layout_marginRight="20dip">
    <TextView android:text="账号：" android:id="@+id/logindialog_textView1"
        android:layout_width="fill_parent" android:layout_height="wrap_content"
        android:textSize="15dip" android:layout_marginLeft="20dip"
        android:layout_marginRight="20dip"></TextView>
    <EditText android:layout_height="wrap_content"
        android:id="@+id/logindialog_usernameEditText" android:text="a"
        android:layout_width="match_parent" android:textSize="15dip"
        android:layout_marginLeft="20dip" android:layout_marginRight="20dip"></EditText>
    <TextView android:text="密码：" android:id="@+id/logindialog_textView1"
        android:layout_width="fill_parent" android:layout_height="wrap_content"
        android:textSize="15dip" android:layout_marginLeft="20dip"
        android:layout_marginRight="20dip"></TextView>
    <EditText android:layout_height="wrap_content"
        android:id="@+id/logindialog_passwordEditText" android:text=""
        android:layout_width="match_parent" android:textSize="15dip"
        android:layout_marginLeft="20dip" android:layout_marginRight="20dip"></EditText>
```

```xml
<LinearLayout android:layout_height="wrap_content"
    android:layout_width="fill_parent" android:orientation="horizontal"
    android:id="@+id/linearLayout1" android:gravity="center">
    <Button android:layout_height="wrap_content" android:id="@+id/logindialog_loginButton"
        android:layout_width="100dip" android:text="登录"></Button>
    <Button android:layout_height="wrap_content" android:id="@+id/logindialog_exitAppButton"
        android:layout_width="100dip" android:text="退出"></Button>
</LinearLayout>
</LinearLayout>
```

系统配置文件 AndroidManifest.xml 的代码如下：

```xml
<?xml version="1.0" encoding="utf-8"?>
<manifest xmlns:android="http://schemas.android.com/apk/res/android"
    package="loginDialog2.test.run" android:versionCode="1"
    android:versionName="1.0">
    <application android:icon="@drawable/icon" android:label="@string/app_name">
        <activity android:name=".Main" android:label="@string/app_name">
            <intent-filter>
                <action android:name="android.intent.action.MAIN" />
                <category android:name="android.intent.category.LAUNCHER" />
            </intent-filter>
        </activity>
        <activity android:name=".LoginDialog" android:label="@string/app_name"
            android:theme="@android:style/Theme.Dialog">
        </activity>
    </application>
</manifest>
```

需要注意的是关键代码：

```xml
<activity android:name=".LoginDialog" android:label="@string/app_name"
    android:theme="@android:style/Theme.Dialog">
</activity>
```

上述代码的功能是设置 LoginDialog 的外观样式为对话框。程序初始运行效果如图 4.26 所示。当单击"登录"按钮弹出对话框效果的界面，如图 4.27 所示。

图 4.26 初始运行效果

图 4.27 弹出对话框外观的布局

在对话框中输入字符 a，然后单击"登录"按钮出现效果如图 4.28 所示。

图 4.28 得到字符 a

提示
本示例仅仅是为了演示回调方法 onActivityResult 的使用，真实的项目中并不使用回调 onActivityResult 来实现登录功能。

4.3 显式启动其他应用程序的 Activity

下面实现两个应用程序之间的互相调用。

新建一个名称为 application2 的 Android 项目，项目不需要特殊的设置，布局文件 main.xml 的代码如下：

```xml
<?xml version="1.0" encoding="utf-8"?>
<LinearLayout xmlns:android="http://schemas.android.com/apk/res/android"
    android:orientation="vertical" android:layout_width="fill_parent"
    android:layout_height="fill_parent">
    <TextView android:layout_width="fill_parent"
        android:layout_height="wrap_content" android:text="this is application2" />
</LinearLayout>
```

再新建一个名称为 application1 的 Android 项目，在 main.xml 布局文件中加入一个按钮，代码如下：

```xml
<?xml version="1.0" encoding="utf-8"?>
<LinearLayout xmlns:android="http://schemas.android.com/apk/res/android"
    android:orientation="vertical" android:layout_width="fill_parent"
    android:layout_height="fill_parent">
    <Button android:text="Button" android:id="@+id/button1"
        android:layout_width="wrap_content" android:layout_height="wrap_content"></Button>
</LinearLayout>
```

在名称为 application1 的 Android 项目中创建 Activity 对象文件 Main.java 的代码如下：

```java
public class Main extends Activity {
    private Button button1;

    @Override
    public void onCreate(Bundle savedInstanceState) {
        super.onCreate(savedInstanceState);
        setContentView(R.layout.main);

        button1 = (Button) this.findViewById(R.id.button1);

        button1.setOnClickListener(new OnClickListener() {
            public void onClick(View arg0) {
                Intent newIntent = new Intent();
```

```
            newIntent.setClassName("application2.test.run",
                    "application2.test.run.Main");
            Main.this.startActivity(newIntent);
        }
    });
    
    }
}
```

其中代码：

```
newIntent.setClassName("application2.test.run",
        "application2.test.run.Main");
```

上述代码中第一个参数是目标对象所在的包名，而第 2 个参数是目标对象所在类的完整类路径。
首先运行 application2 项目，目的是将 application2 项目安装到 AVD 设备中。
再运行 application1 项目，运行后的 application1 项目界面如图 4.29 所示。
单击 application1 项目界面中的按钮 Button 后弹出 application2 的界面效果如图 4.30 所示。

图 4.29　application1 项目界面　　　图 4.30　application2 被调用出来

4.4　发送文本短信的简单示例

新建名称为 SimpleTextSMS 的 Android 项目，更改 main.xml 的代码如下：

```xml
<?xml version="1.0" encoding="utf-8"?>
<LinearLayout xmlns:android="http://schemas.android.com/apk/res/android"
    android:orientation="vertical" android:layout_width="fill_parent"
    android:layout_height="fill_parent">
    <TextView android:text="对方手机号：" android:id="@+id/textView1"
        android:layout_width="wrap_content" android:layout_height="wrap_content"></TextView>
    <EditText android:id="@+id/editText1" android:text="13888888888"
        android:layout_height="wrap_content" android:layout_width="match_parent"></EditText>
    <TextView android:text="文本短信正文：" android:id="@+id/textView2"
        android:layout_width="wrap_content" android:layout_height="wrap_content"></TextView>
    <EditText android:id="@+id/editText2" android:text="你好，全同学！"
        android:layout_height="wrap_content" android:layout_width="match_parent"></EditText>
    <Button android:text="Button" android:id="@+id/button1"
        android:layout_width="wrap_content" android:layout_height="wrap_content"></Button>
</LinearLayout>
```

文件 Main.java 的代码如下：

```java
public class Main extends Activity {
    private EditText editText1;
    private EditText editText2;
    private Button button1;
    private String sendPhoneNum = "";
    private String sendText = "";

    @Override
    public void onCreate(Bundle savedInstanceState) {
        super.onCreate(savedInstanceState);
        setContentView(R.layout.main);

        editText1 = (EditText) this.findViewById(R.id.editText1);
        editText2 = (EditText) this.findViewById(R.id.editText2);
        button1 = (Button) this.findViewById(R.id.button1);

        sendPhoneNum = editText1.getText().toString();
        sendText = editText2.getText().toString();
        button1.setOnClickListener(new OnClickListener() {
            public void onClick(View arg0) {
                Main.this.sendText = Main.this.sendText
                        + "--由高洪岩Android手机客户端发送！1949-10-1";
                PendingIntent pi = PendingIntent.getActivity(Main.this, 0,
                        new Intent(Main.this, Main.class), 0);
                SmsManager sms = SmsManager.getDefault();
                Log.v("!", "" + Main.this.sendPhoneNum + " "
                        + Main.this.sendText);
                sms.sendTextMessage("tel:" + Main.this.sendPhoneNum, null,
                        Main.this.sendText, pi, null);

            }
        });
    }
}
```

文件AndroidManifest.xml的配置如下：

```xml
<?xml version="1.0" encoding="utf-8"?>
<manifest xmlns:android="http://schemas.android.com/apk/res/android"
    package="SimpleTextSMS.test.run" android:versionCode="1"
    android:versionName="1.0">
    <application android:icon="@drawable/icon" android:label="@string/app_name">
        <activity android:name=".Main" android:label="@string/app_name">
            <intent-filter>
                <action android:name="android.intent.action.MAIN" />
                <category android:name="android.intent.category.LAUNCHER" />
            </intent-filter>
        </activity>
    </application>
    <uses-permission android:name="android.permission.SEND_SMS"></uses-permission>
</manifest>
```

将代码布署到真机上,正确发送了短信到目的手机号中。

4.5 Notification 通知的使用

通知在 Android 系统中的使用率非常高,效果就是在 Android 系统主界面上方的状态条中显示出相关的信息提示。

4.5.1 Notification 通知的初入

新建名称为 notification_test1 的 Android 项目,布局文件 main.xml 的代码如下:

```xml
<?xml version="1.0" encoding="utf-8"?>
<LinearLayout xmlns:android="http://schemas.android.com/apk/res/android"
    android:orientation="vertical" android:layout_width="fill_parent"
    android:layout_height="fill_parent">
    <Button android:text="启动通知" android:id="@+id/button1"
        android:layout_width="wrap_content" android:layout_height="wrap_content"></Button>
</LinearLayout>
```

布局文件 second.xml 的代码如下:

```xml
<?xml version="1.0" encoding="utf-8"?>
<LinearLayout xmlns:android="http://schemas.android.com/apk/res/android"
    android:orientation="vertical" android:layout_width="fill_parent"
    android:layout_height="fill_parent">
    <TextView android:layout_width="fill_parent"
        android:layout_height="wrap_content" android:text="我是 second" />
</LinearLayout>
```

文件 Main.java 的代码如下:

```java
public class Main extends Activity {
    private Button button1;
    private int count = 0;

    @Override
    public void onCreate(Bundle savedInstanceState) {
        super.onCreate(savedInstanceState);
        setContentView(R.layout.main);
        button1 = (Button) this.findViewById(R.id.button1);
        button1.setOnClickListener(new OnClickListener() {
            public void onClick(View arg0) {
                count++;
                // 取得 Notification 的通知管理服务对象 NotificationManager
                NotificationManager nm = (NotificationManager) Main.this
                        .getSystemService(Context.NOTIFICATION_SERVICE);
                // 设置在 android 最上方状态的图片和文本及显示时间
                Notification n = new Notification(R.drawable.icon, "我是通知的小提示",
```

```
                    System.currentTimeMillis());
                // 创建 Intent 的目的是当展开通知看到详细内容时
                // 单击其中的界面去往哪个 Activity 对象
                Intent intent = new Intent(Main.this, Second.class);
                // 往 Intent 对象中存放数据
                intent.putExtra("username", "username" + count);
                intent.putExtra("password", "password" + count);
                // 用 PendingIntent 对象关联 Intent 对象
                // 第 2 个参数是请求的代码
                // 第 3 个参数的功能是更新以前使用的 Intent 对象中的数据
                PendingIntent contentIntent = PendingIntent.getActivity(
                        Main.this, 100, intent,
                        PendingIntent.FLAG_UPDATE_CURRENT);
                // 设置通知的具体标题和内容
                n.setLatestEventInfo(Main.this, "我是标题", "我是正文", contentIntent);
                // 发送通知
                nm.notify(1, n);
            }
        });
    }
}
```

别忘在 AndroidManifest.xml 文件中配置 Second.java 对象，还要创建 Second.java 文件，代码如下：

```
public class Second extends Activity {
    @Override
    public void onCreate(Bundle savedInstanceState) {
        super.onCreate(savedInstanceState);
        setContentView(R.layout.second);
        Log.v("!username=", "" + this.getIntent().getStringExtra("username"));
        Log.v("!password=", "" + this.getIntent().getStringExtra("password"));
    }
}
```

项目运行后的初始界面如图 4.31 所示。

单击"启动通知"按钮在状态条出现 Notification 通知小提示，如图 4.32 所示。

图 4.31　初始运行效果

图 4.32　出现通知小提示文本和图标

拖曳通知进行展开，如图 4.33 所示。

单击通知信息界面的内部转到 Second.java 对象中，如图 4.34 所示。

图 4.33　展开的通知信息

图 4.34　到达 Second.java 界面

并在 LogCat 中打印出 Intent 中的数据，如图 4.35 所示。

```
V  431   !username=                    username1
V  431   !password=                    password1
```

图 4.35　打印 Intent 中的数据

4.5.2　自动隐藏状态条的图标

上面实验中单击了通知并转到 Second.java 对象后，状态条还显示刚才的通知图标，如何实现单击通知后，状态条中的图标自动消失呢？很简单，更改代码如下：

```
Notification n = new Notification(R.drawable.icon, "我是通知的小提示",
       System.currentTimeMillis());
   n.flags = Notification.FLAG_AUTO_CANCEL;
```

也就是对 Notification 对象设置 1 个自动取消的 flag 标记，运行项目，当再单击通知中的内容后转到 Second.java 时，状态条中的图标自动消失了，如图 4.36 所示。

图 4.36　自动消失的图标

4.5.3　每个通知对象拥有自己的 Intent 对象

在 4.5.1 和 4.5.2 节中使用了如下代码：

```
nm.notify(1, n);
```

第 1 个参数是标识这个通知的唯一 id，如果每次执行的 id 不同，则创建出新的通知对象。新建名称为 notification_test2 的 Android 项目，Main.java 的代码如下：

```java
public class Main extends Activity {
    private Button button1;
    private int count = 0;

    @Override
    public void onCreate(Bundle savedInstanceState) {
        super.onCreate(savedInstanceState);
        setContentView(R.layout.main);
        button1 = (Button) this.findViewById(R.id.button1);
        button1.setOnClickListener(new OnClickListener() {
            public void onClick(View arg0) {
```

```
                count++;
                NotificationManager nm = (NotificationManager) Main.this
                        .getSystemService(Context.NOTIFICATION_SERVICE);
                Notification n = new Notification(R.drawable.icon, "我是通知的小提示",
                        System.currentTimeMillis());
                n.flags = Notification.FLAG_AUTO_CANCEL;
                Intent intent = new Intent(Main.this, Second.class);
                intent.putExtra("username", "username" + intent);
                PendingIntent contentIntent = PendingIntent.getActivity(
                        Main.this, 100, intent,
                        PendingIntent.FLAG_UPDATE_CURRENT);
                n.setLatestEventInfo(Main.this, "我是标题", "我是正文", contentIntent);
                nm.notify(count, n);
            }
        });

    }
}
```

布局文件 main.xml 的代码如下：

```
<?xml version="1.0" encoding="utf-8"?>
<LinearLayout xmlns:android="http://schemas.android.com/apk/res/android"
    android:orientation="vertical" android:layout_width="fill_parent"
    android:layout_height="fill_parent">
    <Button android:text="启动通知1" android:id="@+id/button1"
        android:layout_width="wrap_content" android:layout_height="wrap_content"></Button>
</LinearLayout>
```

文件 Second.java 的代码如下：

```
public class Second extends Activity {
    @Override
    public void onCreate(Bundle savedInstanceState) {
        super.onCreate(savedInstanceState);
        setContentView(R.layout.second);
        Log.v("!username=", "" + this.getIntent().getStringExtra("username"));
    }
}
```

项目运行后单击 3 次 "启动通知 1" 按钮，出现效果如图 4.37 所示。

上面的实验仅仅是视觉上有 3 个通知对象的效果，但所有通知关联的 Intent 对象却是最后 1 个通知所产生的，这很明显不符合每个通知拥有自己 Intent 对象的情况，单击所有的通知内容，验证是 1 个 Intent 对象，在 LogCat 中打印的日志内容如图 4.38 所示。

图 4.37　出现 3 个不同的通知对象　　　　图 4.38　不同的通知对象拥有相同的 Intent 对象

如何解决呢？很简单，更改代码如下：

```
PendingIntent contentIntent = PendingIntent.getActivity(
        Main.this, count, intent,
        PendingIntent.FLAG_UPDATE_CURRENT);
```

对 getActivity()对象的 requestCode 请求码设置唯一就解决这个问题了，再次运行项目，出现的日志内容如图 4.39 所示。

图 4.39　每个通知拥有不同的 Intent 对象

4.5.4　设置状态栏中通知的数量显示

相同唯一标识的通知对象数量的显示是通过 Notification 对象的 number 属性来进行处理的，使用起来比较简单，将 Main.java 的代码更改如下：

```
button1.setOnClickListener(new OnClickListener() {
    public void onClick(View arg0) {
        count++;
        NotificationManager nm = (NotificationManager) Main.this
                .getSystemService(Context.NOTIFICATION_SERVICE);
        Notification n = new Notification(R.drawable.icon, "我是通知的小提示",
                System.currentTimeMillis());
        n.flags = Notification.FLAG_AUTO_CANCEL;
        Intent intent = new Intent(Main.this, Second.class);
        intent.putExtra("username", "username" + count);
        PendingIntent contentIntent = PendingIntent.getActivity(
                Main.this, count, intent,
                PendingIntent.FLAG_UPDATE_CURRENT);
        n.setLatestEventInfo(Main.this, "我是标题", "我是正文", contentIntent);
        n.number = count;
        nm.notify(1, n);
    }
```

});

程序运行后，单击 3 次"启动通知 1"按钮，在状态栏显示出通知数量为 3，如图 4.40 所示。

图 4.40　显示通知数量

4.5.5　取消通知

更改 main.xml 的代码如下：

```xml
<?xml version="1.0" encoding="utf-8"?>
<LinearLayout xmlns:android="http://schemas.android.com/apk/res/android"
    android:orientation="vertical" android:layout_width="fill_parent"
    android:layout_height="fill_parent">
    <Button android:text="启动通知1" android:id="@+id/button1"
        android:layout_width="wrap_content" android:layout_height="wrap_content"></Button>
    <Button android:text="取消通知1" android:id="@+id/button2"
        android:layout_width="wrap_content" android:layout_height="wrap_content"></Button>
</LinearLayout>
```

更改 Main.java 的代码如下：

```java
public class Main extends Activity {
    private Button button1;
    private Button button2;
    private int count = 0;
    private NotificationManager nm;

    @Override
    public void onCreate(Bundle savedInstanceState) {
        super.onCreate(savedInstanceState);
        setContentView(R.layout.main);
        button1 = (Button) this.findViewById(R.id.button1);
        button2 = (Button) this.findViewById(R.id.button2);

        button1.setOnClickListener(new OnClickListener() {
            public void onClick(View arg0) {
                count++;
                nm = (NotificationManager) Main.this
                        .getSystemService(Context.NOTIFICATION_SERVICE);
                Notification n = new Notification(R.drawable.icon, "我是通知的小提示",
                        System.currentTimeMillis());
                n.flags = Notification.FLAG_AUTO_CANCEL;
                Intent intent = new Intent(Main.this, Second.class);
                intent.putExtra("username", "username" + count);
                PendingIntent contentIntent = PendingIntent.getActivity(
                        Main.this, count, intent,
                        PendingIntent.FLAG_UPDATE_CURRENT);
```

```
                    n.setLatestEventInfo(Main.this, "我是标题", "我是正文", contentIntent);
                    n.number = count;
                    nm.notify(9999, n);
                }
            });

            button2.setOnClickListener(new OnClickListener() {
                public void onClick(View arg0) {
                    nm.cancel(9999);
                }
            });

        }
}
```

程序运行后创建两个通知，如图 4.41 所示。

单击"取消通知 1"按钮，取消通知后的界面如图 4.42 所示。

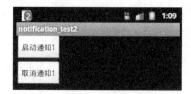   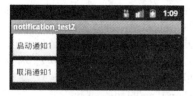

　　图 4.41　创建 1 个通知　　　　　　　　图 4.42　取消了通知

4.5.6　设置振动模式和发出提示音和 LED 灯

为了让用户能对通知的到达有比较明显的察觉，通知对象还可以使用振动的功能，新建名称为 moreTestTest 的 Android 项目，设置 Main.java 的代码如下：

```java
public class Main extends Activity {
    private Button button1;
    private Button button2;
    private int count = 0;
    private NotificationManager nm;

    @Override
    public void onCreate(Bundle savedInstanceState) {
        super.onCreate(savedInstanceState);
        setContentView(R.layout.main);
        button1 = (Button) this.findViewById(R.id.button1);
        button2 = (Button) this.findViewById(R.id.button2);

        button1.setOnClickListener(new OnClickListener() {
            public void onClick(View arg0) {
                count++;
                nm = (NotificationManager) Main.this
                        .getSystemService(Context.NOTIFICATION_SERVICE);

                Notification n = new Notification(R.drawable.icon, "我是通知的小提示",
```

```
                    System.currentTimeMillis());
            n.flags |= n.FLAG_AUTO_CANCEL;
            n.vibrate = new long[] { 0, 5000, 1000, 5000, 1000, 5000 };
            // 0, 5000, 1000, 5000, 1000, 5000
            // 分别代表不休息，振动 5 秒，休息 1 秒，振动 5 秒，休息 1 秒，振动 5 秒

            Intent intent = new Intent(Main.this, Second.class);
            intent.putExtra("username", "username" + count);
            PendingIntent contentIntent = PendingIntent.getActivity(
                    Main.this, count, intent, 0);
            n.setLatestEventInfo(Main.this, "我是标题", "我是正文", contentIntent);
            n.number = count;
            nm.notify(count, n);

        }
    });

    button2.setOnClickListener(new OnClickListener() {
        public void onClick(View arg0) {
            count++;
            nm = (NotificationManager) Main.this
                    .getSystemService(Context.NOTIFICATION_SERVICE);
            Notification n = new Notification(R.drawable.icon, "我是通知的小提示",
                    System.currentTimeMillis());
            n.ledARGB = Color.GREEN;// LED 绿色灯闪烁
            n.defaults = n.DEFAULT_LIGHTS;// 设置默认行为
            n.flags |= n.FLAG_AUTO_CANCEL;
            n.flags |= n.FLAG_SHOW_LIGHTS;

            n.sound = Uri.parse("android.resource://"
                    + Main.this.getPackageName() + "/" + R.raw.aqmm);

            Intent intent = new Intent(Main.this, Second.class);
            intent.putExtra("username", "username" + count);
            PendingIntent contentIntent = PendingIntent.getActivity(
                    Main.this, count, intent,
                    PendingIntent.FLAG_UPDATE_CURRENT);
            n.setLatestEventInfo(Main.this, "我是标题", "我是正文", contentIntent);
            nm.notify(count, n);
        }
    });

  }
}
```

在<manifest>标签中加入下述振动权限代码即可：

```
<uses-permission android:name="android.permission.VIBRATE"></uses-permission>
```

4.5.7 自定义通知布局内容

创建通知自定义布局文件 mynotificationview.xml 的代码如下：

```xml
<?xml version="1.0" encoding="utf-8"?>
<LinearLayout xmlns:android="http://schemas.android.com/apk/res/android"
    android:orientation="horizontal" android:layout_width="fill_parent"
    android:layout_height="fill_parent">
    <ImageView android:layout_width="wrap_content" android:src="@drawable/a"
        android:layout_height="wrap_content"
android:id="@+id/imageView1"></ImageView>
    <ImageView android:layout_width="wrap_content" android:src="@drawable/b"
        android:layout_height="wrap_content"
android:id="@+id/imageView2"></ImageView>
    <ImageView android:layout_width="wrap_content" android:src="@drawable/c"
        android:layout_height="wrap_content"
android:id="@+id/imageView3"></ImageView>
</LinearLayout>
```

更改 Main.java 的代码如下：

```java
public class Main extends Activity {
    private Button button1;
    private int count = 0;
    private NotificationManager nm;

    @Override
    public void onCreate(Bundle savedInstanceState) {
        super.onCreate(savedInstanceState);
        setContentView(R.layout.main);
        button1 = (Button) this.findViewById(R.id.button1);

        button1.setOnClickListener(new OnClickListener() {
            public void onClick(View arg0) {
                count++;
                nm = (NotificationManager) Main.this
                        .getSystemService(Context.NOTIFICATION_SERVICE);

                Notification n = new Notification(R.drawable.icon, "我是通知的小提示",
                        System.currentTimeMillis());
                n.flags = n.flags | n.FLAG_ONGOING_EVENT;// 一直显示在通知栏中

                RemoteViews remoteViews = new RemoteViews(Main.this
                        .getPackageName(), R.layout.mynotificationview);
                n.contentView = remoteViews;

                Intent intent = new Intent(Main.this, Second.class);
                intent.putExtra("username", "username" + count);
                PendingIntent contentIntent = PendingIntent.getActivity(
                        Main.this, count, intent, 0);
                n.contentIntent = contentIntent;
```

```
                    nm.notify(count, n);
                }
            });
        }
    }
```

程序运行后的效果如图 4.43 所示。

图 4.43　一直显示在通知栏中的自定义布局

4.5.8　Notification.FLAG_INSISTENT 和 Notification.FLAG_ONGOING_EVENT 的使用

在 Notification 对象中还有 2 个常量比较重要，它们是 Notification.FLAG_INSISTENT 和 Notification.FLAG_ONGOING_EVENT 的使用，使用 Notification.FLAG_INSISTENT 常量的含义是如果没有人为的干预（比如下滑通知栏查看通知）则通知的某些特效（比如振动）就一直持续下去，而使用 Notification.FLAG_ONGOING_EVENT 常量的含义是通知一直在运行的状态，也就是在通知栏中呈现出"正在进行的"的状态。

新建名称为 insistent_notification 的 Android 项目，Main.java 的核心代码如下：

```
public class Main extends Activity {
    private Button button1;
    private Button button2;

    @Override
    public void onCreate(Bundle savedInstanceState) {
        super.onCreate(savedInstanceState);
        setContentView(R.layout.main);

        button1 = (Button) this.findViewById(R.id.button1);
        button2 = (Button) this.findViewById(R.id.button2);

        button1.setOnClickListener(new OnClickListener() {
            public void onClick(View arg0) {
                NotificationManager nm = (NotificationManager) Main.this
                        .getSystemService(Context.NOTIFICATION_SERVICE);

                Notification n = new Notification(R.drawable.icon, "我是通知的小提示",
                        System.currentTimeMillis());
                n.flags |= n.FLAG_AUTO_CANCEL;
                n.flags |= n.FLAG_INSISTENT;
```

```
                    n.vibrate = new long[] { 0, 2000, 1000 };
                    Intent intent = new Intent(Main.this, Second.class);
                    PendingIntent contentIntent = PendingIntent.getActivity(
                            Main.this, 1, intent, 0);
                    n.setLatestEventInfo(Main.this, "我是标题", "我是正文", contentIntent);
                    nm.notify(100, n);
                }
            });

            button2.setOnClickListener(new OnClickListener() {
                public void onClick(View arg0) {
                    NotificationManager nm = (NotificationManager) Main.this
                            .getSystemService(Context.NOTIFICATION_SERVICE);

                    Notification n = new Notification(R.drawable.icon, "我是通知的小提示",
                            System.currentTimeMillis());
                    n.flags |= n.FLAG_ONGOING_EVENT;
                    n.vibrate = new long[] { 0, 2000, 1000 };
                    Intent intent = new Intent(Main.this, Second.class);
                    PendingIntent contentIntent = PendingIntent.getActivity(
                            Main.this, 2, intent, 0);
                    n.setLatestEventInfo(Main.this, "我是标题", "我是正文", contentIntent);
                    nm.notify(200, n);
                }
            });

        }
    }
```

上面的示例运行的效果是，单击 Button1 则振动一直在进行，单击 Button2 时只振动一次，而通知则在通知栏中呈现"正在进行的"状态。

4.6 Activity 的 4 种启动方式

在 Android 的系统中有一个 Activity 栈，在这个栈中用后进先出 LIFO 的原则来弹出 Activity 对象，这个 Activity 栈也叫做 Task，这个 Task 就是为了完成某个工作的一系列 Activity 的集合，而这些 Activity 又被组织成了堆栈的形式。当一个 Activity 启动时，就会把它压入该 Task 的堆栈，而当用户在该 Activity 中按返回键，或者代码中 finish 掉时，就会将它从该 Task 的堆栈中弹出。

在本书第 1 章已经和大家介绍了 Activity 的生命周期，主要执行的有 onCreate()和 onStart()及 onResume()等，但这些回调函数在具有多个 Activity 项目开发中一定会频繁地被调用，因为 Activity 一直在创建和销毁，主要就是由于 Activity 跳转造成的，还有可能存在调用其他程序可复用的 Activity 对象，这时就有可能会希望跳转到原来某个 Activity 实例，而不是产生大量重复的 Activity。这时，就需要为 Activity 配置特定的加载模式，而不是使用默认的加载模式。

Activity 加载模式有以下 4 种：

- standard 模式。
- singleTop 模式。
- singleTask 模式。
- singleInstance 模式。

默认的加载模式为 standard 标准模式。下面分别进行详细介绍。

4.6.1 standard 模式

设置<activity>的启动模式可以在 AndroidManifest.xml 配置文件中通过<activity>标签的属性 android:launchMode 来设置。

新建名称为 intentFlag_test1 的 Android 项目，main.xml 布局文件的代码如下：

```xml
<?xml version="1.0" encoding="utf-8"?>
<LinearLayout xmlns:android="http://schemas.android.com/apk/res/android"
    android:orientation="vertical" android:layout_width="fill_parent"
    android:layout_height="fill_parent">
    <TextView android:id="@+id/textView1" android:layout_width="fill_parent"
        android:layout_height="wrap_content" android:text="@string/hello" />
    <Button android:text="启动 Main" android:id="@+id/button1"
        android:layout_width="wrap_content" android:layout_height="wrap_content"></Button>
</LinearLayout>
```

文件 Main.java 的代码如下：

```java
public class Main extends Activity {
    private Button button1;
    private TextView textView1;

    @Override
    public void onCreate(Bundle savedInstanceState) {
        super.onCreate(savedInstanceState);
        setContentView(R.layout.main);

        button1 = (Button) this.findViewById(R.id.button1);
        textView1 = (TextView) this.findViewById(R.id.textView1);
        textView1.setText("" + this.hashCode());

        if (this.getIntent().getExtras() == null) {
            Log.v("!第一次运行打印自己的 hashCode：", "hashCode=" + this.hashCode());
        } else {
            Log.v("!不是第一次运行打印传过来的 hashCode：", "hashCode="
                    + this.getIntent().getIntExtra("hashCode", 9999999));
        }

        button1.setOnClickListener(new OnClickListener() {
            public void onClick(View arg0) {
                Intent intent = new Intent(Main.this, Main.class);
                intent.putExtra("hashCode", Main.this.hashCode());
```

```
                    Main.this.startActivity(intent);
                }
            });

        }
    }
```

程序运行后出现界面如图 4.44 所示。

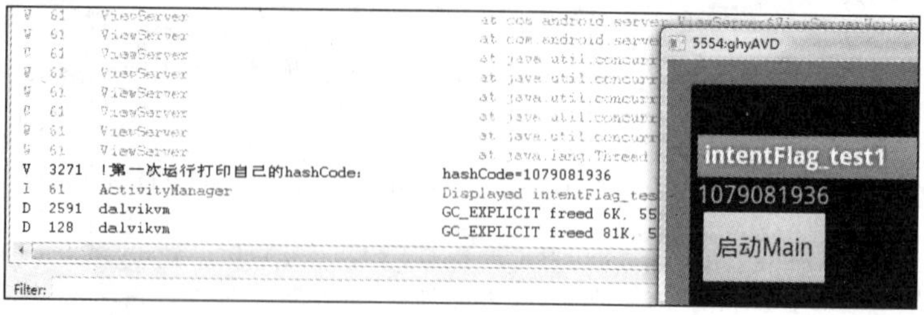

图 4.44　初次启动

记住这个尾数：1936，由于是初次运行，所以 getIntent()方法并没有取到值，在 LogCat 中打印自己的 hashCode。

单击"启动 Main"按钮继续启动 Main.java 的 Activity 对象，出现界面如图 4.45 所示。

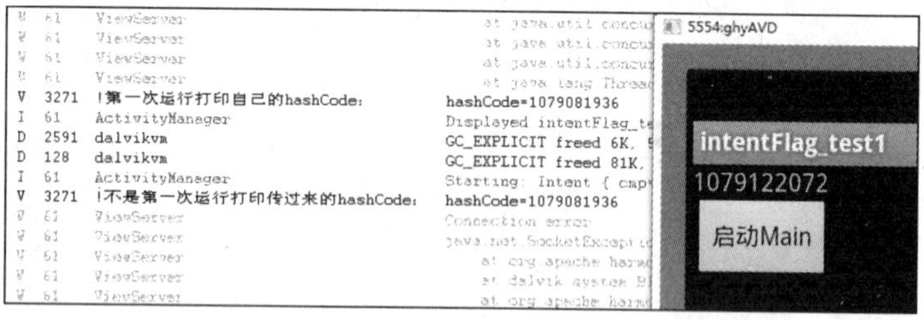

图 4.45　第 1 次单击的界面

继续记住当前 Activity 对象 hashCode 的尾数：2072，在 LogCat 打印出传过来的 hashCode 尾数值 1936。

再继续单击"启动 Main"按钮，继续启动 Main.java 的 Activity 对象，并且把 hashCode 尾数值 2072 继续传递，出现界面如图 4.46 所示。

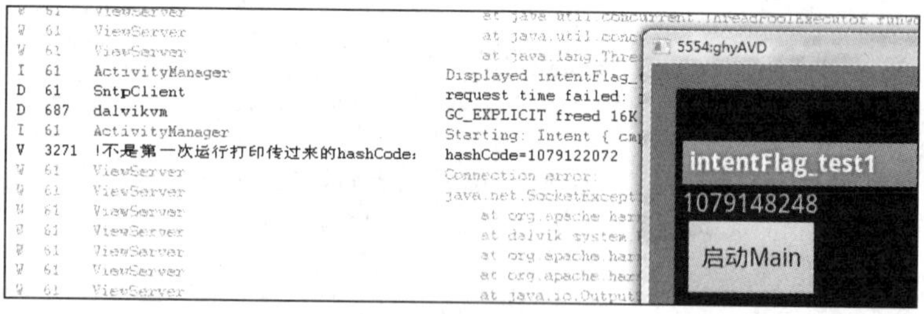

图 4.46　第 2 次单击的界面

从图 4.46 中可以看到，成功从 Intent 对象中取到尾数为 2072 的 hashCode，证明在标准启动模式下可以取到 Intent 中的数据，当前 Activity 对象的 hashcode 的尾数为：8248。

这时单击 back 按钮，会出现如图 4.47 所示的界面。
出现重复的 2072 这个 hashcode，继续按下 Back 按钮，出现界面如图 4.48 所示。

图 4.47　第 1 次回退 back　　　　　图 4.48　第 2 次按下 back 回退按钮

从这个测试项目可以发现，在标准模式下，Activity 栈中重复地创建了 Main.java 这个 Activity 对象，并且 Activity 对象是以 LIFO 的原则进行弹出，而且可以正常地从 Intent 对象中取出数据，这就是标准启动模式，标准启动模式具有创建多个 Activity 对象的特点，按栈中 Activity 的顺序来进行彼此之间的导航，Activity 就像糖葫芦一样被放在 Task 栈中。

4.6.2　singleTop 模式

与标准启动模式比较类似的是 singleTop 模式，它也是具有启动多种 Activity 对象实例的特点，但它还有一个特殊点，就是如果栈顶是目标 Activity 对象则不必重新创建 Activity 的实例。

 如果栈顶是目标 Activity 对象，则把 Intent 对象传递给 onNewIntent 回调函数。

创建名称为 intentFlag_test2 的 Android 项目，Main.java 的代码如下：

```java
public class Main extends Activity {
    private Button button1;
    private TextView textView1;


    @Override
    public void onCreate(Bundle savedInstanceState) {
        super.onCreate(savedInstanceState);
        setContentView(R.layout.main);
        
        Log.v("!","执行了 onCreate");
        
        button1 = (Button) this.findViewById(R.id.button1);
        textView1 = (TextView) this.findViewById(R.id.textView1);
        textView1.setText("" + this.hashCode());
        
        if (this.getIntent().getExtras() == null) {
```

```
                    Log.v("!第一次运行打印自己的 hashCode：", "hashCode=" + this.hashCode());
            } else {
                    Log.v("!不是第一次运行打印传过来的 hashCode：", "hashCode="
                            + this.getIntent().getIntExtra("hashCode", 9999999));
            }

        button1.setOnClickListener(new OnClickListener() {
                public void onClick(View arg0) {
                    Intent intent = new Intent(Main.this, Main.class);
                    intent.putExtra("hashCode", Main.this.hashCode());
                    Main.this.startActivity(intent);
                }
        });

    }
}
```

由于需要改变 Activity 对象的启动模式，则必须在配置文件 AndroidManifest.xml 中更改 Activity 的启动模式，代码如下：

```
<?xml version="1.0" encoding="utf-8"?>
<manifest xmlns:android="http://schemas.android.com/apk/res/android"
    package="intentFlag_test2.test.run" android:versionCode="1"
    android:versionName="1.0">
    <application android:icon="@drawable/icon" android:label="@string/app_name">
        <activity android:name=".Main" android:label="@string/app_name"
            android:launchMode="singleTop">
            <intent-filter>
                <action android:name="android.intent.action.MAIN" />
                <category android:name="android.intent.category.LAUNCHER" />
            </intent-filter>
        </activity>
    </application>
</manifest>
```

启动程序界面如图 4.49 所示。

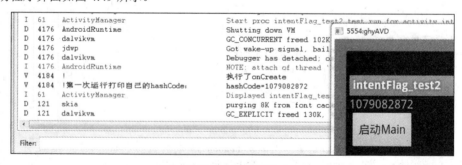

图 4.49　初次启动

多次单击"启动 Main"按钮，出现界面如图 4.50 所示。

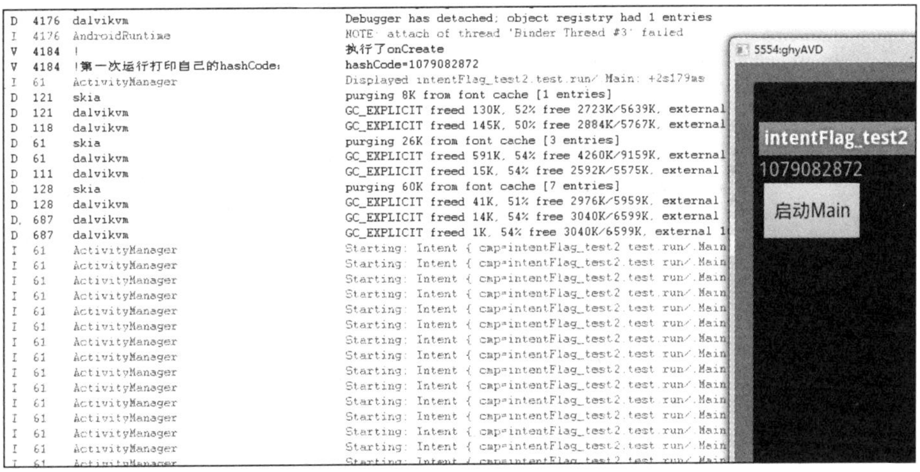

图 4.50　并没有重复创建的 Activity 对象

虽然多次单击"启动 Main"按钮，但并没有创建新的 Activity 实例，并且也从未从 Intent 对象中取出数据并打印，这就是 singleTop 模式的特点：如果栈顶是目标 Activity 对象就不必创建新的对象。

那如何取出 Intent 中的数据呢？使用 onNewIntent 回调函数，更改后的 Main.java 代码如下：

```java
public class Main extends Activity {
    private Button button1;
    private TextView textView1;

    @Override
    protected void onNewIntent(Intent intent) {
        super.onNewIntent(intent);
        Log.v("!onNewIntent 中取出来的 hashCode 值：", ""
                + intent.getIntExtra("hashCode", 9999999));
    }

    @Override
    public void onCreate(Bundle savedInstanceState) {
        super.onCreate(savedInstanceState);
        setContentView(R.layout.main);

        Log.v("!", "执行了 onCreate");

        button1 = (Button) this.findViewById(R.id.button1);
        textView1 = (TextView) this.findViewById(R.id.textView1);
        textView1.setText("" + this.hashCode());

        if (this.getIntent().getExtras() == null) {
            Log.v("!第一次运行打印自己的 hashCode：", "hashCode=" + this.hashCode());
        } else {
            Log.v("!不是第一次运行打印传过来的 hashCode：", "hashCode="
                    + this.getIntent().getIntExtra("hashCode", 9999999));
        }
```

```
button1.setOnClickListener(new OnClickListener() {
    public void onClick(View arg0) {
        Intent intent = new Intent(Main.this, Main.class);
        intent.putExtra("hashCode", Main.this.hashCode());
        Main.this.startActivity(intent);
    }
});
```

程序重新运行后的效果如图 4.51 所示。

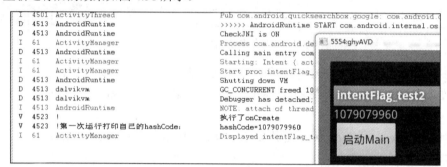

图 4.51　加入回调后的初次运行

多次单击"启动 Main"按钮出现效果如图 4.52 所示。

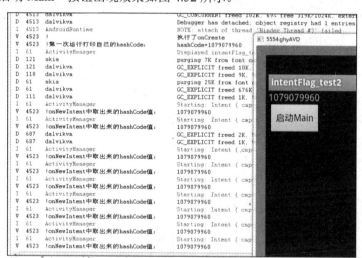

图 4.52　成功从回调中取出值

那如果栈顶的 Activity 不是目标 Activity 时会如何呢？

创建名称为 intentFlag_test2_1 的 Android 项目，文件 Main.java 的代码如下：

```
public class Main extends Activity {
    private Button button1;
    private TextView textView1;
```

```java
    @Override
    protected void onNewIntent(Intent intent) {
        super.onNewIntent(intent);
        Log.v("!onNewIntent 中取出来的 hashCode 值：", ""
                + intent.getIntExtra("hashCode", 9999999));
    }

    @Override
    public void onCreate(Bundle savedInstanceState) {
        super.onCreate(savedInstanceState);
        setContentView(R.layout.main);

        Log.v("!", "执行了 onCreate");

        button1 = (Button) this.findViewById(R.id.button1);
        textView1 = (TextView) this.findViewById(R.id.textView1);
        textView1.setText("" + this.hashCode());

        if (this.getIntent().getExtras() == null) {
            Log.v("!Intent 中的數據為空：", "hashCode=" + this.hashCode());
        } else {
            Log.v("!Intent 中的數據不為空：", "hashCode="
                    + this.getIntent().getIntExtra("hashCode", 9999999));
        }

        button1.setOnClickListener(new OnClickListener() {
            public void onClick(View arg0) {
                Intent intent = new Intent(Main.this, Second.class);
                intent.putExtra("hashCode", Main.this.hashCode());
                Main.this.startActivity(intent);
            }
        });

    }
}
```

布局文件 main.xml 的代码如下：

```xml
<?xml version="1.0" encoding="utf-8"?>
<LinearLayout xmlns:android="http://schemas.android.com/apk/res/android"
    android:orientation="vertical" android:layout_width="fill_parent"
    android:layout_height="fill_parent">
    <TextView android:id="@+id/textView1" android:layout_width="fill_parent"
        android:layout_height="wrap_content" android:text="@string/hello" />
    <Button android:text="启动 Second" android:id="@+id/button1"
        android:layout_width="wrap_content" android:layout_height="wrap_content"></Button>
</LinearLayout>
```

文件 Second.java 的代码如下：

```java
public class Second extends Activity {
```

```java
    private Button button2;

    @Override
    public void onCreate(Bundle savedInstanceState) {
        super.onCreate(savedInstanceState);
        setContentView(R.layout.second);
        Log.v("!", "Second 的 onCreate()方法被调用 并且收到 Intent 传遇来的值："
                + this.getIntent().getIntExtra("hashCode", 8888));
        button2 = (Button) this.findViewById(R.id.button2);
        button2.setOnClickListener(new OnClickListener() {
            public void onClick(View v) {
                Intent intent = new Intent(Second.this, Main.class);
                intent.putExtra("hashCode", 98765);
                Second.this.startActivity(intent);
            }
        });
    }
}
```

配置文件 AndroidManifest.xml 的代码如下：

```xml
<?xml version="1.0" encoding="utf-8"?>
<manifest xmlns:android="http://schemas.android.com/apk/res/android"
    package="intentFlag_test2_1.test.run" android:versionCode="1"
    android:versionName="1.0">
    <application android:icon="@drawable/icon" android:label="@string/app_name">
        <activity android:name=".Main" android:label="@string/app_name"
            android:launchMode="singleTop">
            <intent-filter>
                <action android:name="android.intent.action.MAIN" />
                <category android:name="android.intent.category.LAUNCHER" />
            </intent-filter>
        </activity>
        <activity android:name=".Second" android:label="@string/app_name">
        </activity>
    </application>
</manifest>
```

程序运行后如图 4.53 所示。

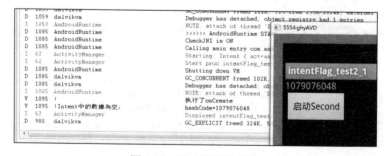

图 4.53　再一次运行项目

这时单击"启动 Second"按钮，出现效果如图 4.54 所示。

图 4.54 进入 Second

再单击 second.xml 中的 "我是 Second,单击我返回 Main" 按钮,LogCat 打印出的效果如图 4.55 所示。

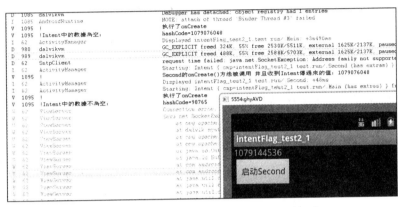

图 4.55 Main.java 不在栈顶所以重新创建出来

通过本示例可以发现,当 Activity 被设置为 singleTop 加载模式时,如果堆栈的顶部已经存在了目标 Activity 的对象,那么它便不会重新创建,而是调用 onNewIntent 回调方法,如果目的 Activity 在栈中存在,但不是在顶部,那么该 Activity 依然要重新创建。

4.6.3 singleTask 模式

在上节介绍的 singleTop 模式中,目标 Activity 只有在栈顶时才可以复用 Activity 对象,不在栈顶则创建 Activity 对象的情况比较苛刻,相对来讲,singleTask 模式的限制就比较宽松,它实现的效果是目标 Activity 对象在栈中就可以复用,而不需要必须在栈顶,如果匹配这条原则,则还要把这个 Activity 上方的所有的 Activity 进行销毁,这点需要留意。

创建名称为 intentFlag_test3 的 Android 项目,在本项目中需要具有 3 个 Activity 对象,分别是 Main、Second 及 Third 对象,它们的代码如下:

文件 Main.java 代码如下:

```
public class Main extends Activity {
    private Button button1;
    private TextView textView1;

    @Override
    protected void onNewIntent(Intent intent) {
        super.onNewIntent(intent);
        Log.v("!在 onNewIntent 中从 Third 取出来的 hashCode 值:", ""
```

```java
                                + intent.getIntExtra("hashCode", 9999999));
        }

        @Override
        public void onCreate(Bundle savedInstanceState) {
            super.onCreate(savedInstanceState);
            setContentView(R.layout.main);

            Log.v("!", "执行了 Main 的 onCreate");

            button1 = (Button) this.findViewById(R.id.button1);
            textView1 = (TextView) this.findViewById(R.id.textView1);
            textView1
                    .setText("" + this.hashCode() + " task id=" + this.getTaskId());

            button1.setOnClickListener(new OnClickListener() {
                public void onClick(View arg0) {
                    Intent intent = new Intent(Main.this, Second.class);
                    Main.this.startActivity(intent);
                }
            });

        }
}
```

文件 main.xml 的代码如下：

```xml
<?xml version="1.0" encoding="utf-8"?>
<LinearLayout xmlns:android="http://schemas.android.com/apk/res/android"
    android:orientation="vertical" android:layout_width="fill_parent"
    android:layout_height="fill_parent">
    <TextView android:id="@+id/textView1" android:layout_width="fill_parent"
        android:layout_height="wrap_content" android:text="@string/hello" />
    <Button android:text="启动 Second" android:id="@+id/button1"
        android:layout_width="wrap_content" android:layout_height="wrap_content"></Button>
</LinearLayout>
```

文件 Second.java 的代码如下：

```java
public class Second extends Activity {
    private Button button2;
    private TextView textView2;

    @Override
    public void onCreate(Bundle savedInstanceState) {
        super.onCreate(savedInstanceState);
        setContentView(R.layout.second);
        textView2 = (TextView) this.findViewById(R.id.textView2);
        textView2
                .setText("" + this.hashCode() + " task id=" + this.getTaskId());
        button2 = (Button) this.findViewById(R.id.button2);
        button2.setOnClickListener(new OnClickListener() {
```

```java
                    public void onClick(View v) {
                            Intent intent = new Intent(Second.this, Third.class);
                            Second.this.startActivity(intent);
                    }
            });
    }

    @Override
    protected void onDestroy() {
        super.onDestroy();
        Log.v("!", "Second 被 onDestroy 销毁了");
    }
}
```

文件 second.xml 的代码如下:

```xml
<?xml version="1.0" encoding="utf-8"?>
<LinearLayout xmlns:android="http://schemas.android.com/apk/res/android"
    android:orientation="vertical" android:layout_width="fill_parent"
    android:layout_height="fill_parent">
    <TextView android:text="TextView" android:id="@+id/textView2"
        android:layout_width="wrap_content" android:layout_height="wrap_content"></TextView>
    <Button android:text="我是 Second，去启动 Third" android:id="@+id/button2"
        android:layout_width="wrap_content" android:layout_height="wrap_content"></Button>
</LinearLayout>
```

文件 Third.java 的代码如下:

```java
public class Third extends Activity {
    private Button button3;
    private TextView textView3;

    @Override
    public void onCreate(Bundle savedInstanceState) {
        super.onCreate(savedInstanceState);
        setContentView(R.layout.third);
        textView3 = (TextView) this.findViewById(R.id.textView3);
        textView3
                .setText("" + this.hashCode() + " task id=" + this.getTaskId());

        button3 = (Button) this.findViewById(R.id.button3);
        button3.setOnClickListener(new OnClickListener() {
            public void onClick(View v) {
                Intent intent = new Intent(Third.this, Main.class);
                intent.putExtra("hashCode", 98765);
                Third.this.startActivity(intent);
            }
        });
    }

    @Override
    protected void onDestroy() {
```

```java
        super.onDestroy();
        Log.v("!", "Third 被 onDestroy 销毁了");
    }
}
```

文件 third.xml 的代码如下：

```xml
<?xml version="1.0" encoding="utf-8"?>
<LinearLayout xmlns:android="http://schemas.android.com/apk/res/android"
    android:orientation="vertical" android:layout_width="fill_parent"
    android:layout_height="fill_parent">
    <TextView android:text="TextView" android:id="@+id/textView3"
        android:layout_width="wrap_content" android:layout_height="wrap_content"></TextView>
    <Button android:text="我是 Third,点我回到 Main" android:id="@+id/button3"
        android:layout_width="wrap_content" android:layout_height="wrap_content"></Button>
</LinearLayout>
```

文件 AndroidManifest.xml 的代码如下：

```xml
<?xml version="1.0" encoding="utf-8"?>
<manifest xmlns:android="http://schemas.android.com/apk/res/android"
    package="intentFlag_test3.test.run" android:versionCode="1"
    android:versionName="1.0">



    <application android:icon="@drawable/icon" android:label="@string/app_name">
        <activity android:name=".Main" android:label="@string/app_name"
            android:launchMode="singleTask">
            <intent-filter>
                <action android:name="android.intent.action.MAIN" />
                <category android:name="android.intent.category.LAUNCHER" />
            </intent-filter>
        </activity>
        <activity android:name=".Second" android:label="@string/app_name">
        </activity>
        <activity android:name=".Third" android:label="@string/app_name">
        </activity>

    </application>
</manifest>
```

运行项目，结果如图 4.56 所示。

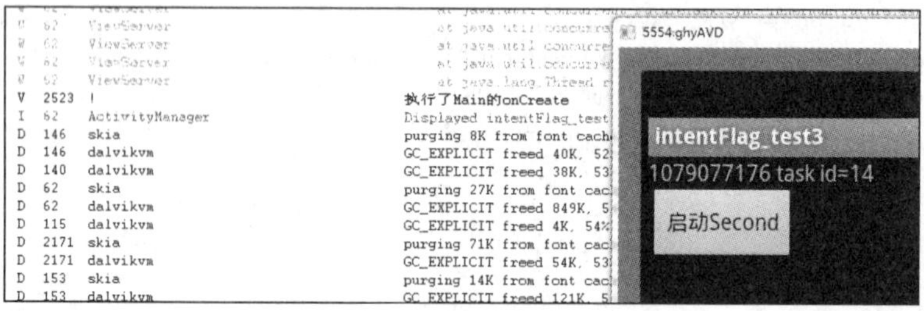

图 4.56　程序初始运行效果

单击"启动 Second"按钮，启动 Second.java 的 Activity 对象，效果如图 4.57 所示。
再单击 Second 上的按钮，切换到 Third 界面上，效果如图 4.58 所示。

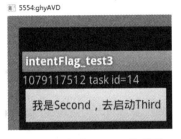 

图 4.57　启动 Second 的界面　　　　　　图 4.58　切换到 Third 界面上

这时当单击 Third 界面上的按钮时，LogCat 打印出来的信息如图 4.59 所示。

图 4.59　没有创建新的对象重回 Main 对象

提示

从图 4.58 中可以发现，在 Activity 栈中 Main 上面的所有 Activity 被销毁，并且调用原来 Main 对象的 onNewIntent()回调方法。

4.6.4　singleInstance 模式

启动模式 singleInstance 的功能是将当前的 Activity 单独放入一个 task 对象中，并且这个 task 中只有一个 Activity 对象，在这里需要说明的是，Activity 栈在 task 中是不允许更改其顺序的，但多个 task 中却可以更改在系统中的顺序，比如可以把一个 task 放入前台，或把一个 task 放入后台。

新建名称为 intentFlag_test4 的项目，需要将 intentFlag_test3 项目中的代码进行复用，但 Main.java 文件还需要做少许改动，代码如下：

```
public class Main extends Activity {
    private Button button1;
    private TextView textView1;

    @Override
    protected void onNewIntent(Intent intent) {
        super.onNewIntent(intent);
```

```java
            Log.v("!在 onNewIntent 中从 Third 取出来的 hashCode 值：", ""
                    + intent.getIntExtra("hashCode", 9999999) + " task id="
                    + this.getTaskId());
    }

    @Override
    public void onCreate(Bundle savedInstanceState) {
        super.onCreate(savedInstanceState);
        setContentView(R.layout.main);

        Log.v("!", "执行了 Main 的 onCreate");

        button1 = (Button) this.findViewById(R.id.button1);
        textView1 = (TextView) this.findViewById(R.id.textView1);
        textView1
                .setText("" + this.hashCode() + " task id=" + this.getTaskId());

        button1.setOnClickListener(new OnClickListener() {
            public void onClick(View arg0) {
                Intent intent = new Intent(Main.this, Second.class);
                Main.this.startActivity(intent);
            }
        });

    }
}
```

AndroidManifest.xml 文件中的代码更改如下：

```xml
<?xml version="1.0" encoding="utf-8"?>
<manifest xmlns:android="http://schemas.android.com/apk/res/android"
    package="intentFlag_test4.test.run" android:versionCode="1"
    android:versionName="1.0">
    <application android:icon="@drawable/icon" android:label="@string/app_name">
        <activity android:name=".Main" android:label="@string/app_name"
            android:launchMode="singleInstance">
            <intent-filter>
                <action android:name="android.intent.action.MAIN" />
                <category android:name="android.intent.category.LAUNCHER" />
            </intent-filter>
        </activity>
        <activity android:name=".Second" android:label="@string/app_name">
        </activity>
        <activity android:name=".Third" android:label="@string/app_name">
        </activity>
    </application>
</manifest>
```

项目运行后的效果如图 4.60 所示。

Intent 对象 第 4 章

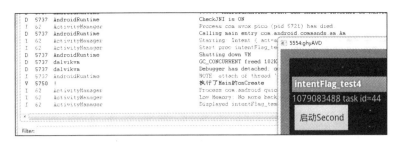

图 4.60 初始运行结果

从图 4.60 中可以看到，Main.java 被放在 Task 值为 44 的栈中，单击"启动 Second"出现如图 4.61 所示的界面。

从图 4.61 中可以看到 Second.java 对象被放在 id 值为 45 的 Task 对象中。

继续单击"我是 Second，去启动 Third"按钮，出现如图 4.62 所示的效果。

图 4.61 Second 另起 Task 炉灶   图 4.62 Third 和 Second 在一起

Third 和 Second 在一起，都放在 id 值为 45 的 Task 对象中。

再单击"我是 Third,点我回到 Main"按钮，出现如图 4.63 所示的效果。

图 4.63 回到 id 值为 44 的 task 中

那么现在 Activity 栈中的顺序为 Second、Third 和 Main，对象 Main 在栈顶，前面介绍过 Activity 栈中的 Activity 对象并不允许更改顺序，但在这里由于是两个 Task 对象，其中 Second 和 Third 在一个 Task 中，而 Main 单独在一个 Task 中，并且这个 Task 中只允许有一个 Main 对象，则系统是可以更改 Task 的顺序的。

4.7 Activity 常用 flag 标记的学习

通过前面的学习，知道了 Activity 是以栈的方式来进行管理，即是在 AndroidManifest.xml 文件中设置 Activity 的启动模式，还可以通过使用代码的方式来管理 Activity 栈中的 Activity 对象，那就是使用 Intent 对象的 setFlag()方法。下面分别介绍这些方法的使用。

4.7.1 FLAG_ACTIVITY_CLEAR_TOP 标记

标记 FLAG_ACTIVITY_CLEAR_TOP 的功能是假设 Activity 栈中已经有 A、B、C 这 3 个 Activity 对象,现在 C 是最后 1 个显示的,所以 C 在 Activity 栈的最上方,这时使用这个标记由 C 转到 B 时,立即将 B 上方的 Activity(C)销毁,而 B 则使用回调 onNewIntent() 取得 Intent 中的数据。

新建名称为 flag_test1 的 Android 项目,创建 3 个 Activity,分别是 Main.java、Second.java 和 Third.java,在这 3 个 Activity 中重写 7 个生命周期方法,并用 Log.v() 打印到 LogCat 标识出当前的 Activity 信息。

文件 Main.java 的核心代码如下:

```java
public class Main extends Activity {
    private TextView main_textView1;
    private Button main_button1;

    @Override
    public void onCreate(Bundle savedInstanceState) {
        super.onCreate(savedInstanceState);

        Log.v("!Main", "onCreate");

        setContentView(R.layout.main);

        main_textView1 = (TextView) this.findViewById(R.id.main_testview1);
        main_textView1.setText("我是 Main:" + this.hashCode());

        main_button1 = (Button) this.findViewById(R.id.main_button1);
        main_button1.setOnClickListener(new OnClickListener() {
            public void onClick(View arg0) {
                Intent intent = new Intent(Main.this, Second.class);
                Main.this.startActivity(intent);
            }
        });

    }
```

文件 Second.java 的核心代码如下:

```java
public class Second extends Activity {
    private TextView second_textView1;
    private Button second_button1;

    @Override
    protected void onNewIntent(Intent intent) {
        super.onNewIntent(intent);
        Log.v("Second onNewIntent username:", ""
                + intent.getStringExtra("username") + " hashCode:"
                + this.hashCode());
    }
```

```java
@Override
public void onCreate(Bundle savedInstanceState) {
    super.onCreate(savedInstanceState);
    setContentView(R.layout.second);

    Log.v("!Second", "onCreate");

    second_textView1 = (TextView) this.findViewById(R.id.second_textView1);
    second_textView1.setText("我是 Second:" + this.hashCode());

    second_button1 = (Button) this.findViewById(R.id.second_button1);
    second_button1.setOnClickListener(new OnClickListener() {
        public void onClick(View arg0) {
            Intent intent = new Intent();
            intent.setClass(Second.this, Third.class);
            Second.this.startActivity(intent);
        }
    });
}
```

文件 Third.java 的核心代码如下:

```java
public class Third extends Activity {
    private TextView third_textView1;
    private Button third_button1;

    @Override
    public void onCreate(Bundle savedInstanceState) {
        super.onCreate(savedInstanceState);
        setContentView(R.layout.third);

        Log.v("!Third", "onCreate");
        third_button1 = (Button) this.findViewById(R.id.third_button1);
        third_button1.setOnClickListener(new OnClickListener() {
            public void onClick(View arg0) {
                Intent intent = new Intent();
                intent.putExtra("username", "gaohongyan");
                intent.setClass(Third.this, Second.class);
                intent.setFlags(Intent.FLAG_ACTIVITY_CLEAR_TOP
                        | Intent.FLAG_ACTIVITY_SINGLE_TOP);
                Third.this.startActivity(intent);
            }
        });
        third_textView1 = (TextView) this.findViewById(R.id.third_textView1);
        third_textView1.setText("我是 Third:" + this.hashCode());
    }
}
```

> 提 示
>
> 如果在 Third.java 文件中只使用 1 个 flag 标记，即
>
> intent.setFlags(Intent.*FLAG_ACTIVITY_CLEAR_TOP*);
>
> 则由 Third.java 返回 Second.java 时，在栈中的 Second.java 对象要销毁，并且重新再创建 1 个新的 Second.java 对象并放入栈顶，这时如果想取得 Intent 中的值，就得用 Second.java 文件中的 onCreate()回调方法通过 this.getIntent()方式进行取得。FLAG_ACTIVITY_SINGLE_TOP 标记的作用是当 Activity 正在运行，并且是 history stack 的 top 时，不会创建新的 Activity。

运行项目后，在 LogCat 打印日志，如图 4.64 所示。

图 4.64　初始打印日志

单击图 4.64 中的按钮，打印出的日志效果如图 4.65 所示。

再单击图 4.65 中的 Button 按钮，出现如图 4.66 所示的效果。

图 4.65　单击 main.xml 中的按钮后效果

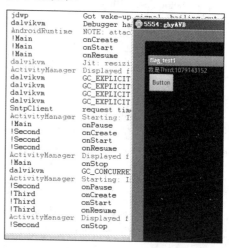

图 4.66　单击 second.xml 中按钮后的效果

再单击图 4.66 中的 Button 按钮，转到 Second.java 对象中，LogCat 打印出的日志效果如图 4.67 所示。

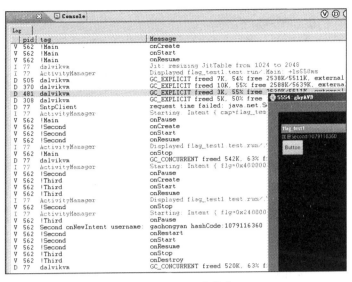

图 4.67 最后日志信息

从截图可以看到，Second.java 对象都是以 6360 结尾的 hashCode，证明是 1 个对象，并且 onNewIntent()回调方法比 onRestart()函数执行要早。

4.7.2 FLAG_ACTIVITY_CLEAR_WHEN_TASK_RESET 标记

标记 FLAG_ACTIVITY_CLEAR_WHEN_TASK_RESET 的功能有些类似于数据还原技术中的"还原点"，不过在 Android 中，这个点叫做"删除点"。例如有两个项目 X 和 Y，X 项目中有两个 Activity 对象 A 和 B，由 A 转到 B 时，此时在 B 中以不加该标签的方式启动 Y 项目中名称为 C 的 Activity 后，按下 Home 键回到桌面，这时再由桌面启动这个 X 项目后看到的还是 Y 项目的 C，但使用这个标记后，重新进入项目看到的就是 X 项目的 B。而发生删除的情况是由桌面返回应用程序时。

新建一个名称为 flag_test2_ext 的 Android 项目，文件 Main.java 的核心代码如下：

```
public class Main extends Activity {
    private EditText editText1;

    @Override
    public void onCreate(Bundle savedInstanceState) {
        super.onCreate(savedInstanceState);
        setContentView(R.layout.main);
        Log.v("!flag_test2_ext.test.run Main", "onCreate");

        editText1 = (EditText) this.findViewById(R.id.editText1);
        if (this.getIntent().getExtras() != null) {
            editText1.setText(this.getIntent().getExtras()
                    .getString("username"));
        }
    }
}
```

这个 flag_test2_ext 是一个"第三方"的支持项目，主项目名称为 flag_test2，有两个 Activity

对象，其中 Main.java 的核心代码如下：

```java
public class Main extends Activity {
    private Button main_button;

    @Override
    public void onCreate(Bundle savedInstanceState) {
        super.onCreate(savedInstanceState);
        setContentView(R.layout.main);
        Log.v("!flag_test2.test.run Main", "onCreate");

        main_button = (Button) this.findViewById(R.id.main_button);
        main_button.setOnClickListener(new OnClickListener() {
            public void onClick(View arg0) {
                Intent intent = new Intent();
                intent.setClass(Main.this, Second.class);
                Main.this.startActivity(intent);
            }
        });
    }
}
```

文件 Second.java 核心代码如下：

```java
public class Second extends Activity {
    private Button second_button;

    @Override
    public void onCreate(Bundle savedInstanceState) {
        super.onCreate(savedInstanceState);
        setContentView(R.layout.second);
        Log.v("!flag_test2.test.run Second", "onCreate");

        second_button = (Button) this.findViewById(R.id.second_button);
        second_button.setOnClickListener(new OnClickListener() {
            public void onClick(View arg0) {
                Intent intent = new Intent();
                //intent.setFlags(Intent.FLAG_ACTIVITY_CLEAR_WHEN_TASK_RESET);
                intent.putExtra("username", "gaohongyan");
                intent.setClassName("flag_test2_ext.test.run",
                        "flag_test2_ext.test.run.Main");
                Second.this.startActivity(intent);
            }
        });
    }
}
```

在 Second.java 代码中有一个注释语句。

先运行 flag_test2_ext 项目，以便把"第三方"项目安装到 AVD 中。再启动 flag_test2 项目之后，由 Main.java 到 Second.java 对象，在 LogCat 打印的信息如图 4.68 所示。

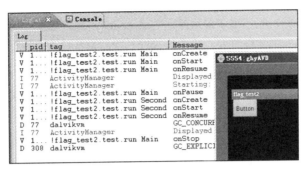

图 4.68　由 Main.java 到 Second.java 对象的日志

这时单击图 4.68 中的 Button 按钮，调用第三方项目的 Activity 界面，单击后日志和显示的界面效果如图 4.69 所示。

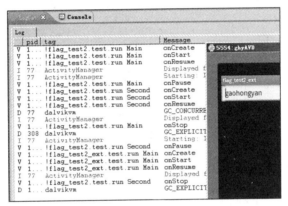

图 4.69　日志和信息

这时单击 Home 键，再由 Home 再次进入 flag_test2 项目，运行结果如图 4.70 所示。

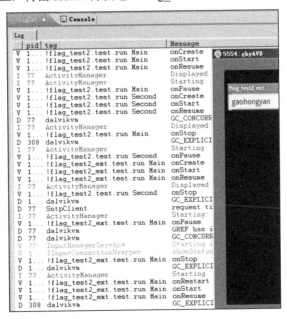

图 4.70　再次进入 flag_test2 项目

从图 4.70 中可以看到，显示出了最后操作的 Activity 对象，如果在重新回到项目不想显示第三方的 Activity，怎么办？把注释打开，再重新运行一次，整个日志结果如图 4.71 所示。

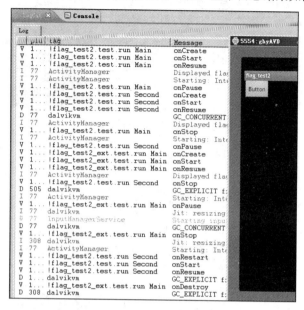

图 4.71 使用标记最后的效果

4.7.3 FLAG_ACTIVITY_EXCLUDE_FROM_RECENTS 标记

标记 FLAG_ACTIVITY_EXCLUDE_FROM_RECENTS 的作用是在当前的 Activity 启动另外一个应用程序的 Activity 界面对象时，目标 Activity 所在的"应用程序"不在历史打开记录中。

新建名称为 flag_test3 的 Android 项目，Main.java 的核心代码如下：

```java
public class Main extends Activity {
    private Button button1;

    @Override
    public void onCreate(Bundle savedInstanceState) {
        super.onCreate(savedInstanceState);
        setContentView(R.layout.main);

        button1 = (Button) this.findViewById(R.id.button1);
        button1.setOnClickListener(new OnClickListener() {
            public void onClick(View arg0) {
                Intent intent = new Intent(Intent.ACTION_VIEW, Uri
                        .parse("http://www.baidu.com"));
                // intent.setFlags(Intent.FLAG_ACTIVITY_EXCLUDE_FROM_RECENTS);
                Main.this.startActivity(intent);
            }
        });

    }
}
```

程序运行后单击按钮弹出浏览器的界面，这时按 Home 按键，应用程序打开历史列表，效果如图 4.72 所示。

在图 4.72 中可以看到，Browser 浏览器应用程序在历史应用程序列表中，加上这个 flag 后便不再出现了，打开注释，再次运行项目，单击 Button 按钮后显示浏览器，再次打开 Home 键，出现效果如图 4.73 所示。

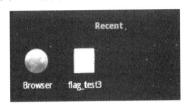
图 4.72　应用程序打开历史列表

图 4.73　浏览器不再出现在历史列表中

4.7.4　FLAG_ACTIVITY_FORWARD_RESULT 标记

假设现在有这么一种 Activity 调用场景，A 启动 B，B 再启动 C，在 C 中将结果返回给 B，B 再将结果返回给 A，这种将数据传来传去的流程略显零乱，能不能使 C 直接将结果返回给 A 呢？当然可以，FLAG_ACTIVITY_FORWARD_RESULT 标记就是用来做这件事情的。

新建名称为 flag_test4 的 Android 项目，创建 3 个 Activity 对象，Main.java、Second.java 和 Third.java 对象。

文件 Main.java 的核心代码如下：

```java
public class Main extends Activity {
    private Button main_button;

    @Override
    protected void onActivityResult(int requestCode, int resultCode, Intent data) {
        super.onActivityResult(requestCode, resultCode, data);
        Log.v("Main", "onActivityResult" + " get username:"
                + data.getStringExtra("username"));
    }

    @Override
    public void onCreate(Bundle savedInstanceState) {
        super.onCreate(savedInstanceState);
        setContentView(R.layout.main);
        Log.v("Main", "onCreate");
        main_button = (Button) this.findViewById(R.id.main_button);
        main_button.setOnClickListener(new OnClickListener() {
            public void onClick(View arg0) {
                Intent intent = new Intent(Main.this, Second.class);
                Main.this.startActivityForResult(intent, 1000);
            }
        });
```

        }
}

Main.java 由于要取得 result 结果,所以以 startActivityForResult 的方式启动 Second.java。文件 Second.java 的核心代码如下:

```java
public class Second extends Activity {
    private Button second_button;
    private Button closeSelf;

    @Override
    public void onCreate(Bundle savedInstanceState) {
        super.onCreate(savedInstanceState);
        setContentView(R.layout.second);

        closeSelf = (Button) this.findViewById(R.id.closeSelf);

        second_button = (Button) this.findViewById(R.id.second_button);
        second_button.setOnClickListener(new OnClickListener() {
            public void onClick(View arg0) {
                Intent intent = new Intent(Second.this, Third.class);
                intent.addFlags(Intent.FLAG_ACTIVITY_FORWARD_RESULT);
                Second.this.startActivity(intent);
            }
        });

        closeSelf.setOnClickListener(new OnClickListener() {
            public void onClick(View v) {
                Second.this.finish();
            }
        });

    }
}
```

Second.java 以 startActivity 的方式启动 Third.java,在 Intent 中带上参数 FLAG_ACTIVITY_FORWARD_RESULT,代表 Third.java 返回的 result 传给 Main.java 对象,但返回 result 数据给 Main.java 对象的前提是 Second.java 和 Third.java 都是在 finish()关闭的情况下。

文件 Third.java 的核心代码如下:

```java
public class Third extends Activity {
    private Button third_button;

    @Override
    public void onCreate(Bundle savedInstanceState) {
        super.onCreate(savedInstanceState);
        setContentView(R.layout.third);

        third_button = (Button) this.findViewById(R.id.third_button);
```

```
           third_button.setOnClickListener(new OnClickListener() {
                public void onClick(View arg0) {
                    Intent intent = new Intent();
                    intent.putExtra("username", "usernameValue");
                    Third.this.setResult(1000, intent);
                    Third.this.finish();
                }
            });
        }
    }
```

程序运行后出现 main.java 界面，如图 4.74 所示。

单击图 4.74 中的按钮，转到 second.xml 布局文件中，显示效果如图 4.75 所示。

图 4.74　显示 main.xml 布局

图 4.75　显示 second.xml 布局文件

单击图 4.75 中的"this is second!goto third"按钮，显示 third.xml 布局文件，效果如图 4.76 所示。

单击图 4.76 中的按钮，调用 setResult() 方法及 finish() 关闭自己后，显示 second.xml 布局界面，效果如图 4.77 所示。

图 4.76　显示 third.xml 布局文件

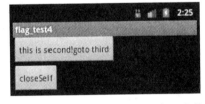
图 4.77　次显示 second.xml 布局文件

单击"closeSelf"按钮关闭 Second.java 后，Third.java 给 Main.java 传递了数据，在 LogCat 打印的信息如图 4.78 所示。

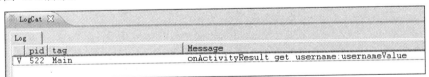
图 4.78　Main.java 取得了数据

另外需要注意的是，如果返回的 Intent 是跨多个 Activity 时，在中途的每 1 个 Activity 都要使用标记 flag，并且结合 startActivity() 方法进行功能的传递，如果在中途有多个 setResult() 方法，则最后一个为有效的，如图 4.79 所示。

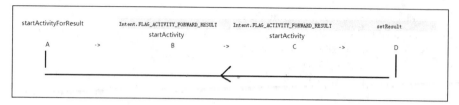

图 4.79 跨多个 Activity 时回传数据的办法

4.7.5 FLAG_ACTIVITY_NEW_TASK 标记

标记 FLAG_ACTIVITY_NEW_TASK 的作用是将 Activity 对象放入新启动的 Task 中，常用在多个应用程序间 Activity 的交互，比如 X 项目中有 Activity 对象 A 和 B，这时在 B 中不用此标记启动 Y 项目的 C，按下 Home 键，再回到 X 项目时显示的还是对象 C，而如果用到此标记，则显示 X 项目的 B，因为 C 被放置在另外一个 Task 对象中。

由于 Activity 被放入其他的 Task 中，所以使用 FLAG_ACTIVITY_NEW_TASK 标记的 Activity 并不销毁，而 FLAG_ACTIVITY_CLEAR_WHEN_TASK_RESET 标记却销毁，这是这两个标记明显区别的地方。

新建名称为 flag_test5_ext 的 Android 项目，Main.java 的代码如下：

```java
public class Main extends Activity {
    private TextView main_textview;

    @Override
    public void onCreate(Bundle savedInstanceState) {
        super.onCreate(savedInstanceState);
        setContentView(R.layout.main);
        main_textview = (TextView) this.findViewById(R.id.main_textview);
        main_textview.setText("ext Main task id=" + this.getTaskId());
    }
}
```

新建名称为 flag_test5 的 Android 项目，文件 Main.java 的代码如下：

```java
public class Main extends Activity {
    private Button button1;
    private TextView main_textview;

    @Override
    public void onCreate(Bundle savedInstanceState) {
        super.onCreate(savedInstanceState);
        setContentView(R.layout.main);

        main_textview = (TextView) this.findViewById(R.id.main_textview);
        main_textview.setText("main task id:" + Main.this.getTaskId());

        button1 = (Button) this.findViewById(R.id.button1);
        button1.setOnClickListener(new OnClickListener() {
            public void onClick(View arg0) {
```

```java
                Intent intent = new Intent();
                intent.setClass(Main.this, Second.class);
                Main.this.startActivity(intent);
            }
        });
    }
}
```

文件 Second.java 的代码如下：

```java
public class Second extends Activity {
    private TextView second_textview;
    private Button button2;

    @Override
    public void onCreate(Bundle savedInstanceState) {
        super.onCreate(savedInstanceState);
        setContentView(R.layout.second);

        button2 = (Button) this.findViewById(R.id.button2);

        second_textview = (TextView) this.findViewById(R.id.second_textview);
        second_textview.setText("second task id:" + this.getTaskId());

        button2.setOnClickListener(new OnClickListener() {
            public void onClick(View arg0) {
                Intent intent = new Intent();
                intent.setClassName("flag_test5_ext.test.run",
                        "flag_test5_ext.test.run.Main");
                // intent.setFlags(Intent.FLAG_ACTIVITY_NEW_TASK);
                Second.this.startActivity(intent);
            }
        });
    }
}
```

注意有一行注释掉的程序代码。

首先运行项目 flag_test5_ext，再退出，目的是在 AVD 中安装这个项目，再运行 flag_test5 项目，出现如图 4.80 所示的界面。

从图 4.80 中可以看到，Main.java 运行在 id 为 12 的 task 中，单击图 4.80 中的 Button 按钮，出现如图 4.81 所示的界面。

从图 4.81 中也可以看到，Main.java 和 Second.java 同在 id 为 12 的 Task 对象中，单击图 4.81 中的按钮出现如图 4.82 所示的界面。

  图 4.80 初始运行界面   图 4.81 Second.java 显示效果

  从图 4.82 中可以看到，这 3 个 Activity 同是运行在 id 为 12 的 Task 中，并且 flag_test5_ext 项目中的 Main.java 对象在 Activity 栈顶，测试到这还没有到关键的步骤！按下 Home 键后，再重新启动 flag_test5 项目，效果如图 4.83 所示。

 图 4.82 打开另外一个项目中的 Activity 图 4.83 重新启动 flag_test5 项目的结果

 提 示 从图 4.82 中可以看到，将 id 为 12 的 Task 栈顶 Activity 对象显示了出来，那如果想要返回 flag_test5 项目后只显示 flag_test5 项目中的 Activity 栈顶对象该如何实现呢？很简单，打开前面代码的注释就可以了！

  将这两个项目从 AVD 中彻底退出，还原测试环境。运行 flag_test5 项目，结果如图 4.84 所示。单击图 4.84 中的按钮后，运行效果如图 4.85 所示。

  图 4.84 最新版的初始运行效果   图 4.85 最新版显示 Second.java 界面

  在图 4.85 中单击按钮启动 flag_test5_ext 项目中的 Main.java 对象，运行后的界面如图 4.86 所示。

  从图 4.86 中可以看到，两个项目中的 Activity 对象在不同的 Task 对象中，一个 id 为 14，另外一个 id 为 15。按下 Home 键，再重新进入 flag_test5 项目，运行界面如图 4.87 所示。

图 4.86 最新版 flag_test5_ext 项目中的 Main.java 对象 图 4.87 最新版显示 Second.java 界面

  另外，需要说明的是，有两个项目 A 和 B，在 A 项目中有 A1 和 A2 的 Activity 对象，而 B 项目中有 B1 和 B2 的 Activity 对象，一共在两个项目中有 4 个 Activity 对象，如果从 A 项目中启动 A1 和 A2，再由 A2 使用 Intent.FLAG_ACTIVITY_NEW_TASK 标记启动 B 项目的 B2，这时系统中有两个 Task，第 1 个 task 中有 A1 和 A2，第 2 个 task 中有 B2，如果在桌面启动 B 项目，将会把 B2 显示出来，因为 B2 在 Task 的栈顶。

4.7.6 FLAG_ACTIVITY_NO_ANIMATION 标记

如果在 Intent 中使用标记 FLAG_ACTIVITY_NO_ANIMATION，并且用 startActivity()方式来启动另外一个 Activity 则没有默认的过渡动画，使用起来比较简单。新建名称为 flag_test6 的 Android 项目，将文件 Main.java 的代码更改为如下即可：

```java
public class Main extends Activity {
    private Button button1;

    @Override
    public void onCreate(Bundle savedInstanceState) {
        super.onCreate(savedInstanceState);
        setContentView(R.layout.main);

        button1 = (Button) this.findViewById(R.id.button1);
        button1.setOnClickListener(new OnClickListener() {
            public void onClick(View arg0) {
                Intent intent = new Intent(Main.this, Second.class);
                intent.addFlags(Intent.FLAG_ACTIVITY_NO_ANIMATION);
                Main.this.startActivity(intent);
            }
        });
    }
}
```

4.7.7 FLAG_ACTIVITY_NO_HISTORY 标记

标记 FLAG_ACTIVITY_NO_HISTORY 的作用是把欲启动的 Activity 不放入栈中。

新建名称为 flag_test7 的 Android 项目，Main.java 的代码如下：

```java
public class Main extends Activity {
    private Button button1;

    @Override
    public void onCreate(Bundle savedInstanceState) {
        super.onCreate(savedInstanceState);
        setContentView(R.layout.main);
        button1 = (Button) this.findViewById(R.id.button1);
        button1.setOnClickListener(new OnClickListener() {
            public void onClick(View arg0) {
                Intent intent = new Intent();
                intent.setClass(Main.this, Second.class);
                intent.setFlags(Intent.FLAG_ACTIVITY_NO_HISTORY);
                Main.this.startActivity(intent);
            }
        });
    }
}
```

Second.java 的代码如下：

```java
public class Second extends Activity {
    private Button button2;

    @Override
    public void onCreate(Bundle savedInstanceState) {
        super.onCreate(savedInstanceState);
        setContentView(R.layout.second);
        button2 = (Button) this.findViewById(R.id.button2);
        button2.setOnClickListener(new OnClickListener() {
            public void onClick(View arg0) {
                Intent intent = new Intent();
                intent.setClass(Second.this, Third.class);
                Second.this.startActivity(intent);
            }
        });
    }
}
```

Third.java 的代码如下：

```java
public class Third extends Activity {
    @Override
    public void onCreate(Bundle savedInstanceState) {
        super.onCreate(savedInstanceState);
        setContentView(R.layout.third);
    }
}
```

程序运行后初始效果如图 4.88 所示。

单击"这是 Main，点我到 Second"按钮后，Second.java 对象被启动，但并没有组织到 Activity 栈中，出现如图 4.89 所示的界面。

图 4.88　初始运行效果　　　　　　图 4.89　Second 对象被显示

单击"这是 Second，并且不在 Activity 栈中，点我到 Third"按钮切换到 Third 对象，运行效果如图 4.90 所示。

由于 Second 不在 Activity 栈中，所以在 Third 界面上按下 Back 按钮返回 Main 界面，效果如图 4.91 所示。

Intent 对象 第 4 章

图 4.90 Third 显示的界面

图 4.91 在 Third 按下 Back 按钮直接返回 Main

4.7.8 FLAG_ACTIVITY_NO_USER_ACTION 标记

标记 FLAG_ACTIVITY_NO_USER_ACTION 的功能暂时先不做讨论，但这个标记与一个回调函数有关，它是 onUserLeaveHint()，所以本小节先来介绍 onUserLeaveHint()回调函数的使用。

回调函数 onUserLeaveHint()的主要作用就是用户按下了 Home 键时它被调用，它也是 Activity 对象生命周期的一部分，但它与 onPause()及 onStop()回调函数有什么区别呢？一起来做一个实验。

本实验的主要目的是当用户按下 Home 键时，当前的应用程序变为在后台运行，并且在通知栏上显示一个图标，单击这个图标后返回这个应用程序。

新建名称为 flag_test8_1 的 Android 项目，文件 Main.java 的代码如下：

```java
public class Main extends Activity {
    private TextView textView1;

    @Override
    protected void onPause() {
        super.onPause();
        Log.v("!", "调用了 onPause()方法");

        NotificationManager nm = (NotificationManager) this
                .getSystemService(Context.NOTIFICATION_SERVICE);

        Notification n = new Notification(R.drawable.icon, "出现通知", System
                .currentTimeMillis());
        n.flags = Notification.FLAG_AUTO_CANCEL;

        Intent intent = new Intent(this, Main.class);
        intent.setFlags(Intent.FLAG_ACTIVITY_CLEAR_TOP);

        PendingIntent contentIntent = PendingIntent.getActivity(this, 1,
                intent, 0);
        n.setLatestEventInfo(this, "我是标题", "我是正文", contentIntent);

        nm.notify(100, n);

    }

    @Override
    public void onCreate(Bundle savedInstanceState) {
        super.onCreate(savedInstanceState);
        setContentView(R.layout.main);
        Log.v("!", "调用了 onCreate()方法");
```

```
        textView1 = (TextView) this.findViewById(R.id.textView1);
        textView1.setText("" + this.getTaskId());
    }
}
```

此程序在运行时，按下 Home 键的确可以在通知栏出现图标，效果如图 4.92 所示。

但现在出现一个问题，按下 Home 键后当前的应用程序变成在后台运行，然后通过单击通知栏的图标再重新进入这个项目，这可以理解，但根据 Activity 的生命周期，在项目中按下 Back 键完全退出应用程序时也会在通知栏出现图标，这样的情况使用 onPause()回调方法就显得不太合适了，那么怎么办呢？可使用 onUserLeaveHint()回调函数解决这个问题。

图 4.92 通知栏出现图标

将 Main.java 的代码更改如下：

```
public class Main extends Activity {
    private TextView textView1;

    @Override
    protected void onUserLeaveHint() {
        super.onUserLeaveHint();
        Log.v("!", "调用了 onUserLeaveHint()方法");

        NotificationManager nm = (NotificationManager) this
                .getSystemService(Context.NOTIFICATION_SERVICE);

        Notification n = new Notification(R.drawable.icon, "出现通知", System
                .currentTimeMillis());
        n.flags = Notification.FLAG_AUTO_CANCEL;

        Intent intent = new Intent(this, Main.class);
        intent.setFlags(Intent.FLAG_ACTIVITY_CLEAR_TOP);

        PendingIntent contentIntent = PendingIntent.getActivity(this, 1,
                intent, 0);
        n.setLatestEventInfo(this, "我是标题", "我是正文", contentIntent);

        nm.notify(100, n);

    }

    @Override
    protected void onPause() {
        //省略
    }

    @Override
```

```
    public void onCreate(Bundle savedInstanceState) {
        //省略
    }
}
```

程序运行后单击 Home 键,LogCat 打印出的结果如图 4.93 所示。

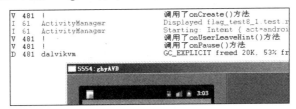

图 4.93　使用 onUserLeaveHint()按下 Home 键的效果

下一步下拉通知栏,重新进入项目,按下 Back 键彻底退出当前应用程序,我们看到并没有在通知栏上显示出图标。

回调函数 onUserLeaveHint()可以按下 Home 键时被触发,也可以在切换 Activity 时被触发,新建名称为 flag_test8_2 的 Android 项目,文件 Main.java 的代码如下:

```
public class Main extends Activity {
    private Button button1;

    @Override
    protected void onUserLeaveHint() {
        super.onUserLeaveHint();
        Log.v("!", "执行了 onUserLeaveHint");
    }

    @Override
    public void onCreate(Bundle savedInstanceState) {
        super.onCreate(savedInstanceState);
        setContentView(R.layout.main);
        button1 = (Button) this.findViewById(R.id.button1);
        button1.setOnClickListener(new OnClickListener() {
            public void onClick(View arg0) {
                Intent intent = new Intent(Main.this, Second.class);
                Main.this.startActivity(intent);
            }
        });
    }
}
```

程序运行后单击按钮在 LogCat 打印日志,结果如图 4.94 所示。

图 4.94 切换 Activity 时回调函数被调用

那有没有办法想要实现按下 Home 键添加通知栏图标,而在切换 Activity 时又不使回调函数 onUserLeaveHint()被调用呢?可以引用本章的主角 FLAG_ACTIVITY_NO_USER_ACTION 标记。

将 flag_test8_2 项目的 Main.java 核心代码更改如下:

```
button1.setOnClickListener(new OnClickListener() {
    public void onClick(View arg0) {
        Intent intent = new Intent(Main.this, Second.class);
        intent.setFlags(Intent.FLAG_ACTIVITY_NO_USER_ACTION);
        Main.this.startActivity(intent);

    }
});
```

程序运行后就可以达到预期的效果了。

4.7.9　FLAG_ACTIVITY_REORDER_TO_FRONT 标记

标记 FLAG_ACTIVITY_REORDER_TO_FRONT 的作用是将当前 Activity 栈中存在的 Activity 对象放入栈顶,比如有 A、B、C、D 这 4 个 Activity,现在的 D 在栈顶,在 D 中呼叫 B 后,则栈的顺序为 ACDB。

新建名称为 flag_test9 的 Android 项目,本示例用 3 个 Activity 对象来进行实验,Main.java 的代码如下:

```java
public class Main extends Activity {
    private Button button1;
    @Override
    public void onCreate(Bundle savedInstanceState) {
        super.onCreate(savedInstanceState);
        setContentView(R.layout.main);

        button1 = (Button) this.findViewById(R.id.button1);
        button1.setOnClickListener(new OnClickListener() {
            public void onClick(View arg0) {
                Intent intent = new Intent(Main.this, Second.class);
                Main.this.startActivity(intent);
            }
        });
    }
}
```

文件 Second.java 的代码如下：

```java
public class Second extends Activity {
    private Button button1;

    @Override
    public void onCreate(Bundle savedInstanceState) {
        super.onCreate(savedInstanceState);
        setContentView(R.layout.second);

        button1 = (Button) this.findViewById(R.id.button1);
        button1.setOnClickListener(new OnClickListener() {
            public void onClick(View arg0) {
                Intent intent = new Intent(Second.this, Third.class);
                Second.this.startActivity(intent);
            }
        });
    }
}
```

文件 Third.java 的代码如下：

```java
public class Third extends Activity {
    private Button button1;

    @Override
    public void onCreate(Bundle savedInstanceState) {
        super.onCreate(savedInstanceState);
        setContentView(R.layout.third);

        button1 = (Button) this.findViewById(R.id.button1);
        button1.setOnClickListener(new OnClickListener() {
            public void onClick(View arg0) {
                Intent intent = new Intent(Third.this, Second.class);
                intent.setFlags(Intent.FLAG_ACTIVITY_REORDER_TO_FRONT);
                Third.this.startActivity(intent);
            }
        });
    }
}
```

程序运行后，经过 Main.java 到 Second.java，再由 Second.java 到 Third.java，到这个步骤，出现的界面如图 4.95 所示。

图 4.95 Third.java 显示界面

系统中 Activity 栈中的顺序从低到高为 Main、Second、Third，此时由 Third.java 使用 FLAG_ACTIVITY_REORDER_TO_FRONT 标记转到 Second.java，栈的顺序从低到高变为 Main、Third、Second，按下 Back 按钮后依次出现 Third 和 Main，再按下 Back 按钮后退出应用程序。

第 5 章 ContentProvider、SharedPreferences 和 SQLite 持久化存储

在 Android 中可以持久化的技术主要分为 File 流、ContentProvider 和 SharedPreferences 以及 SQLite 技术，本章将分别介绍这几种主流的持久化技术的使用方案。

本章中应该着重掌握如下技术点：

- 如何将数据进行持久化，比如使用 File 对象的 IO 流，使用 SharedPreferences 及 SQLite 数据库等
- 跨进程的 ContentProvider 的使用

5.1 在 Android 中使用 File 对象实现文件基本操作

在使用 Java SE 平台开发 CS 结构的软件中，File 的 IO 输入输出流的使用率是非常高的，通过使用 IO 输入输出流可以对存储介质上的文件进行读写操作，下面就实现一个在 Android 平台中使用 File 对象操作文件的功能。

创建名称为 ioTest 的 Android 项目，Main.java 的代码如下：

```java
public class Main extends Activity {

    @Override
    public void onCreate(Bundle savedInstanceState) {
        super.onCreate(savedInstanceState);
        setContentView(R.layout.main);

        try {
            File file = new File("ghy.txt");
            Log.v("path=", file.getAbsolutePath());
            file.createNewFile();
        } catch (IOException e) {
            // TODO Auto-generated catch block
            e.printStackTrace();
        }

    }
}
```

在代码中可以看到创建了一个名称为 ghy.txt 的文件,并且打印保存文件路径的名称,但运行这个项目时却出现了异常,如图 5.1 所示。

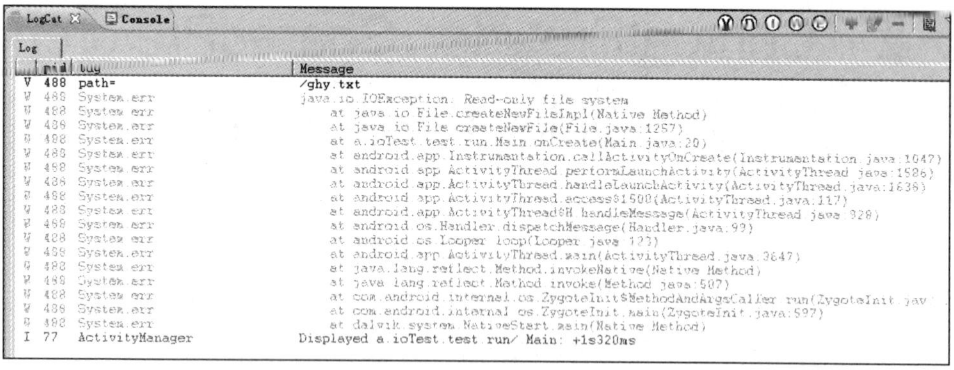

图 5.1 打印文件的绝对路径时出现异常

从出错结果可以看到,创建的文件是在只读的文件系统上,这个只读的存储路径是在系统的根目录下,如图 5.2 所示。

Linux 对权限的要求比较高,不允许随便存储文件,那这个文件到底存储到哪儿不出错呢?请看图 5.3 所示。

图 5.2 根目录是只读的

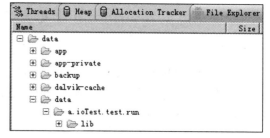

图 5.3 存储到这个目录不出错

从图 5.3 中可以看到,不出异常的路径是在 "/data/data/包名" 中,所以要更改代码如下:

```
public class Main extends Activity {

    @Override
    public void onCreate(Bundle savedInstanceState) {
        super.onCreate(savedInstanceState);
        setContentView(R.layout.main);

        try {
            File file = new File("/data/data/a.ioTest.test.run/ghy.txt");
            Log.v("path=", file.getAbsolutePath());
            file.createNewFile();
        } catch (IOException e) {
            // TODO Auto-generated catch block
            e.printStackTrace();
        }
```

```
        }
    }
```

再次运行项目，没有报错，而且还出现创建的文件 ghy.txt，如图 5.4 所示。

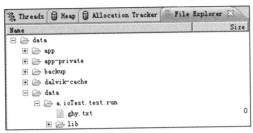

图 5.4 正确创建出文件 ghy.txt

但图 5.4 中的 ghy.txt 文件大小为 0，下面将对 ghy.txt 文件加入文本内容，更改代码如下：

```java
public class Main extends Activity {

    @Override
    public void onCreate(Bundle savedInstanceState) {
        super.onCreate(savedInstanceState);
        setContentView(R.layout.main);

        try {
            File file = new File("/data/data/a.ioTest.test.run/ghy.txt");
            file.createNewFile();
            FileOutputStream fos = new FileOutputStream(file);
            fos.write("我是中国人".getBytes());
            fos.close();
        } catch (IOException e) {
            // TODO Auto-generated catch block
            e.printStackTrace();
        }

    }
}
```

再次运行项目，文件大小变为 15，如图 5.5 所示。

但 ghy.txt 文件内容如何查看？很简单，选中 ghy.txt 后再单击 导出按钮将文件导出到桌面，打开 ghy.txt 文件查看内容，如图 5.6 所示。

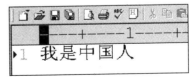

图 5.5 为 ghy.txt 文件添加内容　　　　图 5.6 打开 ghy.txt 确认里面有文本

前面的步骤都是创建文件和写文件的操作，那如何读文件呢？很简单，更改代码如下：

```java
public class Main extends Activity {

    @Override
    public void onCreate(Bundle savedInstanceState) {
        super.onCreate(savedInstanceState);
        setContentView(R.layout.main);

        try {
            File file = new File("/data/data/a.ioTest.test.run/ghy.txt");
            FileInputStream fis = new FileInputStream(file);
            InputStreamReader isrRef = new InputStreamReader(fis);
            char[] charArray = new char[2];
            int readLength = isrRef.read(charArray);
            StringBuffer sbRef = new StringBuffer();
            while (readLength != -1) {
                sbRef.append(charArray, 0, readLength);
                readLength = isrRef.read(charArray);
            }
            Log.v("ghy.txt 文件内容为：", sbRef.toString());
            fis.close();
        } catch (IOException e) {
            // TODO Auto-generated catch block
            e.printStackTrace();
        }

    }
}
```

读取的内容在 LogCat 打印出来，如图 5.7 所示。

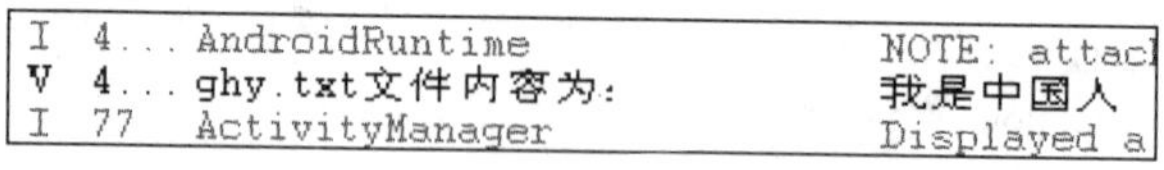

```
I  4... AndroidRuntime              NOTE: attac
V  4... ghy.txt文件内容为：         我是中国人
I  77   ActivityManager             Displayed a
```

图 5.7 从 ghy.txt 中读出的文本

到此，使用 File 类在 Android 平台中操作文件的过程就结束了，操作的代码和在 Java SE 平台中操作的代码一模一样，没有什么区别。

5.2 在 Android 中使用 Android 平台自带对象实现文件的基本操作

虽然在 Java SE 平台上提供了 File 对象来进行文件的基本操作，但在 Android 平台上可以使用 Android 自己独有的操作文件对象，使用这些对象也可以对文件进行读写，而且使用起来比较方便。

5.2.1 使用 openFileOutput 和 openFileInput 读写文件

在上一节中使用 File 类来进行文件的存储，并且要指定文件所在的路径：

File file = new File("/data/data/a.ioTest.test.run/ghy.txt");

其实在 Android 中已经提供了方便的在存储项目中自定义文件的方法，即 openFileOutput 和 openFileInput。本示例就来演示它们的使用方法。

新建 Android 项目，名称为 FileOperate_5_3。更改文件 Main.java 的代码如下：

```java
public class Main extends Activity {
    private Button button1;
    private Button button2;

    @Override
    public void onCreate(Bundle savedInstanceState) {
        super.onCreate(savedInstanceState);
        setContentView(R.layout.main);

        button1 = (Button) this.findViewById(R.id.button1);
        button2 = (Button) this.findViewById(R.id.button2);

        button1.setOnClickListener(new OnClickListener() {
            public void onClick(View arg0) {
                try {
                    FileOutputStream fosRef = Main.this.openFileOutput(
                            "ghy.txt", Context.MODE_PRIVATE);
                    fosRef.write("我又是中国人".getBytes());
                    fosRef.close();
                    // ////////////

                    StringBuffer sbRef = new StringBuffer();
                    FileInputStream fisRef = Main.this.openFileInput("ghy.txt");
                    InputStreamReader isrRef = new InputStreamReader(fisRef);
                    char[] charArray = new char[2];
                    int readLength = isrRef.read(charArray);
                    while (readLength != -1) {
                        sbRef.append(charArray, 0, readLength);
                        readLength = isrRef.read(charArray);
                    }
                    Log.v("读入的值：", new String(sbRef.toString()));
                    fisRef.close();
                    isrRef.close();
                } catch (FileNotFoundException e) {
                    // TODO Auto-generated catch block
                    e.printStackTrace();
                } catch (IOException e) {
                    // TODO Auto-generated catch block
                    e.printStackTrace();
                }
            }
        });

        button2.setOnClickListener(new OnClickListener() {
```

```java
                public void onClick(View arg0) {
                    try {
                        FileOutputStream fosRef = Main.this.openFileOutput(
                                "ghy.txt", Context.MODE_APPEND);
                        fosRef.write("  我不是中国人我是追加的".getBytes());
                        fosRef.close();

                        // ////////////

                        StringBuffer sbRef = new StringBuffer();
                        FileInputStream fisRef = Main.this.openFileInput("ghy.txt");
                        InputStreamReader isrRef = new InputStreamReader(fisRef);
                        char[] charArray = new char[2];
                        int readLength = isrRef.read(charArray);
                        while (readLength != -1) {
                            sbRef.append(charArray, 0, readLength);
                            readLength = isrRef.read(charArray);
                        }
                        Log.v("读入的值: ", new String(sbRef.toString()));
                        fisRef.close();
                        isrRef.close();

                    } catch (FileNotFoundException e) {
                        // TODO Auto-generated catch block
                        e.printStackTrace();
                    } catch (IOException e) {
                        // TODO Auto-generated catch block
                        e.printStackTrace();
                    }

                }
            });

        }
    }
```

程序初始运行效果如图 5.8 所示。

单击上面的按钮正确创建 ghy.txt 文件，并且在 LogCat 面板中打印 ghy.txt 文件的内容，如图 5.9 所示。

图 5.8　程序初始运行效果

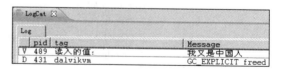

图 5.9　单击上面按钮的执行效果

单击下面的按钮对 ghy.txt 文件追加文本内容，如图 5.10 所示。

ContentProvider、SharedPreferences 和 SQLite 持久化存储 第 5 章

```
D  77    dalvikvm              GC_CONCURRENT freed 455K, 63% free 39
V  489   读入的值：             我又是中国人 我不是中国人我是追加的
```

图 5.10 打印追加后的文本内容

但 ghy.txt 存储到哪里了呢？在图 5.11 所示中可以看到该文件的存储位置。

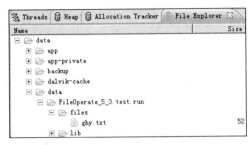

图 5.11 文件 ghy.txt 被保存到 files 目录中了

5.2.2 读取 assets 目录中的文件

Android 中的文件夹 assets 存放的是二进制的文件格式，比如音频、视频、图片等，但 assets 目录中的文件不会被 R.java 文件索引，如果想读取 assets 目录中的文件还需要借助 AssetManager 对象。

新建名称为 assetsTest 的 Android 项目，在 assets 文件夹中加入 ghy.txt 文件，文件内容如图 5.12 所示。

图 5.12 在 assets 文件夹中加入的 ghy.txt 文件内容

更改 Main.java 的代码如下：

```java
public class Main extends Activity {
    @Override
    public void onCreate(Bundle savedInstanceState) {
        super.onCreate(savedInstanceState);
        setContentView(R.layout.main);

        try {
            StringBuffer sbRef = new StringBuffer();
            AssetManager amRef = this.getAssets();
            InputStream isRef = amRef.open("ghy.txt");
            InputStreamReader isrRef = new InputStreamReader(isRef);

            char[] charArray = new char[2];
            int readLength = isrRef.read(charArray);
            while (readLength != -1) {
                sbRef.append(charArray, 0, readLength);
                readLength = isrRef.read(charArray);
```

```
            }
            Log.v("取入的文件内容为：", sbRef.toString());
            isrRef.close();
            isRef.close();

        } catch (IOException e) {
            // TODO Auto-generated catch block
            e.printStackTrace();
        }
    }
}
```

程序运行后取出 assets 目录中 ghy.txt 文件内容，如图 5.13 所示。

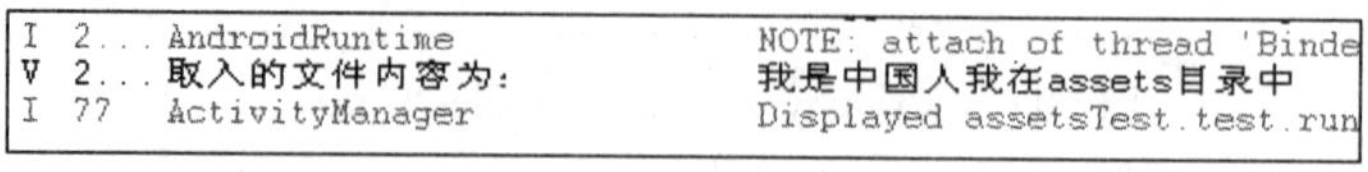

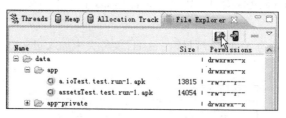

图 5.13　读出 assets 目录中的 ghy.txt 文件内容

但这个 ghy.txt 文件保存到哪儿呢？进入 File Explorer 面板，选中/data/app 目录中的 APK 文件，如图 5.14 所示。

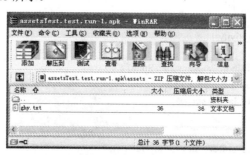

图 5.14　选中 apk 文件准备导出

导出 assetsTest.test.run-1.apk 文件到桌面，用 rar 打开这个 APK 文件后进入 assets 文件夹，看到了 ghy.txt 文件，如图 5.15 所示。

图 5.15　APK 文件中 assets 目录下有 ghy.txt 文件

由于 assets 文件夹中的文件是被打包进 APK 文件中的，所以 assets 目录中的文件只能读，不能写。

5.2.3　读取 res/raw 文件夹中已经存在的 TXT 和 PNG 文件

目录 res 的子目录 raw 的特点是文件能被 R.java 文件索引，目录里面的文件不被编译成二进制的格式，目录中的文件存储到 APK 文件中。而目录 xml 主要的功能就是存放 XML 文件，并且这些 XML 文件是要被编译成二进制的文件格式，有助于程序的运行效率。

创建名称为 FileOperate_5_5 的 Android 项目，并且在 res 目录下创建 raw 子目录，项目的目录结构如图 5.16 所示。

图 5.16　项目结构

还要在 raw 目录中添加 ghyghy.txt 文件和 1 个 PNG 图片，TXT 文件的内容如图 5.17 所示。

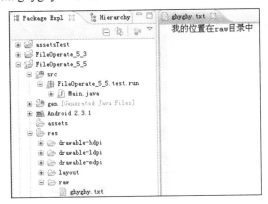

图 5.17　ghyghy.txt 文件内容

提示　一定要设置 ghyghy.txt 文件的编码为 utf-8，不然打印出来的汉字是乱码。

更改文件 Main.java 的代码如下：

```
public class Main extends Activity {
    @Override
    public void onCreate(Bundle savedInstanceState) {
        super.onCreate(savedInstanceState);
        setContentView(R.layout.main);
```

```java
try {
    InputStream isRef = this.getResources().openRawResource(
            R.raw.ghyghy);
    InputStreamReader isrRef = new InputStreamReader(isRef);
    StringBuffer sbRef = new StringBuffer();
    char[] charArray = new char[2];
    int readLength = isrRef.read(charArray);
    while (readLength != -1) {
        sbRef.append(charArray, 0, readLength);
        readLength = isrRef.read(charArray);
    }
    isRef.close();
    isrRef.close();

    Log.v("----", "" + sbRef.toString());

    InputStream isPNGRef = this.getResources().openRawResource(
            R.raw.ghy);
    Bitmap bitmpa = BitmapFactory.decodeStream(isPNGRef);
    ((ImageView) this.findViewById(R.id.imageView1))
            .setImageBitmap(bitmpa);

} catch (NotFoundException e) {
    // TODO Auto-generated catch block
    e.printStackTrace();
} catch (IOException e) {
    // TODO Auto-generated catch block
    e.printStackTrace();
}

}
```

程序运行后在 LogCat 打印的结果如图 5.18 所示。

```
D  6... dalvikvm           Debugger has detached; obj
V  6... ----               我的位置在raw目录中
I  77  ActivityManager     Displayed FileOperate_5_5.
```

图 5.18 LogCat 打印 raw 目录 ghyghy.txt 文件内容

在 AVD 中也打印出了图片，如图 5.19 所示。

图 5.19 显示出了图片

5.2.4 读取 res/xml 文件夹中已经存在的 XML 文件

创建名称为 FileOperate_5_6 的 Android 项目，在 xml 文件夹中新建 userinfo.xml 文件，XML 文件的代码如图 5.20 所示。

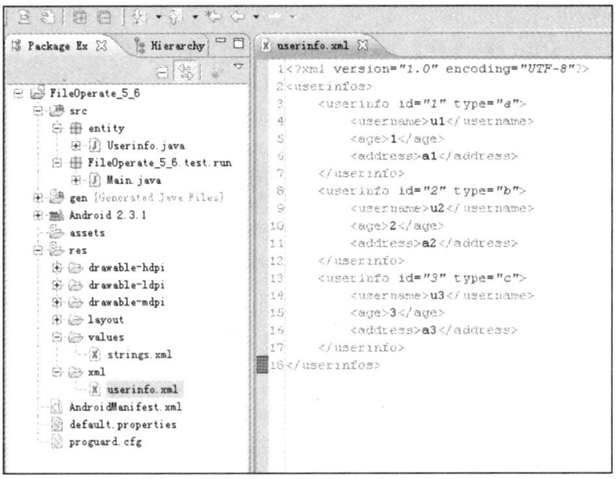

图 5.20 userinfo.xml 文件代码内容

文件 Main.java 解析 XML 文件的代码如下：

```
public class Main extends Activity {
    @Override
    public void onCreate(Bundle savedInstanceState) {
        super.onCreate(savedInstanceState);
        setContentView(R.layout.main);

        List<Userinfo> listUserinfo = new ArrayList<Userinfo>();

        try {
            XmlResourceParser parser = this.getResources().getXml(
                    R.xml.userinfo);
            int eventType = parser.getEventType();
            Userinfo userinfo = null;
            while (eventType != XmlResourceParser.END_DOCUMENT) {

                switch (eventType) {
                case XmlResourceParser.START_TAG:
                    Log.v("START_TAG", parser.getName());
                    if (parser.getName().equals("userinfo")) {
                        userinfo = new Userinfo();
                        userinfo.setId(parser.getAttributeValue(null, "id"));
                        userinfo
                                .setType(parser.getAttributeValue(null, "type"));
                    }
                    if (parser.getName().equals("username")) {
                        userinfo.setUsername(parser.nextText());
                    }
```

```java
                    if (parser.getName().equals("age")) {
                        userinfo.setAge(parser.nextText());
                    }
                    if (parser.getName().equals("address")) {
                        userinfo.setAddress(parser.nextText());
                    }
                    break;
                case XmlResourceParser.END_TAG:
                    Log.v("END_TAG", parser.getName());
                    if (parser.getName().equals("userinfo")) {
                        listUserinfo.add(userinfo);
                    }
                    break;
                }
                eventType = parser.next();
            }

            for (int i = 0; i < listUserinfo.size(); i++) {
                Userinfo userinfoEach = listUserinfo.get(i);
                Log.v("!", "id=" + userinfoEach.getId() + " type="
                        + userinfoEach.getType() + " username="
                        + userinfoEach.getUsername() + " age="
                        + userinfoEach.getType() + " address="
                        + userinfoEach.getAddress());

            }

        } catch (NotFoundException e) {
            // TODO Auto-generated catch block
            e.printStackTrace();
        } catch (XmlPullParserException e) {
            // TODO Auto-generated catch block
            e.printStackTrace();
        } catch (IOException e) {
            // TODO Auto-generated catch block
            e.printStackTrace();
        }

    }
}
```

程序运行结果如图 5.21 所示。

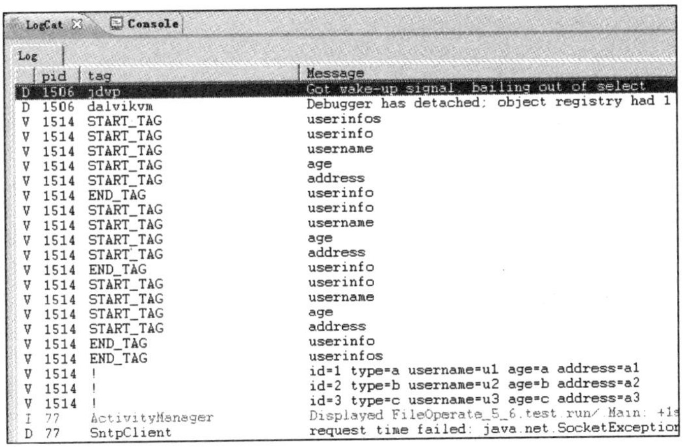

图 5.21　程序运行结果

5.2.5　操作 SD 卡中的文件

前面讲解了操作项目中的文件，本节将实现操作 SD 卡中的文件。

新建名称为 FileOperate_5_4 的 Android 项目，更改文件 Main.java 的代码如下：

```java
public class Main extends Activity {
    @Override
    public void onCreate(Bundle savedInstanceState) {
        super.onCreate(savedInstanceState);
        setContentView(R.layout.main);

        try {
            File sdFile = new File("/sdcard");
            if (sdFile.exists() && sdFile.canWrite()) {
                File txtFile = new File("/sdcard/ghysd.txt");
                txtFile.createNewFile();
                FileOutputStream fosRef = new FileOutputStream(txtFile);
                fosRef.write("我在 SD 卡中".getBytes());
                fosRef.close();
            }

            File readTxtFile = new File("/sdcard/ghysd.txt");
            FileInputStream fisRef = new FileInputStream(readTxtFile);
            InputStreamReader isrRef = new InputStreamReader(fisRef);
            StringBuffer sbRef = new StringBuffer();
            char[] charArray = new char[2];
            int readLength = isrRef.read(charArray);
            while (readLength != -1) {
                sbRef.append(charArray, 0, readLength);
                readLength = isrRef.read(charArray);
            }
            fisRef.close();
            isrRef.close();
```

```
                    Log.v("----", "" + sbRef.toString());
            } catch (IOException e) {
                    e.printStackTrace();
            }
        }
    }
```

程序运行后在 sdcard 目录中创建了文件，文件的内容如图 5.22 所示。

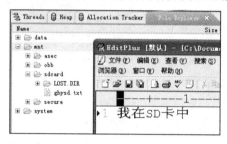

图 5.22　在 SD 卡中创建文件并显示内容

5.3　Linux 中的文件操作权限

操作系统 Linux 和 Windows 在文件访问权限上有本质上的不同，例如图 5.23 所示给出了文件权限的情况。

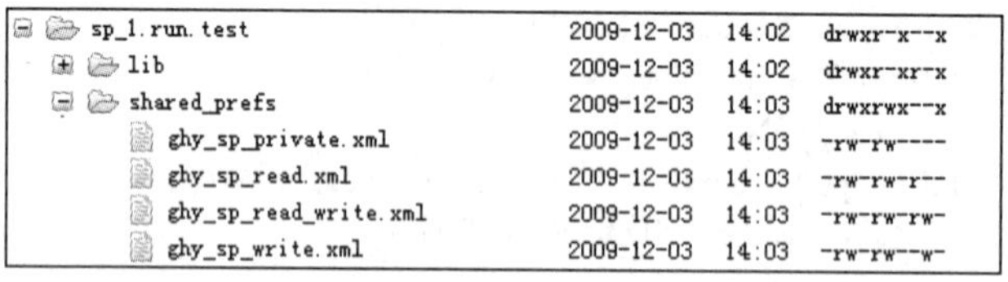

图 5.23　文件权限

图中显示了 4 个文件的权限属性，例如文件 ghy_sp_private.xml 的权限属性值为-rw-rw----，这段权限属性字符串分为 4 段，分别说明如下。

- 第一段是 "-"：其中-（减号）代表此对象是文件，如果是 d 则代表此对象是文件夹。
- 第二段是 "rw-"：r 代表可读，w 代表可写，-代表不允许可执行，此段代表文件的所有者所拥有的权限。
- 第三段是 "rw-"：权限的解释和第二段一样，此段代表文件所有者所在的用户组中其他用户的权限。
- 第四段是 "---"：-代表不允许 r 读，不允许 w 写，不允许 x 可执行，此段表示其他用户组中的用户操作文件的权限，都是-（减号）表示无权限。

5.4 SharedPreferences 的读写权限实验

在 Android 中还可以使用 SharedPreferences 对象来进行数据的持久化，SharedPreferences 对象有些类似于 Web 中的 Cookie 对象，存储的数据结构是 key 和 value 键值对的形式。

创建名称为 sp_1 的 Android 项目，更改文件 Main.java 的代码如下：

```java
public class Main extends Activity {
    private Button button;
    private TextView textView;

    @Override
    public void onCreate(Bundle savedInstanceState) {
        super.onCreate(savedInstanceState);
        setContentView(R.layout.main);

        button = (Button) this.findViewById(R.id.button1);
        textView = (TextView) this.findViewById(R.id.textView1);
        // 私有
        SharedPreferences spRef_private = this.getSharedPreferences(
                "ghy_sp_private", Context.MODE_PRIVATE);
        SharedPreferences.Editor editor_private = spRef_private.edit();
        editor_private.putString("username", "我是高洪岩 private");
        editor_private.commit();
        // 只读
        SharedPreferences spRef_read = this.getSharedPreferences("ghy_sp_read",
                Context.MODE_WORLD_READABLE);
        SharedPreferences.Editor editor_read = spRef_read.edit();
        editor_read.putString("username", "我是高洪岩 read");
        editor_read.commit();
        // 可写
        SharedPreferences spRef_write = this.getSharedPreferences(
                "ghy_sp_write", Context.MODE_WORLD_WRITEABLE);
        SharedPreferences.Editor editor_write = spRef_write.edit();
        editor_write.putString("username", "我是高洪岩 write");
        editor_write.commit();
        // 可读可写
        SharedPreferences spRef_read_write = this.getSharedPreferences(
                "ghy_sp_read_write", Context.MODE_WORLD_READABLE
                        + Context.MODE_WORLD_WRITEABLE);
        SharedPreferences.Editor editor_read_write = spRef_read_write.edit();
        editor_read_write.putString("username", "我是高洪岩 read_write");
        editor_read_write.commit();

        button.setOnClickListener(new OnClickListener() {

            public void onClick(View arg0) {

                textView
```

```
                                .setText("private_username:"
                                        + Main.this.getSharedPreferences(
                                                "ghy_sp_private", 0).getString(
                                                "username", "默认名称 ghy_sp_private")
                                        + "\r\n"
                                        + "read_username:"
                                        + Main.this.getSharedPreferences("ghy_sp_read",
                                                0).getString("username",
                                                "默认名称 ghy_sp_read")
                                        + "\r\n"
                                        + "write_username:"
                                        + Main.this.getSharedPreferences(
                                                "ghy_sp_write", 0).getString(
                                                "username", "默认名称 ghy_sp_write")
                                        + "\r\n"
                                        + "read_write_username:"
                                        + Main.this.getSharedPreferences(
                                                "ghy_sp_read_write", 0).getString(
                                                "username", "默认名称 ghy_sp_read_write"));
                    }
                });
    }
}
```

在上面的代码中，通过使用 SharedPreferences 对象创建的数据具有以下 4 种权限属性：

- Context.MODE_PRIVATE 私有数据。
- Context.MODE_WORLD_READABLE 只读数据。
- Context.MODE_WORLD_WRITEABLE 可写数据。
- Context.MODE_WORLD_WRITEABLE + Context.MODE_WORLD_READABLE 可读写数据。

文件 main.xml 的代码如下：

```xml
<?xml version="1.0" encoding="utf-8"?>
<LinearLayout xmlns:android="http://schemas.android.com/apk/res/android"
    android:orientation="vertical" android:layout_width="fill_parent"
    android:layout_height="fill_parent">
    <TextView android:text="TextView" android:id="@+id/textView1"
        android:layout_width="wrap_content" android:layout_height="wrap_content"></TextView>
    <Button android:text="Button" android:id="@+id/button1"
        android:layout_width="wrap_content" android:layout_height="wrap_content"></Button>
</LinearLayout>
```

程序运行后创建的文件存放在 File Explorer 面板中的/data/data 路径，如图 5.24 所示。

在图 5.24 中继续找文件夹 sp_1.run.test/shared_prefs，在文件夹下有 4 个创建的 XML 文件，如图 5.25 所示。

图 5.24　存放在/data/data 路径中　　图 5.25　创建的 4 个 XML 文件

程序运行后单击界面中的 Button 按钮，取出 XML 文件中的数据，如图 5.26 所示。

图 5.26　正确取出 4 个 XML 文件中的数据

从上面的示例中可以看出，当前的应用程序可以无限制地取出 SharedPreferences 中的数据，具有完全的操作权限，那么如果是两个项目呢？

下面再创建另外一个 Android 项目，名称为 sp_2，更改文件 Main.java 的代码如下：

```java
public class Main extends Activity {

    private TextView textView;

    @Override
    public void onCreate(Bundle savedInstanceState) {
        super.onCreate(savedInstanceState);
        setContentView(R.layout.main);

        try {
            Context otherContext = createPackageContext("sp_1.run.test", 0);

            textView = (TextView) this.findViewById(R.id.textView1);
            textView.setText("private_username:"
                    + otherContext.getSharedPreferences("ghy_sp_private", 0)
                            .getString("username", "默认名称 ghy_sp_private")
                    + "\r\n"
                    + "read_username:"
                    + otherContext.getSharedPreferences("ghy_sp_read", 0)
                            .getString("username", "默认名称 ghy_sp_read")
                    + "\r\n"
                    + "write_username:"
                    + otherContext.getSharedPreferences("ghy_sp_write", 0)
                            .getString("username", "默认名称 ghy_sp_write")
                    + "\r\n"
                    + "read_write_username:"
                    + otherContext.getSharedPreferences("ghy_sp_read_write", 0)
                            .getString("username", "默认名称 ghy_sp_read_write"));
```

```
                } catch (NameNotFoundException e) {
                    // TODO Auto-generated catch block
                    e.printStackTrace();
                }
            }
        }
```

程序运行后出现的效果如图 5.27 所示。

图 5.27 sp2 项目打印的结果

从图 5.27 中可以看到，Context.MODE_PRIVATE 私有数据和 Context.MODE_WORLD_WRITEABLE 可写数据在其他的应用程序中是不能被访问的。

5.5 Uri 对象的匹配

Uri 是通用资源标志符，即 Universal Resource Identifier，它的功能是定义数据的位置，Uri 通常情况下由 5 部分组成，分别是：

[scheme:][//authority][path][?query][#fragment]

scheme 代表模式，也就是访问数据的类别。

authority 代表主机名。

path 代表资源的路径。

query 是参数。

fragment 是资源的片段，它用于在 HTML 语言中进行锚点定位。

如果想掌握 ContentProvider 对象的使用，必须要掌握 Uri 的匹配,这样就可以在 ContentProvider 对象中自定义 Uri 类型。

新建名称为 UriTest 的 Android 项目，将 Main.java 文件的代码更改如下：

```
public class Main extends Activity {

    // NO_MATCH 表示不匹配任何路径的返回代码
    private static final UriMatcher uriMatcherRef = new UriMatcher(
            UriMatcher.NO_MATCH);

    // 添加匹配路径
    static {
        // 第 1 个参数是 authority
        // 第 2 个参数是 path
```

```java
        // 第 3 个参数是匹配路径成功后的代码值
        uriMatcherRef.addURI("accp", "student", 1000);
        uriMatcherRef.addURI("accp", "studentNo/#", 2000);
        uriMatcherRef.addURI("accp", "select", 3000);
        // 这里的#代表匹配任意数字，另外还可以用*来匹配任意文本
    }

    @Override
    public void onCreate(Bundle savedInstanceState) {
        super.onCreate(savedInstanceState);
        setContentView(R.layout.main);

        // match 方法返回的匹配值，没有匹配返回-1
        // 如果匹配成功返回 addURI 的第 3 个参数值
        Uri uri1 = Uri.parse("//accp/student");
        Log.v("====", "" + uriMatcherRef.match(uri1));

        Uri uri10 = Uri.parse("abc://accp/student");
        Log.v("====", "" + uriMatcherRef.match(uri10));

        Uri uri2 = Uri.parse("//accp/studentNo/2000");
        Log.v("====", "" + uriMatcherRef.match(uri2) + "   "
                + ContentUris.parseId(uri2));

        Uri uri3 = Uri.parse("//accp/select?studentNo=3001&class=3002");
        Log.v("====", "" + uriMatcherRef.match(uri3) + "   "
                + uri3.getQueryParameter("studentNo") + "   "
                + uri3.getQueryParameter("class"));

        Uri uri4 = Uri.parse("http://localhost:8081/testProject");
        uri4 = uri4.withAppendedPath(uri4, "4000");
        Log.v("====", "" + uri4.getPath() + "  参数为值："
                + ContentUris.parseId(uri4));

    }
}
```

程序运行后的效果如图 5.28 所示。

```
V  5561  ====                             1000
V  5561  ====                             1000
V  5561  ====                             2000   2000
V  5561  ====                             3000   3001   3002
V  5561  ====                             /testProject/4000  参数为值: 4000
```

图 5.28　程序运行结果

5.6　ContentProvider 对象的初步使用

为了以后完全掌握 ContentProvider 对象的使用，本节的目标是初步掌握 ContentProvider 对象

的 Uri 匹配，只有 Uri 匹配正确，才可以正确地操作数据。

　　ContentProvider 对象是无界面的，主要提供多个应用程序之间互相传输数据的方式，通过使用 Uri 对象就能找到对方的数据操作接口进而进行数据的管理与操作。

　　新建名称为 homeCP 的 Android 项目，再创建名称为 GHYContentProvider.java 的 ContentProvider 对象，代码如下：

```java
public class GHYContentProvider extends ContentProvider {

    private static final UriMatcher uriMatcherRef = new UriMatcher(
            UriMatcher.NO_MATCH);

    static {
        uriMatcherRef.addURI("com.gaohongyan.www", "insert", 1000);
        uriMatcherRef.addURI("com.gaohongyan.www", "delete/#", 2000);
    }

    @Override
    public String getType(Uri arg0) {
        Log.v("！ ", "调用了 getType()方法");
        switch (uriMatcherRef.match(arg0)) {
        case 1000:
            Log.v("！ ", "匹配了 insert");
            return "vnd.android.cursor.item/insert";
        case 2000:
            Log.v("！ ", "匹配了 delete");
            return "vnd.android.cursor.item/delete";
        default:
            throw new IllegalArgumentException();
        }
    }

    @Override
    public int delete(Uri arg0, String arg1, String[] arg2) {
        // TODO Auto-generated method stub
        return 0;
    }

    @Override
    public Uri insert(Uri arg0, ContentValues arg1) {
        // TODO Auto-generated method stub
        return null;
    }

    @Override
    public boolean onCreate() {
        // TODO Auto-generated method stub
        return false;
    }
```

```java
    @Override
    public Cursor query(Uri arg0, String[] arg1, String arg2, String[] arg3,
            String arg4) {
        // TODO Auto-generated method stub
        return null;
    }

    @Override
    public int update(Uri arg0, ContentValues arg1, String arg2, String[] arg3) {
        // TODO Auto-generated method stub
        return 0;
    }

}
```

方法 getType()返回的字符串和 AndroidManifest.xml 文件中<activity>子标签<data>的 android:mimeType 属性值一样,即可成功隐式匹配 Activity。

布局文件 main.xml 的代码如下:

```xml
<?xml version="1.0" encoding="utf-8"?>
<LinearLayout xmlns:android="http://schemas.android.com/apk/res/android"
    android:orientation="vertical" android:layout_width="fill_parent"
    android:layout_height="fill_parent">
    <Button android:text="gotoInsert" android:id="@+id/button1"
        android:layout_width="wrap_content" android:layout_height="wrap_content"></Button>
    <Button android:text="gotoDelete" android:id="@+id/button2"
        android:layout_width="wrap_content" android:layout_height="wrap_content"></Button>
</LinearLayout>
```

文件 Main.java 的代码如下:

```java
public class Main extends Activity {
    private Button button1;
    private Button button2;

    @Override
    public void onCreate(Bundle savedInstanceState) {
        super.onCreate(savedInstanceState);
        setContentView(R.layout.main);

        button1 = (Button) this.findViewById(R.id.button1);
        button1.setOnClickListener(new OnClickListener() {
            @Override
            public void onClick(View arg0) {
                Intent intent = new Intent("insertAction", Uri
                        .parse("content://com.gaohongyan.www/insert"));
```

```
                    Main.this.startActivity(intent);
            }
        });

        button2 = (Button) this.findViewById(R.id.button2);
        button2.setOnClickListener(new OnClickListener() {
            @Override
            public void onClick(View arg0) {
                Intent intent = new Intent("deleteAction", Uri
                        .parse("content://com.gaohongyan.www/delete/999"));
                Main.this.startActivity(intent);
            }
        });
    }
}
```

> content://是协议的固定写法，专用于 ContentProvider 技术。另外，一定要在 Uri 字符串 delete 后面写上 999 参数值，不写会出现匹配不成功的结果。

布局文件 insert.xml 的代码如下：

```xml
<?xml version="1.0" encoding="utf-8"?>
<LinearLayout xmlns:android="http://schemas.android.com/apk/res/android"
    android:orientation="vertical" android:layout_width="fill_parent"
    android:layout_height="fill_parent">
    <EditText android:text="insert" android:layout_width="match_parent"
        android:layout_height="wrap_content" android:id="@+id/editText1"></EditText>
</LinearLayout>
```

对应的 Activity 文件 Insert.java 的代码如下：

```java
public class Insert extends Activity {
    /** Called when the activity is first created. */
    @Override
    public void onCreate(Bundle savedInstanceState) {
        super.onCreate(savedInstanceState);
        setContentView(R.layout.insert);
    }
}
```

布局文件 delete.xml 的代码如下：

```xml
<?xml version="1.0" encoding="utf-8"?>
<LinearLayout xmlns:android="http://schemas.android.com/apk/res/android"
    android:orientation="vertical" android:layout_width="fill_parent"
    android:layout_height="fill_parent">
    <EditText android:text="delete" android:layout_width="match_parent"
        android:layout_height="wrap_content" android:id="@+id/editText1"></EditText>
</LinearLayout>
```

对应的 Activity 文件 Delete.java 的代码如下:

```java
public class Delete extends Activity {
    /** Called when the activity is first created. */
    @Override
    public void onCreate(Bundle savedInstanceState) {
        super.onCreate(savedInstanceState);
        setContentView(R.layout.delete);
    }
}
```

项目配置文件的代码如下:

```xml
<?xml version="1.0" encoding="utf-8"?>
<manifest xmlns:android="http://schemas.android.com/apk/res/android"
    package="homeCP.test.run" android:versionCode="1" android:versionName="1.0">

    <application android:icon="@drawable/icon" android:label="@string/app_name">
        <activity android:name=".Main" android:label="@string/app_name">
            <intent-filter>
                <action android:name="android.intent.action.MAIN" />
                <category android:name="android.intent.category.LAUNCHER" />
            </intent-filter>
        </activity>
        <activity android:name=".Delete" android:label="@string/app_name">
            <intent-filter>
                <action android:name="deleteAction"></action>
                <category android:name="android.intent.category.DEFAULT" />
                <data android:mimeType="vnd.android.cursor.item/delete"></data>
            </intent-filter>
        </activity>

        <activity android:name=".Insert" android:label="@string/app_name">
            <intent-filter>
                <action android:name="insertAction"></action>
                <category android:name="android.intent.category.DEFAULT" />
                <data android:mimeType="vnd.android.cursor.item/insert"></data>
            </intent-filter>
        </activity>
        <provider android:name="ghycontentprovider.GHYContentProvider"
            android:authorities="com.gaohongyan.www"></provider>

    </application>
</manifest>
```

在 Main.java 文件中的下述代码中:

```java
button1.setOnClickListener(new OnClickListener() {
    public void onClick(View arg0) {
        Intent intent = new Intent("insertAction", Uri
```

```
                        .parse("content://com.gaohongyan.www/insert"));
                Main.this.startActivity(intent);
            }
        });
```

其中的 com.gaohongyan.www 主机名字符串必须与 AndroidManifest.xml 文件中的<provider>节点 android:authorities 属性值一样，android:authorities 属性值的作用是标识一个 ContentProvider 对象在系统中的唯一。

程序运行后的初始效果如图 5.29 所示。

单击 gotoInsert 按钮进入 Insert 界面，如图 5.30 所示。

再单击 gotoDelete 按钮进入 delete 界面，如图 5.31 所示。

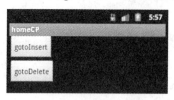  

　　图 5.29　初始运行效果　　　　　图 5.30　进入 insert 界面　　　　图 5.31　进入 delete 界面

至此已经在本示例中实现了 Uri 匹配，ContentProvider 初步使用及自定义<data>数据类型的使用。此示例的执行顺序为：

（1）启动项目，将 AndroidManifest.xml 文件中的所有对象注册到 Android 的 AVD 虚拟机中。

（2）单击 Button1 执行如下代码：

```
button1.setOnClickListener(new OnClickListener() {
    public void onClick(View arg0) {
        Intent intent = new Intent("insertAction", Uri
                .parse("content://com.gaohongyan.www/insert"));
        Main.this.startActivity(intent);
    }
});
```

到系统中去匹配<action android:name=" " />的值为 insertAction 的 Activity，这时已经有满足条件，但是还要继续匹配 1 个 Uri。

（3）由于这个 Uri 是一个 content://协议，一个 ContentProvider 对象，所以将 Uri 协议的主机名 com.gaohongyan.www 字符串提取出来，到 AndroidManifest.xml 文件中找有没有<provider>标签的属性 android:authorities 值为 com.gaohongyan.www 的 ContentProvider 对象。

（4）本示例中存在 android:authorities 属性值为 com.gaohongyan.www 的 ContentProvider 对象，所以自动调用 ContentProvider 对象的 public String getType(Uri arg0)方法返回 1 个 MIME 的字符串，并且这个返回的 MIME 字符串要与<activity>标签子标签<data>中声明的值一样，参见如下代码：

```
<activity android:name=".Delete" android:label="@string/app_name">
    <intent-filter>
        <action android:name="deleteAction"></action>
        <category android:name="android.intent.category.DEFAULT" />
        <data android:mimeType="vnd.android.cursor.item/delete"></data>
```

```
        </intent-filter>
    </activity>

    <activity android:name=".Insert" android:label="@string/app_name">
        <intent-filter>
            <action android:name="insertAction"></action>
            <category android:name="android.intent.category.DEFAULT" />
            <data android:mimeType="vnd.android.cursor.item/insert"></data>
        </intent-filter>
    </activity>
```

（5）经过 action 及 data 的成功匹配后，指定的 Activity 对象即被显示在前台。

其实本示例也是隐式调用 Activity 对象的另外一种实现。

5.7 SQLite 数据库的使用

Android 中的数据库持久化方案采用的是 SQLite，它是一种文件型数据库，支持常用的函数，使用起来比较方便，文件的体积非常小巧。

5.7.1 使用 Navicat_for_SQLite 工具创建 SQLite 数据库及表

SQLite 数据库也有自己的 UI 软件，名称是 Navicat_for_SQLite.rar。软件 Navicat_for_SQLite.rar 中的文件列表如图 5.32 所示。

双击名称为 navicat.exe 的可执行文件，运行效果如图 5.33 所示。

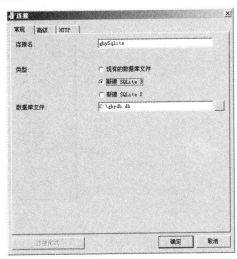

图 5.32 软件 Navicat_for_SQLite.rar 中的文件列表　　图 5.33 连接数据库界面

输入连接数据库的"连接名"，类型选择"新建 SQLite 3"版本，因为是新的示例，以前从未创建数据库，所以新建一个版本为 SQLite3 的数据库，这个新建的数据库文件放在 C 盘下，数据库文件名为 ghydb.db，配置完成后单击"确定"按钮进行数据库的创建，成功创建数据库后的文件名如图 5.34 所示。

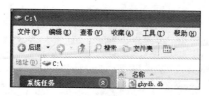

图 5.34　生成的数据库文件 ghydb.db

单击"确定"按钮后自动进入如图 5.35 所示的软件主界面。

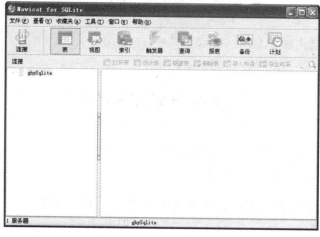

图 5.35　软件主界面

双击左边"连接"列表中的"ghySQLite"数据库，准备新建数据表，如图 5.36 所示。

图 5.36　新建数据表

添加 3 个字段，并且设置主键，如图 5.37 所示。

ContentProvider、SharedPreferences 和 SQLite 持久化存储 第 5 章

图 5.37　设置主键添加其他字段

设置主键自增的属性，界面如图 5.38 所示。

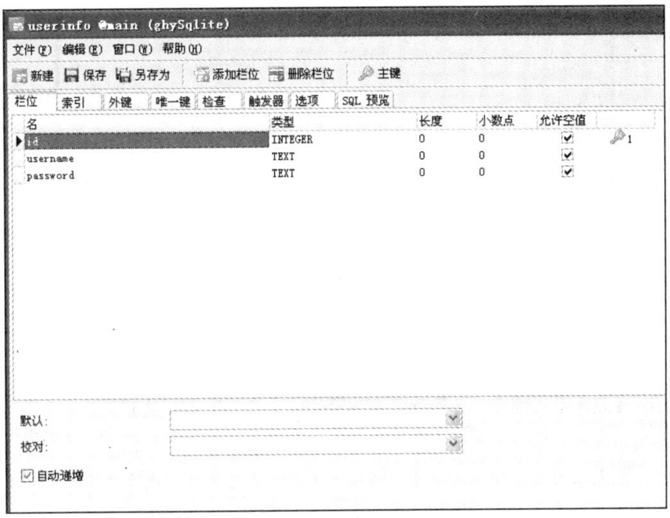

图 5.38　设置主键的自增属性

最后保存数据表，名称为 userinfo，在 userinfo 数据表上单击鼠标右键打开表，如图 5.39 所示。添加两条数据记录，如图 5.40 所示。

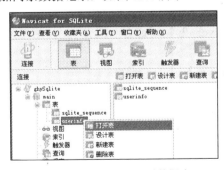

图 5.39　打开 userinfo 数据表

图 5.40　添加两条记录

添加两条记录后，别忘了单击界面下方工具条中的提交铵钮✓将事务提交，将两条记录保存进数据库，再单击主界面中的"查询"按钮，如图 5.41 所示。

再单击"查询"按钮下面的"新建查询"按钮，如图 5.42 所示。

　　图 5.41　单击靠右侧的"查询"按钮　　　图 5.42　单击"新建查询"按钮

显示出 SQL 编辑器后，输入查询语句，得到记录总数为两条的结果，如图 5.43 所示。

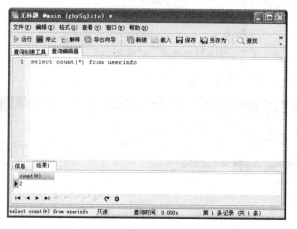

图 5.43　查询到两条数据记录

到此，使用 Navicat_for_SQLite 工具创建 SQLite 数据库及表的过程就介绍完毕。

5.7.2　使用 SQLiteDatabase 对象的常用方法操作数据库

Android 中提供了 SQLiteDatabase 类来操作 SQLite 数据库，SQLiteDatabase 类中的方法比较简单，使用起来和 JDBC 差不多，都是针对 SQL 语言的字符串进行数据库的操作。

首先，创建名称为 SQLiteDatabaseMethodTest 的 Android 项目，然后进行下述操作。

1．使用 execSQL 方法实现 insert 操作与事务提交（commit）与回滚（rollback）的处理

前面创建的 ghy.db 数据库中的 userinfo 数据表有两条记录，本节将实现一个事务与 insert 操作的联合示例，在 res 目录下创建 raw 目录，并且把 ghydb.db 复制到这个目录中，如图 5.44 所示。

图 5.44　复制 ghydb.db 数据库到 raw 目录中

先了解一下 Android 中数据库事务相关的知识点。

在 SQLiteDatabase 类中涉及事务的 API 有 3 个：

（1）public void beginTransaction()

功能：开启一个事务，在 Android 中的 SQLite 数据库中事务是可以嵌套的，如果在实现所有的数据库操作后，调用 setTransactionSuccessful() 方法就证明数据库操作没有异常，setTransactionSuccessful()方法的主要功能是设置数据库操作成功的标志，通过执行这个标志 setTransactionSuccessful()方法代码 SQLite 数据库才可以确定到底是将数据库提交还是回滚。

（2）public void endTransaction()

功能：如果在执行 endTransaction()方法之前没有调用 setTransactionSuccessful()方法，执行 endTransaction()方法时就是回滚操作，反之就是提交操作。

（3）setTransactionSuccessful()

功能：用于设置数据库操作结果的标志。

执行 insert 语句的 execSQL 方法声明如下：

```
public void execSQL (String sql)
```

这个方法是执行一个没有返回值的 SQL 语句，比如执行 insert 插入记录的情况下就可以使用这个 public void execSQL (String sql)方法。

但在通常的情况下 SQLite 数据库是存放在项目中的 res/raw 目录下，在运行时这个目录中的文件只能读，不能写入，因为是打包进 APK 文件中，所以在项目运行后必须通过代码的方式复制到项目私有目录中，这个目录是在 AVD 的/data/data/包名/下，所以新建一个名称为 CopyDBTools.java 的 java 文件，代码实现的主要功能就是判断/data/data/包名/文件夹下有没有数据库文件，如果没有则复制过去，如果有就判断一个 boolean 标志，true 代表覆盖旧的数据库，false 代表不覆盖数据库，CopyDBTools.java 的代码如下：

```java
public class CopyDBTools {
    public static void beginCopyDB(Context context, String dbname,
            String copyToPath, boolean isReWrite) {

        try {
            File file = new File(copyToPath + dbname);

            if (file.exists() == false) {
                // 不存在
                Log.v("!", "数据库不存在 COPY 新的去");
                InputStream isRef = context.getResources().openRawResource(
                        R.raw.ghydb);
                FileOutputStream fosRef = new FileOutputStream(file);
                byte[] byteArray = new byte[2];
                int readLength = isRef.read(byteArray);
                while (readLength != -1) {
                    fosRef.write(byteArray, 0, readLength);
                    readLength = isRef.read(byteArray);
                }
```

```
                    fosRef.close();
                    isRef.close();
                } else {
                    // 存在
                    if (isReWrite == true) {
                        Log.v("!", "数据库存在并且删除旧的写新的");
                        if (file.delete() == true) {
                            Log.v("覆盖的情况：", "删除旧的数据库成功");
                            InputStream isRef = context.getResources()
                                    .openRawResource(R.raw.ghydb);
                            FileOutputStream fosRef = new FileOutputStream(file);
                            byte[] byteArray = new byte[2];
                            int readLength = isRef.read(byteArray);
                            while (readLength != -1) {
                                fosRef.write(byteArray, 0, readLength);
                                readLength = isRef.read(byteArray);
                            }
                            fosRef.close();
                            isRef.close();
                        }
                    }
                }
            } catch (NotFoundException e) {
                // TODO Auto-generated catch block
                e.printStackTrace();
            } catch (FileNotFoundException e) {
                // TODO Auto-generated catch block
                e.printStackTrace();
            } catch (IOException e) {
                // TODO Auto-generated catch block
                e.printStackTrace();
            }
        }
    }
```

还要更改 Main.java 的代码如下：

```
public class Main extends Activity {
    @Override
    public void onCreate(Bundle savedInstanceState) {
        super.onCreate(savedInstanceState);
        setContentView(R.layout.main);
        CopyDBTools.beginCopyDB(this, "ghy.db",
                "/data/data/SQLiteDatabaseMethodTest.test.run/", true);
    }
}
```

项目中的 ghydb.db 文件在 raw 目录中，如图 5.45 所示。

运行这个项目，ghydb.db 文件被成功地复制到了项目私有访问目录中，如图 5.46 所示。

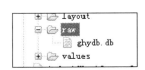

图 5.45　ghydb.db 在 raw 目录中　　　图 5.46　ghydb.db 成功复制到私有目录中

有了数据库文件 ghydb.db 就可以实现本节的主题：事务与插入。

更改 Main.java 的代码如下：

```java
public class Main extends Activity {
    @Override
    public void onCreate(Bundle savedInstanceState) {
        super.onCreate(savedInstanceState);
        setContentView(R.layout.main);
        CopyDBTools.beginCopyDB(this, "ghydb.db",
                "/data/data/SQLiteDatabaseMethodTest.test.run/", false);

        SQLiteDatabase database = SQLiteDatabase.openOrCreateDatabase(
                "/data/data/SQLiteDatabaseMethodTest.test.run/ghydb.db", null);
        boolean isSuccessTransaction = false;
        try {
            database.beginTransaction();
            database
                    .execSQL("insert into userinfo(username,password) values('a','aa')");
            database
                    .execSQL("insert into userinfo(username,password) values('b','bb')");
            database.setTransactionSuccessful();
            Log.v("设置事务成功标志", "设置事务成功标志");
            isSuccessTransaction = true;
        } finally {
            if (isSuccessTransaction == true) {
                Log.v("事务提交", "事务提交");
            } else {
                Log.v("事务回滚", "事务回滚");
            }
            database.endTransaction();
        }
        database.close();

    }
}
```

运行项目，从 LogCat 打印出来的结果来看事务成功提交了，如图 5.47 所示。

```
V  3891   设置事务成功标志    设置事务成功标志
V  3891   事务提交            事务提交
```

图 5.47 成功 insert 并实现事务提交

将数据库 ghydb.db 从 AVD 设备导出到 C 盘根目录下，将旧的 ghydb.db 文件覆盖掉，然后进入 Navicat_for_SQLite.rar 工具打开数据表 userinfo，查看新添加的 2 条记录，如图 5.48 所示。

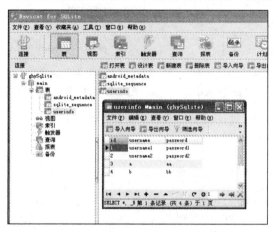

图 5.48 查看新添加的 2 条记录

上面的测试提交是成功了，但还没有实现回滚，我们继续进行下面的测试。

将前面代码第 2 个 insert 语句改成如下形式：

```
database
    .execSQL("insert into userinfozzzzzz(username,password) values('b','bb')");
```

数据库中并没有 userinfozzzzzz 数据表，执行 SQL 时肯定出错，运行这个项目，在 LogCat 中打印出错信息，如图 5.49 所示。

```
V  5662  事务回滚          事务回滚
D  5662  AndroidRuntime    Shutting down VM
V  5662  dalvikvm          threadid=1: thread exiting with uncaught exception (group=0x40015560)
E  5662  AndroidRuntime    FATAL EXCEPTION: main
E  5662  AndroidRuntime    java.lang.RuntimeException: Unable to start activity ComponentInfo{SQLiteDatabaseMethodTe
E  5662  AndroidRuntime        at android.app.ActivityThread.performLaunchActivity(ActivityThread.java:1622)
E  5662  AndroidRuntime        at android.app.ActivityThread.handleLaunchActivity(ActivityThread.java:1638)
E  5662  AndroidRuntime        at android.app.ActivityThread.access$1500(ActivityThread.java:117)
E  5662  AndroidRuntime        at android.app.ActivityThread$H.handleMessage(ActivityThread.java:928)
E  5662  AndroidRuntime        at android.os.Handler.dispatchMessage(Handler.java:99)
E  5662  AndroidRuntime        at android.os.Looper.loop(Looper.java:123)
E  5662  AndroidRuntime        at android.app.ActivityThread.main(ActivityThread.java:3647)
E  5662  AndroidRuntime        at java.lang.reflect.Method.invokeNative(Native Method)
E  5662  AndroidRuntime        at java.lang.reflect.Method.invoke(Method.java:507)
E  5662  AndroidRuntime        at com.android.internal.os.ZygoteInit$MethodAndArgsCaller.run(ZygoteInit.java:839)
E  5662  AndroidRuntime        at com.android.internal.os.ZygoteInit.main(ZygoteInit.java:597)
E  5662  AndroidRuntime        at dalvik.system.NativeStart.main(Native Method)
E  5662  AndroidRuntime    Caused by: android.database.sqlite.SQLiteException: no such table: userinfozzzzzz : insert
E  5662  AndroidRuntime        at android.database.sqlite.SQLiteDatabase.native_execSQL(Native Method)
E  5662  AndroidRuntime        at android.database.sqlite.SQLiteDatabase.execSQL(SQLiteDatabase.java:1743)
E  5662  AndroidRuntime        at SQLiteDatabaseMethodTest.test.run.Main.onCreate(Main.java:25)
E  5662  AndroidRuntime        at android.app.Instrumentation.callActivityOnCreate(Instrumentation.java:1047)
E  5662  AndroidRuntime        at android.app.ActivityThread.performLaunchActivity(ActivityThread.java:1586)
E  5662  AndroidRuntime        ... 11 more
W  78    ActivityManager   Force finishing activity SQLiteDatabaseMethodTest.test.run/.Main
W  78    ActivityManager   Force finishing activity test.test.run/.Main
```

图 5.49 没有 userinfozzzzzz 数据表出错了

再把 AVD 中的 ghydb.db 数据库导出到 C 盘覆盖旧的数据库，再次查看数据表 userinfo 中的数据还是 4 条记录，如图 5.50 所示。

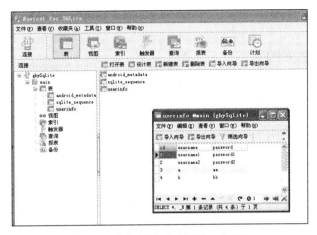

图 5.50　事务回滚成功还是 4 条记录

当然，在 Android 中还可以加入监听事务的机制，更改 Main.java 的代码如下：

```
public class Main extends Activity {
    @Override
    public void onCreate(Bundle savedInstanceState) {
        super.onCreate(savedInstanceState);
        setContentView(R.layout.main);
        CopyDBTools.beginCopyDB(this, "ghydb.db",
                "/data/data/SQLiteDatabaseMethodTest.test.run/", false);

        SQLiteDatabase database = SQLiteDatabase.openOrCreateDatabase(
                "/data/data/SQLiteDatabaseMethodTest.test.run/ghydb.db", null);
        boolean isSuccessTransaction = false;
        try {
            database
                    .beginTransactionWithListener(new SQLiteTransactionListener() {
                        public void onRollback() {
                            Log.v("在事务监听中监听到回滚了！", "！");
                        }

                        public void onCommit() {
                            Log.v("在事务监听中监听到提交了！", "！");
                        }

                        public void onBegin() {
                            Log.v("在事务监听中监听到开启了！", "！");
                        }
                    });

            database
                    .execSQL("insert into userinfo(username,password) values('a','aa')");
            database
                    .execSQL("insert into userinfozzzzzz(username,password) values('b','bb')");
            database.setTransactionSuccessful();
            Log.v("设置事务成功标志", "设置事务成功标志");
```

```
            isSuccessTransaction = true;
        } finally {
            if (isSuccessTransaction == true) {
                Log.v("事务提交", "事务提交");
            } else {
                Log.v("事身回滚", "事务回滚");
            }
            database.endTransaction();
        }
        database.close();
    }
}
```

程序运行后在 LogCat 中的打印效果如图 5.51 所示。

图 5.51　加入事务监听的代码

2. SQLiteStatement compileStatement (String sql)方法的使用

compileStatement(String sql) 方法用于返回一个预编译 SQL 语句的对象 SQLiteStatement，在这个 SQLiteStatement 对象中可以用 "?"（问号）的方式传递参数，但这个方法不能返回多行数据，只能返回 1 行 1 列的结果值，也可以执行 insert 语句，但如果真的返回多行多列的结果集时，则只能取得第 1 行第 1 列的值。

将 Main.java 的代码更改如下：

```
            // 省略
            // 计算插入前一共有多少条记录
            SQLiteStatement stmt0 = database
                    .compileStatement("select count(*) from userinfo");
            Log.v("计算插入前一共有多少条记录： ", "" + stmt0.simpleQueryForLong());
            // //////////////////////////////////////////

            // 只有 insert 不处理其他功能
```

```
Log.v("execute()方法开始", "!");
for (int i = 0; i < 5; i++) {
    SQLiteStatement stmt1 = database
            .compileStatement("insert into userinfo(username,password) values(?,?)");
    stmt1.bindString(1, "usernameA" + (i + 1));
    stmt1.bindString(2, "passwordA" + (i + 1));
    stmt1.execute();
}
Log.v("execute()方法结束", "!");
// ////////////////////////////////////////

// 不光 insert 还查看 insert 后的主键值
Log.v("executeInsert()方法开始", "!");
for (int i = 5; i <= 10; i++) {
    SQLiteStatement stmt2 = database
            .compileStatement("insert into userinfo(username,password) values(?,?)");
    stmt2.bindString(1, "usernameA" + (i + 1));
    stmt2.bindString(2, "passwordA" + (i + 1));
    Log.v("插入后的主键 ID 值为：", "" + stmt2.executeInsert());
}
Log.v("executeInsert()方法结束", "!");
// ////////////////////////////////////////

// 返回一共有多少条记录
SQLiteStatement stmt3 = database
        .compileStatement("select count(*) from userinfo");
Log.v("一共有多少条记录：", "" + stmt3.simpleQueryForLong());
// ////////////////////////////////////////

// 返回 id 是 7 的 username 的值是多少
SQLiteStatement stmt4 = database
        .compileStatement("select username from userinfo where id=7");
Log.v("id 为 7 的 username 值为：", "" + stmt4.simpleQueryForString());
// ////////////////////////////////////////
// 省略
```

程序运行后的效果如图 5.52 所示。

图 5.52　程序运行后的打印结果

数据表 userinfo 中的数据如图 5.53 所示。

图 5.53　userinfo 表中的数据

3. int delete (String table, String whereClause, String[] whereArgs)方法的使用

方法 delete()具有删除记录的功能，本示例实现删除 id 为 8 和 9 的数据记录，更改 Main.java 的代码如下：

```
// 省略
    database.delete("userinfo", "id=? or id=?",
            new String[] { "8", "9" });
// 省略
```

删除后的数据表 userinfo 如图 5.54 所示。

图 5.54　删除 id 为 8 和 9 后的数据

4. execSQL(String sql)和 execSQL(String sql, Object[] bindArgs)方法的使用

execSQL 方法的功能是执行没有返回结果值的语句，比如 delete 和 insert 或 create 创建对象类型的 SQL 语句。

更改 Main.java 的代码如下：

```
// 省略
```

```
database
        .execSQL("insert into userinfo(username,password) values('cc','ccc')");
database
        .execSQL("insert into userinfo(username,password) values('dd','ddd')");

database.execSQL(
        "insert into userinfo(username,password) values(?,?)",
        new String[] { "xx", "xxx" });
database.execSQL(
        "insert into userinfo(username,password) values(?,?)",
        new String[] { "yy", "yyy" });
// 省略
```

程序运行后在 userinfo 表中添加了 4 条记录, 如图 5.55 所示。

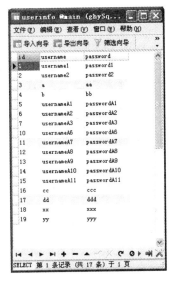

图 5.55　添加了 4 条记录

5. final String getPath ()方法的使用

getPath()方法的作用是返回数据库的存放路径, 在 Main.java 文件中使用 getPath()函数的代码如下:

```
Log.v("数据库存放路径", "" + database.getPath());
```

程序运行后效果如图 5.56 所示。

```
V  1...  数据库存放路径         /data/data/SQLiteDatabaseMethodTest.test.run/ghydb.db
V  1...  设置事务成功标志        设置事务成功标志
V  1...  事务提交              事务提交
```

图 5.56　打印数据库存放路径

6. SQLiteDatabase create (SQLiteDatabase.CursorFactory factory)创建内存数据库

create()方法可以在内存中创建数据表及动态操作表中的数据, 更改 Main.java 的代码如下:

```
public class Main extends Activity {
    @Override
```

```
    public void onCreate(Bundle savedInstanceState) {
        super.onCreate(savedInstanceState);
        setContentView(R.layout.main);
        boolean isSuccessTransaction = false;
        SQLiteDatabase database2 = SQLiteDatabase.create(null);
        try {
            database2.beginTransaction();
            database2
                    .execSQL("create    table    abc(\"id\"    INTEGER    PRIMARY    KEY
AUTOINCREMENT,\"uu\"    TEXT,\"pp\"    TEXT)");
            database2.execSQL("insert into abc(uu,pp) values('uu1','pp1')");
            Cursor cursorRef = database2.query("abc",
                    new String[] { "uu", "pp" }, "", new String[] {}, "", "",
                    "");
            while (cursorRef.moveToNext()) {
                Log.v("uu:", cursorRef.getString(0) + " pp="
                        + cursorRef.getString(1));
            }
            database2.setTransactionSuccessful();
            Log.v("设置事务成功标志", "设置事务成功标志");
            isSuccessTransaction = true;
        } finally {
            if (isSuccessTransaction == true) {
                Log.v("事务提交", "事务提交");
            } else {
                Log.v("事务回滚", "事务回滚");
            }
            database2.endTransaction();
        }
        database2.close();

    }
}
```

程序运行后在 LogCat 中的打印结果如图 5.57 所示。

V	1...	uu:	uu1 pp=pp1
V	1...	设置事务成功标志	设置事务成功标志
V	1...	事务提交	事务提交

图 5.57　打印动态创建数据表中的数据

7. long insert (String table, String nullColumnHack, ContentValues values)方法的使用

insert()方法的功能是往数据库中插入数据，该方法需要 3 个参数：

- 参数 table：对哪个表进行 insert 操作。
- 参数 values：往表中插入什么数据，用 ContentValues 对象封装。
- 参数 nullColumnHack：由于在 SQL 中不允许所有的列都是空，这样的数据是没有意义的，所以当传入第 3 个参数 ContentValues 的值为 null 或 size()为 0 时，使用本参数往指定的列名字段中插入 null 值，这样至少有 1 个列有值了，虽然这个值是 null。

更改 Main.java 的代码如下：

```
// 省略
            Log.v("新插入的主键 id 值为：", ""
                + database.insert("userinfo", "username",
                    new ContentValues()));
            Log.v("新插入的主键 id 值为：", ""
                + database.insert("userinfo", "username", null));
// 省略
```

程序运行后在 LogCat 中打印出日志记录，如图 5.58 所示。

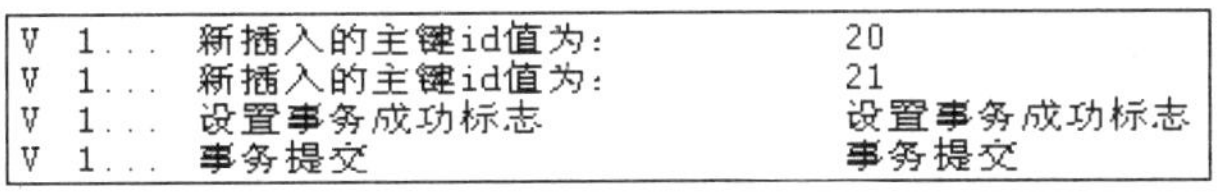

图 5.58 使用 insert 方法插入记录

查看数据库 ghydb.db 中的数据表 userinfo 中的数据，如图 5.59 所示。

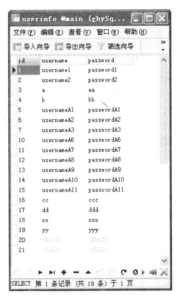

图 5.59 userinfo 表中新加两条空记录

8. Cursor query (String table, String[] columns, String selection, String[] selectionArgs, String groupBy, String having, String orderBy, String limit)方法的使用

query()方法用于实现查询功能，参数解释如下。

- 参数 table：要查询表的名称。
- 参数 columns：要查询列名的数组，如果传空值将要返回全部的列。
- 参数 selection：要查询的条件，但不包含 where 关键字。
- 参数 selectionArgs：查询条件的参数值，如果用 "?" 号的形式进行查询，则必须用这个参数对 "?" 问号值进行填充。
- 参数 groupBy：分组语句，不包括 group by 关键字。

- 参数 having：对分组进行查询条件的过滤，不包含 having 关字键。
- 参数 orderBy：对某个字段进行排序，不包括 order by 关键字。
- 参数 limit：限定返回的行数。

更改数据表 userinfo 的结构及数据表中的数据，如图 5.60 所示。

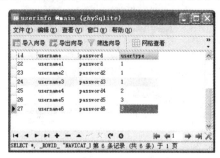

图 5.60 更改 userinfo 表结构及表数据

我们看到，数据表 userinfo 中一共有 6 条记录。

现在，更改 Main.java 的代码如下：

```
// 省略
    Cursor cursor1 = database.query("userinfo", new String[] { "id",
            "username", "password" }, "id=? or id=? or id=?",
            new String[] { "22", "23", "24" }, "", "", "id desc", "");
    Log.v("普通查询开始", "----------------");
    while (cursor1.moveToNext()) {
        String id = cursor1.getString(0);
        String username = cursor1.getString(1);
        String password = cursor1.getString(2);
        Log.v("数据为：", "id=" + id + " username=" + username
                + " password=" + password);
    }
    Log.v("普通查询结束", "----------------");

    Cursor cursor2 = database.query("userinfo", new String[] {
            "usertype", "count(*) as count" }, "", new String[] {},
            "usertype", "count(*) >=2", "id desc", "");
    Log.v("分组查询开始", "----------------");
    while (cursor2.moveToNext()) {
        String usertype = cursor2.getString(0);
        String count = cursor2.getString(1);
        Log.v("数据为：", "usertype=" + usertype + " count=" + count);
    }
    Log.v("分组查询结束", "----------------");

    //下面的 limit 参数是从结果集中返回的记录数
    //这个功能可以实现分页，但关于复杂版分页功能请参看后面的章节
    Cursor cursor3 = database.query("userinfo", new String[] { "id",
            "username" }, "", new String[] {}, "", "", "id desc", "4");
    Log.v("分页查询开始", "----------------");
```

```
        while (cursor3.moveToNext()) {
            String id = cursor3.getString(0);
            String username = cursor3.getString(1);
            Log.v("数据为：", "id=" + id + " username=" + username);
        }
        Log.v("分页查询结束", "----------------");
// 省略
```

程序运行结果如图 5.61 所示。

```
V  632  普通查询开始           ----------------
V  632  数据为：              id=24 username=username3 password=password3
V  632  数据为：              id=23 username=username2 password=password2
V  632  数据为：              id=22 username=username1 password=password
V  632  普通查询结束           ----------------
V  632  分组查询开始
V  632  数据为：              usertype=3 count=2
V  632  数据为：              usertype=1 count=3
V  632  分组查询结束           ----------------
V  632  分页查询开始
V  632  数据为：              id=27 username=username6
V  632  数据为：              id=26 username=username5
V  632  数据为：              id=25 username=username4
V  632  数据为：              id=24 username=username3
V  632  分页查询结束           ----------------
V  632  设置事务成功标志        设置事务成功标志
V  632  事务提交              事务提交
```

图 5.61   query 方法的执行结果

9. Cursor query (boolean distinct, String table, String[] columns, String selection, String[] selectionArgs, String groupBy, String having, String orderBy, String limit)方法的使用

重载的 query()方法的第 1 个参数用来设置是否对"查询结果集"中的"整行"重复记录进行去重复操作。更改数据表 userinfo 的数据如图 5.62 所示，从图中可以看到除了 id 列外，有重复的数据。

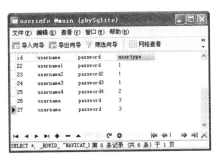

图 5.62   userinfo 数据表内容

如果要去掉重复的数据，更改文件 Main.java 的代码如下：

```
// 省略
        Cursor cursor1 = database.query(true, "userinfo", new String[] {
                "username", "password" }, "", new String[] {},"","",
                "id desc", "");
        Log.v("普通查询第 1 种开始", "----------------");
        while (cursor1.moveToNext()) {
            String id = cursor1.getString(0);
            String username = cursor1.getString(1);
            Log.v("数据为：", "id=" + id + " username=" + username);
        }
        Log.v("普通查询第 1 种结束", "----------------");
```

```
Cursor cursor2 = database.query(true, "userinfo", new String[] {
        "id", "username", "password" }, "", new String[] {}, "",
        "", "id desc", "");
Log.v("普通查询第 2 种开始", "---------------");
while (cursor2.moveToNext()) {
    String id = cursor2.getString(0);
    String username = cursor2.getString(1);
    String password = cursor2.getString(2);
    Log.v("数据为: ", "id=" + id + " username=" + username
            + " password=" + password);
}
Log.v("普通查询第 2 种结束", "---------------");
// 省略
```

程序运行结果如图 5.63 所示。

```
V 1... 普通查询第1种开始      ---------------
V 1... 数据为:             id=username username=password
V 1... 数据为:             id=username4 username=password4
V 1... 数据为:             id=username3 username=password3
V 1... 数据为:             id=username2 username=password2
V 1... 数据为:             id=username1 username=password
V 1... 普通查询第1种结束      ---------------
V 1... 普通查询第2种开始      ---------------
V 1... 数据为:             id=27 username=username password=password
V 1... 数据为:             id=26 username=username password=password
V 1... 数据为:             id=25 username=username4 password=password4
V 1... 数据为:             id=24 username=username3 password=password3
V 1... 数据为:             id=23 username=username2 password=password2
V 1... 数据为:             id=22 username=username1 password=password
V 1... 普通查询第2种结束      ---------------
V 1... 设置事务成功标志       设置事务成功标志
V 1... 事务提交            事务提交
```

图 5.63　去重复的测试结果

可以看到，去掉了重复的数据。

10. Cursor rawQuery (String sql, String[] selectionArgs)方法的使用

查询方法 query 的简化版是 rawQuery 方法，该方法只有两个参数。

更改 Main.java 的代码如下：

```
// 省略
Cursor cursor1 = database
        .rawQuery(
                "select id,username,password from userinfo where id=? or id=?",
                new String[] { "26", "27" });
Log.v("rawQuery 开始", "---------------");
while (cursor1.moveToNext()) {
    String id = cursor1.getString(0);
    String username = cursor1.getString(1);
    Log.v("数据为: ", "id=" + id + " username=" + username);
}
Log.v("rawQuery 结束", "---------------");
// 省略
```

程序运行后，显示查询的结果，如图 5.64 所示。

```
V  2... rawQuery开始
V  2... 数据为：                    id=26 username=username
V  2... 数据为：                    id=27 username=username
V  2... rawQuery结束
V  2... 设置事务成功标志             设置事务成功标志
V  2... 事务提交                    事务提交
```

图 5.64 rawQuery 查询的结果

11. long replace (String table, String nullColumnHack, ContentValues initialValues)方法的使用

replace 方法的功能是替换数据表中的某一行的数据，类似于 update 的功能，但功能上还有区别。更改 Main.java 的代码如下：

```
// 省略
ContentValues contentValuesRef = new ContentValues();
contentValuesRef.put("id", "27");
contentValuesRef.put("username", "zzzzzzzzzzz");
contentValuesRef.put("password", "yyyyyyyyyyy");

long returnLongValue = database.replace("userinfo", "username",
                        contentValuesRef);
Log.v("更新 row 的 id 是：", "" + returnLongValue);
// 省略
```

程序运行后替换掉了数据表中的数据，如图 5.65 所示。

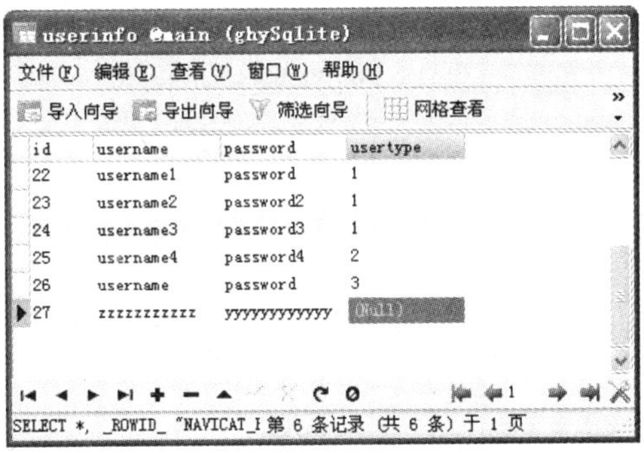

图 5.65 replace 更新 id 为 27 的数据

字段 usertype 并没有被赋值，所以它的值被替换成 null 了，如果想实现在不指定某些字段的情况下，更新数据还保留原有的值，该怎么办呢？可以用 update()方法来实现。

12. int update (String table, ContentValues values, String whereClause, String[] whereArgs)方法的使用

更改 Main.java 的代码如下：

```
// 省略
ContentValues contentValuesRef = new ContentValues();
contentValuesRef.put("username", "123");
contentValuesRef.put("password", "456");
```

```
            long returnLongValue = database.update("userinfo",
                        contentValuesRef, "id=?", new String[] { "26" });
            Log.v("受影响的行数是：", "" + returnLongValue);
            // 省略
```

程序运行后的数据表数据如图 5.66 所示。

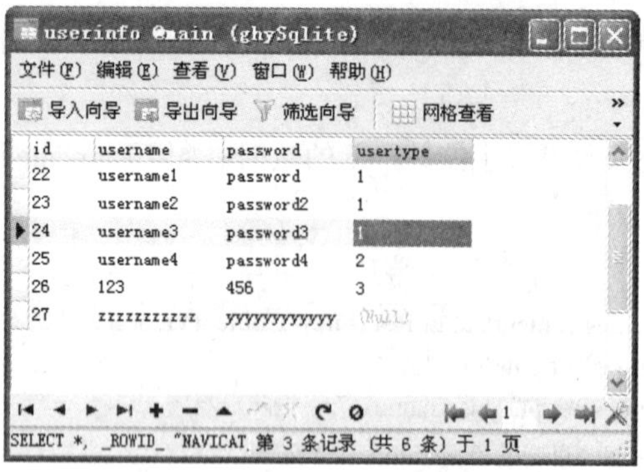

图 5.66 update 更新了数据行

从图 5.66 中可以看到，使用 update()方法后，username 和 password 的值被更新了，而 usertype 原有的值也被保留了下来。

13. 多表联接查询

在数据库 ghydb.db 中新建一个数据表 usertype，表结构如图 5.67 所示。

再次更改 userinfo 数据表，数据表的内容如图 5.68 所示。

图 5.67 usertype 数据表的结构

图 5.68 更改 userinfo 数据表的内容

更改 Main.java 的代码如下：

```
            // 省略
            Cursor cursor1 = database
                    .rawQuery(
                            "select  userinfo.id,username,password,typename  from  userinfo,usertype where userinfo.usertype=usertype.id",
                            new String[] {});
            Log.v("rawQuery 开始", "-----------------");
            while (cursor1.moveToNext()) {
```

```
            String id = cursor1.getString(0);
            String username = cursor1.getString(1);
            String password = cursor1.getString(2);
            String typename = cursor1.getString(3);
            Log.v("数据为: ", "id=" + id + " username=" + username
                    + " password=" + password + " typename=" + typename);
        }
        Log.v("rawQuery 结束", "----------------");
// 省略
```

程序运行效果如图 5.69 所示。

```
V  5676  rawQuery开始
V  5676  数据为:                    id=22 username=username1 password=password typename=普通用户
V  5676  数据为:                    id=23 username=username2 password=password2 typename=普通用户
V  5676  数据为:                    id=24 username=username3 password=password3 typename=普通用户
V  5676  数据为:                    id=25 username=username4 password=password4 typename=VIP用户
V  5676  数据为:                    id=26 username=123 password=456 typename=管理员
V  5676  数据为:                    id=27 username=zzzzzzzzzzz password=yyyyyyyyyyy typename=管理员
V  5676  rawQuery结束
V  5676  设置事务成功标志             设置事务成功标志
V  5676  事务提交                    事务提交
```

图 5.69　多表查询结果

实现了 usertype 和 userinfo 两个表的联接查询。

14. SQLite 分页功能的实现

新建 SQLite 数据库及数据表 userinfo，表中的数据内容如图 5.70 所示。

图 5.70　数据表 userinfo 中的数据

从图 5.70 可以看到，id 并不连续，本示例就来实现对这个表进行分页处理的功能。
新建名称为 sqlite_page 的 Android 项目，更改 Main.java 的代码如下：

```
// 省略
        // 每一页显示 3 条，显示第 3 页的数据
        // offset 4 代表从查询结果中乎略前 4 行
        // limit 4 代表从乎略的行数此往下显示 4 条
        Cursor cursor = db.rawQuery(
                "select * from userinfo limit 4,4", null);
```

```
            while (cursor.moveToNext()) {
                Log.v("!", "id=" + cursor.getString(0) + " username="
                    + cursor.getString(1));
            }
            // 省略
```

子句 limit 的功能和 MySQL 中的 limit 使用方式及功能一样。

程序运行后的结果如图 5.71 所示。

图 5.71　分页过后查询出来第 2 页的结果

5.7.3　封装数据库操作类

下面要在 ADT 中创建一个操作数据库的封装工具类，新建名称为 sqlite_5_7_1 的 Android 项目，在 res 目录下创建 raw 目录，并且把 ghydb.db 复制到这个目录中，如图 5.72 所示。

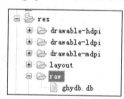

图 5.72　复制 ghydb.db 文件到 raw 目录中

在项目中创建 GetDataBaseFromFile.java 类，类的主要功能是判断指定路径下有没有数据库 ghydb.db 文件，如果没有，则把项目中的 ghydb.db 数据库文件复制到指定目录中，再创建数据库的对象并返回，进而对数据库进行 CURD 方面的操作，代码如下：

```java
public class GetDataBaseFromFile {

    private String copyToPath = "/data/data/sqlite_5_7_1.test.run/";
    private String dbname = "ghydb.db";
    private boolean isReWrite = false;

    private Context context;

    public GetDataBaseFromFile(Context context) {
        super();
        this.context = context;
```

```java
}
public SQLiteDatabase getSQLiteDatabase() {

    SQLiteDatabase database = null;

    try {
        File file = new File(copyToPath + dbname);

        if (file.exists() == false) {
            // 不存在
            Log.v("!", "数据库不存在COPY新的去");
            InputStream isRef = context.getResources().openRawResource(
                    R.raw.ghydb);
            FileOutputStream fosRef = new FileOutputStream(file);
            byte[] byteArray = new byte[2];
            int readLength = isRef.read(byteArray);
            while (readLength != -1) {
                fosRef.write(byteArray, 0, readLength);
                readLength = isRef.read(byteArray);
            }
            fosRef.close();
            isRef.close();
        } else {
            // 存在
            if (isReWrite == true) {
                Log.v("!", "数据库存在并且删除旧的写新的");
                if (file.delete() == true) {
                    Log.v("覆盖的情况：", "删除旧的数据库成功");
                    InputStream isRef = context.getResources()
                            .openRawResource(R.raw.ghydb);
                    FileOutputStream fosRef = new FileOutputStream(file);
                    byte[] byteArray = new byte[2];
                    int readLength = isRef.read(byteArray);
                    while (readLength != -1) {
                        fosRef.write(byteArray, 0, readLength);
                        readLength = isRef.read(byteArray);
                    }
                    fosRef.close();
                    isRef.close();
                }
            }
        }

        database = SQLiteDatabase.openOrCreateDatabase(copyToPath + dbname,
                null);
        database.beginTransaction();

    } catch (NotFoundException e) {
        // TODO Auto-generated catch block
```

```java
                e.printStackTrace();
            } catch (FileNotFoundException e) {
                // TODO Auto-generated catch block
                e.printStackTrace();
            } catch (IOException e) {
                // TODO Auto-generated catch block
                e.printStackTrace();
            }
        return database;

    }
}
```

还要创建文件 GetDataBase.java，并且使用 ThreadLocal 来管理数据库对象，代码如下：

```java
public class GetDataBase {

    private static ThreadLocal tl = new ThreadLocal();

    public static SQLiteDatabase getDataBase(Context context) {
        SQLiteDatabase SQLiteDatabaseRef;
        Object object = tl.get();
        if (object == null) {
            Log.v("===========", "ThreadLocal 中无对象");
            GetDataBaseFromFile getDataBaseRef = new GetDataBaseFromFile(
                    context);
            SQLiteDatabaseRef = getDataBaseRef.getSQLiteDatabase();
            tl.set(SQLiteDatabaseRef);
        } else {
            Log.v("===========", "ThreadLocal 中有对象");
            SQLiteDatabaseRef = (SQLiteDatabase) object;
        }
        return SQLiteDatabaseRef;
    }

    public static void commit() {
        Log.v("===========", "设置正确的提交标志！");
        ((SQLiteDatabase) tl.get()).setTransactionSuccessful();
    }

    public static void close() {
        ((SQLiteDatabase) tl.get()).endTransaction();
        ((SQLiteDatabase) tl.get()).close();
        tl.set(null);
    }

}
```

还有最重要的 CURD 增、删、改、查工具类，继续创建名称为 DBOperate.java 的类，代码如下：

```java
public class DBOperate {

    private Context context;

    public DBOperate(Context context) {
        super();
        this.context = context;
    }

    private ContentValues getContentValuesFromHashMap(HashMap valueMap) {
        ContentValues returnContentValuesRef = new ContentValues();
        Iterator iterator = valueMap.keySet().iterator();
        while (iterator.hasNext()) {
            String key = "" + iterator.next();
            String value = "" + valueMap.get(key);
            returnContentValuesRef.put(key, value);
        }
        return returnContentValuesRef;
    }

    public long insert(String tableName, HashMap valueMap)
            throws SQLiteException {
        SQLiteDatabase dataBaseRef = GetDataBase.getDataBase(context);
        return dataBaseRef.insertOrThrow(tableName, "",
                getContentValuesFromHashMap(valueMap));
    }

    public long delete(String tableName, String where, String[] whereValue)
            throws SQLiteException {
        SQLiteDatabase dataBaseRef = GetDataBase.getDataBase(context);
        return dataBaseRef.delete(tableName, where, whereValue);
    }

    public long update(String tableName, HashMap valueMap, String where,
            String[] whereValue) throws SQLiteException {
        SQLiteDatabase dataBaseRef = GetDataBase.getDataBase(context);
        return dataBaseRef.update(tableName,
                getContentValuesFromHashMap(valueMap), where, whereValue);
    }

    public List select(String sql, String[] selectValue) throws SQLiteException {
        ArrayList returnList = new ArrayList();

        SQLiteDatabase dataBaseRef = GetDataBase.getDataBase(context);
        Cursor cursorRef = dataBaseRef.rawQuery(sql, selectValue);

        while (cursorRef.moveToNext()) {
            HashMap rowMap = new HashMap();
            int selectColCount = cursorRef.getColumnCount();
            for (int i = 0; i < selectColCount; i++) {
```

```java
                    String[] colName = cursorRef.getColumnNames();
                    for (int j = 0; j < colName.length; j++) {
                        int colIndex = cursorRef.getColumnIndex(colName[j]);
                        String colValue = cursorRef.getString(colIndex);
                        rowMap.put(colName[j], colValue);
                    }
                }
                returnList.add(rowMap);
            }

            return returnList;
        }
    }
```

更改文件 Main.java 的代码如下:

```java
public class Main extends Activity {

    @Override
    public void onCreate(Bundle savedInstanceState) {
        super.onCreate(savedInstanceState);
        setContentView(R.layout.main);

        try {
            Log.v("--------------", ""
                    + GetDataBase.getDataBase(this.getApplicationContext())
                            .hashCode());
            Log.v("--------------", ""
                    + GetDataBase.getDataBase(this.getApplicationContext())
                            .hashCode());
            Log.v("--------------", ""
                    + GetDataBase.getDataBase(this.getApplicationContext())
                            .hashCode());

            DBOperate dboRef = new DBOperate(this.getApplicationContext());

            List returnList = dboRef.select("select * from userinfo", null);
            for (int i = 0; i < returnList.size(); i++) {
                Map rowMap = (Map) returnList.get(i);
                Log.v("+++++++++++++++++", "id:" + rowMap.get("id")
                        + "  username:" + rowMap.get("username")
                        + "  password:" + rowMap.get("password"));
            }

            GetDataBase.commit();

        } catch (Exception exception) {
            Log.v("###########", "报异常了!");
        } finally {
```

```
            GetDataBase.close();
        }
    }
}
```

本示例是实现一个查询的功能，程序运行后，LogCat 打印的信息如图 5.73 所示。

```
V  464  ============      ThreadLocal中无对象
V  464  !                 数据库不存在COPY新的去
V  464  ----------------  1079096512
V  464  ============      ThreadLocal中有对象
V  464  ----------------  1079096512
V  464  ============      ThreadLocal中有对象
V  464  ----------------  1079096512
V  464  ============      ThreadLocal中有对象
V  464  +++++++++++++++   id:22   username:username1   password:password1
V  464  +++++++++++++++   id:23   username:username2   password:password2
V  464  +++++++++++++++   id:24   username:username3   password:password3
V  464  +++++++++++++++   id:25   username:username4   password:password4
V  464  +++++++++++++++   id:26   username:123   password:456
V  464  +++++++++++++++   id:27   username:zzzzzzzzzzz   password:yyyyyyyyyyy
V  464  ============      设置正确的提交标志!
```

图 5.73　使用封装工具类的查询功能

再实现一个 insert 数据的功能，更改 Main.java 的代码如下：

```
        // 省略
        DBOperate dboRef = new DBOperate(this.getApplicationContext());

        HashMap valueMap = new HashMap();
        valueMap.put("username", "username3");
        valueMap.put("password", "password3");
        dboRef.insert("userinfo", valueMap);

        HashMap valueMap2 = new HashMap();
        valueMap2.put("username", "username4");
        dboRef.insert("userinfo", valueMap2);

        List returnList = dboRef.select("select * from userinfo", null);
        for (int i = 0; i < returnList.size(); i++) {
            Map rowMap = (Map) returnList.get(i);
            Log.v("+++++++++++++++", "id:" + rowMap.get("id")
                    + "  username:" + rowMap.get("username")
                    + "  password:" + rowMap.get("password"));
        }

        GetDataBase.commit();

    } catch (Exception exception) {
        Log.v("###########", "报异常了!");
    } finally {
        GetDataBase.close();
    }

    }
}
```

程序运行结果如图 5.74 所示。

```
V  689  ============              ThreadLocal中无对象
V  689  ----------------          1079097024
V  689  ============              ThreadLocal中有对象
V  689  ----------------          1079097024
V  689  ============              ThreadLocal中有对象
V  689  ----------------          1079097024
V  689  ============              ThreadLocal中有对象
V  689  ============              ThreadLocal中有对象
V  689  ============              ThreadLocal中有对象
V  689  ++++++++++++++            id:22  username:username1  password:password
V  689  ++++++++++++++            id:23  username:username2  password:password2
V  689  ++++++++++++++            id:24  username:username3  password:password3
V  689  ++++++++++++++            id:25  username:username4  password:password4
V  689  ++++++++++++++            id:26  username:123        password:456
V  689  ++++++++++++++            id:27  username:zzzzzzzzzz password:yyyyyyyyyy
V  689  ++++++++++++++            id:28  username:username3  password:password3
V  689  ++++++++++++++            id:29  username:username4  password:null
V  689  ============              设置正确的提交标志！
```

图 5.74 插入两条记录后的数据表内容

再来实现一个删除功能，更改 Main.java 的代码如下：

```
// 省略
    DBOperate dboRef = new DBOperate(this.getApplicationContext());

    Log.v("---------删除影响行数----------",""
            + dboRef.delete("userinfo", "id=?", new String[] { "22" }));

    List returnList = dboRef.select("select * from userinfo", null);
    for (int i = 0; i < returnList.size(); i++) {
        Map rowMap = (Map) returnList.get(i);
        Log.v("++++++++++++++++", "id:" + rowMap.get("id")
                + "  username:" + rowMap.get("username")
                + "  password:" + rowMap.get("password"));
    }

    GetDataBase.commit();

} catch (Exception exception) {
    Log.v("###########", "报异常了!");
} finally {
    GetDataBase.close();
}
    }
}
```

程序运行效果如图 5.75 所示。

```
V  778  ============              ThreadLocal中无对象
V  778  ----------------          1079098152
V  778  ============              ThreadLocal中有对象
V  778  ----------------          1079098152
V  778  ============              ThreadLocal中有对象
V  778  ----------------          1079098152
V  778  ============              ThreadLocal中有对象
V  778  ----------删除影响行数----  1
V  778  ============              ThreadLocal中有对象
V  778  ++++++++++++++            id:23  username:username2  password:password2
V  778  ++++++++++++++            id:24  username:username3  password:password3
V  778  ++++++++++++++            id:25  username:username4  password:password4
V  778  ++++++++++++++            id:26  username:123        password:456
V  778  ++++++++++++++            id:27  username:zzzzzzzzzz password:yyyyyyyyyy
V  778  ++++++++++++++            id:28  username:username3  password:password3
V  778  ++++++++++++++            id:29  username:username4  password:null
V  778  ============              设置正确的提交标志！
```

图 5.75 删除后的数据表内容

现在数据表中的数据个数是 7 条，id 分别是 23 到 29，下面来实现一个事务回滚的效果，更改 Main.java 的代码如下：

```java
// 省略
DBOperate dboRef = new DBOperate(this.getApplicationContext());

HashMap valueMap = new HashMap();
valueMap.put("username", "username3");
valueMap.put("password", "password3");
dboRef.insert("userinfo", valueMap);

HashMap valueMap2 = new HashMap();
valueMap2.put("username", "username4");
dboRef.insert("userinfo22222", valueMap2);

List returnList = dboRef.select("select * from userinfo", null);
for (int i = 0; i < returnList.size(); i++) {
    Map rowMap = (Map) returnList.get(i);
    Log.v("+++++++++++++++", "id:" + rowMap.get("id")
        + "  username:" + rowMap.get("username")
        + "  password:" + rowMap.get("password"));
}

GetDataBase.commit();

} catch (Exception exception) {
    Log.v("############", "报异常了!");
} finally {
    GetDataBase.close();
}
```

程序运行后的效果如图 5.76 所示。

没有数据表 userinfo22222，第 1 条 insert 语句应该回滚，所以程序中还是 7 条记录，数据表内容如图 5.77 所示。

图 5.76　报异常了

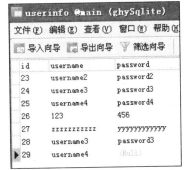

图 5.77　7 条记录没有变

再来实现一个混合的事务测试，代码中有 insert 和 delete 还有 update 混合使用，然后故意使代码出错，抛出异常，看看数据表是否回滚，Main.java 的代码如下：

```java
// 省略
DBOperate dboRef = new DBOperate(this.getApplicationContext());

// 删除 id 为 23 的数据
Log.v("---------删除影响行数-----------", ""
        + dboRef.delete("userinfo", "id=?", new String[] { "23" }));

// 插入新的记录
HashMap valueMap = new HashMap();
valueMap.put("username", "username3");
valueMap.put("password", "password3");
dboRef.insert("userinfo", valueMap);

// 更新 id 为 24 的数据
HashMap valueUpdateMap = new HashMap();
valueUpdateMap.put("username", "usernameZZZZZZ");
valueUpdateMap.put("password", "passwordZZZZZZ");
dboRef.update("userinfo", valueUpdateMap, "id=?",
        new String[] { "24" });

// 故意异常代码
HashMap valueMap2 = new HashMap();
valueMap2.put("username", "username4");
dboRef.insert("userinfo22222", valueMap2);

List returnList = dboRef.select("select * from userinfo", null);
for (int i = 0; i < returnList.size(); i++) {
    Map rowMap = (Map) returnList.get(i);
    Log.v("+++++++++++++++++", "id:" + rowMap.get("id")
            + "   username:" + rowMap.get("username")
            + "   password:" + rowMap.get("password"));
}

GetDataBase.commit();

} catch (Exception exception) {
    Log.v("###########", "报异常了！");
} finally {
    GetDataBase.close();
}
```

程序运行后事务回滚，数据表 userinfo 的内容如图 5.78 所示。

图 5.78 内容不变的 userinfo 数据表

继续实现一个修改的示例，修改 id 为 23 的记录的 username 和 password 字段的值，将文件 Main.java 的代码更改如下：

```
// 省略
DBOperate dboRef = new DBOperate(this.getApplicationContext());

HashMap valueMap = new HashMap();
valueMap.put("username", "usernameZZZZZZ");
valueMap.put("password", "passwordZZZZZZ");

dboRef.update("userinfo", valueMap, "id=?", new String[] { "23" });

GetDataBase.commit();

} catch (Exception exception) {
    Log.v("###########", "报异常了!");
} finally {
    GetDataBase.close();
}
    }
  }
}
```

程序运行效果如图 5.79 所示。

图 5.79 更新了数据

5.7.4 使用 DBOperate 对象将数据表中的数据显示在 ListView 中

上面的示例仅仅是在 LogCat 中打印出信息，有些时候需要将数据库中的数据以列表的方式显示，而使用 DBOperate 对象查询出来的数据类型正是 List，这个 List 中的每一个元素都是 Map 对象，这种数据结构可以和 SimpleAdapter 对象完全美的结合，下面的示例就实现这样的功能。

新建名称为 sqlite_5_7_3 的 Android 项目，更改 ghydb.db 数据库中的 userinfo 数据表数据如图 5.80 所示。

图 5.80　更改 userinfo 数据表内容

将前面的 dbtools 包中的所有类复制到当前项目的 src 资源路径中，更改文件 Main.java 的代码如下：

```
// 省略
    DBOperate dboRef = new DBOperate(this.getApplicationContext());

    List returnList = dboRef.select("select username from userinfo",
            null);

    SimpleAdapter arrayAdapter = new SimpleAdapter(this, returnList,
            android.R.layout.simple_list_item_1,
            new String[] { "username" },
            new int[] { android.R.id.text1 });

    listView1.setAdapter(arrayAdapter);

    GetDataBase.commit();

} catch (Exception exception) {
    Log.v("############", "报异常了!");
} finally {
    GetDataBase.close();
}
```

程序运行后正确地在 ListView 控件中显示出列表，如图 5.81 所示。

图 5.81　正确显示列表

5.8 ContentProvider 对象的使用

本示例将要实现两个应用程序之间互相传输数据的示例，传输的媒体使用 ContentProvider 对象。

5.8.1 创建数据提供者 ContentProvider 对象

更改 userinfo 数据表如图 5.82 所示。

需要把这个 ghydb.db 数据库文件复制到 raw 目录下，如图 5.83 所示。

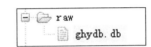

图 5.82　更改 userinfo 数据表内容　　　图 5.83　复制 ghydb.db 数据库文件

新建名称为 test 的 Android 项目，在项目中新建名称为 ContentProvider 的对象，代码如下：

```
public class GHYContentProvider extends ContentProvider {

    private static final UriMatcher matcher = new UriMatcher(
            UriMatcher.NO_MATCH);

    // 添加欲匹配的 Uri 路径
    static {
        matcher.addURI("com.gaohongyan.www", "selectById", 10);
        matcher.addURI("com.gaohongyan.www", "selectById/#", 1);
        matcher.addURI("com.gaohongyan.www", "selectAll", 2);
        matcher.addURI("com.gaohongyan.www", "updateById", 3);
        matcher.addURI("com.gaohongyan.www", "deleteById", 4);
        matcher.addURI("com.gaohongyan.www", "insert", 5);
    }

    // 为<Activity>的<data>子标签与 Intent 对象
    // 和 ContentProvider 提供的界面联合使用做准备
    // 由于本项目只提供了 1 个操作 ContentProvider 的界面
    // 所以 getType 方法只匹配 oneRow/selectById
    @Override
    public String getType(Uri arg0) {
        Log.v("!", "自动执行了 getType()方法");
        switch (matcher.match(arg0)) {
        case 1:
            Log.v("匹配了：", "oneRow/selectById");
```

```java
                    return "oneRow/selectById";
            default:
                throw new IllegalArgumentException();
            }
    }

    @Override
    public Cursor query(Uri arg0, String[] arg1, String arg2, String[] arg3,
            String arg4) {
        Log.v("!", "进入 GHYContentProvider 的 query 方法");
        switch (matcher.match(arg0)) {
        case 10:
        case 1:
            Log.v("!", "匹配值为 1 的 selectById");
            String getSelectByIdValue = "" + ContentUris.parseId(arg0);
            SQLiteDatabase databaseById = SQLiteDatabase.openOrCreateDatabase(
                    "/data/data/test.test.run/ghydb.db", null);
            Cursor cursorById = databaseById.rawQuery(
                    "select * from userinfo where id=?",
                    new String[] { getSelectByIdValue });
            MatrixCursor MatrixCursorRef = new MatrixCursor(new String[] {
                    "id", "username", "password" });
            while (cursorById.moveToNext()) {
                String id = cursorById.getString(0);
                String username = cursorById.getString(1);
                String password = cursorById.getString(2);
                Log.v("!GHYContentProvider 的 query 打印 selectById: ", "id=" + id
                        + " username=" + username + " password=" + password);
                MatrixCursorRef.addRow(new Object[] { id, username, password });
            }
            databaseById.close();
            return MatrixCursorRef;
        case 2:
            Log.v("!", "匹配值为 2 的 selectAll");
            SQLiteDatabase databaseAll = SQLiteDatabase.openOrCreateDatabase(
                    "/data/data/test.test.run/ghydb.db", null);
            Cursor cursorAll = databaseAll.rawQuery("select * from userinfo",
                    new String[] {});
            MatrixCursor MatrixCursorAllRef = new MatrixCursor(new String[] {
                    "id", "username", "password" });
            while (cursorAll.moveToNext()) {
                String id = cursorAll.getString(0);
                String username = cursorAll.getString(1);
                String password = cursorAll.getString(2);
                Log.v("!GHYContentProvider 的 query 打印 selectAll: ", "id=" + id
                        + " username=" + username + " password=" + password);
                MatrixCursorAllRef
                        .addRow(new Object[] { id, username, password });
            }
            databaseAll.close();
```

```java
            return MatrixCursorAllRef;
        default:
            throw new IllegalArgumentException();
    }
}

@Override
public int delete(Uri arg0, String arg1, String[] arg2) {
    Log.v("!", "进入 GHYContentProvider 的 delete 方法");
    switch (matcher.match(arg0)) {
    case 4:
        Log.v("!", "匹配值为 4 的 deleteByid");
        SQLiteDatabase databaseDeleteById = SQLiteDatabase
                .openOrCreateDatabase("/data/data/test.test.run/ghydb.db",
                        null);
        int deleteRowNum = databaseDeleteById
                .delete("userinfo", arg1, arg2);
        databaseDeleteById.close();
        return deleteRowNum;
    default:
        throw new IllegalArgumentException();
    }

}

@Override
public Uri insert(Uri arg0, ContentValues arg1) {
    Log.v("!", "进入 GHYContentProvider 的 insert 方法");
    switch (matcher.match(arg0)) {
    case 5:
        Log.v("!", "匹配值为 5 的 insert");
        SQLiteDatabase databaseDeleteById = SQLiteDatabase
                .openOrCreateDatabase("/data/data/test.test.run/ghydb.db",
                        null);
        long insertRowNum = databaseDeleteById.insert("userinfo", "", arg1);

        Uri uriRef = null;
        if (insertRowNum > 0) {
            uriRef = Uri.parse("content://com.gaohongyan.www/selectById/"
                    + insertRowNum);

        }
        databaseDeleteById.close();
        return uriRef;
    default:
        throw new IllegalArgumentException();
    }
}

@Override
```

```java
    public boolean onCreate() {
        Log.v("!", "执行了 GHYContentProvider 的 onCreate()方法");
        CopyDBTools.beginCopyDB(this.getContext(), "ghydb.db",
                "/data/data/test.test.run/", false);
        // 返回 true 代表当前的 ContentProvider 创建成功
        // 反之失败
        return true;
    }

    @Override
    public int update(Uri arg0, ContentValues arg1, String arg2, String[] arg3) {
        switch (matcher.match(arg0)) {
        case 3:
            Log.v("!", "匹配值为 3 的 updateById");
            SQLiteDatabase database = SQLiteDatabase.openOrCreateDatabase(
                    "/data/data/test.test.run/ghydb.db", null);
            int updateRowNum = database.update("userinfo", arg1, arg2, arg3);
            database.close();
            return updateRowNum;
        default:
            throw new IllegalArgumentException();
        }

    }
}
```

在 test 项目中提供了界面 Activity 来对 ContentProvider 进行操作，这个界面 Activity 是供其他 Android 项目来进行调用的，这个 Activity 的名称为 SelectById.java，代码如下：

```java
public class SelectById extends Activity {
    private Button selectbyid_button1;
    private EditText selectbyid_editText1;
    private TextView selectbyid_textView1;

    @Override
    public void onCreate(Bundle savedInstanceState) {
        super.onCreate(savedInstanceState);
        setContentView(R.layout.selectbyid);

        selectbyid_button1 = (Button) this
                .findViewById(R.id.selectbyid_button1);
        selectbyid_editText1 = (EditText) this
                .findViewById(R.id.selectbyid_editText1);
        selectbyid_textView1 = (TextView) this
                .findViewById(R.id.selectbyid_textView1);

        selectbyid_button1.setOnClickListener(new OnClickListener() {
            public void onClick(View arg0) {

                String findId = selectbyid_editText1.getText().toString();
```

```java
                ContentResolver crRef = SelectById.this.getContentResolver();
                Cursor cursor = crRef.query(Uri
                        .parse("content://com.gaohongyan.www/selectById/"
                                + findId), null, "", null, "");

                String returnShowString = "";
                while (cursor.moveToNext()) {
                    String id = cursor.getString(0);
                    String username = cursor.getString(1);
                    String password = cursor.getString(2);
                    returnShowString = "id=" + id + " username=" + username
                            + " password=" + password;
                }
                selectbyid_textView1.setText(returnShowString);
            }
        });

    }
}
```

Activity 对象 SelectById.java 对应的布局文件为 selectbyid.xml，代码如下：

```xml
<?xml version="1.0" encoding="utf-8"?>
<LinearLayout xmlns:android="http://schemas.android.com/apk/res/android"
    android:orientation="vertical" android:layout_width="fill_parent"
    android:layout_height="fill_parent">
    <EditText android:text="1" android:layout_height="wrap_content"
        android:layout_width="match_parent" android:id="@+id/selectbyid_editText1"></EditText>
    <Button android:text="Button" android:id="@+id/selectbyid_button1"
        android:layout_width="wrap_content" android:layout_height="wrap_content"></Button>
    <TextView android:text="TextView" android:id="@+id/selectbyid_textView1"
        android:layout_width="wrap_content" android:layout_height="wrap_content"></TextView>
</LinearLayout>
```

系统项目文件 AndroidManifest.xml 的代码如下：

```xml
<?xml version="1.0" encoding="utf-8"?>
<manifest xmlns:android="http://schemas.android.com/apk/res/android"
    package="test.test.run" android:versionCode="1" android:versionName="1.0">


    <application android:icon="@drawable/icon" android:label="@string/app_name">
        <activity android:name=".SelectById" android:label="@string/app_name">
            <intent-filter>
                <action android:name="selectById" />
                <category android:name="android.intent.category.DEFAULT" />
                <data android:mimeType="oneRow/selectById"></data>
            </intent-filter>
        </activity>
        <provider android:name="extcontentprovider.GHYContentProvider"
            android:authorities="com.gaohongyan.www"></provider>
```

```xml
        </application>
</manifest>
```

5.8.2 创建 ContentProvider 对象的使用者

新建名称为 callCP 的 Android 项目，布局文件 main.xml 的代码如下：

```xml
<?xml version="1.0" encoding="utf-8"?>
<LinearLayout xmlns:android="http://schemas.android.com/apk/res/android"
    android:orientation="vertical" android:layout_width="fill_parent"
    android:layout_height="fill_parent">
    <Button android:text="selectById" android:id="@+id/selectById"
        android:layout_width="wrap_content" android:layout_height="wrap_content"></Button>
    <Button android:text="showById" android:id="@+id/dataFromDB"
        android:layout_width="wrap_content" android:layout_height="wrap_content"></Button>
    <Button android:text="ShowAll" android:id="@+id/dataFromDBAll"
        android:layout_width="wrap_content" android:layout_height="wrap_content"></Button>
    <Button android:text="updateById" android:id="@+id/updateUserinfo_button1"
        android:layout_width="wrap_content" android:layout_height="wrap_content"></Button>
    <Button android:text="insert" android:id="@+id/insert_button1"
        android:layout_width="wrap_content" android:layout_height="wrap_content"></Button>
    <Button android:text="deleteById" android:id="@+id/deleteById_button1"
        android:layout_width="wrap_content" android:layout_height="wrap_content"></Button>
</LinearLayout>
```

更改 Main.java 的代码如下：

```java
public class Main extends Activity {
    private Button selectById;
    private Button dataFromDB;
    private Button dataFromDBAll;
    private Button updateUserinfo_button1;
    private Button insert_button1;
    private Button deleteById_button1;

    @Override
    public void onCreate(Bundle savedInstanceState) {
        super.onCreate(savedInstanceState);
        setContentView(R.layout.main);

        selectById = (Button) this.findViewById(R.id.selectById);
        dataFromDB = (Button) this.findViewById(R.id.dataFromDB);
        dataFromDBAll = (Button) this.findViewById(R.id.dataFromDBAll);
        insert_button1 = (Button) this.findViewById(R.id.insert_button1);
        deleteById_button1 = (Button) this
                .findViewById(R.id.deleteById_button1);
        updateUserinfo_button1 = (Button) this
                .findViewById(R.id.updateUserinfo_button1);

        // 界面版操作 ContentProvider 对象
```

```java
selectById.setOnClickListener(new OnClickListener() {
    public void onClick(View arg0) {
        Intent intent = new Intent("selectById", Uri
                .parse("content://com.gaohongyan.www/selectById/1"));
        Main.this.startActivity(intent);
    }
});

// 代码版操作 ContentProvider 对象
dataFromDB.setOnClickListener(new OnClickListener() {
    public void onClick(View arg0) {
        ContentResolver crRef = Main.this.getContentResolver();
        Cursor cursorRef = crRef.query(Uri
                .parse("content://com.gaohongyan.www/selectById/1"),
                null, "", null, "");
        while (cursorRef.moveToNext()) {
            String id = cursorRef.getString(0);
            String username = cursorRef.getString(1);
            String password = cursorRef.getString(2);
            Log.v("!返回 Cursor 在 Main.java:  ", "id=" + id + " username="
                    + username + " password=" + password);
        }

    }
});

// 代码版操作 ContentProvider 对象
dataFromDBAll.setOnClickListener(new OnClickListener() {
    public void onClick(View arg0) {
        ContentResolver crRef = Main.this.getContentResolver();
        Cursor cursorRef = crRef.query(Uri
                .parse("content://com.gaohongyan.www/selectAll"), null,
                "", null, "");
        while (cursorRef.moveToNext()) {
            String id = cursorRef.getString(0);
            String username = cursorRef.getString(1);
            String password = cursorRef.getString(2);
            Log.v("!返回 Cursor 在 Main.java:  ", "id=" + id + " username="
                    + username + " password=" + password);
        }

    }
});

updateUserinfo_button1.setOnClickListener(new OnClickListener() {

    public void onClick(View arg0) {
        ContentValues cvRef = new ContentValues();
        cvRef.put("username", "zzzz");
        cvRef.put("password", "xxxx");
```

```
                    ContentResolver crRef = Main.this.getContentResolver();
                    crRef.update(Uri
                            .parse("content://com.gaohongyan.www/updateById"),
                            cvRef, "id=?", new String[] { "1" });
                }
            });

            insert_button1.setOnClickListener(new OnClickListener() {
                public void onClick(View arg0) {
                    ContentValues cvRef = new ContentValues();
                    cvRef.put("username", "newUsername");
                    cvRef.put("password", "newPassword");
                    ContentResolver crRef = Main.this.getContentResolver();
                    crRef.insert(Uri.parse("content://com.gaohongyan.www/insert"),
                            cvRef);

                }
            });

            deleteById_button1.setOnClickListener(new OnClickListener() {
                public void onClick(View arg0) {
                    ContentResolver crRef = Main.this.getContentResolver();
                    crRef.delete(Uri
                            .parse("content://com.gaohongyan.www/deleteById"),
                            "id=?", new String[] { "3" });
                }
            });

        }
    }
```

5.8.3 调用 ContentProvider 对象的应用运行效果

程序初始运行效果如图 5.84 所示。

图 5.84 程序初始运行效果

1. 根据 id 查询数据——Activity 方式

单击"selectById"根据 id 查询数据的按钮,单击后进入名称为 test 项目提供的 Activity 界面,运行效果如图 5.85 所示。

单击 Button 按钮在下方的 TextView 直接显示根据 id 查询出来的数据,如图 5.86 所示。

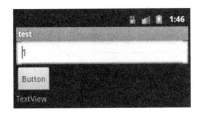
图 5.85　进入 test 项目提供的 Activity

图 5.86　根据 id 找到数据

2. 根据 id 查询数据——代码方式

返回主界面,单击 showById 按钮,用代码的方式直接操作 test 项目中的 ContentProvider 对象,LogCat 打印的结果如图 5.87 所示。

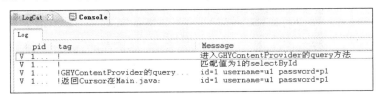

图 5.87　根据 id 用代码方式操作 ContentProvider 中的数据

3. 查询全部数据

返回主界面,单击 ShowAll 按钮,用代码的方式返回 ContentProvider 对象 userinfo 表的所有数据,LogCat 打印的结果如图 5.88 所示。

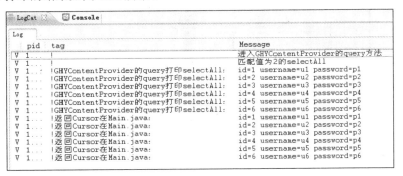

图 5.88　查询 userinfo 表中所有数据

4. 根据 id 更改数据

返回主界面,单击 updateById 按钮,根据 id 更新 userinfo 表中的数据,LogCat 打印的结果如图 5.89 所示。

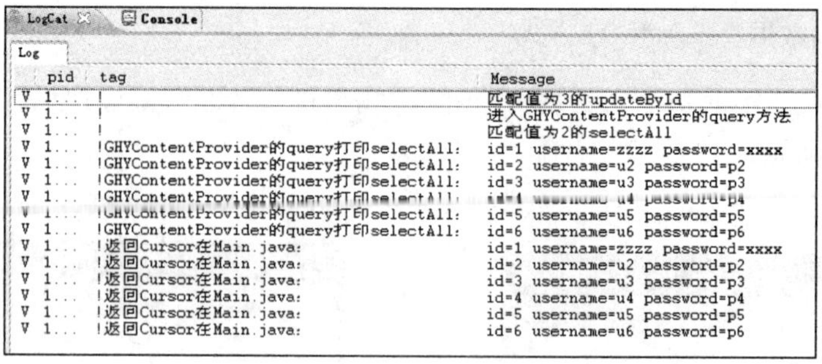

图 5.89 更新后的数据表内容

5. 插入记录

返回主界面，单击 insert 按钮插入新的记录并且查询出来，LogCat 打印的结果如图 5.90 所示。

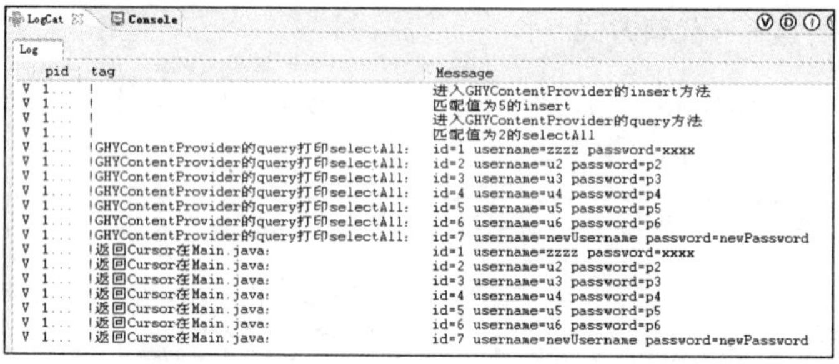

图 5.90 插入新记录后的 userinfo 数据表内容

6. 删除记录

返回主界面，单击 deleteById 按钮，根据 id 删除数据，LogCat 打印的结果如图 5.91 所示。

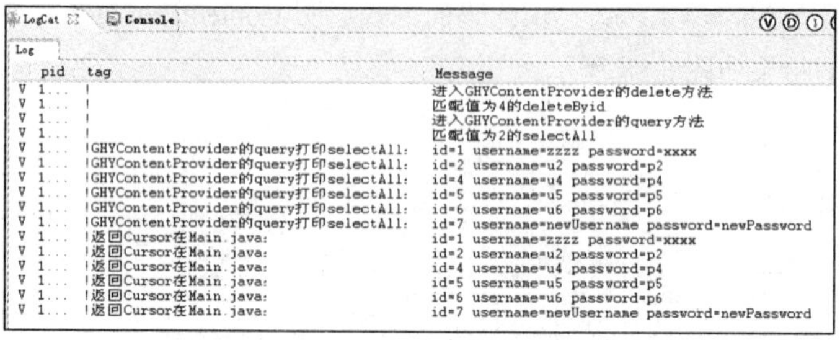

图 5.91 根据 id 删除后的数据表内容

到此已经在两个项目 callCP 和 test 中通过 ContentProvider 实现了数据共享，如图 5.92 所示。

图 5.92 显示出两个项目

使用 ContentProvider 可以使两个模块进行有效的解耦（解耦的含义是使两个软件模块之间的关系降到最低），而 Uri 对象的使用真正解决了系统间通过物理路径直接访问资源的缺点，其实在

Android 系统中，大多数的应用都是使用 ContentProvider 来完成的，ContentProvider 对象结合 SQLite 数据库可以使操作 SQLite 数据的接口更加规范和标准，从而有效地提高系统间代码的移植性。

5.9 Application 全局数据存储对象的使用

Application 对象有些像 Web 技术中的 Application 对象，在此对象中可以存储整个项目中共用的资源，在 Android 中也可以做到这样的效果，使用的对象就是 Application 类，它的继承关系如图 5.93 所示。

图 5.93　Application 类继承关系

使用 Application 对象的过程中一定要确保 Application 在系统中是唯一的，也就是单例模式，所以需要程序员写代码去维护 Application 单例，使用 Application 类的方式也很简单，只需要在 AndroidManifest.xml 文件中注册 Application 就可以了。

Application 类的实验代码在名称为 applicationTest 的项目中。

首先创建一个 Application 的子类 MyApplication.java，代码如下：

```
public class MyApplication extends Application {

    private static MyApplication application;

    public static MyApplication getApplication() {
        return application;
    }

    @Override
    public void onCreate() {
        super.onCreate();
        Log.v("!", "MyApplication onCreate 被调用了!");
        application = this;

        editText = new EditText(this.getApplicationContext());
        editText.setText("我是在 application 自动创建的");
    }

    // 以下为业务方法
    private EditText editText;
```

```java
    public EditText getEditText() {
        return editText;
    }

    public void setEditText(EditText editText) {
        this.editText = editText;
    }

    public static void setApplication(MyApplication application) {
        MyApplication.application = application;
    }
}
```

还需要在 AndroidManifest.xml 文件中进行注册,代码如下:

```xml
<application android:icon="@drawable/icon" android:label="@string/app_name"
    android:name="extpackage.MyApplication">
```

还需要创建两个 Activity,其中 Main.java 代码如下:

```java
public class Main extends Activity {

    private Button button1;
    private Button button2;

    @Override
    public void onCreate(Bundle savedInstanceState) {
        super.onCreate(savedInstanceState);
        setContentView(R.layout.main);

        button1 = (Button) this.findViewById(R.id.button1);
        button2 = (Button) this.findViewById(R.id.button2);

        button1.setOnClickListener(new OnClickListener() {
            public void onClick(View arg0) {
                Log.v("!", "取出的 text 值: "
                        + MyApplication.getApplication().getEditText()
                            .getText().toString());
            }
        });

        button2.setOnClickListener(new OnClickListener() {
            public void onClick(View arg0) {
                Intent intent = new Intent(Main.this, Second.class);
                Main.this.startActivity(intent);
            }
        });
```

        }
    }

文件 Second.java 代码如下：

```java
public class Second extends Activity {

    private EditText editText1;

    @Override
    public void onCreate(Bundle savedInstanceState) {
        super.onCreate(savedInstanceState);
        setContentView(R.layout.second);

        editText1 = (EditText) this.findViewById(R.id.editText1);
        editText1.setText(""
                + MyApplication.getApplication().getEditText().getText()
                        .toString());
    }
}
```

启动项目时自动实例化了 MyApplication.java 对象，如图 5.94 所示。

图 5.94　自动实例化 MyApplication

程序运行后界面如图 5.95 所示。

单击上面的按钮打印出 MyApplication 初始化 EditText 的 text 属性值，如图 5.96 所示。

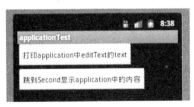

图 5.95　显示 main.xml 界面　　　图 5.96　在 Main.java 中打印 EditText 的 text 属性

单击下面的按钮转到 Second.java 后也正确显示出 text 属性值，如图 5.97 所示。

图 5.97　在 Second.java 中也显示出了 text 属性值

使用 Application 的优势是在 onCreate() 自动被调用时可以取得上下文 Context 对象，以方便后面的数据处理及对象的初始化工作。

第 6 章 Broadcast、Service 服务及 Handle 对象

本章介绍 Android 的核心组件 Service，通过 Service 这种技术可以在系统的后台进行一些隐藏性任务的执行，前台用户并不会体会到程序在运行，这种技术通常用在一些计划任务中，Android 的 4 大组件 Service 在实现这一功能时的使用率非常高，而要实现一个功能完善的软件不可能不用到 Service。

本章的知识点非常重要，也是学习 Android 必须掌握的技术，可以着重关注以下的知识点：

- startService 和 bindService 的区别和使用
- 定时服务 AlarmManager 的使用
- 串行化 Parcelable 接口的使用
- 重点：AIDL 的使用
- Handler 对象的使用

6.1 使用 Broadcast 的种类

前面已经介绍过创建广播 Broadcast 对象的两种形式，一种是静态的形式，需要在 AndroidManifest.xml 文件中注册广播 Broadcast 对象，另一种是动态的形式，也就是使用程序代码来动态创建广播 Broadcast 对象，本章使用 Broadcast 对象与 Service 对象进行联合开发，来实现一个简化版的 Mp3 播放器。

广播接收者的生命周期非常短，仅仅在执行方法 onReceiver()时存在，该方法执行完毕后，广播接收者也即被系统所销毁，所以不要试图在广播接收者中执行回调函数。

6.1.1 多 BroadcastReceiver 同时匹配 Intent 的情况

首先介绍有多个 BroadcastReceiver 对象同时匹配 Intent 的情况。新建名称为 Broadcast 的 Android 项目，创建两个广播接收者 BroadcastReceiver 对象，第一个 GHYBroadcastReceiver1.java 对象的代码如下：

```
public class GHYBroadcastReceiver1 extends BroadcastReceiver {

    @Override
    public void onReceive(Context arg0, Intent arg1) {
        Log.v("!", "执行了 GHYBroadcastReceiver1 的 onReceive 方法");
```

另外一个 GHYBroadcastReceiver2.java 广播接收者对象的代码如下：

```java
public class GHYBroadcastReceiver2 extends BroadcastReceiver {

    @Override
    public void onReceive(Context arg0, Intent arg1) {
        Log.v("!", "执行了 GHYBroadcastReceiver2 的 onReceive 方法");
    }

}
```

文件 Main.java 的代码如下：

```java
public class Main extends Activity {
    @Override
    public void onCreate(Bundle savedInstanceState) {
        super.onCreate(savedInstanceState);
        setContentView(R.layout.main);

        Intent intent = new Intent();
        intent.setAction("ghyAction");
        this.sendBroadcast(intent);

    }
}
```

系统配置文件 AndroidManifest.xml 的代码如下：

```xml
<?xml version="1.0" encoding="utf-8"?>
<manifest xmlns:android="http://schemas.android.com/apk/res/android"
    package="Broadcast.test.run" android:versionCode="1"
    android:versionName="1.0">
    <application android:icon="@drawable/icon" android:label="@string/app_name">
        <activity android:name=".Main" android:label="@string/app_name">
            <intent-filter>
                <action android:name="android.intent.action.MAIN" />
                <category android:name="android.intent.category.LAUNCHER" />
            </intent-filter>
        </activity>
        <receiver android:name="ghypackage.GHYBroadcastReceiver1">
            <intent-filter>
                <action android:name="ghyAction"></action>
            </intent-filter>
        </receiver>
        <receiver android:name="ghypackage.GHYBroadcastReceiver2">
            <intent-filter>
                <action android:name="ghyAction"></action>
            </intent-filter>
```

```
            </receiver>
        </application>
</manifest>
```

程序运行后在 LogCat 中打印的结果如图 6.1 所示。

```
I  490  AndroidRuntime         NOTE: attach of thread 'Binder Thread #3' fai
V  499  !                      执行了GHYBroadcastReceiver1的onReceive方法
V  499  !                      执行了GHYBroadcastReceiver2的onReceive方法
I  77   ActivityManager        Displayed Broadcast.test.run/.Main: +1s467ms
```

图 6.1 同时打印

6.1.2 用广播实现程序开机运行的效果

创建名称为 autoRun 的 Android 项目，新建两个 Activity 对象 Main.java 和 Second.java 文件，代码默认即可。

创建自定义广播 GhyBroadcastReceiver.java 对象，核心代码如下：

```java
public class GhyBroadcastReceiver extends BroadcastReceiver {

    @Override
    public void onReceive(Context arg0, Intent arg1) {
        Log.v("!", "onReceive");
        Intent intent = new Intent(arg0, Second.class);
        intent.setFlags(Intent.FLAG_ACTIVITY_NEW_TASK);
        arg0.startActivity(intent);
    }

}
```

文件 AndroidManifest.xml 的代码如下：

```xml
<?xml version="1.0" encoding="utf-8"?>
<manifest xmlns:android="http://schemas.android.com/apk/res/android"
    package="autoRun.test.run" android:versionCode="1" android:versionName="1.0">

    <application android:icon="@drawable/icon" android:label="@string/app_name">
        <activity android:name=".Main" android:label="@string/app_name">
            <intent-filter>
                <action android:name="android.intent.action.MAIN" />
                <category android:name="android.intent.category.LAUNCHER" />
            </intent-filter>
        </activity>

        <activity android:name=".Second" android:label="@string/app_name">
        </activity>

        <receiver android:name="ghybr.GhyBroadcastReceiver">
            <intent-filter>
                <action android:name="android.intent.action.BOOT_COMPLETED"></action>
            </intent-filter>
```

```xml
        </receiver>

    </application>
    <permission android:name="android.permission.RECEIVE_BOOT_COMPLETED"></permission>
</manifest>
```

装有 Android 系统的真机重新启动后将会自动运行 Second.java 界面。

6.1.3 sendStickyBroadcast 函数的使用

比如有这种情况，ActivityA 发送广播到 ActivityB，但 BroadcastReceiver 是在 ActivityB 中用代码进行注册的，ActivityA 发送出去的广播 ActivityB 是接收不到的，如果遇到这种情况该怎么办呢？使用 sendStickyBroadcast 方法就解决了。

新建名称为 test 的 Android 项目，Main.java 的核心代码如下：

```java
public class Main extends Activity {
    private Button button1;
    private Button button2;
    private int count = 0;

    @Override
    public void onCreate(Bundle savedInstanceState) {
        super.onCreate(savedInstanceState);
        setContentView(R.layout.main);

        // 添加 android.permission.BROADCAST_STICKY 权限
        button1 = (Button) this.findViewById(R.id.button1);
        button1.setOnClickListener(new OnClickListener() {
            public void onClick(View arg0) {
                count++;
                Intent intent = new Intent("sendMyBroadcastReceiver");
                intent.putExtra("username", "username" + count);
                Main.this.sendStickyBroadcast(intent);
            }
        });

        button2 = (Button) this.findViewById(R.id.button2);
        button2.setOnClickListener(new OnClickListener() {
            public void onClick(View arg0) {
                Intent intent = new Intent(Main.this, Second.class);
                Main.this.startActivity(intent);
            }
        });
    }
}
```

文件 Second.java 的核心代码如下：

```java
class MyBroadcastReceiver extends BroadcastReceiver {
    @Override
```

```
            public void onReceive(Context arg0, Intent arg1) {
                Log.v("!", "username=" + arg1.getStringExtra("username"));
            }
        }

        public class Second extends Activity {
            @Override
            public void onCreate(Bundle savedInstanceState) {
                super.onCreate(savedInstanceState);
                setContentView(R.layout.main);

                IntentFilter filter = new IntentFilter();
                filter.addAction("sendMyBroadcastReceiver");
                MyBroadcastReceiver myBroadcastReceiverRef = new MyBroadcastReceiver();
                this.registerReceiver(myBroadcastReceiverRef, filter);
            }
        }
```

在 AndroidManifest.xml 文件中添加权限代码：

`<uses-permission android:name="android.permission.BROADCAST_STICKY"></uses-permission>`

程序运行后单击两次"单击我 2 次发送 2 次 sendStickyBroadcast 广播"按钮，如图 6.2 所示。这时再单击"到 Second.java"按钮，在 LogCat 控制台打印出最后一次 Intent 中的数据，如图 6.3 所示。

图 6.2　单击 2 次上面的按钮

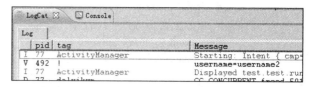

图 6.3　打印最后 1 次 Intent 中的数据

因为在 Second.java 文件中的 onCreate()方法中才注册了 MyBroadcastReceiver.java 广播接收者，此示例在于 Main.java 先发广播，然后等 Second.java 启动后再接收。

6.2　Service 服务

Android 中的服务与 Activity 不同，它是不能与用户交互、也不能自己启动且运行在后台的程序，当我们退出应用时，Service 进程并没有结束，它仍然在后台运行。那什么时候会用到 Service 呢？例如我们播放音乐的时候，有可能想边听音乐边做其他事情，但在退出播放音乐的应用，如果不用 Service 就听不到歌了，所以这时候就得用到 Service。再举一个例子，当一个应用的数据是通过网络获取的，并且获取的时候并不确定，这时候也可以用 Service 在后台定时执行指定的任务，而不用每次打开应用的时候再去获取。想要获取启动的 Service 实例，可以使用 bindService()方法来实现。

Service 的 onCreate 和 onStartCommand 是运行在主线程里的，所以，如果里面有处理耗时的任务，请开启新的线程来进行处理或使用 IntentService 对象。

在学习 Service 之前，先来掌握一下 Service 的生命周期，参看图 6.4 所示。

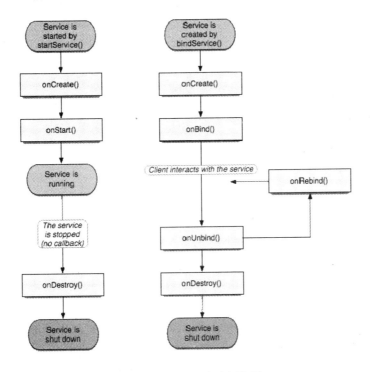

图 6.4　Service 的生命周期图

从图 6.4 中可以看到，启动服务有两种方式：startService()和 bindService()。需要注意的是，使用 startService()方法启动 Service 时回调函数 onStartCommand()有可能被调用多次，而其他回调函数只被调用 1 次。

注意，要学习 Service 服务，一定先要掌握以下 4 个函数的基本概念。

- startService()方法：启动服务，也就是创建服务，在内存中生成服务的实例对象。
- stopService()方法：停止服务，使用本方法用来销毁内存中的 Service 实例对象，如果内存中的 Service 对象曾经被 bindService()方法关联绑定过，那么要想销毁内存中的 Service，则先要 unbindService()反绑定服务。
- bindService()方法：具有创建服务和绑定服务（与服务进行通信）的能力，使得与 Service 对象取得连接，进而进行数据上的通信，想要通信就得在 Service 的 onBind()方法返回给客户端一个 IBind 接口实例，IBind 接口允许客户端调用服务的方法，比如得到 Service 运行的状态或执行其他业务方法操作。
- unbindService()方法：断开与 Service 的通信。

启动 Service 有两种方式：一种是用 startService()方法，另外一种就是前面介绍过的 bindService()

方法，它们之间的区别如下：

（1）使用 startService()启动服务
- 用 startService()方法启动的服务 onCreate()和 onStartCommand()回调方法依次被调用。
- 如果 Service 启动后多次重复调用 startService()方法则 onStartCommand()方法被执行多次，也就是说，如果 Service 还没有运行，则 Android 先调用 Service 的 onCreate()方法，然后调用 onStartCommand()。如果 Service 已经运行，则只调用 onStartCommand()。所以一个 Service 的 onStartCommand 方法可能会重复调用多次，可以用 Intent 封装要传递的数据给 onStartCommand()回调方法，然后在 Service 中进行获取。
- 使用 stopService()方法的功能是停止服务，并且回调 Service 类中的 onDestroy()方法。
- 用 startService()方法启动的服务，当 Activity 退出时系统中的服务并不销毁，还在内存中执行，也就是说，用 startService()方法启动的服务在内存中只有一个实例，并不随着 Activity 的退出而退出。

（2）使用 bindService()启动服务并绑定服务
- 用 bindService()方法启动的服务自动依次调用 onCreate()和 onBind()方法。
- 当运行中的 Service 宿主对象 Activity 对象退出时，当前的 Service 也被销毁，并且自动调用 onUnbind()方法和 onDestroy()方法。
- 当在 Service 的回调方法 onBind()返回 null 时，不调用 ServiceConnection 对象的 onServiceConnected()方法，当 onBind 方法返回一个 IBinder 对象时，则自动调用 ServiceConnection 对象中的 onServiceConnected()方法。
- 传递给 bindService()方法的 Intent 对象在 IBinder onBind(Intent arg0)方法中进行处理，而传递给 unbindService()方法的 Intent 对象在 onUnbind(Intent intent)方法中进行处理。

6.2.1　用 startService 启动 Service 方式与生命周期

新建名称为 beginService 的 Android 项目，用 Eclipse 向导创建自定义的 Service 类 GhyService.java，它的父类是 Service，创建后的代码如下：

```java
public class GhyService extends Service {

    private BroadcastReceiver stopServiceReceiver = new BroadcastReceiver() {
        @Override
        public void onReceive(Context arg0, Intent arg1) {
            GhyService.this.stopSelf();
            GhyService.this.unregisterReceiver(stopServiceReceiver);
            Log.v("!", "关闭了用 startService 方式启动的 Service");
        }
    };

    @Override
    public void onCreate() {
        super.onCreate();
        Log.v("!", "调用了 onCreate 方法");
```

```
            IntentFilter intentFilter = new IntentFilter("closeService");
            GhyService.this.registerReceiver(stopServiceReceiver, intentFilter);
        }

        @Override
        public void onDestroy() {
            super.onDestroy();
            Log.v("!", "调用了 onDestroy 方法");
        }

        @Override
        public int onStartCommand(Intent intent, int flags, int startId) {
            Log.v("!", "调用了 onStartCommand 方法");
            return super.onStartCommand(intent, flags, startId);
        }

        // onBind 必须要重写
        @Override
        public IBinder onBind(Intent arg0) {
            return null;
        }
    }
```

在上面的代码中可以发现，onBind()方法必须被重写，因为它在父类 Service 中是抽象的，如图 6.5 所示。

| abstract IBinder | onBind (Intent intent) Return the communication channel to the service. |

图 6.5 抽的 onBind()方法

更改布局文件 main.xml 的代码如下：

```xml
<?xml version="1.0" encoding="utf-8"?>
<LinearLayout xmlns:android="http://schemas.android.com/apk/res/android"
    android:orientation="vertical" android:layout_width="fill_parent"
    android:layout_height="fill_parent">
    <Button android:text="用 startService() 方式启动 Service" android:id="@+id/button1"
        android:layout_width="wrap_content" android:layout_height="wrap_content"></Button>
    <Button android:text="结束用 startService() 方式启动的 Service" android:id="@+id/button2"
        android:layout_width="wrap_content" android:layout_height="wrap_content"></Button>
</LinearLayout>
```

更改名称为 Main.java 的 Activity 代码如下：

```java
public class Main extends Activity {
    private Button button1;
    private Button button2;

    @Override
```

```java
    public void onCreate(Bundle savedInstanceState) {
        super.onCreate(savedInstanceState);
        setContentView(R.layout.main);

        button1 = (Button) this.findViewById(R.id.button1);
        button2 = (Button) this.findViewById(R.id.button2);
        button1.setOnClickListener(new OnClickListener() {
            public void onClick(View arg0) {
                Intent intent = new Intent(Main.this, GhyService.class);
                Main.this.startService(intent);
            }
        });

        button2.setOnClickListener(new OnClickListener() {
            public void onClick(View arg0) {
                Intent intent = new Intent("closeService");
                Main.this.sendBroadcast(intent);
            }
        });

    }
}
```

程序运行后出现界面图 6.6 所示。

单击"用 startService()方式启动 Service"按钮，在 LogCat 中打印出的信息如图 6.7 所示。

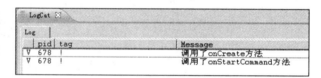

图 6.6 程序初次运行界面　　图 6.7 单击用 startService()方式启动 Service 出现信息

从界面中可以看到，GhyService 执行了两个回调函数：onCreate()和 onStartCommand()，再多次单击"用 startService()方式启动 Service"按钮，出现如图 6.8 所示的结果。

图 6.8 再多次单击用 startService()方式启动 Service 出现信息

从图 6.8 中可以看到，当用 startService()方式启动服务后多次调用 startService()方法时是不执行 onCreate()回调方法的，onCreate()这个方法只被执行 1 次，只是多次执行 onStartCommand()方法，这时单击 AVD 面板中的 后退按钮退出当前的 Activity，也就是本应用程序，再次查看 LogCat

面板中的数据,如图 6.9 所示。

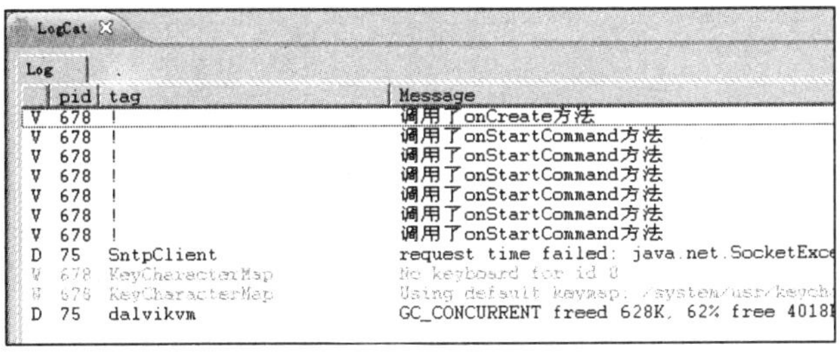

图 6.9 退出应用程序

从图 6.9 中可以看到,用 startService()方法启动的 Service,当宿主程序 Activity 对象退出时,Service 并不销毁,也就是用 startService()方式启动的 Service 的生命周期并不和宿主 Activity 一致,那 GhyService 怎么退出呢?其实在 Main.java 代码中已经给出了答案,就是在 GhyService.java 中配置一个 BroadcastReceiver 广播接收者对象,然后通过 Activity 对象发出广播告诉 Service 退出并销毁,单击 AVD 应用程序列表中的 beginService 应用程序,如图 6.10 所示。

图 6.10 重新进入 beginService

重新进入 beginService 后,LogCat 并没有打印出相关的数据,如图 6.11 所示。

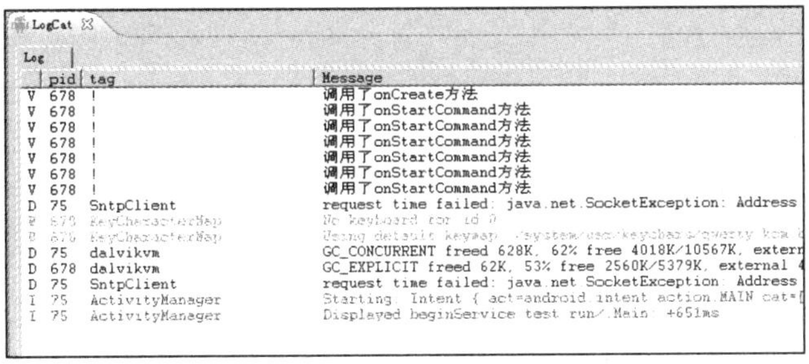

图 6.11 重新进入 beginService 后的 LogCat 日志

这时单击"结束用 startService()方式启动的 Service"按钮发送广播使 GhyService.java 服务退出并销毁,LogCat 打印出相关的日志信息,如图 6.12 所示。

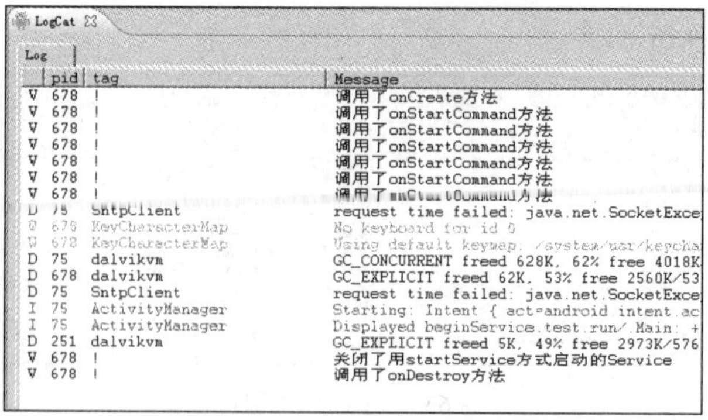

图 6.12　停止用 startService()方式启动的 GhyService 服务

上面的示例是使用广播的方式使 GhyService.java 服务退出，其实还有更直接的方法 stopService()来停止 Service，将 Main.java 代码中的 button2 的 onClick 事件代码更改如下：

```
button2.setOnClickListener(new OnClickListener() {
    public void onClick(View arg0) {
        Intent intent = new Intent(Main.this, GhyService.class);
        // Main.this.sendBroadcast(intent);
        Main.this.stopService(intent);
    }
});
```

还要更改 GhyService.java 文件 onCreate() 的代码，屏蔽掉动态注册广播接收者 BroadcastReceiver 的代码，更改后的回调函数 onCreate()如下：

```
@Override
public void onCreate() {
    super.onCreate();
    Log.v("!", "调用了 onCreate 方法");
    //IntentFilter intentFilter = new IntentFilter("closeService");
    //GhyService.this.registerReceiver(stopServiceReceiver, intentFilter);
}
```

再重新运行本项目，单击"启动服务"按钮，再单击 3 次"启动服务"按钮，然后再单击后退按钮退出应用程序，再重新进入 beginService 服务，单击"停止服务"按钮，LogCat 中的日志信息如图 6.13 所示。

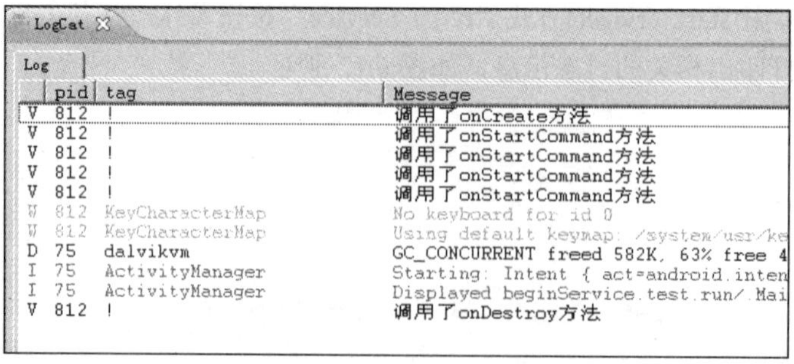

图 6.13　用 stopService()方法停止服务

上面的示例是用人为的因素来停止 Service 的运行，有时候需要程序自己关闭 Service，比如下载成功后，或音乐播放结束后等情况。用代码的方式来进行 Service 自关闭的实现思路是用 Intent 夹带参数的方式去启动 Service，示例代码如下：

```
selfStop.setOnClickListener(new OnClickListener() {
    public void onClick(View arg0) {
        Intent selfStopIntent = new Intent(Main.this, GhyService.class);
        selfStopIntent.putExtra("selfStop", "yes");
        Main.this.startService(selfStopIntent);
    }
});
```

然后在自定义的 Service 类中的 onStartCommand() 方法中进行判断，以便得知是否要自关闭当前的 Service 服务，回调函数 onStartCommand() 的代码如下：

```
@Override
public int onStartCommand(Intent intent, int flags, int startId) {
    Log.v("执行了 onStartCommand 方法", "执行了 onStartCommand 方法 flags=" + flags);
    try {
        if (intent.getStringExtra("selfStop") != null
                && intent.getStringExtra("selfStop").equals("yes")) {
            Log.v("关闭前时间：", "" + new Date().toLocaleString());
            Thread.sleep(3000);
            this.stopSelf();
            Log.v("关闭后时间：", "" + new Date().toLocaleString());
        }
    } catch (InterruptedException e) {
        // TODO Auto-generated catch block
        e.printStackTrace();
    }

    return super.onStartCommand(intent, flags, startId);
}
```

这样 Service 程序也就具有自关闭的功能了。

6.2.2 用 bindService 启动 Service 的方式与生命周期

用 startService() 方法启动的 Service 只能通过 Intent 对象进行数据的传递，却不能调用 Service 类中相关的方法，虽然间接地通过广播 BroadCast 可以实现，但还是走了一些弯路，有没有办法能使 Activity 客户端直接调用 Service 类中的业务方法呢？当然可以！使用 bindService() 方法。

新建名称为 bindServiceTest 的 Android 项目，用 Eclipse 的向导新建自定义的 GhyService.java 服务类，它的父类是 Service，生成的默认代码如下：

```
public class GhyService extends Service {

    @Override
    public IBinder onBind(Intent arg0) {
        // TODO Auto-generated method stub
```

```
            return null;
    }

}
```

从程序中可以看到有默认的回调方法 onBind()，它的返回值类型为 IBinder 对象，其主要的作用是 Service 与外界进行交互的一种手段，而 IBinder 对象数据类型是接口，其在 Android 中的声明如图 6.14 所示。

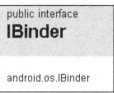

图 6.14　IBinder 接口的声明

所以需要 Android 开发的程序员来实现这个接口，而用 bindService()方式启动 Service 生命周期的回调函数有：onCreate()、onBind()、onUnbind()、onDestroy()、onRebind()，可将这些函数在 GhyService.java 代码中进行重写，完整的 GhyService.java 代码如下：

```java
public class GhyService extends Service {

    public class IBinderImple extends Binder {
        public GhyService getGhyService() {
            Log.v("!", "执行了 IBinderImple 的 getGhyService 方法");
            return GhyService.this;
        }
    }

    @Override
    public void onCreate() {
        super.onCreate();
        Log.v("!", "调用了 onCreate 方法");
    }

    @Override
    public void onDestroy() {
        super.onDestroy();
        Log.v("!", "调用了 onDestroy 方法");
    }

    @Override
    public void onRebind(Intent intent) {
        super.onRebind(intent);
        Log.v("!", "调用了 onRebind 方法");
    }

    @Override
    public boolean onUnbind(Intent intent) {
        Log.v("!", "调用了 onUnbind 方法");
        return super.onUnbind(intent);
    }

    @Override
    public IBinder onBind(Intent arg0) {
```

```
            Log.v("!", "调用了 onBind 方法");
            return new IBinderImple();
        }
    }
```

在 GhyService.java 类中继承了 Binder 类，而 Binder 类是 IBinder 接口的实现类，其主要的目的是通过 IBinder 对象在 Activity 对象中操作 Service 类。

Activity 对象 Main.java 的代码如下：

```
public class Main extends Activity {
    private Button button1;
    private Button button2;
    private GhyService ghyServiceRef;

    private ServiceConnection serviceConnection = new ServiceConnection() {
        public void onServiceConnected(ComponentName arg0, IBinder arg1) {
            Log.v("onServiceConnected", "onServiceConnected");
            ghyServiceRef = ((GhyService.IBinderImple) arg1).getGhyService();
        }

        public void onServiceDisconnected(ComponentName arg0) {
            Log.v("onServiceDisconnected", "onServiceDisconnected");
            ghyServiceRef = null;
        }
    };

    @Override
    public void onCreate(Bundle savedInstanceState) {
        super.onCreate(savedInstanceState);
        setContentView(R.layout.main);

        button1 = (Button) this.findViewById(R.id.button1);
        button2 = (Button) this.findViewById(R.id.button2);

        button1.setOnClickListener(new OnClickListener() {
            public void onClick(View arg0) {
                Intent intent = new Intent(Main.this, GhyService.class);
                Main.this.bindService(intent, serviceConnection,
                        Service.BIND_AUTO_CREATE);
            }
        });
        button2.setOnClickListener(new OnClickListener() {
            public void onClick(View arg0) {
                if (ghyServiceRef != null) {
                    unbindService(serviceConnection);
                    ghyServiceRef = null;
                }
            }
```

```
        });
    }

    @Override
    public void onStop() {
        super.onStop();
        if (ghyServiceRef != null) {
            unbindService(serviceConnection);
            ghyServiceRef = null;
        }
    }
}
```

在 Main.java 代码中声明了 ServiceConnection 对象，它的作用是收到从 GhyService.java 文件中 onBind()方法返回的 Binder 对象，进而通过 Binder 对象取得 GhyService.java 对象的实例，从而能调用 GhyService.java 对象的一些业务方法。

运行项目出现如图 6.15 所示的界面。

单击"用 bindService 方式启动 Service"按钮，LogCat 控制台打印出的日志信息如图 6.16 所示。

图 6.15　初次运行程序

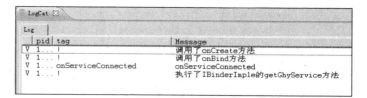

图 6.16　单击用 bindService 方式启动 Service 按钮

再重复单击"用 bindService 方式启动 Service"按钮，LogCat 日志结果没有变化，还是和图 6.16 中的一样，证明重复执行 bindService()方法不触发任何的回调函数。

这时单击后退按钮退出当前的应用程序，在 LogCat 中打印出的日志内容如图 6.17 所示。

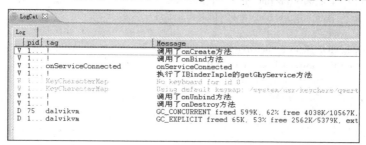

图 6.17　退出用 bindService 方式启动的 Service

从上面的实验可以观察到，通过使用 bindService()方法启动 Service 的生命周期和 Activity 对象一致，也就是说使用 bindService()方法启动 Service 后，Service 就和调用 bindService()方法的进程同生共死，当调用 bindService()方法的进程结束了，那么它 bind（绑定）的 Service 也要跟着被结束，所以好的编程习惯是在 Activity 的 onStop()回调函数中加上反绑定 Service 的代码，例如：

```
public void onStop() {
    super.onStop();
    if (ghyServiceRef != null) {
        unbindService(serviceConnection);
        ghyServiceRef = null;
    }
}
```

来看看正确执行生命周期函数 onUnbind()和 onDestroy()的方法。将 LogCat 中的内容清空，重新进入 bindService 应用程序，启动服务后再单击"停止用 bindService 方式启动的 Service"按钮停止 GhyService.java 服务，LogCat 控制台打印出的日志内容如图 6.18 所示。

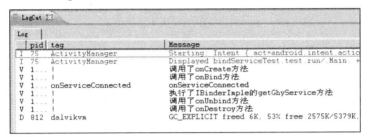

图 6.18　用按钮停止 GhyService.java 服务

程序测试到此，还有一个回调函数 onRebind()方法没有被执行，下面介绍该函数的使用。

6.2.3　回调函数 onRebind()的调用时机

回调函数 onRebind()重绑定的调用时机就像它的名字一样，是在客户端重新绑定 Service 时进行调用，注意，这里有一个限制就是反绑定 Service 时 onUnbind ()函数必须返回为 true。

新建名称为 onRebind_Test 的 Android 项目，创建自定义的 Service 类 GhyService.java，代码如下：

```
public class GhyService extends Service {

    public class IBinderImple extends Binder {
        public GhyService getGhyService() {
            Log.v("!", "执行了 IBinderImple 的 getGhyService 方法");
            return GhyService.this;
        }
    }

    @Override
    public void onCreate() {
        super.onCreate();
        Log.v("!", "调用了 onCreate 方法");
    }

    @Override
    public void onDestroy() {
        super.onDestroy();
        Log.v("!", "调用了 onDestroy 方法");
    }
```

```java
@Override
public int onStartCommand(Intent intent, int flags, int startId) {
    Log.v("!", "调用了 onStartCommand 方法");
    return super.onStartCommand(intent, flags, startId);
}

@Override
public void onRebind(Intent intent) {
    super.onRebind(intent);
    Log.v("!", "调用了 onRebind 方法");
}

@Override
public boolean onUnbind(Intent intent) {
    Log.v("!", "调用了 onUnbind 方法");
    // 按照 android doc 说明必须返回 true
    return true;
}

@Override
public IBinder onBind(Intent arg0) {
    Log.v("!", "调用了 onBind 方法");
    return new IBinderImple();
}

}
```

项目中 Activity 文件 Main.java 的代码如下：

```java
public class Main extends Activity {

    private Button button1;
    private Button button2;
    private Button button3;
    private GhyService ghyServiceRef;

    private ServiceConnection serviceConnection = new ServiceConnection() {
        public void onServiceConnected(ComponentName arg0, IBinder arg1) {
            Log.v("onServiceConnected", "onServiceConnected");
            ghyServiceRef = ((GhyService.IBinderImple) arg1).getGhyService();
        }

        public void onServiceDisconnected(ComponentName arg0) {
            Log.v("onServiceDisconnected", "onServiceDisconnected");
            ghyServiceRef = null;
        }
    };

    @Override
```

```java
public void onCreate(Bundle savedInstanceState) {
    super.onCreate(savedInstanceState);
    setContentView(R.layout.main);

    button1 = (Button) this.findViewById(R.id.button1);
    button2 = (Button) this.findViewById(R.id.button2);
    button3 = (Button) this.findViewById(R.id.button3);

    button1.setOnClickListener(new OnClickListener() {
        public void onClick(View arg0) {
            Intent intent = new Intent(Main.this, GhyService.class);
            // 要想退出 Activity 时 Service 不关闭
            // 必须先 startService 然后再 bindService
            Main.this.startService(intent);
            Main.this.bindService(intent, serviceConnection,
                    Service.BIND_AUTO_CREATE);
        }
    });
    button2.setOnClickListener(new OnClickListener() {
        public void onClick(View arg0) {
            Intent intent = new Intent(Main.this, GhyService.class);
            Main.this.bindService(intent, serviceConnection,
                    Service.BIND_AUTO_CREATE);
        }
    });

    button3.setOnClickListener(new OnClickListener() {
        public void onClick(View arg0) {
            Intent intent = new Intent(Main.this, GhyService.class);
            ghyServiceRef.stopService(intent);
            Main.this.finish();
        }
    });
}

@Override
public void onStop() {
    super.onStop();//////////////**************此代码不能简略
    unbindService(serviceConnection);
}
}
```

在文件 AndroidManifest.xml 中注册 Service,配置代码如下:

```xml
<service android:name="service.GhyService"></service>
```

程序运行后的初始界面如图 6.19 所示。

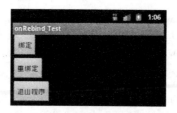

图 6.19 初始界面

单击"绑定"按钮，在 LogCat 中打印出日志信息，如图 6.20 所示。

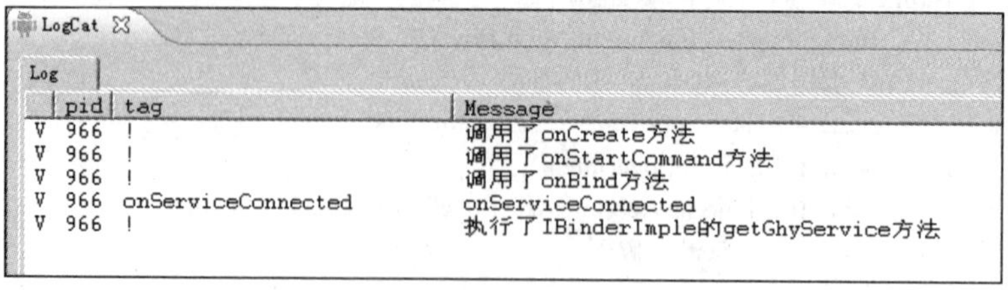

图 6.20 单击绑定按钮后的日志

再单击 AVD 面板中的 Back 按钮，打印出的日志信息如图 6.21 所示。

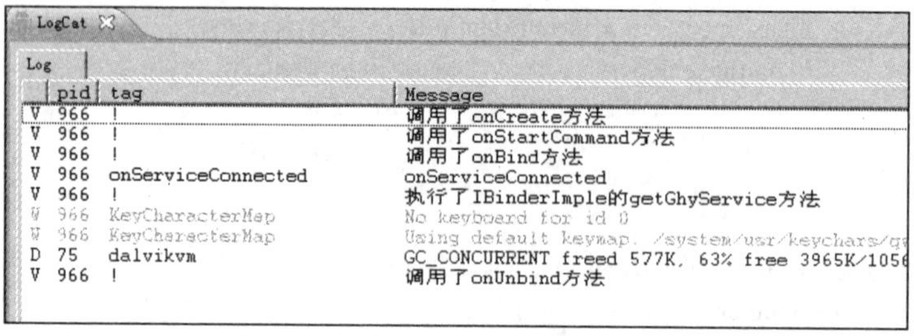

图 6.21 按下 Back 按钮后的日志信息

从图 6.21 中可以看到 onUnbind()方法执行了，再次进入这个项目，单击"重绑定"按钮出现日志信息，如图 6.22 所示。

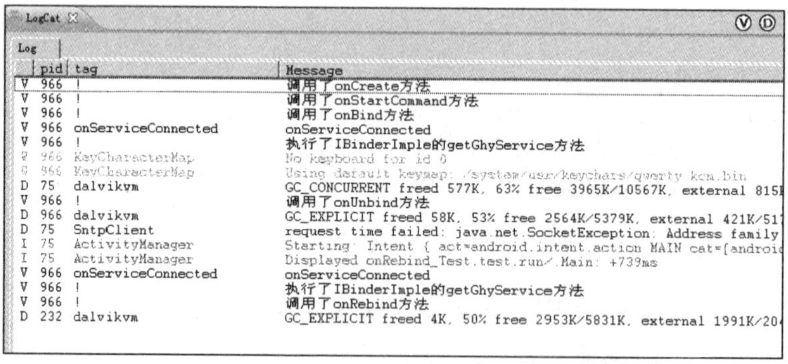

图 6.22 onRebind()方法被加载

总结：onRebind()方法执行的时机是 Service 在内存中已经存在，然后使用 bindService()方法再次与 Service 取得通信，这时 onRebind()方法被调用。

6.2.4　ServiceConnection 对象的 onServiceDisconnected()方法调用时机

类 ServiceConnection 中的 onServiceDisconnected()方法在正常情况下是不被调用的，它的调用时机是当 Service 服务被异外销毁时，例如内存的资源不足时这个方法才被自动调用。

6.3　Service 相关示例及知识点

Service 有很多相关的技术，本节就常用的技术作为示例一一列举，以使读者进一步加深对 Service 技术的使用。

6.3.1　定时服务 AlarmManager 的使用

对象 AlarmManager 的功能有些像"计划任务"，即在某一个时间执行某一个任务，大多数闹钟功能都使用这个对象设计，但这个对象在设备重启后会被取消，所以建议在系统启动时使用广播 BroadCast 再重新注册一下。

本示例将演示对象 AlarmManagerr 的基本使用方法。创建名称为 AlarmManagerTest1 项目，文件 Main.java 的代码如下：

```java
public class Main extends Activity {
    private Button button1;
    private Button button2;
    private Button button3;
    private Button button3_1;

    private AlarmManager am;

    @Override
    public void onCreate(Bundle savedInstanceState) {
        super.onCreate(savedInstanceState);
        setContentView(R.layout.main);

        button1 = (Button) this.findViewById(R.id.button1);
        button2 = (Button) this.findViewById(R.id.button2);
        button3 = (Button) this.findViewById(R.id.button3);
        button3_1 = (Button) this.findViewById(R.id.button3_1);

        button1.setOnClickListener(new OnClickListener() {
            public void onClick(View arg0) {
                am = (AlarmManager) Main.this
                        .getSystemService(Context.ALARM_SERVICE);
                Intent intent = new Intent("AlarmManagerTest1SendBroadCase");
                intent.putExtra("buttonType", "button1");
                intent.putExtra("button1Time", "" + System.currentTimeMillis());

                PendingIntent pi = PendingIntent.getBroadcast(Main.this, 1,
```

```java
                    intent, PendingIntent.FLAG_UPDATE_CURRENT);
            am.set(AlarmManager.RTC_WAKEUP,
                    System.currentTimeMillis() + 5000, pi);
        }
    });

    button2.setOnClickListener(new OnClickListener() {
        public void onClick(View arg0) {
            am = (AlarmManager) Main.this
                    .getSystemService(Context.ALARM_SERVICE);
            Intent intent = new Intent("AlarmManagerTest1SendBroadCase");
            intent.putExtra("buttonType", "button2");

            Calendar calendarRef = Calendar.getInstance();
            long getTime = calendarRef.getTimeInMillis();
            calendarRef.setTimeInMillis(System.currentTimeMillis());
            calendarRef.add(Calendar.SECOND, 60);
            intent.putExtra("button2Time", "" + getTime);

            PendingIntent pi = PendingIntent.getBroadcast(Main.this, 2,
                    intent, PendingIntent.FLAG_UPDATE_CURRENT);

            am.set(AlarmManager.RTC_WAKEUP, calendarRef.getTimeInMillis(),
                    pi);
        }
    });

    button3.setOnClickListener(new OnClickListener() {
        public void onClick(View arg0) {
            am = (AlarmManager) Main.this
                    .getSystemService(Context.ALARM_SERVICE);
            long getTime = SystemClock.elapsedRealtime();
            Intent intentButton3 = new Intent(
                    "AlarmManagerTest1SendBroadCase");
            intentButton3.putExtra("buttonType", "button3");
            intentButton3.putExtra("button3Time", "" + getTime);

            PendingIntent pi = PendingIntent.getBroadcast(Main.this, 3,
                    intentButton3, PendingIntent.FLAG_UPDATE_CURRENT);

            am.setRepeating(AlarmManager.ELAPSED_REALTIME_WAKEUP, getTime,
                    3000, pi);
        }
    });

    button3_1.setOnClickListener(new OnClickListener() {
        public void onClick(View arg0) {
            am = (AlarmManager) Main.this
                    .getSystemService(Context.ALARM_SERVICE);
            Intent intentButton3 = new Intent(
```

```java
                    "AlarmManagerTest1SendBroadCase");
            PendingIntent pi = PendingIntent.getBroadcast(Main.this, 3,
                    intentButton3, PendingIntent.FLAG_UPDATE_CURRENT);
            am.cancel(pi);
        }
    });
    }
}
```

广播类 GhyBroadcastReceiver.java 的核心代码如下：

```java
public class GhyBroadcastReceiver extends BroadcastReceiver {

    @Override
    public void onReceive(Context arg0, Intent arg1) {
        long getTime = System.currentTimeMillis();

        String buttonType = arg1.getStringExtra("buttonType");
        if (buttonType.equals("button1")) {
            Log.v("!", "按钮 1 延迟秒数："
                    + (getTime - Long.parseLong(arg1
                            .getStringExtra("button1Time"))) / 1000);
        }
        if (buttonType.equals("button2")) {
            Log.v("!", "按钮 2 延迟秒数："
                    + (getTime - Long.parseLong(arg1
                            .getStringExtra("button2Time"))) / 1000);
        }
        if (buttonType.equals("button3")) {
            Log.v("!", "每隔 3 秒执行一次："
                    + (getTime - Long.parseLong(arg1
                            .getStringExtra("button3Time"))) / 1000);
        }
    }
}
```

程序运行结果如图 6.23 所示。

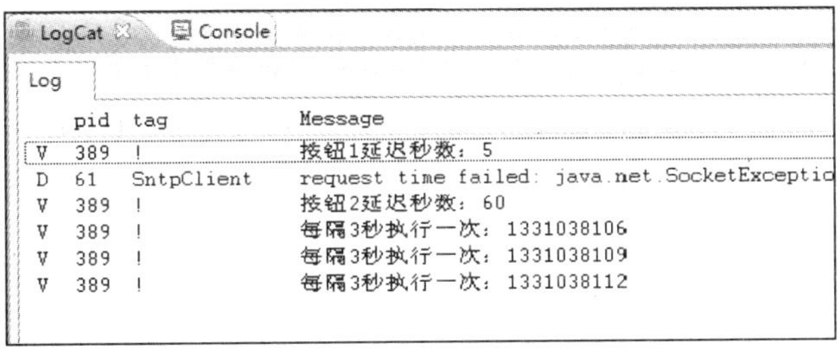

图 6.23　程序运行效果

常量 AlarmManager.RTC_WAKEUP 的作用是在指定的时间唤醒设备并执行 Intent 意图，而常量 AlarmManager.ELAPSED_REALTIME_WAKEUP 的作用是当设备启动后所经过的时间之后再触发指定的意图。

6.3.2 判断 Service 是否在运行中

创建名称为 serviceIsRun 的项目，Main.java 文件的核心代码如下：

```java
public class Main extends Activity {
    private Button button1;
    private Button button2;
    private Button button3;

    @Override
    public void onCreate(Bundle savedInstanceState) {
        super.onCreate(savedInstanceState);
        setContentView(R.layout.main);

        button1 = (Button) this.findViewById(R.id.button1);
        button2 = (Button) this.findViewById(R.id.button2);
        button3 = (Button) this.findViewById(R.id.button3);

        button1.setOnClickListener(new OnClickListener() {
            public void onClick(View arg0) {
                Intent intent = new Intent(Main.this, GhyService.class);
                Main.this.startService(intent);
            }
        });

        button2.setOnClickListener(new OnClickListener() {
            public void onClick(View arg0) {
                Intent intent = new Intent(Main.this, GhyService.class);
                Main.this.stopService(intent);
            }
        });

        button3.setOnClickListener(new OnClickListener() {
            public void onClick(View arg0) {

                ActivityManager am = (ActivityManager) Main.this
                        .getSystemService(Context.ACTIVITY_SERVICE);
                List<RunningServiceInfo> serviceList = am
                        .getRunningServices(Integer.MAX_VALUE);
                boolean isRun = false;
                for (int i = 0; i < serviceList.size(); i++) {
                    if (serviceList.get(i).service.getClassName().equals(
                            "extservice.GhyService")) {
                        isRun = true;
                        break;
```

```
                }
            }
            if (isRun == true) {
                Log.v("!", "运行中");
            } else {
                Log.v("!", "不运行中");
            }
        }
    });
  }
}
```

程序运行后即可查询 Service 的运行状态,如图 6.24 所示。

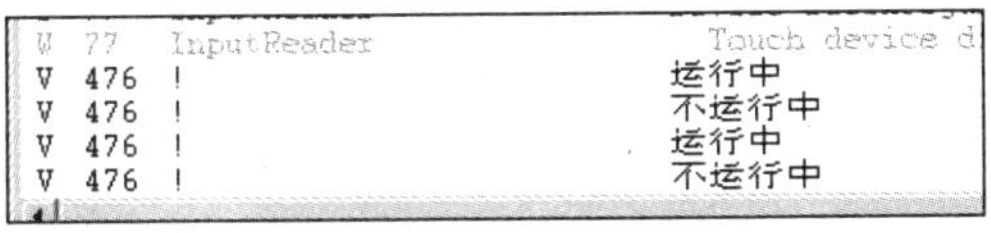

图 6.24 运行效果

6.3.3 方法 onStartCommand 的返回值实验

方法 onStartCommand()的返回值为 int 类型,主要的作用是当 Service 进程被意外 kill 掉时,Service 服务下一步要做哪些行为,主要有 3 种值。

- START_STICKY: Service 被异外终止时不调用 onDestroy()回调,并且终止后自动重启 Service 服务,只执行 Service 对象的 onCreate()生命周期方法。
- START_NOT_STICKY: Service 被异外终止时不调用 onDestroy()回调,并且不自动重启服务。
- START_REDELIVER_INTENT: Service 被异外终止时不调用 onDestroy()回调,并且终止后自动重启 Service 服务,还要执行 Service 对象的 onCreate()和 onStartCommand()生命周期方法,并且从 Intent 中能取到值。

新建名称为 testService 的 Android 项目,创建名称为 Main.java 的 Activity 对象,核心代码如下:

```
public class Main extends Activity {
    private Button button1;

    @Override
    public void onCreate(Bundle savedInstanceState) {
        super.onCreate(savedInstanceState);
        setContentView(R.layout.main);

        Log.v("!", "Main onCreate");

        button1 = (Button) this.findViewById(R.id.button1);
        button1.setOnClickListener(new OnClickListener() {
            public void onClick(View arg0) {
```

```java
                    Intent intent = new Intent(Main.this, GhyService.class);
                    intent.putExtra("username", "gaohongyan");
                    Main.this.startService(intent);
                }
            });
        }
    }
```

创建名称为 GhyService.java 的 Service 对象,核心代码如下:

```java
public class GhyService extends Service {

    @Override
    public boolean onUnbind(Intent intent) {
        Log.v("!", "GhyService onUnbind");
        return super.onUnbind(intent);
    }

    @Override
    public void onRebind(Intent intent) {
        super.onRebind(intent);
        Log.v("!", "GhyService onRebind");
    }

    @Override
    public void onCreate() {
        super.onCreate();
        Log.v("!", "GhyService onCreate");
    }

    @Override
    public int onStartCommand(Intent intent, int flags, int startId) {
        Log.v("!", "GhyService onStartCommand username="
                + intent.getStringExtra("username"));
        return Service.START_STICKY;
    }

    @Override
    public IBinder onBind(Intent arg0) {
        Log.v("!", "GhyService onBind");
        return null;
    }

    @Override
    public void onDestroy() {
        super.onDestroy();
        Log.v("!", "GhyService onDestroy");
    }

}
```

从上面的代码可以看到，onStartCommand 方法返回的常量为 Service.START_STICKY，运行项目，在 LogCat 中打印日志信息，如图 6.25 所示。

从图 6.25 中可以看到 Activity 的 onCreate()方法被调用，单击 Button 按钮启动 Service，LogCat 的日志内容如图 6.26 所示。

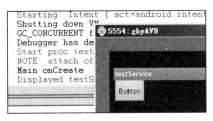

图 6.25　初始运行效果

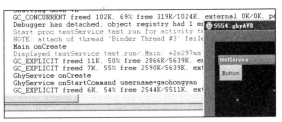

图 6.26　启动 Service 的日志

这时进入 CMD 控制台，找到 androidSDK 所在的文件夹，进入 androidSDK 的如下路径：

C:\android\安装完成后\android-sdk_r09-windows\android-sdk-windows\platform-tools

进入上方的路径后，输入 adb shell 后执行 ps 命令，查看一下当前系统中运行的进程列表，在列表中发现当前运行的进程，如图 6.27 所示。

图 6.27　发现运行的进程

从图 6.27 中可以看到，当前进程的 pid 为 742，继续输入命令 kill 742 结束这个进程，再次查看控制台日志，如图 6.28 所示。

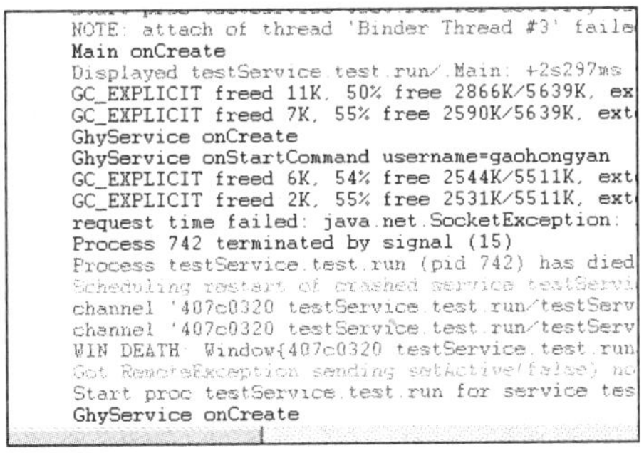

图 6.28　重新创建了 Service 但未执行 onStartCommand()

继续更改代码，将 onStartCommand()方法的代码改成如下形式：

```
public int onStartCommand(Intent intent, int flags, int startId) {
    Log.v("!", "GhyService onStartCommand username="
            + intent.getStringExtra("username"));
    return Service.START_NOT_STICKY;
}
```

重新运行此项目，在 LogCat 中打印的日志信息如图 6.29 所示。

单击 Button 按钮出现如图 6.30 所示。

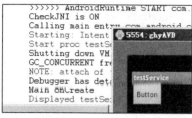
图 6.29　第 2 次初始运行

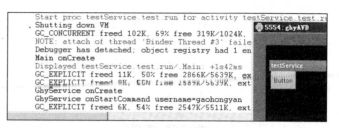
图 6.30　第 2 次按下 Button

这时再 kill 掉当前的进程，查看一下 LogCat 中的内容，如图 6.31 所示。

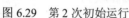

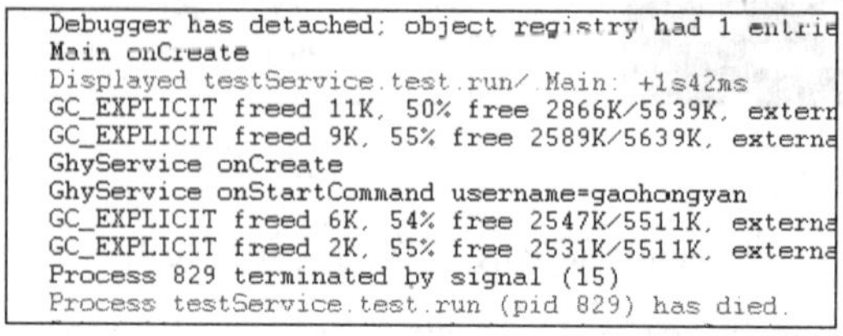

图 6.31　第 2 次并没有重新创建 Service

继续做第 3 个常量的实验，将 onStartCommand()方法的代码更改如下：

```
public int onStartCommand(Intent intent, int flags, int startId) {
    Log.v("!", "GhyService onStartCommand username="
        + intent.getStringExtra("username"));
    return Service.START_REDELIVER_INTENT;
}
```

运行项目，按流程去操作 Button->adb shell->ps->kill xxxx，在 LogCat 中打印出的日志如图 6.32 所示。

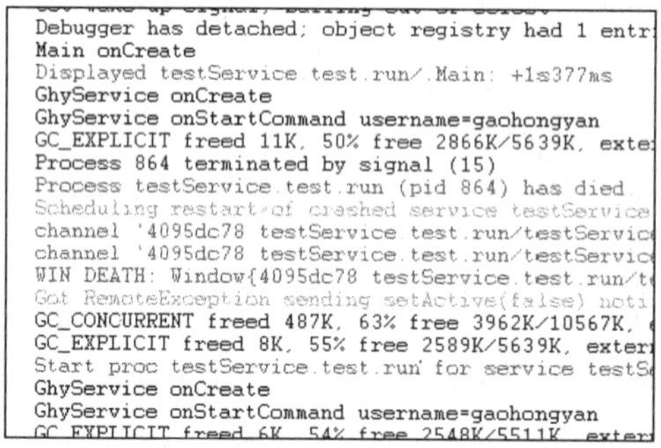

图 6.32　第 3 次实验完整流程

这次 Service 正确地重启并且调用了 onStartCommand()方法，还从 Intent 中取到了 username 的值。

6.3.4　Parcelable 接口串行化的使用

Android 序列化对象主要有两种方法，一种是实现 Serializable 接口，另外一种是实现 Parcelable 接口。实现 Serializable 接口是 Java SE 本身就支持的，而 Parcelable 是 Android 特有的功能，效率比实现 Serializable 接口高，而且还可以用在 IPC 中。接口 Parcelable 的作用是把数据进行打包以利于后期的传输,传输到目的地址后再分解出来,相当于实体的作用,如果你想在 Android 中传递自定义数据类型这是其中的一种手段。

如果某个类实现了这个 Parcelable 接口,那么它的对象实例可以写入到 Parcel 中,并且能够从中恢复,而且这个类必须要有一个 static 的字段,并且字段变量的名称一定为 CREATOR,这个变量是某个实现了 Parcelable.Creator 接口的类的对象实例。

接口 Parcelable 的作用其实就是为了数据的串行化,新建一个名称为 testParcelable 的实例来在两个 Activity 中传递 Parcelable 对象。

创建一个实现 Parcelable 接口的实体 Userinfo.java,完整的代码如下:

```java
package entity;

import android.os.Parcel;
import android.os.Parcelable;
import android.util.Log;

public class Userinfo implements Parcelable {

    public Userinfo() {
        Log.v("!", "Userinfo()");
    }

    public Userinfo(Parcel parcel) {
        Log.v("!", "Userinfo(Parcel parcel)");
        this.username = parcel.readString();//2 先读 username
        this.age = parcel.readInt();//2 再读 age

    }

    private String username;
    private int age;

    public String getUsername() {
        Log.v("!", "Userinfo getUsername");
        return username;
    }

    public void setUsername(String username) {
        Log.v("!", "Userinfo setUsername");
        this.username = username;
    }

    public int getAge() {
        Log.v("!", "Userinfo getAge");
        return age;
    }

    public void setAge(int age) {
        Log.v("!", "Userinfo setAge");
```

```java
            this.age = age;
        }

        public int describeContents() {
            return 0;
        }

        public void writeToParcel(Parcel arg0, int arg1) {
            Log.v("!", "writeToParcel");
            arg0.writeString(username);//1 先写 username
            arg0.writeInt(age);//1 再写 age
        }

        public static final Creator<Userinfo> CREATOR = new Creator<Userinfo>() {
            public Userinfo createFromParcel(Parcel arg0) {
                Log.v("!", "createFromParcel");
                return new Userinfo(arg0);
            }

            public Userinfo[] newArray(int arg0) {
                Log.v("!", "newArray");
                return new Userinfo[arg0];
            }

        };

    }
```

方法 writeToParcel() 的作用是把 Userinfo 的 username 和 age 值存入 Parcel 对象中，也就是将你的对象序列化为一个 Parcel 对象，传递到目的地址后再调用 createFromParcel() 方法，从 Parcel 对象中将值取出来再放入 Userinfo 的 username 和 age 属性字段中，也就是使用 createFromParcel() 方法将 Parcel 对象反序列化为 Userinfo 对象。而且在 Userinfo.java 类中必须要有一个实例化的静态常量内部对象 CREATOR，它实现了接口 Parcelable.Creator，修饰符 public static final 都不能少，内部对象 CREATOR 的名称也不能改变，必须全部大写。

大体的流程为：通过 writeToParcel() 方法将 Userinfo 对象映射成 Parcel 对象，再通过 createFromParcel() 方法将 Parcel 对象映射成 Userinfo 对象。在这里可以将 Parcel 看成是一个实体，通过 writeToParcel() 方法把 Userinfo 对象的值写到 Parcel 对象里面，再通过 createFromParcel() 方法从 Parcel 对象里读取数据再生成 1 个 Userinfo 对象，只不过这个过程需要程序来实现。

注意，写入的顺序和读取的顺序一定要一致。

文件 Main.java 的核心代码如下：

```java
public class Main extends Activity {
    @Override
    public void onCreate(Bundle savedInstanceState) {
        super.onCreate(savedInstanceState);
        setContentView(R.layout.main);
```

```java
        Userinfo userinfo = new Userinfo();
        userinfo.setUsername("高洪岩");
        userinfo.setAge(100);

        Intent intent = new Intent(this, Second.class);
        intent.putExtra("userinfoKey", userinfo);
        this.startActivity(intent);

    }
}
```

文件 Second.java 的核心代码如下：

```java
public class Second extends Activity {
    @Override
    public void onCreate(Bundle savedInstanceState) {
        super.onCreate(savedInstanceState);
        setContentView(R.layout.second);

        Userinfo userinfo = this.getIntent().getParcelableExtra("userinfoKey");

        Log.v("!", "second getUsername=" + userinfo.getUsername());
        Log.v("!", "second getAge=" + userinfo.getAge());

    }
}
```

程序运行后在 Main 和 Second 之间传递自定义数据类型 Userinfo，在 LogCat 中打印出相关的日志信息，如图 6.33 所示。

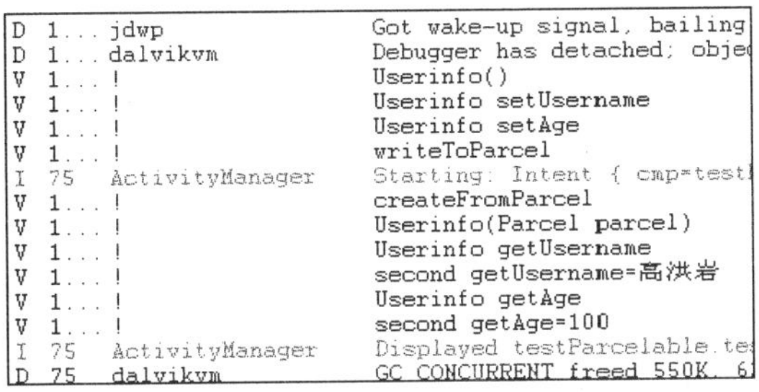

图 6.33 在 Second 中取出从 Main 传递过来的数据

6.3.5 使用 AIDL 技术跨进程传递 Parcelable 对象

有了前面 Parcelable 对象的学习基础，那么在实现跨进程传递数据对象时就非常容易了，比如在不同的进程中访问其他进程中 Service 的对象。

在 Android 中想要跨进程传递对象，必须使用 AIDL（Android Interface Definition Language）服务，什么是 AIDL 服务呢？为了使其他的应用程序也可以访问本应用程序提供的服务，Android 系统采用了远程过程调用（Remote Procedure Call，RPC）方式来实现。与很多其他的基于 RPC 的解决方案一样，Android 使用一种接口定义语言 IDL（Interface Definition Language）来公开服务的接口。因此，可以将这种跨进程访问的服务称为 AIDL 服务。

AIDL 是一种 IDL 语言，用于生成可以在 Android 设备上两个进程之间进行通信（IPC）的代码。如果在一个进程中（例如 1 个 Activity）要调用另一个进程中（例如 Service）对象的操作，就可以使用 AIDL 生成可序列化的参数。但服务 AIDL 传递数据的数据类型是有一些限制的，例如常用的基本数据类型 String 和 int 等是支持的，集合框架 List 和 Map 也是支持的，如果想实现一些自定义的数据类型则必须要实现 Parcelable 接口，比如实体类，也支持 List<实体>这样的用法。

新建名称为 service_1 的 Android 项目，本项目是提供业务服务的，称为服务端项目。

新建实体类 School.java 文件，代码如下：

```java
package aidlpackage.entity;

import android.os.Parcel;
import android.os.Parcelable;

public class School implements Parcelable {

    private String name;
    private String type;

    public School() {
    }

    public School(Parcel source) {
        super();
        this.setName(source.readString());
        this.setType(source.readString());
    }

    public String getName() {
        return name;
    }

    public void setName(String name) {
        this.name = name;
    }

    public String getType() {
        return type;
    }

    public void setType(String type) {
        this.type = type;
    }

    public int describeContents() {
        return 0;
    }
```

```java
    public void writeToParcel(Parcel dest, int flags) {
        dest.writeString(name);
        dest.writeString(type);
    }

    public static final Parcelable.Creator<School> CREATOR = new Parcelable.Creator<School>() {
        public School createFromParcel(Parcel source) {
            return new School(source);
        }

        public School[] newArray(int size) {
            return new School[size];
        }
    };
}
```

由于 School.java 文件是一个自定义的实体类，所以再新建一个名称为 School.aidl 的文件来声明这个自定义的数据类型 School.java 对象，School.aidl 文件的内容如下：

```
package aidlpackage.entity;
parcelable School;
```

关键字 package 声明 School.java 类在哪个包中，parcelable 是在系统中注册这个类名称。

再新建 1 个名称为 Userinfo.java 的文件，代码如下：

```java
package aidlpackage.entity;

import java.util.ArrayList;
import java.util.HashMap;
import java.util.List;
import java.util.Map;

import android.os.Parcel;
import android.os.Parcelable;
import android.util.Log;

public class Userinfo implements Parcelable {

    private String username;// 1
    private int age;// 2
    private List<String> stuList = new ArrayList<String>();// 3
    private Map<String, String> stuMap = new HashMap<String, String>();// 4
    private List<School> schoolList = new ArrayList<School>();// 5
    private School schoolInfo = new School();// 6

    public List<School> getSchoolList() {
        return schoolList;
    }

    public void setSchoolList(List<School> schoolList) {
```

```java
        this.schoolList = schoolList;
    }

    public School getSchoolInfo() {
        return schoolInfo;
    }

    public void setSchoolInfo(School schoolInfo) {
        this.schoolInfo = schoolInfo;
    }

    public List getStuList() {
        return stuList;
    }

    public void setStuList(List stuList) {
        this.stuList = stuList;
    }

    public Map getStuMap() {
        return stuMap;
    }

    public void setStuMap(Map stuMap) {
        this.stuMap = stuMap;
    }

    public Userinfo() {
        super();
    }

    public Userinfo(Parcel parcel) {
        super();
        this.setUsername(parcel.readString());// 1
        this.setAge(parcel.readInt());// 2
        this.setStuList(parcel.readArrayList(List.class.getClassLoader()));// 3
        this.setStuMap(parcel.readHashMap(Map.class.getClassLoader()));// 4
        this.setSchoolList(parcel.readArrayList(School.class.getClassLoader()));// 5
        this.setSchoolInfo((School) parcel.readParcelable(School.class
                .getClassLoader()));// 6
    }

    public static Parcelable.Creator<Userinfo> getCreator() {
        return CREATOR;
    }

    public String getUsername() {
        return username;
    }
```

```java
    public void setUsername(String username) {
        this.username = username;
    }

    public int getAge() {
        return age;
    }

    public void setAge(int age) {
        this.age = age;
    }

    public int describeContents() {
        return 0;
    }

    public void writeToParcel(Parcel arg0, int arg1) {
        Log.v("!", "writeToParcel");

        arg0.writeString(username);// 1
        arg0.writeInt(age);// 2
        arg0.writeList(stuList);// 3
        arg0.writeMap(stuMap);// 4
        arg0.writeList(schoolList);// 5
        arg0.writeParcelable(schoolInfo, arg1);// 6

    }

    public static final Parcelable.Creator<Userinfo> CREATOR = new Parcelable.Creator<Userinfo>() {
        public Userinfo createFromParcel(Parcel parcel) {
            Log.v("!", "createFromParcel");
            Userinfo userinfo = new Userinfo(parcel);
            return userinfo;
        }

        public Userinfo[] newArray(int size) {
            Log.v("!", "newArray");
            return new Userinfo[size];
        }
    };

}
```

对应的 Userinfo.aidl 文件的代码如下:

```
package aidlpackage.entity;
parcelable Userinfo;
```

再创建一个名称为 GhyService.java 的 Service 文件, 代码如下:

```
package extservice;
```

```java
import java.util.ArrayList;
import java.util.HashMap;
import java.util.LinkedHashMap;
import java.util.List;
import java.util.Map;

import aidlpackage.gaohongyanService;
import aidlpackage.entity.School;
import aidlpackage.entity.Userinfo;
import android.app.Service;
import android.content.Intent;
import android.os.IBinder;
import android.os.RemoteException;
import android.util.Log;

public class GhyService extends Service {

    private class GhyBinder extends gaohongyanService.Stub {

        public boolean getBoolean() throws RemoteException {
            return true;
        }

        public int getInt() throws RemoteException {
            return 99;
        }

        public String getString() throws RemoteException {
            return "gaohongyan";
        }

        public Userinfo getUserinfo() throws RemoteException {
            Userinfo userinfo = new Userinfo();
            userinfo.setUsername("gaohongyan username");
            userinfo.setAge(10000);

            List stuList = new ArrayList();
            stuList.add("list1");
            stuList.add("list2");
            stuList.add("list3");

            userinfo.setStuList(stuList);

            Map stuMap = new HashMap();
            stuMap.put("key1", "value1");
            stuMap.put("key2", "value2");
            stuMap.put("key3", "value3");

            userinfo.setStuMap(stuMap);
```

```java
        List<School> schoolList = new ArrayList<School>();
        School s1 = new School();
        s1.setName("学校名称 1");
        s1.setType("学校类别 1");

        School s2 = new School();
        s2.setName("学校名称 2");
        s2.setType("学校类别 2");

        schoolList.add(s1);
        schoolList.add(s2);

        userinfo.setSchoolList(schoolList);

        School schoolRef = new School();
        schoolRef.setName("学校名称");
        schoolRef.setType("学校类别");
        userinfo.setSchoolInfo(schoolRef);

        return userinfo;
    }

    public List<Userinfo> getUserinfoList() throws RemoteException {
        List<Userinfo> listUserinfo = new ArrayList<Userinfo>();
        for (int i = 0; i < 10; i++) {
            Userinfo userinfo = new Userinfo();
            userinfo.setUsername("username" + (i + 1));
            userinfo.setAge(i + 1);
            listUserinfo.add(userinfo);
        }
        return listUserinfo;
    }

    public Map getMap() throws RemoteException {
        Map returnMap = new LinkedHashMap();
        returnMap.put("key1", "value1");
        returnMap.put("key2", "value2");
        returnMap.put("key3", "value3");
        returnMap.put("key4", "value4");
        returnMap.put("key5", "value5");
        returnMap.put("key6", "value6");
        return returnMap;
    }

    public void setUserinfo(Userinfo userinfo) throws RemoteException {
        Log.v("!", "setUserinfo getUserinfo username="
                + userinfo.getUsername());

    }
```

```java
            public void setUserinfoList(List<Userinfo> userinfoList)
                    throws RemoteException {
                Log.v("!", "setUserinfoList userinfoList.get(0).getUsername()="
                        + userinfoList.get(0).getUsername());
            }
        };
    }

    @Override
    public void onCreate() {
        super.onCreate();
        Log.v("!GhyService", "onCreate");
    }

    @Override
    public void onRebind(Intent intent) {
        super.onRebind(intent);
        Log.v("!GhyService", "onRebind");
    }

    @Override
    public void onDestroy() {
        super.onDestroy();
        Log.v("!GhyService", "onDestroy");
    }

    @Override
    public int onStartCommand(Intent intent, int flags, int startId) {
        Log.v("!GhyService", "onStartCommand");
        return super.onStartCommand(intent, flags, startId);
    }

    @Override
    public IBinder onBind(Intent arg0) {
        Log.v("!GhyService", "onBind");
        return new GhyBinder();
    }
}
```

把服务 Service 在 AndroidManifest.xml 文件中进行注册。

```xml
<service android:name="extservice.GhyService">
    <intent-filter>
        <action android:name="xiaoXueAction"></action>
    </intent-filter>
</service>
```

最为主要的还要创建一个名称为 gaohongyanService.aidl 的文件，这个文件就是操作业务的接口，只不过是 aidl 扩展名，代码如下：

```
package aidlpackage;

import aidlpackage.entity.Userinfo;

interface gaohongyanService{

String getString();

int getInt();

boolean getBoolean();

Userinfo getUserinfo();

List<Userinfo> getUserinfoList();

Map getMap();

void setUserinfo(in Userinfo userinfo);

void setUserinfoList(in List<Userinfo> userinfoList);

}
```

写 AIDL 业务接口时不需要添加修饰符，完成后的项目文件结构如图 6.34 所示。

启动项目以使本项目中的 Service 在系统中进行注册，为其他应用程序提供服务。

再新建一个名称为 service_2_caller 的 Android 项目，本项目称为客户端项目。

将 service_1 项目中的 aidlpackage 包中的所有内容复制到 src 路径下，完成后的项目文件结构如图 6.35 所示。

图 6.34　服务器端的项目文件结构

图 6.35　service_2_caller 项目结构

文件 Main.java 的代码如下：

```java
public class Main extends Activity {

    private ServiceConnection connection = new ServiceConnection() {
        public void onServiceConnected(ComponentName arg0, IBinder arg1) {
            try {
                gaohongyanService ghyServiceRef = gaohongyanService.Stub
                        .asInterface(arg1);
                Log.v("!", "getString=" + ghyServiceRef.getString());
                Log.v("!", "getInt=" + ghyServiceRef.getInt());
                Log.v("!", "getBoolean=" + ghyServiceRef.getBoolean());

                Userinfo userinfo = ghyServiceRef.getUserinfo();
                Log.v("!", "userinfo username=" + userinfo.getUsername());
                Log.v("!", "userinfo age=" + userinfo.getAge());

                List stuList = userinfo.getStuList();
                for (int i = 0; i < stuList.size(); i++) {
                    Log.v("!", "stuList each value=" + stuList.get(i));
                }

                Map stuMap = userinfo.getStuMap();
                Iterator iteratorMap = stuMap.keySet().iterator();
                while (iteratorMap.hasNext()) {
                    String key = "" + iteratorMap.next();
                    Log.v("!", "stuMap key=" + key + " value="
                            + stuMap.get(key));

                }

                List<School> schoolList = userinfo.getSchoolList();
                for (int i = 0; i < schoolList.size(); i++) {
                    Log.v("!", "each list school name="
                            + schoolList.get(i).getName() + " type="
                            + schoolList.get(i).getType());
                }

                Log.v("!", "school name=" + userinfo.getSchoolInfo().getName()
                        + " type=" + userinfo.getSchoolInfo().getType());

                List<Userinfo> getUserinfoList = ghyServiceRef
                        .getUserinfoList();
                for (int i = 0; i < getUserinfoList.size(); i++) {
                    Userinfo eachUserinfo = getUserinfoList.get(i);
                    Log.v("!", "eachUserinfo username="
                            + eachUserinfo.getUsername()
                            + "   eachUserinfo age=" + eachUserinfo.getAge());
                }
```

```java
            Map map = ghyServiceRef.getMap();
            Iterator iterator = map.keySet().iterator();
            while (iterator.hasNext()) {
                String mapKey = "" + iterator.next();
                Log.v("!", "map key=" + mapKey + " value="
                        + map.get(mapKey));
            }

            Userinfo userinfoParam = new Userinfo();
            userinfoParam.setUsername("单实体 useranme 值");
            ghyServiceRef.setUserinfo(userinfoParam);

            List<Userinfo> userinfoList = new ArrayList<Userinfo>();
            Userinfo userinfoListElement = new Userinfo();
            userinfoListElement.setUsername("多实体 list.get(0)的 useranme 值");
            userinfoList.add(userinfoListElement);
            ghyServiceRef.setUserinfoList(userinfoList);

        } catch (RemoteException e) {
            // TODO Auto-generated catch block
            e.printStackTrace();
        }

    }

    public void onServiceDisconnected(ComponentName arg0) {
    }
};

private Button button1;

@Override
public void onCreate(Bundle savedInstanceState) {
    super.onCreate(savedInstanceState);
    setContentView(R.layout.main);

    Intent intent = new Intent("xiaoXueAction");
    Main.this.bindService(intent, connection, Service.BIND_AUTO_CREATE);

    button1 = (Button) this.findViewById(R.id.button1);
    button1.setOnClickListener(new View.OnClickListener() {
        public void onClick(View arg0) {
            Main.this.unbindService(connection);
        }
    });
}
```

程序运行结果如图 6.36 所示。

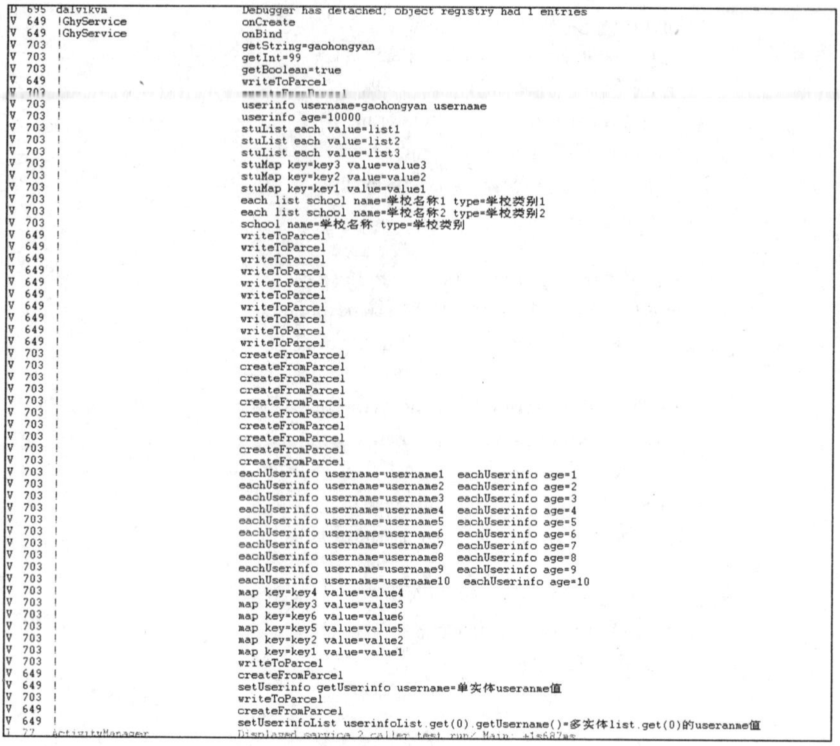

图 6.36 程序运行结果

在开发 AIDL 项目时需要注意以下几个问题：

- 在负责业务的 AIDL 描述文件中，如果引用自定义的实体也要必须显式的调用 import 指令进行实体对象的引用。
- 在 AIDL 文件中所有非 Java 原始类型参数必须加上标记:in、out、inout。
- 如果使用 ADT 进行 Android 的开发，会自动生成一个和 AIDL 文件名同名的接口 java 文件，存放在 gen 路径下。
- 接口前不用加访问权限修饰符 public、private、protected 等，也不能用 final 和 static。
- 两个项目中的 AIDL 文件一定要一模一样，方法的顺序也要一样，建议使用 copy 法。

而一个类要使用 Parcelable 功能，大体实现如下 5 个步骤：

01　实现 Parcelable 接口。

02　实现 writeToParcel(Parcel out)方法。

03　实现 readFromParcel(Parcel in)方法。

04　添加一个静态字段 CREATOR。

05　创建若干个 AIDL 文件声明业务及自定义实体对象。

6.4 Handle 对象的使用

对象 Handle 的主要作用是可以发送和处理消息队列,在 Android 中模仿了 Windows 操作系统中的 Message 原理来实现组件间的解耦,它可以接受子线程发送的 Message 对象,并用此 Message 对象中封装的数据在主线程中更新 UI 界面。

需要注意的是,在 UI 线程中启动 Handler 对象时,Handler 与调用者 Activity 处于同一线程,也就是通常所说的 UI 线程。如果 Handler 里面做耗时的动作,UI 线程会阻塞,另外由于 Android 的 UI 线程不是安全的,并且这些操作必须在 UI 线程中执行,如果不是在 UI 线程中操作 View 对象则系统报出异常。每个 Handler 实例都会绑定到创建它的线程中(一般是位于主线程)。

在 Android 中进行与 UI 通信的开发时,经常会使用 Handler 对象来控制 UI 程序的界面,它的作用可以理解为与其他线程协同工作,接收其他线程的消息并通过接收到的消息更新 UI 界面。

现在有这么一种情况,在一个 UI 界面上有一个按钮,当单击这个按钮的时候会进行网络连接,并把网络上的数据取下来显示到 UI 界面中一个 TextView 里,这时出现一个问题,就是如果这个网络连接的延迟过大,或根本连接不上,可能用时数秒甚至更长,那么程序的界面将处于一种假死状态,这样的效果很明显不符合体验性好的软件标准,这时理论上可以创建一个线程,在线程中取得网络上的数据,但下一步出现了问题!在用户自定义的线程中将取到的数据去更新 UI 则会报出异常,这个情况在第二章已经介绍过此实验,因为 Android 是单线程模型,不允许程序员在自定义的线程类中直接操作 UI 界面,为了解决这个问题,Android 开发了 Handler 对象,由它来负责与子线程进行通信,从而让子线程与主线程之间建立起协作的桥梁,当然也就可以传递数据(大多使用 Message 对象传递),使 Android 的 UI 更新问题得到解决。

6.4.1 Handler 对象的初步使用

本示例就模拟从网络下载数据再显示到 UI 界面上的效果。新建名称为 handler1 的 Android 项目,文件 Main.java 的核心代码如下:

```java
public class Main extends Activity {
    private Button button1;

    private Handler handler = new Handler() {
        @Override
        public void handleMessage(Message msg) {
            super.handleMessage(msg);
            Log.v("!", "Activity print status="
                    + msg.getData().getString("status") + " thread name="
                    + Thread.currentThread().getName());
        }
    };

    @Override
    public void onCreate(Bundle savedInstanceState) {
        super.onCreate(savedInstanceState);
```

```java
        setContentView(R.layout.main);

        Log.v("!", "Activity Thread name=" + Thread.currentThread().getName());

        button1 = (Button) this.findViewById(R.id.button1);
        button1.setOnClickListener(new OnClickListener() {
            public void onClick(View arg0) {
                GhyThread ghyThreadRef = new GhyThread(handler);
                ghyThreadRef.start();
            }
        });

    }
}
```

自定义线程类 GhyThread.java 的核心代码如下:

```java
public class GhyThread extends Thread {

    public GhyThread(Handler handler) {
        super();
        this.handler = handler;
    }

    private Handler handler;

    @Override
    public void run() {
        super.run();
        try {
            int i = 0;
            while (i < 10) {
                i++;
                Log.v("!", "GhyThread threadName="
                        + this.currentThread().getName() + " i=" + i);
                Thread.sleep(1000);
            }
            Bundle bundle = new Bundle();
            bundle.putString("status", "end");

            Message message = new Message();
            message.setData(bundle);
            handler.sendMessage(message);
        } catch (InterruptedException e) {
            // TODO Auto-generated catch block
            e.printStackTrace();
        }
    }
}
```

程序运行后的效果如图 6.37 所示。

```
I  75    ActivityManager  Start proc handler1.test.run for activity handl
V  459   !                Activity Thread name=main
I  75    ActivityManager  Displayed handler1.test.run/.Main: +1s295ms
V  459   !                GhyThread threadName=Thread-10 i=1
V  459   !                GhyThread threadName=Thread-10 i=2
V  459   !                GhyThread threadName=Thread-10 i=3
V  459   !                GhyThread threadName=Thread-10 i=4
V  459   !                GhyThread threadName=Thread-10 i=5
V  459   !                GhyThread threadName=Thread-10 i=6
V  459   !                GhyThread threadName=Thread-10 i=7
V  459   !                GhyThread threadName=Thread-10 i=8
V  459   !                GhyThread threadName=Thread-10 i=9
V  459   !                GhyThread threadName=Thread-10 i=10
V  459   !                Activity print status=end thread name=main
```

图 6.37 运行效果

从图 6.37 中可以看到，方法 handleMessage() 是运行在 main 主线程中的，也就是 Handler 被绑定到了主线程中。

为了进一步演示 Handler 绑定到主线程中的情况，新建一个名称为 HandlerBindUIThread 项目，Activity 文件 Main.java 的代码如下：

```java
public class Main extends Activity {

    private Runnable run = new Runnable() {
        public void run() {
            try {
                Log.v("!", "run thread is=" + Thread.currentThread().getId()
                        + " thread name=" + Thread.currentThread().getName());
                Thread.sleep(10000);
            } catch (InterruptedException e) {
                // TODO Auto-generated catch block
                e.printStackTrace();
            }
        }
    };

    @Override
    public void onCreate(Bundle savedInstanceState) {
        super.onCreate(savedInstanceState);
        Log.v("!", "onCreate thread is=" + Thread.currentThread().getId()
                + " thread name=" + Thread.currentThread().getName());
        Log.v("!", "begin");
        long beginTime = System.currentTimeMillis();
        Handler hanlder = new Handler();
        hanlder.post(run);
        setContentView(R.layout.main);
        long endTime = System.currentTimeMillis();
        Log.v("!", "耗时： " + (endTime - beginTime) / 1000);
    }
}
```

程序运行的结果如图 6.38 所示。

```
jdwp                      got wake-up signal, bailing out of select
dalvikvm                  Debugger has detached; object registry h
                          onCreate thread is=1 thread name=main
                          begin
                          耗时：0
                          run thread is=1 thread name=main
ActivityManager           Launch timeout has expired, giving up wa
```

图 6.38　程序运行结果

这是打印出来的结果，真正的运行流程是先打印出图 6.38 所示的日志信息，然后项目挂起 10 秒钟后再显示出界面，从图 6.38 中还可以看到都是在线程名称为 main 中运行，即属于同步的方式运行，具有"阻塞"的特点，有没有办法实现异步方式运行呢？也就是新开启一个线程运行，并且不耽误 Activity 界面的显示。这只要将 Main.java 的代码更改为如下形式就可以实现这种要求。

```java
public class Main extends Activity {

    private Runnable run = new Runnable() {
        public void run() {
            try {
                Log.v("!", "run thread is=" + Thread.currentThread().getId()
                        + " thread name=" + Thread.currentThread().getName());
                Thread.sleep(10000);
                Log.v("!", "run end!");
            } catch (InterruptedException e) {
                // TODO Auto-generated catch block
                e.printStackTrace();
            }
        }
    };

    @Override
    public void onCreate(Bundle savedInstanceState) {
        super.onCreate(savedInstanceState);
        Log.v("!", "onCreate thread is=" + Thread.currentThread().getId()
                + " thread name=" + Thread.currentThread().getName());
        Log.v("!", "begin");
        long beginTime = System.currentTimeMillis();
        // Handler hanlder = new Handler();
        // hanlder.post(run);
        Thread thread = new Thread(run);
        thread.start();
        setContentView(R.layout.main);
        long endTime = System.currentTimeMillis();
        Log.v("!", "耗时： " + (endTime - beginTime) / 1000);
    }
}
```

程序运行后界面也优先显示了出来，如图 6.39 所示。

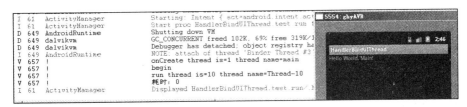

图 6.39 先显示界面

10 秒后打印出了结束日志，结果如图 6.40 所示。

图 6.40 打印结束日志

对象 Handler 的初步使用主要体现在构造函数的方法上，这些函数的功能说明如下。

（1）public Handler()：无参的构造函数，将创建好的 Handler 实例绑定到代码所在的线程的消息队列上，因此一定要确定该线程开启了消息队列，否则程序将发生错误，使用这个构造函数创建的 Handler 实例需要重写 Hanler 类的 handleMessage() 方法，以便在之后的消息处理时调用。

（2）public Handler(Callback callback)：接口 Callback 是 Handler 内部定义的一个接口，因此想要使用这个构造函数创建 Handler 对象，需要自定义一个类实现 Callback 接口，并重写接口中定义的 handleMessage() 方法，这个构造函数其实与无参的构造函数类似，也要确保代码所在的线程开启了消息队列，不同的是在之后处理消息时，将调用接口 Callback 的 handleMessage() 方法，而不是 Handler 对象的 handleMssage() 方法。

（3）public Handler(Looper looper)：表示创建一个 Handler 实例并将其绑定在 Looper 所在的线程上，此时 looper 不能为 null，一般也需要重写 Hanler 类的 handleMessage() 方法。

（4）public Handler(Looper looper,Callback callback)：与（2）和（3）功能相结合。

还有几个知识点需要留意：

（1）调用 Handler 类中以 send 开头的方法可以将 Message 对象压入消息队列中，调用 Handler 类中以 post 开头的方法可以将一个 Runnable 对象包装在一个 Message 对象中，然后再压入消息队列，此时入队的 Message 其 Callback 字段不为 null，值就是这个 Runnable 对象。

（2）调用 Message 对象的 sendToTarget() 方法可以将其本身（Message）压入与其 target 字段（即 handler 对象）所关联的消息队列中。

6.4.2　postDelayed 方法和 removeCallbacks 方法的使用

方法 postDelayed 的作用是延迟多少毫秒后开始运行，而 removeCallbacks 方法是删除指定的 Runnable 对象，使线程对象停止运行。

方法声明如下：

```
public final boolean postDelayed (Runnable r, long delayMillis)
```

其中参数 Runnable r 在 Handler 对象所运行的线程中执行。

创建名称为 handler2 的 Android 项目，Main.java 的核心代码如下：

```java
public class Main extends Activity {
    private Button button1;
    private Button button2;

    private Handler handler = new Handler();

    private int count = 0;

    private Runnable runnableRef = new Runnable() {
        public void run() {
            Log.v("2", Thread.currentThread().getName());
            count++;
            Log.v("!", "count=" + count);
            handler.postDelayed(runnableRef, 1000);
        }
    };

    @Override
    public void onCreate(Bundle savedInstanceState) {
        super.onCreate(savedInstanceState);
        setContentView(R.layout.main);

        Log.v("1", Thread.currentThread().getName());

        button1 = (Button) this.findViewById(R.id.button1);
        button2 = (Button) this.findViewById(R.id.button2);

        button1.setOnClickListener(new OnClickListener() {
            public void onClick(View arg0) {
                Thread thread = new Thread(runnableRef);
                thread.start();
                Log.v("!!!!!!!!!!!!!!", "end");
            }
        });

        button2.setOnClickListener(new OnClickListener() {
            public void onClick(View arg0) {
                handler.removeCallbacks(runnableRef);
            }
        });

    }
}
```

程序运行后单击 button1 按钮开始循环，count 累加 1，运行结果如图 6.41 所示。

Broadcast、Service 服务及 Handle 对象 第 6 章

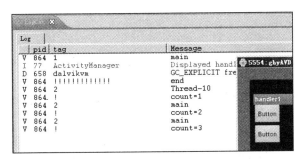

图 6.41 循环加 1 效果

从打印结果可以发现，使用代码：

```
handler.postDelayed(runnableRef, 1000)
```

上述代码运行的 Runnable 并没有新建一个线程，而是运行在 main 线程里。

当单击 button2 按钮时，停止这种累加 1 的功能。

关于循环执行某一个任务还可以使用 Java SE 自带的类来进行处理，新建名称为 TimerTest 项目，文件 Main.java 的代码如下：

```java
public class Main extends Activity {
    private int count = 0;

    @Override
    public void onCreate(Bundle savedInstanceState) {
        super.onCreate(savedInstanceState);
        setContentView(R.layout.main);

        TimerTask task = new TimerTask() {
            @Override
            public void run() {
                Log.v("!", "" + (++count));
            }
        };
        Timer timer = new Timer();
        timer.schedule(task, 1000, 1000);
    }
}
```

打印的效果如图 6.42 所示。

图 6.42 Timer 循环执行某一任务

如果想在 TimerTask 中控制 View 控件，还需要用 Handler 对象以发送消息 Message 的方式来处理 View 的更新。

6.4.3 post 方法的使用

方法 post 是将 Message 对象放入消息队列中，以待后面执行消息队列中的任务。
方法声明如下：

public final boolean post (Runnable r)

其中参数 Runnable r 在 Handler 对象所运行的线程中执行。
新建名称为 handler3 的 Android 项目，文件 Main.java 的代码如下：

```java
public class Main extends Activity {
    private Button button1;
    private int count = 0;

    private Handler handler = new Handler() {
        @Override
        public void handleMessage(Message msg) {
            super.handleMessage(msg);
            Log.v("!", "count=" + msg.getData().getString("count") + "   3"
                    + Thread.currentThread().getName());
        }
    };

    @Override
    public void onCreate(Bundle savedInstanceState) {
        super.onCreate(savedInstanceState);
        setContentView(R.layout.main);

        Log.v("1", "" + Thread.currentThread().getName());

        button1 = (Button) this.findViewById(R.id.button1);
        button1.setOnClickListener(new OnClickListener() {
            public void onClick(View arg0) {
                count = 0;
                handler.post(new Runnable() {
                    public void run() {
                        count++;
                        while (count < 10) {
                            try {
                                Log.v("2", ""
                                        + Thread.currentThread().getName());

                                Bundle bundle = new Bundle();
                                bundle.putString("count", "" + count);

                                Message message = new Message();
                                message.setData(bundle);

                                handler.sendMessage(message);
```

```
                        count++;
                        Thread.sleep(200);
                    } catch (InterruptedException e) {
                        // TODO Auto-generated catch block
                        e.printStackTrace();
                    }
                }
            }
        });

        for (int i = 0; i < 10; i++) {
            try {
                Log.v("!", "i=" + (i + 1));
                Thread.sleep(1000);
            } catch (InterruptedException e) {
                // TODO Auto-generated catch block
                e.printStackTrace();
            }
        }
    }
});
}
}
```

程序运行后单击 Button 按钮出现如图 6.43 所示的效果。

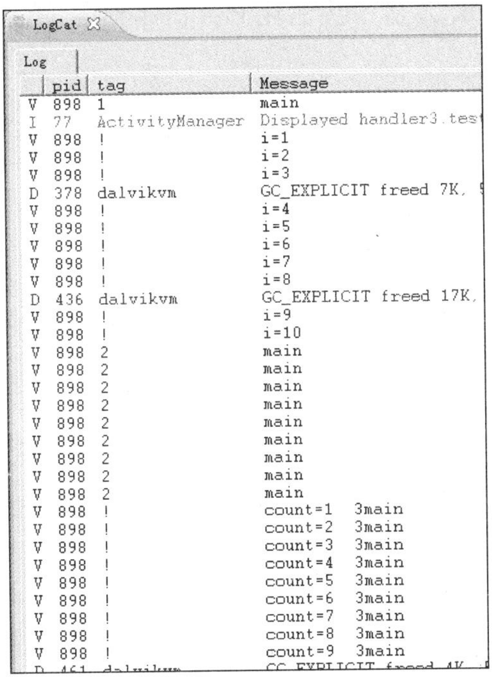

图 6.43　运行效果

从本示例可以发现，Handler 的 post()方法对线程的处理也不是真正创建一个新的线程，而是直接调用了线程的 run 方法。

6.4.4　postAtTime 方法的使用

方法 postAtTime 的作用是实现隔几秒后自动执行，本示例代码在项目 handler4 中。

```
handler.postAtTime(new Runnable() {
    public void run() {
        count++;
        while (count < 10) {
            Bundle bundle = new Bundle();
            bundle.putString("count", "" + count);

            Message message = new Message();
            message.setData(bundle);

            handler.sendMessage(message);

            count++;
        }
    }
}, SystemClock.uptimeMillis() + 5000);
```

本示例实现的效果是隔 5 秒后执行。

6.4.5　在线程对象的 run 方法中实例化 Handler 对象的注意事项

在有些情况下，需要在线程中创建 Handler 对象然后发送消息。

创建名称为 threadUseHandler 的 Android 项目，创建自定义 Handler 对象 GhyHandler.java，代码如下：

```
package exthandler;

import android.os.Handler;
import android.os.Message;
import android.util.Log;

public class GhyHandler extends Handler {
    @Override
    public void handleMessage(Message msg) {
        super.handleMessage(msg);
        Log.v("!", "username=" + msg.getData().getString("username"));
    }
}
```

创建自定义线程类 GhyThread.java，代码如下：

```
package extthread;

import android.os.Bundle;
```

```java
import android.os.Message;
import exthandler.GhyHandler;

public class GhyThread extends Thread {

    @Override
    public void run() {
        super.run();

        GhyHandler handler = new GhyHandler();
        Message message = handler.obtainMessage();
        Bundle bundle = new Bundle();
        bundle.putString("username", "gaohongyan");
        message.setData(bundle);
        handler.sendMessage(message);

    }
}
```

文件 Main.java 的代码如下：

```java
package threadUseHandler.test.run;

import android.app.Activity;
import android.os.Bundle;
import extthread.GhyThread;

public class Main extends Activity {
    @Override
    public void onCreate(Bundle savedInstanceState) {
        super.onCreate(savedInstanceState);
        setContentView(R.layout.main);

        GhyThread ghyThreadRef = new GhyThread();
        ghyThreadRef.start();

    }
}
```

程序运行后出现错误如下：

java.lang.RuntimeException: Can't create handler inside thread that has not called Looper.prepare()

出错的原因是当前的线程 GhyThread.java 并没有创建 Looper 对象，一个线程可以产生一个 Looper 对象，由 Looper 对象来管理线程里的 Message Queue（消息队列），Message Queue 按顺序处理队列中的 Message 对象，每一个线程里可含有一个 Looper 对象以及一个 MessageQueue。

Handler 在创建的时候可以指定 Looper，这样通过 Handler 的 sendMessage()方法发送出去的消息就会添加到指定 Looper 里面的 MessageQueue 里面去，但在不指定 Looper 的情况下，Handler 绑定的是创建它的线程的 Looper，如果这个线程的 Looper 不存在，程序将抛出"java.lang.RuntimeException: Can't create handler inside thread that has not called Looper.prepare()"的

异常，这就是上面代码出错的原因。

介绍到这儿，有必要对一些知识点进行一下总结：

（1）Message 消息，可以理解为线程间通信的数据单元，通过将数据放入 Message 对象中以便达到线程间的通信。例如后台线程在处理数据完毕后需要更新 UI，则可发送一条包含最新数据信息的 Message 给 UI 线程。

（2）Message Queue 消息用来存放通过 Handler 发布的消息，按照先进先出执行。队列中的每一个 Message 都有一个 when 字段，这个字段用来决定 Message 应该何时处理，消息队列中的每一个 Message 根据 when 字段的大小由小到大排列，排在最前面的消息会首先得到处理，因此可以说消息队列并不是一个严格的先进先出的队列。

Message 对象的 target 字段表示关联了哪个线程的消息队列，这个消息就会被压入哪个线程的消息队列中，Message 类用于表示消息。Message 对象可以通过 arg1、arg2、obj 字段和 setData() 携带数据，此外还具有很多字段。when 字段决定 Message 应该何时处理，target 字段用来表示将由哪个 Handler 对象处理这个消息，next 字段表示在消息队列中排在这个 Message 之后的下一个 Message，callback 字段如果不为 null，表示这个 Message 包装了一个 runnable 对象，what 字段表示 code，即这个消息具体是什么类型的消息。每个 what 都在其 handler 的 namespace 中，只需要确保将由同一个 handler 处理的消息的 what 属性不重复就可以。

（3）Handler 是 Message 的主要处理者，负责将 Message 添加到消息队列以及对消息队列中的 Message 进行处理。

（4）Looper 循环器扮演 Message Queue 和 Handler 之间桥梁的角色，循环取出 Message Queue 里面的 Message，并交付给相应的 Handler 进行处理。Looper 类主要用来创建消息队列，每个线程最多只能有一个消息队列，在 Android 中 UI 线程默认具有消息队列，但非 UI 线程在默认情况下是不具备消息队列的，比如自定义的线程类。如果需要在非 UI 线程中开启消息队列，需要调用 Looper.prepare()方法，该方法在执行过程中会创建一个 Looper 对象，而在源代码中的 Looper 构造函数中会创建一个 MessageQueue 实例，此后再为该线程绑定一个 Handler 实例，再调用 Looper.loop() 方法，就可以不断地从消息队列中取出消息和处理消息了。Looper.myLoop()方法可以得到线程的 Looper 对象，如果为 null，说明此时该线程尚未开启消息队列。通过 Loop.getMainLooper()可以获得当前进程的主线程的 Looper 对象。

如果想让该线程具有消息队列和消息循环，需要在线程中首先调用 Looper.prepare()来创建消息队列，然后调用 Looper.loop()进入消息循环，这样该线程就具有了消息处理机制，可以在 Handler 对象中进行消息处理。

下面来看看其实现方法，更改 GhyThread.java 的代码如下：

```java
public class GhyThread extends Thread {

    @Override
    public void run() {
        super.run();

        Looper.prepare();//准备创建 1 个 Looper 对象
```

```
        GhyHandler handler = new GhyHandler();
        Message message = handler.obtainMessage();
        Bundle bundle = new Bundle();
        bundle.putString("username", "gaohongyan");
        message.setData(bundle);
        handler.sendMessage(message);

        Looper.loop();//执行消息队列中的 Message 对象

    }
}
```

程序运行后正确地取出了 username 的值, 如图 6.44 所示。

图 6.44 成功打印 username 的值

6.4.6 以异步方式打开网络图片

创建持有 PNG 图标资源的 Web 项目 pngProject, 布署到 tomcat 中, 项目文件结构如图 6.45 所示。

图 6.45 持有 png 图标的 web 项目

创建 Android 客户端应用程序项目 synchronizedOpenNetPNG, 由于是以异步方式访问远程 PNG 图片资源, 所以创建自定义线程类 OpenNetPNGThread.java, 该类主要的功能就是通过远程 PNG 图片的 URL 返回 Bitmap 位图资源, 核心代码如下:

```java
public class OpenNetPNGThread extends Thread {

    private String pngPath;
    private Handler handler;
    private int imageViewId;

    public OpenNetPNGThread(Handler handler, String pngPath, int imageViewId) {
        super();
        this.pngPath = pngPath;
```

```java
            this.handler = handler;
            this.imageViewId = imageViewId;
        }

        @Override
        public void run() {
            super.run();
            try {
                Log.v("!", "启动线程" + Thread.currentThread().getId() + "  "
                        + Thread.currentThread().getName());
                URL url = new URL(pngPath);
                URLConnection connection = url.openConnection();
                InputStream isRef = connection.getInputStream();
                Bitmap bitmap = BitmapFactory.decodeStream(isRef);

                Bundle bundle = new Bundle();
                bundle.putInt("imageViewId", imageViewId);
                bundle.putParcelable("bitmap", bitmap);

                Message message = handler.obtainMessage();
                message.setData(bundle);
                handler.sendMessage(message);

            } catch (MalformedURLException e) {
                // TODO Auto-generated catch block
                e.printStackTrace();
            } catch (IOException e) {
                // TODO Auto-generated catch block
                e.printStackTrace();
            }

        }
    }
```

创建自定义 Handler 对象 PNGHandler.java，该类主要的作用是从 Message 中取出 Bitmap 资源来对 ImageView 进行更新，核心代码如下：

```java
public class PNGHandler extends Handler {

    private Context context;

    public PNGHandler(Context context) {
        super();
        this.context = context;
    }

    @Override
    public void handleMessage(Message msg) {
        super.handleMessage(msg);

        Bundle bundle = msg.getData();
```

```java
            Bitmap bitmap = bundle.getParcelable("bitmap");
            int imageViewId = bundle.getInt("imageViewId");

            ImageView findImageView = (ImageView) ((Activity) context)
                    .findViewById(imageViewId);
            findImageView.setImageBitmap(bitmap);

        }
    }
```

项目的核心 Activity 对象 Main.java 文件的主要代码如下：

```java
public class Main extends Activity {

    private PNGHandler[] handler = new PNGHandler[5];
    private String[] pngFileName = new String[5];

    private ImageView imageView1;
    private ImageView imageView2;
    private ImageView imageView3;
    private ImageView imageView4;
    private ImageView imageView5;

    private ImageView[] imageViewArray = new ImageView[5];

    @Override
    public void onCreate(Bundle savedInstanceState) {
        super.onCreate(savedInstanceState);
        setContentView(R.layout.main);

        imageView1 = (ImageView) this.findViewById(R.id.imageView1);
        imageView2 = (ImageView) this.findViewById(R.id.imageView2);
        imageView3 = (ImageView) this.findViewById(R.id.imageView3);
        imageView4 = (ImageView) this.findViewById(R.id.imageView4);
        imageView5 = (ImageView) this.findViewById(R.id.imageView5);

        imageViewArray[0] = imageView1;
        imageViewArray[1] = imageView2;
        imageViewArray[2] = imageView3;
        imageViewArray[3] = imageView4;
        imageViewArray[4] = imageView5;

        pngFileName[0] = "http://10.0.2.2:8081/pngProject/a.png";
        pngFileName[1] = "http://10.0.2.2:8081/pngProject/b.png";
        pngFileName[2] = "http://10.0.2.2:8081/pngProject/c.png";
        pngFileName[3] = "http://10.0.2.2:8081/pngProject/d.png";
        pngFileName[4] = "http://10.0.2.2:8081/pngProject/e.png";

        for (int i = 0; i < handler.length; i++) {
            handler[i] = new PNGHandler(this);
        }
```

```
for (int i = 0; i < handler.length; i++) {
    OpenNetPNGThread mythread = new OpenNetPNGThread(handler[i],
            pngFileName[i], imageViewArray[i].getId());
    mythread.start();
        }
    }
}
```

程序运行后的结果如图 6.46 所示。控件 ImageView 显示出 5 张图片资源，如图 6.47 所示。

图 6.46 运行结果   图 6.47 5 张 PNG 资源显示在 ImageView 控件中

6.5 Appwidget 小部件的使用

小部件 Appwidget 是在 Android 操作系统 Home 界面上的小控件，比如图 6.48 中显示的搜索和信息提示。这些外观漂亮的小部件 Appwidget 都是使用图片进行美化的，本节仅仅只是想演示一下 Appwidget 小部件的使用。

在 Android2.3 版本的 SDK 中，Appwidget 小部件仅仅能使用如图 6.49 所示的布局。

```
FrameLayout
LinearLayout
RelativeLayout
```

图 6.48 Home 中的小部件   图 6.49 Android2.3 能使用布局

另外，Appwidget 中使用的控件也会有所限制，只能使用如图 6.50 所列的控件。

```
AnalogClock
Button
Chronometer
ImageButton
ImageView
ProgressBar
TextView
```

图 6.50　Appwidget 能使用的控件

不过没有关系，在 Android SDK 更高的版本中已经支持更多的 Widget 控件，可以在 Appwidget 中使用了。

6.5.1　初入 Appwidget 小部件

类 AppWidgetProvider 继承自 BroadcastReceiver，如图 6.51 所示，也就是说 AppWidgetProvider 具有广播接收者一切的功能，这样系统就可以定时地发送广播来实现一些 Appwidget 界面的更新。

```
public class
AppWidgetProvider
extends BroadcastReceiver

java.lang.Object
  └android.content.BroadcastReceiver
     └android.appwidget.AppWidgetProvider
```

图 6.51　AppWidgetProvider 继承关系

本小节仅仅想实现一个具有两个 Button 按钮的 Appwidget 小部件，新建名称为 zeroAppwidget 的 Android 项目，创建 AppWidgetProvider 类的子类 GhyAppWidgetProvider，代码如下：

```java
public class GhyAppWidgetProvider extends AppWidgetProvider {

    @Override
    public void onUpdate(Context context, AppWidgetManager appWidgetManager,
            int[] appWidgetIds) {
        super.onUpdate(context, appWidgetManager, appWidgetIds);
        // 由于运行在不同的环境中，所以要使用
        // ComponentName 组件的名称来作为目的地址标识
        ComponentName componentName1 = new ComponentName(context, Second.class);
        ComponentName componentName2 = new ComponentName(context, Third.class);
        // Appwidget 中的 View 属于 RemoteView 远程视图
        // RemoteViews 构造方法第 2 个参数指的是
        // Appwidget 关联的布局文件资源 id
        RemoteViews remoteView = new RemoteViews(context.getPackageName(),
                R.layout.myappwidgetlayout);

        // 设置 Appwidget 小部件中第 1 个按钮的 Intent
        Intent intent1 = new Intent();
        intent1.setComponent(componentName1);
```

```
        // 设置 Appwidget 小部件中第 2 个按钮的 Intent
        Intent intent2 = new Intent();
        intent2.setComponent(componentName2);

        // 创建 PendingIntent 和 remoteView 对象的关联
        // 以便单击 Button1 和 Button2 时有相应的动作发生
        PendingIntent pendingIntent1 = PendingIntent.getActivity(context, 1,
                intent1, PendingIntent.FLAG_UPDATE_CURRENT);
        remoteView.setOnClickPendingIntent(R.id.button1, pendingIntent1);

        PendingIntent pendingIntent2 = PendingIntent.getActivity(context, 2,
                intent2, PendingIntent.FLAG_UPDATE_CURRENT);
        remoteView.setOnClickPendingIntent(R.id.button2, pendingIntent2);
        // 更新 RemoteViews 对象
        appWidgetManager.updateAppWidget(appWidgetIds, remoteView);

    }
}
```

继续创建 Appwidget 关联的布局文件 myappwidgetlayout.xml，代码如下：

```xml
<?xml version="1.0" encoding="utf-8"?>
<LinearLayout xmlns:android="http://schemas.android.com/apk/res/android"
    android:orientation="horizontal" android:layout_width="fill_parent"
    android:layout_height="fill_parent">
    <LinearLayout android:orientation="horizontal"
        android:layout_width="fill_parent" android:layout_height="fill_parent">
        <Button android:text="Button" android:id="@+id/button1"
            android:layout_width="wrap_content" android:layout_height="wrap_content"></Button>
        <Button android:text="Button" android:id="@+id/button2"
            android:layout_width="wrap_content" android:layout_height="wrap_content"></Button>
    </LinearLayout>
</LinearLayout>
```

还要在 res/xml 文件夹下创建 Appwidget 的 XML 配置文件 ghy_appwidget_info.xml，此文件主要用来配置 Appwidget 的相关属性，代码如下：

```xml
<appwidget-provider xmlns:android="http://schemas.android.com/apk/res/android"
    android:minWidth="144dip" android:minHeight="72dip"
    android:updatePeriodMillis="0" android:initialLayout="@layout/myappwidgetlayout">
</appwidget-provider>
```

属性 android:updatePeriodMillis 值是 0，代表 Appwidget 控件不周期性地更新，因为使用大于 Android 1.6R1 版本的 SDK 时，系统默认的更新周期为 30 分钟，另外，周期性更新的效果已经不再使用该属性来实现，比如周期性的显示当前的时间。关于如何在 Appwidget 中使用周期性更新视图请参阅后面的章节。

创建两个 Activity 对象 Second.java 和 Third.java 文件，它们对应的布局 layout 内容如图 6.52 所示。

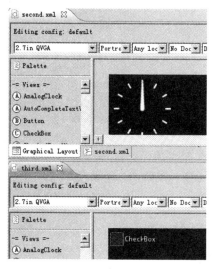

图 6.52　second.xml 和 third.xml 个布局文件中的内容

最后，还要在 AndroidManifest.xml 文件中配置 Appwidget 和 Activity 对象，其中配置 Appwidget 代码如下：

```
<receiver android:name="extappwidgetprovider.GhyAppWidgetProvider">
    <intent-filter>
        <action android:name="android.appwidget.action.APPWIDGET_UPDATE" />
    </intent-filter>
    <meta-data android:name="android.appwidget.provider"
        android:resource="@xml/ghy_appwidget_info" />
</receiver>
```

程序运行后返回 Home，并且在 Home 长按操作出现菜单，如图 6.53 所示。
在图 6.53 中选择"Widgets"小部件菜单项，单击"zeroAppwidget"菜单项，如图 6.54 所示。

图 6.53　出现菜单

图 6.54　单击 zeroAppwidget

Home 上出现自定义的 Appwidget 小部件，如图 6.55 所示。

图 6.55 zeroAppwidget 成功添加到 Home 中

单击不同的按钮跳转到不同的 Activity 界面。

6.5.2 Appwidget 的生命周期

对 zeroAppwidget 项目中的类进行修改代码如下：

```java
public class GhyAppWidgetProvider extends AppWidgetProvider {

    @Override
    public void onDeleted(Context context, int[] appWidgetIds) {
        super.onDeleted(context, appWidgetIds);
        Log.v("!", "onDeleted");
    }

    @Override
    public void onDisabled(Context context) {
        super.onDisabled(context);
        Log.v("!", "onDisabled");
    }

    @Override
    public void onEnabled(Context context) {
        super.onEnabled(context);
        Log.v("!", "onEnabled");
    }

    @Override
    public void onUpdate(Context context, AppWidgetManager appWidgetManager,
            int[] appWidgetIds) {
        super.onUpdate(context, appWidgetManager, appWidgetIds);
        Log.v("!", "onUpdate");
        // 其它代码省略
        appWidgetManager.updateAppWidget(appWidgetIds, remoteView);
    }
}
```

在运行项目前将 Home 中自定义的所有 Appwidget 删除，再重新运行 zeroAppwidget 项目，重新添加一个 zeroAppwidget 控件，查看 LogCat 打印的结果，如图 6.56 所示。

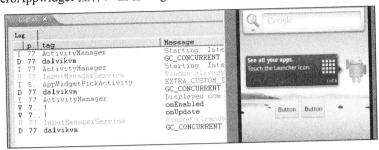

图 6.56　第 1 步生命周期

从图 6.56 中可以看到，第 1 次创建 Appwidget 时，onEnabled 和 onUpdate 方法都被调用。

再添加 1 个 zeroAppwidget 控件，查看 Logcat 内容，如图 6.57 所示。

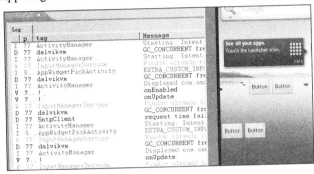

图 6.57　第 2 步生命周期

从图 6.57 中可以看到，重复添加 Appwidget 时仅仅 onUpdate 方法被调用，也就是说 onEnabled 方法仅仅在第一次创建 Appwidget 时调用，每生成 1 个新的 Appwidget 控件时，onUpdate 方法都被调用。

接下来删除一个 Appwidget，然后查看 LogCat 的信息，如图 6.58 所示。

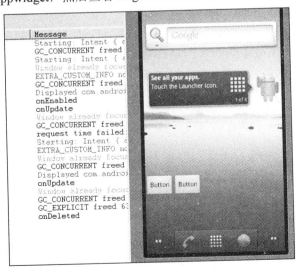

图 6.58　第 3 步生命周期

从图 6.57 中可以看到，每删除一个 Appwidget 都会调用 onDeleted 方法，继续操作，把最后 1 个 Appwidget 也删除，再查看 LogCat 内容，如图 6.59 所示。

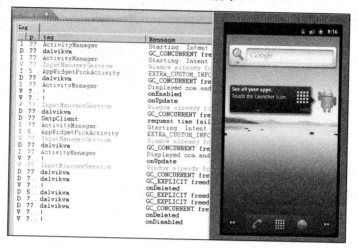

图 6.59　第 4 步生命周期

从图 6.58 中可以发现，Appwidget 全部实例都被删除了，方法 onDisabled 被调用。

6.5.3　Appwidget 的隔时刷新界面的效果——使用 AlarmManager

在应用 Appwidget 技术时，经常用到每隔一段时间刷新界面的功能，本示例使用 AlarmManager 对象来实现这一效果。

新建名称为 myFirstAppwidget 的 Android 项目，创建类型为 AppWidgetProvider 的文件 MyAppWidgetProvider.java，代码如下：

```java
public class MyAppWidgetProvider extends AppWidgetProvider {

    @Override
    public void onEnabled(Context context) {
        super.onEnabled(context);
        Log.v("!", "onEnabled");
    }

    @Override
    public void onDeleted(Context context, int[] appWidgetIds) {
        super.onDeleted(context, appWidgetIds);
        Log.v("!", "onDeleted");
    }

    @Override
    public void onDisabled(Context context) {
        super.onDisabled(context);
        Log.v("!", "onDisabled");
    }

    @Override
```

```java
public void onReceive(Context context, Intent intent) {
    super.onReceive(context, intent);
    Log.v("!", "onReceive");

    RemoteViews remoteViews = new RemoteViews(context.getPackageName(),
            R.layout.mywidgetlayout);
    remoteViews.setTextViewText(R.id.button1, new Date().toLocaleString());
    ComponentName thisWidget = new ComponentName(context,
            MyAppWidgetProvider.class);
    AppWidgetManager manager = AppWidgetManager.getInstance(context);
    int[] appWidgetIds = manager.getAppWidgetIds(thisWidget);
    manager.updateAppWidget(appWidgetIds, remoteViews);
}

@Override
public void onUpdate(Context context, AppWidgetManager appWidgetManager,
        int[] appWidgetIds) {
    super.onUpdate(context, appWidgetManager, appWidgetIds);
    Log.v("!", "onUpdate");

    AlarmManager am = (AlarmManager) context
            .getSystemService(Context.ALARM_SERVICE);
    long getTime = SystemClock.elapsedRealtime();
    Intent intent = new Intent(context, MyAppWidgetProvider.class);

    PendingIntent pi = PendingIntent.getBroadcast(context, 1, intent,
            PendingIntent.FLAG_UPDATE_CURRENT);
    am
            .setRepeating(AlarmManager.ELAPSED_REALTIME_WAKEUP, getTime,
                    1000, pi);

}
}
```

创建 Appwidget 关联的布局文件 mywidgetlayout.xml，代码如下：

```xml
<?xml version="1.0" encoding="utf-8"?>
<LinearLayout xmlns:android="http://schemas.android.com/apk/res/android"
    android:orientation="horizontal" android:layout_width="fill_parent"
    android:layout_height="fill_parent">
    <LinearLayout android:orientation="horizontal"
        android:layout_width="fill_parent" android:layout_height="fill_parent">
        <Button android:text="Button" android:id="@+id/button1"
            android:layout_width="wrap_content" android:layout_height="wrap_content"></Button>
    </LinearLayout>
</LinearLayout>
```

创建 Appwidget 配置文件 my_widget_info.xml，代码如下：

```xml
<appwidget-provider xmlns:android="http://schemas.android.com/apk/res/android"
```

```
            android:minWidth="72dip" android:minHeight="72dip"
            android:updatePeriodMillis="0" android:initialLayout="@layout/mywidgetlayout">
</appwidget-provider>
        <!-- 在 android 中规定 appwidget 的高度或宽度一定是(74*倍数-2) -->
```

还要在文件 AndroidManifest.xml 中配置 Appwidget，代码如下：

```
        <receiver android:name="extwidgetprovider.MyAppWidgetProvider">
            <intent-filter>
                    <action android:name="android.appwidget.action.APPWIDGET_UPDATE" />
            </intent-filter>
            <meta-data android:name="android.appwidget.provider"
                android:resource="@xml/my_widget_info" />
        </receiver>
```

程序运行后的效果如图 6.60 所示。

图 6.60　程序运行效果

在图 6.60 的界面中，每隔 5 秒 Button 的 text 时间即被更新。

6.6　章节 AsyncTask 对象的使用

在 Android 技术中除了使用 Handler、Thread 和 Service 来实现任务的功能，还有一种异步任务对象，它就是 AsyncTask，该类在 JDK 中的声明如图 6.61 所示。

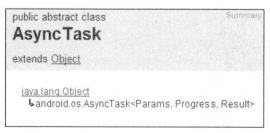

图 6.61　AsyncTask 对象的声明结构

从图 6.61 中可以看到，要想使用它还必须确定 3 个必要的泛型参数类型。

参数 1：指定外部使用 AsyncTask 对象时对 AsyncTask 对象传入的初始化参数类型。

参数 2：指定在执行任务时可以随时返回一些数据，比如进度这些，参数 2 就是定义这个随时返回数据的数据类型。

参数 3：任务执行完了返回的数据类型

虽然上面这样的解释还不足充分证明 AsyncTask 的使用，那就用实验来进行测试吧。

6.6.1 初入 AsyncTask

新建名称为 task1 的 Android 项目，新建 AsyncTask.java 类的子类 MyAsyncTask.java 后就会发现有一个编译错误，如图 6.62 所示。

图 6.62　新建 MyAsyncTask.java 后出现编译错误

因为在图 6.62 中并没有设置 3 个泛型的参数类型，这时改动代码如图 6.63 所示。

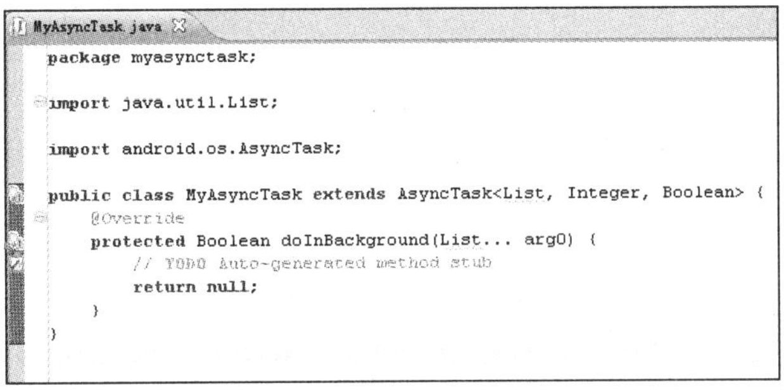

图 6.63　改动后无编译错误的代码

从图 6.63 中可以看到，对 3 个泛型参数的数据类型进行设置后并没有出现编译错误，再重写一下其它的必要方法，完整的代码如下：

```
package myasynctask;

import java.util.List;

import android.os.AsyncTask;

public class MyAsyncTask extends AsyncTask<List, Integer, Boolean> {
    // 把长时间运行的代码放入此方法中运行
    @Override
    protected Boolean doInBackground(List... arg0) {
        return null;
    }

    // 任务执行完了执行下面的代码
    @Override
    protected void onPostExecute(Boolean result) {
        super.onPostExecute(result);
```

```java
            }

            // 可以在下面的方法中中取得当前任务执行的状态
            @Override
            protected void onProgressUpdate(Integer... values) {
                super.onProgressUpdate(values);
            }

}
```

在这里需要注意的是，不能在 doInBackground 方法中操作 View 对象，比如，如下的代码：

```java
public class MyAsyncTask extends AsyncTask<List, Integer, Boolean> {
    // 把长时间运行的代码放入此方法中运行
    @Override
    protected Boolean doInBackground(List... arg0) {
        EditText editText = (EditText) (arg0[0].get(0));
        editText.setText("gaohongyanTextValue");
        return null;
    }

    // 任务执行完了执行下面的代码
    @Override
    protected void onPostExecute(Boolean result) {
        super.onPostExecute(result);
    }

    // 可以在下面的方法中中取得当前任务执行的状态
    @Override
    protected void onProgressUpdate(Integer... values) {
        super.onProgressUpdate(values);
    }

}
```

上面的程序虽然没有编译的错误，但运行后却出现异常如下：

06-01   08:31:48.846:   ERROR/AndroidRuntime(679):   Caused by: android.view.ViewRoot$CalledFromWrongThreadException: Only the original thread that created a view hierarchy can touch its views.

如何去使 AsyncTask 任务运行呢，使用如下的代码：

```java
public class Main extends Activity {
    private Button button1;
    private EditText editText1;

    @Override
    public void onCreate(Bundle savedInstanceState) {
        super.onCreate(savedInstanceState);
        setContentView(R.layout.main);
```

```java
            editText1 = (EditText) this.findViewById(R.id.editText1);

            button1 = (Button) this.findViewById(R.id.button1);
            button1.setOnClickListener(new OnClickListener() {
                @Override
                public void onClick(View arg0) {
                    List paramList = new ArrayList();
                    paramList.add(editText1);
                    new MyAsyncTask().execute(paramList);
                }
            });
        }
    }
```

上例出错的原因也就是在另外的线程中操作了 View 对象,那如何去更新 View 的 UI 界面呢?

6.6.2 使用 AsyncTask 更新 UI 的示例

新建名称为 asyncTaskTestBegin 的 Android 项目,文件 GhyAsyncTask.java 核心代码如下:

```java
public class GhyAsyncTask extends AsyncTask<Object, Integer, Integer> {
    private EditText editText;

    // 此方法是第 1 个被调用的
    @Override
    protected Integer doInBackground(Object... arg0) {
        try {
            editText = (EditText) arg0[2];
            Log.v("!", "从外部传入 2 个参数 arg0[0]=" + arg0[0] + " arg0[1]=" + arg0[1]);
            Log.v("!", "在 doInBackground 方法中做一些耗时的动作");
            for (int i = 0; i < 5; i++) {
                // 每次执行后方法 onProgressUpdate 就被调用
                publishProgress(i + 1, i + 2);
                Thread.sleep(1000);
            }
        } catch (InterruptedException e) {
            // TODO Auto-generated catch block
            e.printStackTrace();
        }

        // 返回 100 代表任务成功运行结束了
        return 100;
    }

    // 任务结束后运行 onPostExecute 方法
    @Override
    protected void onPostExecute(Integer result) {
        super.onPostExecute(result);
        if (result == 100) {
            Log.v("!", "方法 doInBackground 成功运行结束了 并且参数 result 值为" + result);
            Log.v("!", "并进入 onPostExecute 方法中");
```

```
        }
    }

    // publishProgress 方法每一次被调用时
    // onProgressUpdate 就被执行 1 次
    @Override
    protected void onProgressUpdate(Integer... values) {
        super.onProgressUpdate(values);
        Log.v("!", "进入了 onProgressUpdate 方法");
        editText.setText("values[0]=" + values[0] + " values[1]=" + values[1]);
    }

}
```

文件 Main.java 的核心代码如下：

```
public class Main extends Activity {
    private Button button;
    private EditText editText1;

    @Override
    public void onCreate(Bundle savedInstanceState) {
        super.onCreate(savedInstanceState);
        setContentView(R.layout.main);
        editText1 = (EditText) this.findViewById(R.id.editText1);
        button = (Button) this.findViewById(R.id.button1);
        button.setOnClickListener(new OnClickListener() {
            public void onClick(View arg0) {
                new GhyAsyncTask().execute("ghy1", "ghy2", editText1);
            }
        });
    }
}
```

程序运行后 EditText 控件中的文本发生改变，如图 6.64 所示。

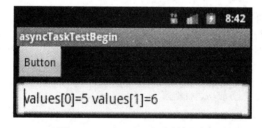

图 6.64　EditText 文本被更改

在 LogCat 打印的结果如图 6.65 所示。

```
706  AndroidRuntime       NOTE: attach of thread 'Binder Thread #3' failed
77   ActivityManager      Displayed asyncTaskTestBegin.test run/.Main: +1s86ms
714  !                    从外部传入2个参数 arg0[0]=ghy1 arg0[1]=ghy2
714  !                    在doInBackground方法中做一些耗时的动作
714  !                    进入了onProgressUpdate方法
714  !                    进入了onProgressUpdate方法
714  !                    进入了onProgressUpdate方法
714  !                    进入了onProgressUpdate方法
714  !                    进入了onProgressUpdate方法
77   SntpClient           request time failed: java.net.SocketException: Address
714  !                    方法doInBackground成功运行结束了 并且参数result值为100
714  !                    并进入onPostExecute方法中
372  dalvikvm             GC_EXPLICIT freed 1K, 53% free 2549K/5379K, external 41
```

图 6.65　LogCat 打印的结果

6.6.3　使用 AsyncTask 时外界无参数与其进行交互的情况

如果有这样的情况就需要把代码更改如下：

```java
public class MyAsyncTask extends AsyncTask<Void, Integer, Boolean> {
    @Override
    protected Boolean doInBackground(Void... arg0) {
        Log.v("!", "进入 doInBackground 方法了");
        return null;
    }
}
```

使用时在 Main.java 代码如下：

```java
public class Main extends Activity {
    @Override
    public void onCreate(Bundle savedInstanceState) {
        super.onCreate(savedInstanceState);
        setContentView(R.layout.main);

        new MyAsyncTask().execute((Void[]) null);
    }
}
```

成功运行在 LogCat 打印字符串如图 6.66 所示。

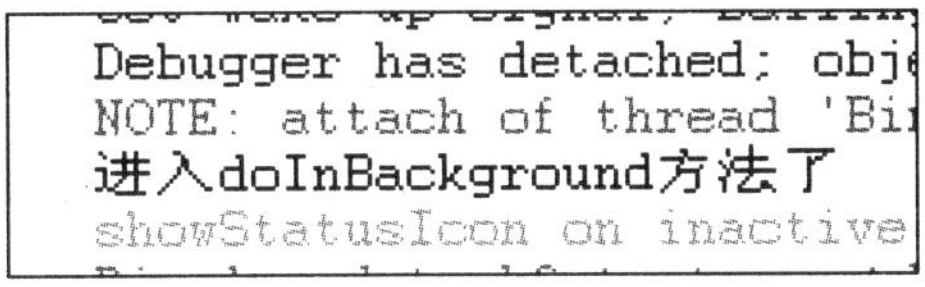

```
Debugger has detached; obj
NOTE: attach of thread 'Bi
进入doInBackground方法了
showStatusIcon on inactive
```

图 6.66　成功打印

此示例的代码在 task2 项目中。

第7章 HTTP 交互、JSON 和 XML

不能和外界通信的 Android 客户端就是一个孤岛,没有任何的生机,所以与服务器端通信就是本章要学习的内容,主要使用 JSON 来进行数据的交互,还可以结合 XML 语言来实现数据的交换。

本章应该着重关注以下内容:

- Gson 框架对 JSON 数据的操作,包括从客户端到服务器端的通信,以及在服务器端如何将 Java 的 Object 对象转成 JSON 字符串
- 如何在 Android 中通过 HTTP 协议访问远程数据

7.1 JSON 介绍

JavaScript 语言的首要目的是为 Web 浏览器提供一种页面脚本语言,用来控制 Web 浏览器中的 DOM 对象。虽然 JavaScript 发展多年,但仍被普遍认为是 Java 的一个子集,但事实上并非如此。它是一种语法类似 C 语言和 Java 并且支持面向对象的语言。

JavaScript 使用了 ECMAScript 语言规范第三版进行了标准化,所以 JavaScript 只是与 C 及 Java 语言语法相似。

JSON 是 JavaScript 面向对象语法的一个子集,也正是由于 JSON 是 JavaScript 的一个子集,因此它可清晰地应用于该语言中。JSON 的全称是 JavaScript Object Notation,它是一个轻量级数据交换格式,程序员可以非常容易地写出符合 JSON 格式的字符串,而且在各种编程语言中都有相应的类库对 JSON 的对象进行解析和生成。JSON 是完全独立的语言,它使用标准的语法格式,来与其他各种编程语言进行数据交换。

7.1.1 Gson 框架与 JSON 字符串交换数据示例

为了演示 JSON 在 Android 中的数据交换,新建名称为 jsonBegin 的 Android 项目,在这个项目中创建实体对象 Userinfo.java,代码如下:

```
package entity;

public class Userinfo {

    private String id;
    private String username;
    private String password;
```

```java
    public String getId() {
        return id;
    }

    public void setId(String id) {
        this.id = id;
    }

    public String getUsername() {
        return username;
    }

    public void setUsername(String username) {
        this.username = username;
    }

    public String getPassword() {
        return password;
    }

    public void setPassword(String password) {
        this.password = password;
    }
}
```

实体 Userinfo.java 有 3 个属性及 get 和 set 方法。

再新建实体对象 ClassEntity.java，代码如下：

```java
public class ClassEntity {

    private String classname;
    private List<Userinfo> userinfoList = new ArrayList();

    public String getClassname() {
        return classname;
    }

    public void setClassname(String classname) {
        this.classname = classname;
    }

    public List getUserinfoList() {
        return userinfoList;
    }

    public void setUserinfoList(List userinfoList) {
        this.userinfoList = userinfoList;
    }
}
```

实体类 ClassEntity 有两个属性，类型分别是 String 和泛型 userinfoList，还有它们对应的 set 和 get 方法。ClassEntity 对象是用泛型 List 集合对 Userinfo.java 实体进行封装并管理。

文件 Main.java 的代码如下：

```java
package jsonBegin.test.run;

import java.lang.reflect.Type;
import java.util.ArrayList;
import java.util.List;

import android.app.Activity;
import android.os.Bundle;
import android.util.Log;

import com.google.gson.Gson;
import com.google.gson.GsonBuilder;
import com.google.gson.reflect.TypeToken;

import entity.ClassEntity;
import entity.Userinfo;

public class Main extends Activity {
    @Override
    public void onCreate(Bundle savedInstanceState) {
        super.onCreate(savedInstanceState);
        setContentView(R.layout.main);

        Userinfo userinfo1 = new Userinfo();// 实体 1
        userinfo1.setId("1");
        userinfo1.setUsername("username1");
        userinfo1.setPassword("password1");

        Userinfo userinfo2 = new Userinfo();// 实体 2
        userinfo2.setId("2");
        userinfo2.setUsername("username2");
        userinfo2.setPassword("password2");

        Userinfo userinfo3 = new Userinfo();// 实体 3
        userinfo3.setId("3");
        userinfo3.setUsername("username3");
        userinfo3.setPassword("password3");

        // 往 listUserinfoBean 中存 Userinfo 对象
        ArrayList listUserinfoBean = new ArrayList();
        listUserinfoBean.add(userinfo1);
        listUserinfoBean.add(userinfo2);
        listUserinfoBean.add(userinfo3);

        // 往 listString 中存 String 对象
        ArrayList listString = new ArrayList();
```

```java
listString.add("gaohongyan1");
listString.add("gaohongyan2");
listString.add("gaohongyan3");

// 新建实体 ClassEntity 对象 ceRef
ClassEntity ceRef = new ClassEntity();
ceRef.setClassname("一年五班");
ceRef.setUserinfoList(listUserinfoBean);

// 创建 google 公司的 Gson 框架对象 gsongRef
// 并且设置解析 JSON 的日期格式为中文格式
Gson gsonRef = new GsonBuilder().setDateFormat("yyyy-MM-dd HH:mm:ss")
        .create();
Log.v("!-1", gsonRef.toJson(userinfo1));// 转成 JSON 字符串
Log.v("!-2", gsonRef.toJson(listUserinfoBean)); // 转成 JSON 字符串
Log.v("!-3", gsonRef.toJson(listString)); // 转成 JSON 字符串
Log.v("!-4", gsonRef.toJson(ceRef)); // 转成 JSON 字符串

// 分割线
Log.v("====================================",
        "====================================");

// 取得实体 JSON 字符串中的属性-开始
// 将 JSON 字符串通过 fromJson 方法转成 Userinfo 对象并打印属性
Userinfo getUserinfo1 = gsonRef.fromJson(gsonRef.toJson(userinfo1),
        Userinfo.class);
Log
        .v("!=====1", getUserinfo1.getId() + " "
                + getUserinfo1.getUsername() + " "
                + getUserinfo1.getPassword());
// 取得实体 JSON 字符串中的属性-结束

// 取得 List 中存 UserinfoBean 属性值-开始
// 由于 List 对象 listUserinfoBean 存放全部是 Userinfo 对象
// 所以在将 JSON 字符串转成 List 对象时必须设置集合框架中数据的类型映射
Type collectionUserinfoType = new TypeToken<ArrayList<Userinfo>>() {
}.getType();
ArrayList<Userinfo> getListUserinfoBean = gsonRef.fromJson(gsonRef
        .toJson(listUserinfoBean), collectionUserinfoType);
for (int i = 0; i < getListUserinfoBean.size(); i++) {
    Log.v("!=====2", "" + getListUserinfoBean.get(i).getId() + " "
            + getListUserinfoBean.get(i).getUsername() + " "
            + getListUserinfoBean.get(i).getPassword());
}
// 取得 List 中存 UserinfoBean 属性值-结束

// 取得 List 中存 String-开始
// List 对象 listString 中存的是 String 数据类型，所以也要设置数据类型映射
Type collectionStringType = new TypeToken<ArrayList<String>>() {
}.getType();
```

```
            List getListString = gsonRef.fromJson(gsonRef.toJson(listString),
                    collectionStringType);
            for (int i = 0; i < getListString.size(); i++) {
                Log.v("!=====3", "" + getListString.get(i));
            }
            // 取得 List 中存 String-结束

            // 取得 ClassEntity 中存属性值-开始
            // 将 JSON 字符串转成 ClassEntity 数据类型
            // 并且一定要将 ClassEntity 对象中的 userinfoList 设置为泛型
            // 不然 Gson 框架不知道 List 中存的是什么类型，也就不能由 JSON 字符串
            // 逆向成存储 Userinfo 对象的 List 对象，代码声明如下：
            // List<Userinfo> userinfoList = new ArrayList();

            ClassEntity getCeRef = gsonRef.fromJson(gsonRef.toJson(ceRef),
                    ClassEntity.class);
            Log.v("!=====4", getCeRef.getClassname());
            List getListUserinfo = getCeRef.getUserinfoList();
            for (int i = 0; i < getListUserinfo.size(); i++) {
                Userinfo userinfo = (Userinfo) getListUserinfo.get(i);
                Log.v("!=====4", "" + userinfo.getId() + " "
                        + userinfo.getUsername() + " " + userinfo.getPassword());
            }
            // 取得 ClassEntity 中存属性值-结束

        }
    }
```

程序运行后在 LogCat 中打印的结果如下：

!-1(1083): {"id":"1","password":"password1","username":"username1"}
!-2(1083):
[{"id":"1","password":"password1","username":"username1"},{"id":"2","password":"password2","username":"username2"},{"id":"3","password":"password3","username":"username3"}]
!-3(1083): ["gaohongyan1","gaohongyan2","gaohongyan3"]
!-4(1083): {"classname":" 一 年 五 班 ","userinfoList":[{"id":"1","password":"password1","username":"username1"},{"id":"2","password":"password2","username":"username2"},{"id":"3","password":"password3","username":"username3"}]}
===(1083): =======================
!=====1(1083): 1 username1 password1
!=====2(1083): 1 username1 password1
!=====2(1083): 2 username2 password2
!=====2(1083): 3 username3 password3
!=====3(1083): gaohongyan1
!=====3(1083): gaohongyan2
!=====3(1083): gaohongyan3
!=====4(1083): 一年五班
!=====4(1083): 1 username1 password1
!=====4(1083): 2 username2 password2
!=====4(1083): 3 username3 password3

7.1.2 在 Android 中通过 HTTP 协议用 JSON 与 Web 项目通信

JSON 字符串有些时候是来自于远程的 Web 项目，然后在 Android 手机终端进行解析，本示例就来实现这个由 Android 终端发起一个 HTTP 请求并把远程返回的 JSON 字符串进行解析，本示例分为两个项目，一个是 Android 项目，另外一个就是 Web 项目。

1. 新建 Web 项目 getJSONString

创建名称为 getJSONString.java 的 Servlet，代码如下：

```java
public class getJSONString extends HttpServlet {

    public void doPost(HttpServletRequest request, HttpServletResponse response)
            throws ServletException, IOException {

        System.out.println("从 android 发送的请求中取出 username="
                + request.getParameter("username"));
        System.out.println("从 android 发送的请求中取出 password="
                + request.getParameter("password"));

        Userinfo userinfo1 = new Userinfo();
        userinfo1.setId("1");
        userinfo1.setUsername("username1");
        userinfo1.setPassword("password1");

        Userinfo userinfo2 = new Userinfo();
        userinfo2.setId("2");
        userinfo2.setUsername("username2");
        userinfo2.setPassword("password2");

        Userinfo userinfo3 = new Userinfo();
        userinfo3.setId("3");
        userinfo3.setUsername("username3");
        userinfo3.setPassword("password3");

        List userinfoList = new ArrayList();
        userinfoList.add(userinfo1);
        userinfoList.add(userinfo2);
        userinfoList.add(userinfo3);

        Gson gsonRef = new Gson();
        System.out.println(gsonRef.toJson(userinfoList));

        response.setContentType("text/html");
        PrintWriter out = response.getWriter();
        out.println(gsonRef.toJson(userinfoList));
        out.flush();
        out.close();

    }
}
```

}
```

实体类 Userinfo.java 的代码如下:

```java
package entity;

public class Userinfo {
    private String id;
    private String username;
    private String password;

    public String getId() {
        return id;
    }

    public void setId(String id) {
        this.id = id;
    }

    public String getUsername() {
        return username;
    }

    public void setUsername(String username) {
        this.username = username;
    }

    public String getPassword() {
        return password;
    }

    public void setPassword(String password) {
        this.password = password;
    }
}
```

将项目布署到 tomcat 下并启动 tomcat 服务,准备被 Android 终端访问。

2. 创建名称为 getJSONStringAndroid 的 Android 项目

把 Web 项目中的 Userinfo.java 文件复制到 Android 项目的 src 路径下,目的是用 Gson 框架解析传递过来的 JSON 字符串,文件 Main.java 的代码如下:

```java
package getJSONStringAndroid.test.run;

import entity.Userinfo;

import java.io.IOException;
import java.io.UnsupportedEncodingException;
import java.lang.reflect.Type;
import java.util.ArrayList;
```
```

```java
import java.util.List;

import org.apache.http.HttpResponse;
import org.apache.http.NameValuePair;
import org.apache.http.ParseException;
import org.apache.http.client.ClientProtocolException;
import org.apache.http.client.entity.UrlEncodedFormEntity;
import org.apache.http.client.methods.HttpPost;
import org.apache.http.impl.client.DefaultHttpClient;
import org.apache.http.message.BasicNameValuePair;
import org.apache.http.protocol.HTTP;
import org.apache.http.util.EntityUtils;

import android.app.Activity;
import android.os.Bundle;
import android.util.Log;

import com.google.gson.Gson;
import com.google.gson.reflect.TypeToken;

public class Main extends Activity {
    @Override
    public void onCreate(Bundle savedInstanceState) {
        super.onCreate(savedInstanceState);
        setContentView(R.layout.main);

        try {
            DefaultHttpClient dhcRef = new DefaultHttpClient();

            HttpResponse response = null;

            HttpPost request = new HttpPost(
                    "http://10.0.2.2:8081/getJSONString/getJSONString");

            List<NameValuePair> params = new ArrayList<NameValuePair>();
            params
                    .add(new BasicNameValuePair("username",
                            "gaohongyanUsername"));
            params
                    .add(new BasicNameValuePair("password",
                            "gaohongyanPassword"));

            request.setEntity(new UrlEncodedFormEntity(params, HTTP.UTF_8));
            response = dhcRef.execute(request);
            String getJSONString = EntityUtils.toString(response.getEntity());

            Gson gsonRef = new Gson();

            Type collectionUserinfoType = new TypeToken<ArrayList<Userinfo>>() {
            }.getType();
```

```java
            List<Userinfo> listUserinfo = gsonRef.fromJson(getJSONString,
                        collectionUserinfoType);
            for (int i = 0; i < listUserinfo.size(); i++) {
                Userinfo userinfo = listUserinfo.get(i);
                Log.v("" + (i + 1), "id=" + userinfo.getId() + " username="
                        + userinfo.getUsername() + " password="
                        + userinfo.getPassword());
            }

        } catch (UnsupportedEncodingException e) {
            // TODO Auto-generated catch block
            e.printStackTrace();
        } catch (ClientProtocolException e) {
            // TODO Auto-generated catch block
            e.printStackTrace();
        } catch (ParseException e) {
            // TODO Auto-generated catch block
            e.printStackTrace();
        } catch (IOException e) {
            // TODO Auto-generated catch block
            e.printStackTrace();
        }

    }
}
```

程序运行结果如图 7.1 所示。

```
V  732  1           id=1 username=username1 password=password1
V  732  2           id=2 username=username2 password=password2
V  732  3           id=3 username=username3 password=password3
```

图 7.1　成功解析远程 Web 回传的 JSON 字符串

7.2　在 Android 中通过 HTTP 协议访问 TXT 文件和 PIC 图片

使用 HTTP 协议不仅可以访问远程 Web 项目中的 servlet，还可以访问一些数据资源，比如图片等。

1. 创建存放有远程资源的 Web 项目

创建名称为 remotePicWeb 的 Web 项目，在 WebRoot 中添加资源文件，如图 7.2 所示。

其中 ghygbk.txt 文件使用 gbk 编码，而 ghyutf8.txt 使用 utf-8 编码，TXT 文件的编码可以使用 EditPlus 软件通过保存 TXT 文件时来进行选择。

将这个 Web 项目部署到 tomcat 中。其中 ghygbk.txt 和 ghyutf8.txt 文件的内容如图 7.3 所示。

图 7.2 添加了两个 TXT 和一个图片的资源

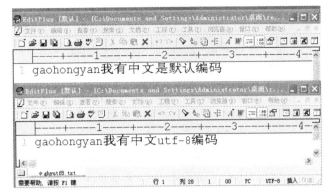

图 7.3 两个 TXT 文件的内容

2. 创建访问远程资源的 Android 项目

创建名称为 remotePic 的 Android 项目，文件 Main.java 的代码如下：

```java
package remotePic.test.run;

import info.monitorenter.cpdetector.io.ASCIIDetector;
import info.monitorenter.cpdetector.io.CodepageDetectorProxy;
import info.monitorenter.cpdetector.io.JChardetFacade;
import info.monitorenter.cpdetector.io.ParsingDetector;
import info.monitorenter.cpdetector.io.UnicodeDetector;

import java.io.BufferedInputStream;
import java.io.File;
import java.io.FileInputStream;
import java.io.IOException;
import java.io.InputStream;
import java.io.InputStreamReader;
import java.net.MalformedURLException;
import java.net.URL;
import java.net.URLConnection;

import android.app.Activity;
import android.graphics.Bitmap;
import android.graphics.BitmapFactory;
import android.os.Bundle;
import android.util.Log;
import android.widget.ImageView;

public class Main extends Activity {
    private ImageView imageView1;
    private ImageView imageView2;

    // 由于 txt 文件字符编码各不相同，所以使用 cpdetector 第三方 jar 包
    // 来取得编码，再根据不同的编码类型正确显示中文字符
    private String getFileChatSet(String filePathAndName) throws IOException {
        URL uriTextRef = new URL(filePathAndName);
```

```java
        URLConnection connectionText = uriTextRef.openConnection();
        InputStream isRef = connectionText.getInputStream();
        BufferedInputStream bisRef = new BufferedInputStream(isRef);

        CodepageDetectorProxy detector = CodepageDetectorProxy.getInstance();
        detector.add(new ParsingDetector(false));
        detector.add(JChardetFacade.getInstance());
        detector.add(ASCIIDetector.getInstance());
        detector.add(UnicodeDetector.getInstance());
        java.nio.charset.Charset charset = null;
        charset = detector.detectCodepage(bisRef, bisRef.available());

        return charset.name();

    }

    @Override
    public void onCreate(Bundle savedInstanceState) {
        super.onCreate(savedInstanceState);
        setContentView(R.layout.main);

        try {
            imageView1 = (ImageView) this.findViewById(R.id.imageView1);
            imageView2 = (ImageView) this.findViewById(R.id.imageView2);
            // 在 ImageView 中显示网络图片-开始
            URL uriRef = new URL("http://10.0.2.2:8081/remotePicWeb/ghy.png");
            URLConnection connection = uriRef.openConnection();
            Bitmap bitmapRef = BitmapFactory.decodeStream(connection
                    .getInputStream());
            imageView1.setImageBitmap(bitmapRef);
            // 在 ImageView 中显示网络图片-结束

            // 在 ImageView 中显示 SDCARD 图片-开始
            FileInputStream fisRef = new FileInputStream(new File(
                    "/sdcard/sdcardpic.png"));
            Bitmap bitmapSDRef = BitmapFactory.decodeStream(fisRef);
            imageView2.setImageBitmap(bitmapSDRef);
            // 在 ImageView 中显示 SDCARD 图片-结束

            // 在 LogCat 打印远程 txt 文件内容-开始
            String txtFile = "http://10.0.2.2:8081/remotePicWeb/ghygbk.txt";

            URL uriText = new URL(txtFile);
            URLConnection connectionText = uriText.openConnection();

            String charSetName = getFileChatSet(txtFile);
            Log.v("当前文件编码==", "" + charSetName);
            InputStream inputStreamRef = connectionText.getInputStream();
            StringBuffer sbRef = new StringBuffer();
            InputStreamReader reader = null;
```

```
if (charSetName.toLowerCase().equals("utf-8")) {
    reader = new InputStreamReader(inputStreamRef, "utf-8");
}
if (charSetName.toLowerCase().equals("gbk")) {
    reader = new InputStreamReader(inputStreamRef, "gbk");
}
char[] charArray = new char[2];
int readLength = reader.read(charArray);
while (readLength != -1) {
    sbRef.append(charArray, 0, readLength);
    readLength = reader.read(charArray);
}

reader.close();
inputStreamRef.close();

if (charSetName.toLowerCase().equals("utf-8")) {
    Log.v("!!utf-8 file=", new String(sbRef.toString().getBytes(),
            "utf-8"));
}
if (charSetName.toLowerCase().equals("gbk")) {
    Log.v("!!gbk file=", new String(sbRef.toString().getBytes(),
            "utf-8"));
}

// 在 LogCat 打印远程 txt 文件内容-结束

} catch (MalformedURLException e) {
    e.printStackTrace();
} catch (IOException e) {
    e.printStackTrace();
}
```

上面的代码使用了第三方 cpdetector 工具包来判断文件的编码类型，需要 4 个 jar 包文件，如图 7.4 所示。

本示例实现了访问 sdcard 卡中的图片，所以还需要在 AVD 中导入图片，如图 7.5 所示。

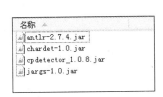

图 7.4 判断 TXT 文件编码的 cpdetector 工具包需要的 jar 文件

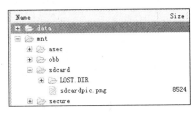

图 7.5 sdcard 卡中的图片资源

项目运行结果如图 7.6 所示。

成功地从 sdcard 卡和远程取得图片并显示出来，并且在 LogCat 中打印出 ghyghk.txt 文件的内容，如图 7.7 所示。

图 7.6　在 AVD 中显示图片资源　　　　图 7.7　成功打印 GBK 编码的 TXT 文件

将文件 Main.java 中访问远程 TXT 文件名的变量代码更改如下：

```
String txtFile = "http://10.0.2.2:8081/remotePicWeb/ghyutf8.txt";
```

再次运行项目，成功地取出 ghyutf8.txt 文件中的内容，如图 7.8 所示。

图 7.8　成功地打印出 utf-8 编码的 TXT 文件

7.3　用 java 语言 DOM 解析 XML

DOM 是 Document Object Model 的缩写，即文档对象模型。前面说过，XML 将数据组织为一棵树，所以 DOM 就是对这棵树的一个对象描述。通俗地说，就是通过解析 XML 文档，为 XML 文档在逻辑上建立一个树模型，树的节点是一个个对象，通过存取这些对象就能够存取 XML 文档的内容。

新建名称为 android_xml_1 的 Android 项目，在 assets 文件夹中创建 userinfo.xml 资源文件，代码如下：

```xml
<?xml version="1.0" encoding="UTF-8"?>
<userinfos>
    <userinfo id="1" type="a">
        <username value="高洪岩1"></username>
        <age value="100"></age>
        <content>我是正文1</content>
    </userinfo>
    <userinfo id="2" type="b">
        <username value="高洪岩2"></username>
        <age value="200"></age>
        <content>我是正文2</content>
```

        </userinfo>
    </userinfos>

更改 Main.java 的代码如下：

```java
package android_xml_1.test.run;

import java.io.IOException;

import javax.xml.parsers.DocumentBuilder;
import javax.xml.parsers.DocumentBuilderFactory;
import javax.xml.parsers.ParserConfigurationException;

import org.w3c.dom.Document;
import org.w3c.dom.Element;
import org.w3c.dom.NodeList;
import org.xml.sax.SAXException;

import android.app.Activity;
import android.os.Bundle;
import android.util.Log;

public class Main extends Activity {
    /** Called when the activity is first created. */
    @Override
    public void onCreate(Bundle savedInstanceState) {
        super.onCreate(savedInstanceState);
        setContentView(R.layout.main);

        try {
            DocumentBuilderFactory factory = DocumentBuilderFactory
                    .newInstance();
            DocumentBuilder builder = factory.newDocumentBuilder();
            Document document = builder.parse(this.getAssets().open(
                    "userinfo.xml"));
            Element rootElement = document.getDocumentElement();
            Log.v("!", "userinfo.xml 根元素名称为：" + rootElement.getNodeName());
            NodeList userinfoNodeList = rootElement
                    .getElementsByTagName("userinfo");
            for (int i = 0; i < userinfoNodeList.getLength(); i++) {
                Element eachUserinfoElement = (Element) userinfoNodeList
                        .item(i);
                Log.v("!", "userinfo 标签的 id 属性值为："
                        + eachUserinfoElement.getAttribute("id") + " type 属性值为："
                        + eachUserinfoElement.getAttribute("type"));

                Log.v("!", "username 标签的 value 属性值为："
                        + eachUserinfoElement.getElementsByTagName("username")
                                .item(0).getAttributes().getNamedItem("value")
                                .getTextContent());
```

```java
                    Log.v("!", "age 标签的 value 属性值为："
                            + eachUserinfoElement.getElementsByTagName("age").item(
                                    0).getAttributes().getNamedItem("value")
                                    .getTextContent());
                    Log.v("!", "content 标签的正文内容为："
                            + eachUserinfoElement.getElementsByTagName("content")
                                    .item(0).getTextContent());
                    Log.v("!", "-----------------------------------");
                }

        } catch (ParserConfigurationException e) {
            // TODO Auto-generated catch block
            e.printStackTrace();
        } catch (SAXException e) {
            // TODO Auto-generated catch block
            e.printStackTrace();
        } catch (IOException e) {
            // TODO Auto-generated catch block
            e.printStackTrace();
        }

    }
}
```

程序运行结果如图 7.9 所示。

```
V 551 !   userinfo.xml根元素名称为: userinfos
V 551 !   userinfo标签的id属性值为: 1 type属性值为: a
V 551 !   username标签的value属性值为: 高洪岩1
V 551 !   age标签的value属性值为: 100
V 551 !   content标签的正文内容为: 我是正文1
V 551 !   -----------------------------------
V 551 !   userinfo标签的id属性值为: 2 type属性值为: b
V 551 !   username标签的value属性值为: 高洪岩2
V 551 !   age标签的value属性值为: 200
V 551 !   content标签的正文内容为: 我是正文2
V 551 !   -----------------------------------
```

图 7.9　运行结果

第 8 章　Activity 活动、Service 服务和 Broadcast 广播彼此调用实验

本章主要实现 Android 核心组件之间的调用，虽然内容较为简单，但凡学习 Android 这几个实验是必须要经历的，所以还需要一步一个脚印地走完这个学习过程。

本章比较重要的学习内容是：

- Activity->BroadCaseReceiver->Activity
- Activity->Service(startService)-> BroadCaseReceiver

8.1　Activity->BroadCaseReceiver->Activity 实验

本小节要实现从 Activity 发起一个广播到 BroadCaseReceiver，然后从 BroadCaseReceiver 再启动一个 Acitivyt 对象。

新建名称为 ActivityBroadcaseActivity 的 Android 项目，文件 Main.java 的代码如下：

```java
public class Main extends Activity {
    private Button button1;

    @Override
    public void onCreate(Bundle savedInstanceState) {
        super.onCreate(savedInstanceState);
        setContentView(R.layout.main);

        button1 = (Button) this.findViewById(R.id.button1);
        button1.setOnClickListener(new OnClickListener() {
            public void onClick(View arg0) {

                Intent intent = new Intent("gotoMyBroadcastReceiver");
                intent.putExtra("username", "高洪岩");

                Main.this.sendBroadcast(intent);

            }
        });

    }
```

}

创建 BroadcastReceiver 广播接收者 MyBroadcastReceiver.java 的代码如下：

```java
public class MyBroadcastReceiver extends BroadcastReceiver {

    @Override
    public void onReceive(Context arg0, Intent arg1) {

        Log.v("!从 Main 取出来的数据值是：", "" + arg1.getStringExtra("username"));

        Intent intent = new Intent(arg0, Second.class);
        intent.setFlags(Intent.FLAG_ACTIVITY_NEW_TASK);
        intent.putExtra("pull_second_Key", "pull_second_Value");

        arg0.startActivity(intent);

    }
}
```

Second.java 的代码如下：

```java
public class Second extends Activity {
    @Override
    public void onCreate(Bundle savedInstanceState) {
        super.onCreate(savedInstanceState);
        setContentView(R.layout.second);

        Log.v("!pull_second_Key=", this.getIntent().getStringExtra(
            "pull_second_Key"));
    }
}
```

程序初始运行界面如图 8.1 所示。

单击按钮后在 LogCat 中打印的日志及界面变化如图 8.2 所示。

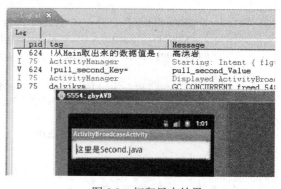

图 8.1 初始界面运行效果

图 8.2 打印日志结果

8.2　Activity->Service(startService)->Activity 实验

本小节要实现一个从 Activity 对象启动一个 Service，再从这个 Service 启动一个 Activity 对象的实例。

新建名称为 ActivityServiceActivity 的 Android 项目，文件 Main.java 的代码如下：

```java
public class Main extends Activity {
    private Button button1;

    @Override
    public void onCreate(Bundle savedInstanceState) {
        super.onCreate(savedInstanceState);
        setContentView(R.layout.main);

        button1 = (Button) this.findViewById(R.id.button1);
        button1.setOnClickListener(new OnClickListener() {
            public void onClick(View arg0) {
                Intent startService = new Intent(Main.this, GhyService.class);
                startService.putExtra("pull_service_key", "pull_service_value");

                Main.this.startService(startService);
            }
        });
    }
}
```

文件 GhyService.java 的代码如下：

```java
public class GhyService extends Service {

    @Override
    public int onStartCommand(Intent intent, int flags, int startId) {

        for (int i = 0; i < 5; i++) {
            Log.v("!", "" + (i + 1));
        }
        Log.v("接收到 pull_service_key", ""
                + intent.getStringExtra("pull_service_key"));

        Intent gotoSecond = new Intent(this.getApplicationContext(),
                Second.class);
        gotoSecond.setFlags(Intent.FLAG_ACTIVITY_NEW_TASK);
        gotoSecond.putExtra("pull_second_key", "pull_second_value");

        this.startActivity(gotoSecond);

        return super.onStartCommand(intent, flags, startId);
    }
}
```

```
        @Override
        public IBinder onBind(Intent arg0) {
            return null;
        }

}
```

文件 Second.java 的代码如下：

```
public class Second extends Activity {
    @Override
    public void onCreate(Bundle savedInstanceState) {
        super.onCreate(savedInstanceState);
        setContentView(R.layout.main);
        Log.v("接收到 pull_second_key", ""
                + this.getIntent().getStringExtra("pull_second_key"));
    }
}
```

程序初始运行结果如图 8.3 所示。

单击按钮后在 LogCat 中打印的信息如图 8.4 所示。

图 8.3　初始运行效果　　　　　　图 8.4　日志打印结果

8.3　Activity->BroadCaseReceiver->Service(startService)实验

本小节实现从 Activity 对象发起一个广播到 BroadCaseReceiver，然后在 BroadCaseReceiver 中启动一个 Service 服务的实例。

新建名称为 ActivityBroadcaseService 的 Android 项目，文件 Main.java 的代码如下：

```
public class Main extends Activity {
    private Button button1;

    @Override
```

```java
public void onCreate(Bundle savedInstanceState) {
    super.onCreate(savedInstanceState);
    setContentView(R.layout.main);

    button1 = (Button) this.findViewById(R.id.button1);
    button1.setOnClickListener(new OnClickListener() {
        public void onClick(View arg0) {

            Intent sendBroadcast = new Intent("sendBroadcast2");
            sendBroadcast.putExtra("pull_broadcast_key",
                    "pull_broadcast_value");

            Main.this.sendBroadcast(sendBroadcast);

        }
    });

}
}
```

自定义 BroadcastReceiver 广播接收者 GhyBroadcastReceiver.java 的代码如下：

```java
public class GhyBroadcastReceiver extends BroadcastReceiver {

    @Override
    public void onReceive(Context arg0, Intent arg1) {
        Log.v("在广播接收到的值 pull_broadcast_key", ""
                + arg1.getStringExtra("pull_broadcast_key"));
        for (int i = 0; i < 5; i++) {
            Log.v("在广播进行的业务", "" + (i + 1));
        }

        Intent intent = new Intent(arg0, service.GhyService.class);
        intent.putExtra("pull_GhyService_key", "pull_GhyService_value");

        arg0.startService(intent);

    }

}
```

自定义 Service 服务类 GhyService.java 的代码如下：

```java
public class GhyService extends Service {

    @Override
    public int onStartCommand(Intent intent, int flags, int startId) {
        Log.v("从 pull_GhyService_key 取出的值为：", ""
                + intent.getStringExtra("pull_GhyService_key"));
        return super.onStartCommand(intent, flags, startId);
    }
```

```
        @Override
        public IBinder onBind(Intent arg0) {
            return null;
        }

    }
```

程序初始运行结果如图 8.5 所示。

单击按钮"发送广播",在 LogCat 中打印的日志内容如图 8.6 所示。

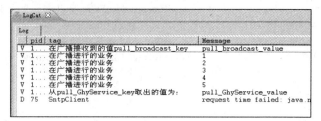

图 8.5　初始运行效果　　　　　　　　　图 8.6　日志打印结果

8.4　Activity->Service(startService)-> BroadCaseReceiver 实验

本小节实现从 Activity 对象启动一个 Service 服务,在这个 Service 服务中发起一个广播。

新建名称为 ActivityServiceBroadcase 的 Android 项目,文件 Main.java 的代码如下:

```
public class Main extends Activity {

    @Override
    public void onCreate(Bundle savedInstanceState) {
        super.onCreate(savedInstanceState);
        setContentView(R.layout.main);

        Intent startService = new Intent(Main.this, GhyService.class);
        startService.putExtra("pull_service_key", "pull_service_value");

        this.startService(startService);
    }
}
```

自定义 Service 文件 GhyService.java 的代码如下:

```
public class GhyService extends Service {

    @Override
    public int onStartCommand(Intent intent, int flags, int startId) {
        Log.v("从 pull_service_key 取出的值为:", ""
                + intent.getStringExtra("pull_service_key"));

        for (int i = 0; i < 5; i++) {
```

```
        Log.v("在服务中做业务：", "" + (i + 1));
    }

    Intent sendBroadcast = new Intent("sendBroadcast3");
    sendBroadcast.putExtra("pull_broadcast_key", "pull_broadcast_value");
    this.sendBroadcast(sendBroadcast);

    return super.onStartCommand(intent, flags, startId);
}

@Override
public IBinder onBind(Intent arg0) {
    return null;
}
}
```

自定义广播接收者 GhyBroadcastReceiver.java 的代码如下：

```
public class GhyBroadcastReceiver extends BroadcastReceiver {

    @Override
    public void onReceive(Context arg0, Intent arg1) {
        Log.v("在广播接收到的值 pull_broadcast_key", ""
                + arg1.getStringExtra("pull_broadcast_key"));
    }

}
```

程序初始运行结果如图 8.7 所示。

单击按钮后在 LogCat 中打印的结果如图 8.8 所示。

图 8.7 初始运行效果

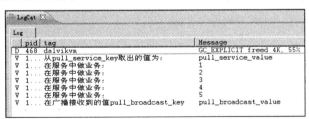

图 8.8 日志打印结果

8.5 Activity->BroadCaseReceiver->Service(bindService)实验

前面都是用 startService()方法来启动 Service 服务组件的，本节将用 bindService()方法来绑定 Service 服务组件。

新建名称为 BroadCastReceiverXml_bindService 的 Android 项目，新建 Service 类 GhyService.java，核心代码如下：

```java
public class GhyService extends Service {

    public String getUsernameFromService() {
        return "高洪岩来自 Service";
    }

    @Override
    public void onCreate() {
        super.onCreate();
        Log.v("!", "GhyService onCreate");
    }

    @Override
    public int onStartCommand(Intent intent, int flags, int startId) {
        Log.v("!", "GhyService onStartCommand");
        return super.onStartCommand(intent, flags, startId);
    }

    @Override
    public boolean onUnbind(Intent intent) {
        Log.v("!", "GhyService onUnbind");
        return true;
    }

    @Override
    public void onRebind(Intent intent) {
        super.onRebind(intent);
        Log.v("!", "GhyService onRebind");
    }

    @Override
    public void onDestroy() {
        super.onDestroy();
        Log.v("!", "GhyService onDestroy");
    }

    public class GhyBinder extends Binder {
        public GhyService getGhyService() {
            Log.v("!", "GhyBinder getGhyService");
            return GhyService.this;
        }
    }

    @Override
    public IBinder onBind(Intent arg0) {
        Log.v("!", "GhyService onBind");
        return new GhyBinder();
    }

}
```

Activity 活动、Service 服务和 Broadcast 广播彼此调用实验 第 8 章

新建广播接收者 BroadCaseReceiver 类 GhyBroadcastReceiver.java，核心代码如下：

```java
public class GhyBroadcastReceiver extends BroadcastReceiver {

    private ServiceConnection connection = new ServiceConnection() {
        public void onServiceDisconnected(ComponentName arg0) {

        }

        public void onServiceConnected(ComponentName arg0, IBinder arg1) {
            GhyService gsRef = ((GhyBinder) arg1).getGhyService();
            Log.v("!", gsRef.getUsernameFromService());
        }
    };

    @Override
    public void onReceive(Context arg0, Intent arg1) {
        Log.v("!", "GhyBroadcastReceiver onReceive username="
                + arg1.getStringExtra("username"));

        Intent intent = new Intent(arg0, GhyService.class);
        arg0.bindService(intent, connection, Service.BIND_AUTO_CREATE);

    }
}
```

Main.java 的核心代码如下：

```java
public class Main extends Activity {
    private Button button1;

    @Override
    public void onCreate(Bundle savedInstanceState) {
        super.onCreate(savedInstanceState);
        setContentView(R.layout.main);

        button1 = (Button) this.findViewById(R.id.button1);
        button1.setOnClickListener(new OnClickListener() {
            public void onClick(View arg0) {

                Intent intent = new Intent("getUsernameGhyBroadcastReceiver");
                intent.putExtra("username", "gaohongyan");
                Main.this.sendBroadcast(intent);
            }
        });
    }
}
```

运行项目单击 Button 按钮，在广播 BroadCaseReceiver 中用 bindService()方法启动 Service，但出现如下异常：

java.lang.RuntimeException: Unable   to   start   receiver   extbroadcastreceiver.GhyBroadcastReceiver:

android.content.ReceiverCallNotAllowedException: IntentReceiver components are not allowed to bind to services

出现此异常的主要原因是由于在 BroadcastReceiver 类中的方法"public void onReceive(Context arg0, Intent arg1)"中的参数 arg0 的对象是一个 android.app.ReceiverRestrictedContext 实例，而这个实例的 bindService()方法在 Android 的源代码中抛出了异常，Android 的源代码如下：

```java
@Override
public boolean bindService(Intent service, ServiceConnection conn, int flags) {
    throw new ReceiverCallNotAllowedException(
        "IntentReceiver components are not allowed to bind to services");
}
```

如何才能实现在 BroadcastReceiver 对象中 bindervice 呢？很简单，不用 arg0 这个 Context 参数即可。

新建名称 BroadCastReceiverXml_bindService_update 的 Android 项目，此项目中的核心代码与 BroadCastReceiverXml_bindService 项目大体一样，只是在这个项目中添加了一个通用工具类，此类名称为 CommonTools.java，主要的作用就是保存一个 Context 持久化参数，以后使用时直接调用即可，代码如下：

```java
package commontools;

import android.content.Context;

public class CommonTools {

    public static Context context;

}
```

对象 GhyBroadcastReceiver.java 的核心代码如下：

```java
@Override
public void onReceive(Context arg0, Intent arg1) {
    Log.v("!", "GhyBroadcastReceiver onReceive username="
        + arg1.getStringExtra("username"));

    Intent intent = new Intent(arg0, GhyService.class);
    CommonTools.context.bindService(intent, connection,
        Service.BIND_AUTO_CREATE);

}
```

项目运行后得到正确的打印结果，如图 8.9 所示。

```
W 1. ActivityManager finishReceiver called but none active
V 7..!           GhyBroadcastReceiver onReceive username=gaohongyan
V 7..!           GhyService onCreate
V 7..!           GhyService onBind
V 7..!           GhyBinder getGhyService
V 7..!           高洪岩来自Service
```

图 8.9　运行结果

第 9 章 UI 控件的美化与动画

本章主要学习在 Android 中如何对界面中的常用控件进行美化，这也是 Android 开发必须要掌握的技术，同 Web 技术中美工和程序分工不同，大多数 Android 开发者既要做一个程序员，还要充当一个优秀的美工，这样开发出来的软件 UI 要表达的意图和程序才能更好地结合，另外 Android 的 UI 界面设计并不像 Web 设计那样单纯的使用 CSS，相反它还会或多或少地用代码来美化装饰界面，有些类似于 Web 程序员使用 js 结合 CSS 美化程序 UI，在美化的同时还要适当地写一写 js 程序等这种情况，所以 Android 程序员掌握用 style 美化界面的同时还要掌握用 java 语言来实现一些视觉动画效果。

学习本章应该着重掌握以下技术点：

- 使用 style 在不同事件下控制控件外观
- 美化常用控件的方法
- 使用 XML 配置文件定义动画

9.1 style 的使用

关于样式知识点的讲解已经在第 3 章中介绍过，并且结合使用 draw9patch 工具实现了美化界面，下面将对 style 及 selector 和在 Android 中的动画进行更加详细地介绍。

在介绍上面的知识点之前，有必要先了解一下 Android 系统中自带 style 样式的相关知识。

在 Android 系统中，样式定义在 "\android-sdk-windows\platforms\android-9\data\res\values" 文件夹中的 styles.xml 文件中，这里面有系统全部的样式定义声明，但有一些样式是隐藏的，它们使用@hide 来作为标记，例如下面的样式代码：

```xml
<!-- @hide -->
<style name="TextAppearance.SearchResult.Title">
    <item name="android:textSize">16sp</item>
</style>
```

样式 TextAppearance.SearchResult.Title 在 ADT 的自动提示中是不显示的，因为是隐藏的。

使用系统自带的样式非常简单，在名称为 systemStyleTest 项目中的 main.xml 代码如下：

```xml
<?xml version="1.0" encoding="utf-8"?>
<LinearLayout xmlns:android="http://schemas.android.com/apk/res/android"
    android:orientation="vertical" android:layout_width="fill_parent"
    android:layout_height="fill_parent">
    <TextView style="@android:style/TextAppearance.Large"
```

```
        android:layout_width="fill_parent" android:layout_height="wrap_content"
        android:text="高洪岩大字体" />
<TextView style="@android:style/TextAppearance.Small"
        android:layout_width="fill_parent" android:layout_height="wrap_content"
        android:text="高洪岩小字体" />
</LinearLayout>
```

上面的代码使用的就是系统自带的样式，显示的外观是大字体和小字体，运行效果如图 9.1 所示。

图 9.1　大字体与小字体

需要注意的是，在 ADT 中将自动提示的样式名称中的 "_"（下划线）改成小字点 "."即可。样式 style 是系统中的资源，在 Android 中使用资源有以下几个知识点：

1. 引用自定义资源：@资源类型/资源名称

这种写法是使用用户自定义的资源名称，例如如下代码：

```
<TextView android:layout_width="fill_parent"
        android:layout_height="wrap_content" android:text="@string/hello"
        android:background="@color/ghyColor" />
```

通过使用@string 和@color 就可以引用对应资源类型的自定义资源名称。

2. 引用系统资源与使用隐藏资源：@android:资源类型/资源名称

在 sdk 文件夹 "android-sdk-windows\platforms\android-9\data\res\values" 中的 colors.xml 配置文件中有系统默认的 color 颜色配置，在项目中可以引用系统资源，例如下述代码：

```
<TextView android:layout_width="fill_parent"
        android:layout_height="wrap_content" android:text="@string/hello"
        android:background="@*android:color/hint_foreground_dark" />
```

代码使用了 colors.xml 文件中的 hint_foreground_dark 样式 style，但由于此样式在 public.xml 并未定义，所以在项目中并不能直接使用，这时使用@*android 的方式来引用隐藏的资源，加入 "*"（星号）的作用是使用系统隐藏的资源，也就是使用非 public 的资源。在 Android 项目中可以使用的资源在路径 "android-sdk-windows\platforms\android-9\data\res\values" 中的 public.xml 文件中。在这里需要说明一下，没在 public.xml 中声明的资源是 Google 不推荐使用的。

9.1.1　style 的概述与定义

定义 style 样式资源可以把 UI 用户界面进行美化及改良，样式可以应用于 1 个或更多控件，也可以应用于 1 个或更多 Activity 对象，还可以应用于整个应用程序。

使用 style 样式非常简单，在 style_1 项目中的 res/values/文件夹下创建一个名称为 style.xml 的文件，样式的文件名是任意的，但为了文件名有意义，尽量给文件名加入 style 的关键字，从而能快速识别 XML 文件资源的类型，内容如下：

```xml
<?xml version="1.0" encoding="utf-8"?>
<resources>
    <style name="ghyStyle1">
    </style>
</resources>
```

在这个样式中并没有对样式添加任何的定义属性，也就是在<style>标签中并没有<item>标签，但<style>标签的属性 name 却代表了这个样式的名称 ghyStyle1，这个名称也在 R.java 文件中进行了注册，也就是样式资源的 id，代码如下：

```java
public static final class style {
    public static final int ghyStyle1=0x7f050000;
}
```

虽然定义了一个名称为 ghyStyle1 的样式，但却没有细节的定义，继续更改 style.xml 中的样式代码如下：

```xml
<?xml version="1.0" encoding="utf-8"?>
<resources>
    <style name="ghyStyle1">
        <item name="android:textSize">40dip</item>
    </style>
</resources>
```

上面代码中的<item>标签的 name 属性值 android:textSize 是来自于 android-sdk-windows\platforms\android-9\data\res\values 文件夹中的 attrs.xml 文件，在此文件中定义了所有 Android 系统自带的属性，其中就有 textSize 属性的声明，如图 9.2 所示。

```
486        </ul>
487        -->
488        <attr name="textSize" format="dimension" />
489
```

图 9.2　textSize 属性在 attrs.xml 文件中的声明

这段样式定义文字的大小为 40dip 单位，<item>标签定义样式的细节信息，name 属性定义样式的名称，而<item>的 body 体定义样式的值。

更加详细的 style 完整语法定义如下：

```xml
<?xml version="1.0" encoding="utf-8"?>
<resources>
    <style name="style_name" parent="@[package:]style/style_to_inherit">
        <item name="[package:]style_property_name">style_value</item>
    </style>
</resources>
```

定义样式时有以下几点需要注意：

（1）样式的存放路径是在 res/values/文件夹下。

（2）元素<resources>是必须具有的标签，是样式 XML 文件的根（root）结点。

（3）元素<style>：style 标签是定义 1 个样式，它有名称为<item>的子结点。<style>的 name 属性可以生成此样式的资源 id，也就是在 R.java 文件中，通过这个 resourceId 就可以将这个样式应用到 View 控件或 Activity 或整个的应用程序中。<style>还有 parent 属性，这个属性定义当前的样式是从哪个样式继承下来，使得样式也可以得到代码的重用。

（4）元素<item>：定义样式的属性，是<style>标签的子标签，具有 name 属性，用于定义样式属性的具体名称。

虽然定义了样式，那如何引用呢？使用如下的语法即可：

```
@[package:]style/style_name
```

9.1.2 style 的使用与继承

在 style_1 项目中将 main.xml 中的<textView>控件应用上一节创建的 ghyStyle1 样式，布局代码如下：

```
<?xml version="1.0" encoding="utf-8"?>
<LinearLayout xmlns:android="http://schemas.android.com/apk/res/android"
    android:orientation="vertical" android:layout_width="fill_parent"
    android:layout_height="fill_parent">
    <TextView style="@style/ghyStyle1" android:layout_width="fill_parent"
        android:layout_height="wrap_content" android:text="@string/hello" />
</LinearLayout>
```

程序运行的结果如图 9.3 所示。

图 9.3　TextView 控件应用 ghyStyle1 样式

下面再来实现一个样式的继承示例，继续更改 style.xml 的样式代码如下：

```
<?xml version="1.0" encoding="utf-8"?>
<resources>
    <style name="ghyParentStyle">
        <item name="android:background">#FF0000</item>
    </style>
    <style name="ghyStyle1" parent="@style/ghyParentStyle">
        <item name="android:textSize">40dip</item>
    </style>
</resources>
```

在 style.xml 文件中定义了一个名称为 ghyParentStyle 的父样式，然后在 ghyStyle1 中进行继承，

AVD 运行后 TextView 控件出现了红色的背景，如图 9.4 所示。

图 9.4 样式的继承示例

9.2 文字颜色 selector 状态列表

与 UI 界面美化有密切关系的是 stateList 状态列表，selector 对控件状态的改变是通过 UI 图形来表达的，比如按钮有按下的状态、默认的状态及屏蔽状态等 UI 图形界面来向用户展示控件的状态。

9.2.1 文字颜色 selector 的概述与定义

文字颜色状态列表 XML 配置文件是存放在 res/color/文件夹下，此信息是来自于 Android 官方的 guide 手册中的内容，说明如图 9.5 所示。

可以在 DOC 文档中的 Application Resources 中的 Resource Types 中的 Color State List 中找到具体的使用方法。另外，创建文字颜色 selector 时只有手动进行配置，使用 ADT 的向导创建 XML 配置文件已经无效，如图 9.6 所示。

图 9.5 selector 的 XML 配置文件的存放位置　　图 9.6 使用 XML 配置文件向导创建不了 selector 对象

需要注意的是，selector 状态列表 XML 配置文件的文件名 filename 就是 resourceId，引用文字颜色状态列表有两种方式，分别是在 Java 文件和 XML 文件中引用，引用方式为：

- Java 引用方式：R.color.filename
- XML 文件引用方式：@[*package*:]color/*filename*

文字颜色状态列表 selector 的完整语法如下：

```
<?xml version="1.0" encoding="utf-8"?>
```

```xml
<selector xmlns:android="http://schemas.android.com/apk/res/android" >
    <item
        android:color="hex_color"
        android:state_pressed=["true" | "false"]
        android:state_focused=["true" | "false"]
        android:state_selected=["true" | "false"]
        android:state_checkable=["true" | "false"]
        android:state_checked=["true" | "false"]
        android:state_enabled=["true" | "false"]
        android:state_window_focused=["true" | "false"] />
</selector>
```

<selector>元素：状态列表的根（root）结点名称为<selector>，必须要有的标签，用它来包含<item>元素。

<item>元素：<item>元素是定义每种状态的详细信息，它是<selector>的子标签，而 android:color 属性是定义每种状态的文本颜色，值为 16 进制的颜色代码，可以加入透明的 alpha 值，比如 Alpha-Red-Green-Blue，取值的格式为#RGB、#ARGB、#RRGGBB 和#AARRGGBB。

属性 android:state_pressed 的取值为 true 和 false，代表控件按下和不按下时的状态匹配。

属性 android:state_focused 值为 true 代表获得了焦点，为 false 代表没有获得焦点。

属性 android:state_selected 为 true 代表控件被选中，为 false 代表控件并没有被选中。

属性 android:state_checkable 值为 true，代表控件能被 checked，为 false 时代表控件不能被 unchecked。

属性 android:state_checked 值为 true，代表控件已经被 checked，为 false 时代表控件并没有被 checked 的状态。

属性 android:state_enabled 值为 true，代表控件可以被使用，为 false 时代表控件是不可用状态。

属性 android:state_window_focused 值为 true，代表当前的窗体获得了焦点，而值为 false 时代表窗体并没有获得焦点。

上面一大段的解释也就是在每一种状态下使用的 android:color 颜色值。那么到底文字颜色 selector 该如何应用呢？很简单，文字颜色 selector 就是用来匹配每种状态 UI 变化的，下节一起做一个实验。

9.2.2 文字颜色 selector 的使用

创建名称为 selector_1 的 Android 项目，在 res/color 文件夹下创建名称为 button_selector_text.xml 的文字颜色配置文件，代码如下：

```xml
<?xml version="1.0" encoding="utf-8"?>
<selector xmlns:android="http://schemas.android.com/apk/res/android">
    <item android:state_pressed="true" android:color="#FF0000" />
    <item android:state_focused="true" android:color="#00FF00" />
    <item android:color="#0000FF" />
</selector>
```

这段文字颜色 selector 的功能就是定义当控件默认状态时的文字颜色为#0000FF（蓝色），而当

android:state_pressed="true"控件被按下时，颜色值为 android:color="#FF0000"（红色），获得焦点时 android:state_focused="true"，控件颜色为 android:color="#00FF00"（绿色）。

关于用 AVD 获得焦点的操作可以用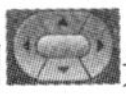来实现。

在文件 main.xml 布局文件中的 Button 控件引用这个文字颜色 selector 资源，代码如下：

```
<Button android:text="Button" android:id="@+id/button1"
    android:layout_width="wrap_content" android:layout_height="wrap_content"
    android:textColor="@color/button_selector_text"></Button>
```

9.3 背景图片 selector 状态列表

在 9.2 节中介绍了文字颜色 selector 美化的实现，本节将实现一个与文字美化同样重要的功能：背景状态美化，也就是用漂亮的背景图片来表达控件的状态信息。

9.3.1 背景图片 selector 状态列表

背景图片 selector 的作用是当控件在不同的状态下显示出不同的背景图片，比如 Button 按钮被按下、抬起，默认时的背景图片都不一样，所以就需要背景图片 selector 的美化了。

想要实现背景图片 selector 的美化，XML 配置文件要存放在 res/drawable/文件夹下，此信息已经在 guide 手册中得以说明，如图 9.7 所示。

```
FILE LOCATION:
    res/drawable/filename.xml
    The filename is used as the resource ID.
```

图 9.7 背景 selector 配置文件 XML 的存放路径

此说明可以在 Android 官方 guide 中的 Application Resources\Resource Types\Drawable 中的 State List 中找到。

如果想要引用 selector 资源，只需要以 XML 文件名 filename 作为 resourceId 即可。引用的方式分别是 java 方式和 XML 方式，如下所示：

Java 引用方式：R.drawable.*filename*

xml 引用方式：@[*package*:]drawable/*filename*

背景美化 selector 文件 mybutton.xml 的完整语法如下：

```
<?xml version="1.0" encoding="utf-8"?>
<selector xmlns:android="http://schemas.android.com/apk/res/android"
    android:constantSize=["true" | "false"]
    android:dither=["true" | "false"]
    android:variablePadding=["true" | "false"] >
    <item
        android:drawable="@[package:]drawable/drawable_resource"
        android:state_pressed=["true" | "false"]
        android:state_focused=["true" | "false"]
```

```
            android:state_selected=/["true" | "false"]
            android:state_checkable=/["true" | "false"]
            android:state_checked=/["true" | "false"]
            android:state_enabled=/["true" | "false"]
            android:state_window_focused=/["true" | "false"] />
</selector>
```

和文字美化 selector 的语法大体相同，仅仅在这里定义的是 android:drawable 属性，并不是 android:color 了。

> 如果既想使用 selector 美化背景又想使用 selector 美化不同状态下的文字颜色，可以在控件中把 android:background 和 android:textColor 结合使用，此效果在下面的内容中进行了实现。

在控件中使用的示例代码如下：

```
android:background="@drawable/mybutton"
```

9.3.2 用 selector 状态列表美化 Button、CheckBox、RadioButton 和 EditText 常用控件

在大多数 UI 控件美化的过程中，都是将文字颜色与背景图片 selector 进行联合使用，本示例就来实现 Button、CheckBox、RadioButton 和 EditText 的控件美化。

新建名称为 moreUI 的 Android 项目来实验美化控件 UI 的功能。

1. 美化 Button 控件

在文件夹 res/drawable-hdpi 下添加 3 个 9Patch 图片资源，Button 背景文件名称如图 9.8 所示。

图 9.8 Button 的 3 个背景图片资源

这 3 个图片代表不同的界面状态，如图 9.9 所示。

图 9.9 3 个不同状态的背景图片

创建 Button 文本颜色的 selector 配置文件 res/color/button_selector_text.xml，代码如下：

```xml
<?xml version="1.0" encoding="utf-8"?>
<selector xmlns:android="http://schemas.android.com/apk/res/android">
    <item android:state_pressed="true" android:color="#FF0000" />
    <item android:state_focused="true" android:color="#00FF00" />
    <item android:color="#0000FF" />
</selector>
```

再创建 Button 不同状态时的背景图片 selector 配置文件 res/drawable/button_selector_drawable.xml，代码如下：

```xml
<?xml version="1.0" encoding="utf-8"?>
<selector xmlns:android="http://schemas.android.com/apk/res/android">
    <item android:state_pressed="true" android:drawable="@drawable/btn_style_one_pressed" />
    <item android:state_focused="true" android:drawable="@drawable/btn_style_one_focused" />
    <item android:drawable="@drawable/btn_style_one_normal" />
</selector>
```

在布局文件 main.xml 中更改代码如下：

```xml
<?xml version="1.0" encoding="utf-8"?>
<LinearLayout xmlns:android="http://schemas.android.com/apk/res/android"
    android:orientation="vertical" android:layout_width="fill_parent"
    android:layout_height="fill_parent">
    <Button android:text="Button" android:id="@+id/button1"
        android:layout_width="fill_parent" android:layout_height="wrap_content"></Button>
    <Button android:textColor="@color/button_selector_text"
        android:background="@drawable/button_selector_drawable" android:text="Button"
        android:id="@+id/button2" android:layout_width="fill_parent"
        android:layout_height="wrap_content"></Button>
</LinearLayout>
```

在 id 为 button2 的 Button 控件中应用文字颜色和背景图片 selector 资源，程序运行后的效果如图 9.10 所示。

可以看到 Button2 默认的状态背景图片为圆角背景，文字颜色为蓝色，当单击 AVD 的方向键将焦点切换到 button2 时，界面又发生变化，如图 9.11 所示。

图 9.10　Button 默认效果

图 9.11　获得焦点的 button2 界面

从图 9.11 中可以看到，button2 获得焦点后文字颜色变成绿色，并且具有蓝色边框的视觉效果。当用鼠标单击 button2 时界面又发生变化，如图 9.12 所示。

图 9.12　按下 button2 时的界面效果

从图 9.12 中可以看到，button2 被按下时背景发生变化，并且文字颜色变成了红色。

2. 美化 CheckBox 控件

在文件夹 res/drawable-hdpi 下添加 4 种不同状态的背景图片，如图 9.13 所示。
这 4 个背景图片的显示效果如图 9.14 所示。

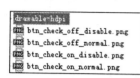

图 9.13　4 个不同状态的 checkbox 背景图片　　图 9.14　4 种 checkbox 不同状态的图片效果

根据文件名可以发现具有 disabled 的 on/off 状态的图片资源。再次更改 main.xml 布局文件，添加代码如下：

```xml
<CheckBox android:checked="true" android:text="是否备份到A 库-我是默认"
    android:id="@+id/checkBox1" android:layout_width="wrap_content"
    android:layout_height="wrap_content"></CheckBox>
<CheckBox android:button="@drawable/checkbox_selector_drawable"
    android:text="是否备份到A 库" android:id="@+id/checkBox1"
    android:layout_width="wrap_content" android:layout_height="wrap_content"></CheckBox>
<CheckBox android:button="@drawable/checkbox_selector_drawable"
    android:text="是否备份到A 库" android:id="@+id/checkBox2"
    android:layout_width="wrap_content" android:layout_height="wrap_content"
    android:enabled="false" android:checked="true"></CheckBox>
<CheckBox android:button="@drawable/checkbox_selector_drawable"
    android:text="是否备份到A 库" android:id="@+id/checkBox2"
    android:layout_width="wrap_content" android:layout_height="wrap_content"
    android:enabled="false" android:checked="false"></CheckBox>
```

添加了 4 个 CheckBox 控件，分别是默认 UI、美化的 CheckBox，还有 android:enabled 值分别是 true 和 false 的两个 CheckBox 控件。

在 res/ drawable 目录中创建名称为 checkbox_selector_drawable.xml 的 selector 配置文件，代码如下：

```xml
<?xml version="1.0" encoding="utf-8"?>
<selector xmlns:android="http://schemas.android.com/apk/res/android">
    <item android:state_checked="true" android:state_enabled="true"
        android:drawable="@drawable/btn_check_on_normal" />
    <item android:state_checked="false" android:state_enabled="true"
        android:drawable="@drawable/btn_check_off_normal" />

    <item android:state_checked="true" android:state_enabled="false"
        android:drawable="@drawable/btn_check_on_disable" />
    <item android:state_checked="false" android:state_enabled="false"
        android:drawable="@drawable/btn_check_off_disable" />
    <item android:drawable="@drawable/btn_check_off_disable" />
</selector>
```

程序运行后的默认效果如图 9.15 所示。

单击第 2 个 CheckBox 后出现 Checked 状态，如图 9.16 所示。

图 9.15　CheckBox 默认运行效果　　图 9.16　单击第 2 个 CheckBox 后 checked 的状态效果

3. 美化 RadionButton 控件

在 res/drawable-hdpi/文件夹下添加两个 RadioButton 状态背景图片，如图 9.17 所示。这两个背景图片的效果如图 9.18 所示。

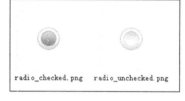

图 9.17　两个 RadioButton 背景图片　　图 9.18　两个 RadioButton 背景图片效果

继续在 res/drawable 文件夹下创建名称为 radiobutton_selector_drawable.xml 的 selector 配置文件，代码如下：

```xml
<?xml version="1.0" encoding="utf-8"?>
<selector xmlns:android="http://schemas.android.com/apk/res/android">
    <item android:state_checked="true" android:drawable="@drawable/radio_checked" />
    <item android:state_checked="false" android:drawable="@drawable/radio_unchecked" />
</selector>
```

更改 main.xml 布局文件，添加如下代码：

```xml
<RadioGroup android:layout_width="fill_parent"
    android:layout_height="wrap_content">
    <RadioButton android:text="我是默认的true" android:id="@+id/radioButton1"
        android:checked="true" android:layout_width="wrap_content"
        android:layout_height="wrap_content"></RadioButton>
    <RadioButton android:text="我是默认的false" android:id="@+id/radioButton2"
        android:checked="false" android:layout_width="wrap_content"
        android:layout_height="wrap_content"></RadioButton>
</RadioGroup>
<RadioGroup android:layout_width="fill_parent"
    android:layout_height="wrap_content">
    <RadioButton android:button="@drawable/radiobutton_selector_drawable"
```

```
                    android:text="我不是默认的true" android:id="@+id/radioButton3"
                    android:checked="true" android:layout_width="wrap_content"
                    android:layout_height="wrap_content"></RadioButton>
            <RadioButton android:button="@drawable/radiobutton_selector_drawable"
                    android:text="我不是默认的false" android:id="@+id/radioButton4"
                    android:checked="false" android:layout_width="wrap_content"
                    android:layout_height="wrap_content"></RadioButton>
    </RadioGroup>
```

程序运行后的 RadioButton 效果如图 9.19 所示。

图 9.19　美化后的 RadioButton 界面效果

4. 美化 EditText 控件

本节美化最后一个控件 EditText，本示例中将 EditText 控件分为两种情况来进行美化，分别是单行和多行。

首先美化单行 EditText 控件，在文件夹下添加两个单行 EditText 控件不同状态时的背景图片，如图 9.20 所示。

这两个单行 EditText 控件不同状态的背景图片界面，如图 9.21 所示。

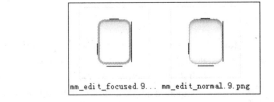

图 9.20　两个单行 EditText 控件不同状态背景图片　图 9.21　两个单行 EditText 控件不同状态背景图片界面效果

在 res/drawable 目录下创建名称为 onerow_edittext_selector_drawable.xml 的 selector 配置文件，代码如下：

```
<?xml version="1.0" encoding="utf-8"?>
<selector xmlns:android="http://schemas.android.com/apk/res/android">
    <item android:state_pressed="true" android:drawable="@drawable/mm_edit_focused" />
```

```
        <item android:state_focused="true" android:drawable="@drawable/mm_edit_focused" />
        <item android:drawable="@drawable/mm_edit_normal" />
</selector>
```

更改 main.xml 代码,添加如下代码:

```
<EditText android:layout_height="wrap_content" android:text="EditText"
    android:id="@+id/editText1" android:layout_width="match_parent"></EditText>
<EditText android:background="@drawable/onerow_edittext_selector_drawable"
    android:layout_height="wrap_content" android:text="EditText"
    android:id="@+id/editText2" android:layout_width="match_parent"></EditText>
```

程序默认运行效果如图 9.22 所示。

当 editText2 控件获得焦点后的背景图片变为如图 9.23 所示。

 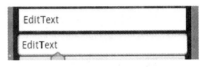

图 9.22 默认的 editText2 的美化为圆角边框   图 9.23 editText2 获得焦点后边框变为绿色

接下来是多行 EditText 控件的美化,添加两个多行 EditText 背景资源文件,如图 9.24 所示。这两个图片的显示效果如图 9.25 所示。

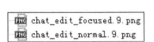 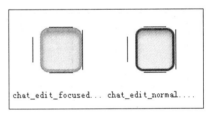

图 9.24 多行 EditText 背景资源文件   图 9.25 两个多行 EditText 背景资源显示效果

在 res/drawable/文件夹下创建名称为 morerow_edittext_selector_drawable.xml 的 selector 配置文件,代码如下:

```
<?xml version="1.0" encoding="utf-8"?>
<selector xmlns:android="http://schemas.android.com/apk/res/android">
    <item android:state_pressed="true" android:drawable="@drawable/chat_edit_focused" />
    <item android:state_focused="true" android:drawable="@drawable/chat_edit_focused" />
    <item android:drawable="@drawable/chat_edit_normal" />
</selector>
```

更改 main.xml 布局文件,添加如下代码:

```
<EditText android:layout_height="wrap_content" android:text="EditText"
    android:id="@+id/editText3" android:layout_width="match_parent"></EditText>
<EditText android:background="@drawable/morerow_edittext_selector_drawable"
    android:layout_height="wrap_content" android:text="EditText"
    android:id="@+id/editText4" android:layout_width="match_parent"
    android:lines="4" android:gravity="top|left"></EditText>
```

程序运行后,默认的 editText4 运行效果如图 9.26 所示。

当 editText4 控件获得焦点后的背景图片变为如图 9.27 所示。

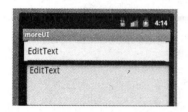
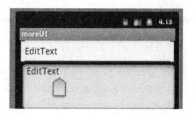

图 9.26　默认的 editText4 多行效果　　　图 9.27　editText4 控件获得焦点后边框为绿色

9.3.3　美化 Option 选项面板

在一些应用程序中经常有设置选项 Option 的列表界面，外观看起来非常美观，如图 9.28 所示。

本小节就从 0 开始美化选项 Option 列表。新建名称为 setupOptionUI 的 Android 项目，添加控件状态图片资源如图 9.29 所示。

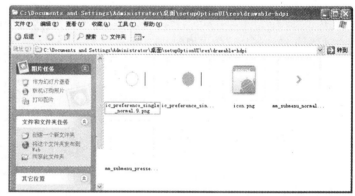

图 9.28　选项 Option 面板的界面外观　　　图 9.29　选项 Option 面板的美化图片

这些图片资源在项目中的列表如图 9.30 所示。

在/res 文件夹下创建名称为 drawable 的子文件夹，如图 9.31 所示。

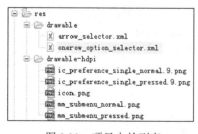
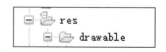

图 9.30　项目中的列表　　　图 9.31　创建 drawable 子文件夹

在 drawable 文件夹中创建名称为 arrow_selector.xml 的 XML 文件，目的是当按下选项列表其中的条目时，使当前条目右边的小箭头变成白色，代码如下：

```
<?xml version="1.0" encoding="utf-8"?>
<selector xmlns:android="http://schemas.android.com/apk/res/android">
```

```xml
    <item android:state_pressed="true" android:drawable="@drawable/mm_submenu_pressed" />
    <item android:drawable="@drawable/mm_submenu_normal" />
</selector>
```

选项 Option 列表是圆角的矩形，圆角的效果是使用图片来进行渲染的，所以还要在 drawable 文件夹中创建 onerow_option_selector.xml 配置文件，这个配置文件是定义每个条目圆角的效果及按下后的效果，代码如下：

```xml
<?xml version="1.0" encoding="utf-8"?>
<selector xmlns:android="http://schemas.android.com/apk/res/android">
    <item android:drawable="@drawable/ic_preference_single_pressed"
        android:state_enabled="true" android:state_pressed="true" />
    <item android:drawable="@drawable/ic_preference_single_normal"
        android:state_enabled="true" android:state_selected="true" />
    <item android:drawable="@drawable/ic_preference_single_normal" />
</selector>
```

在文件夹 values 中创建 color.xml 配置文件，定义文本的字体颜色，color.xml 配置文件的代码如下：

```xml
<?xml version="1.0" encoding="utf-8"?>
<resources>
    <color name="blackText">#000000</color>
</resources>
```

还要在 values 文件夹中创建 style.xml 配置文件，主要对条目进行外边距、高度、背景和对齐等属性的设置，style.xml 样式文件代码如下：

```xml
<?xml version="1.0" encoding="utf-8"?>
<resources>
    <style name="onerow_option_height_style">
        <item name="android:layout_height">35dip</item>
    </style>

    <style name="MainStyle">
        <item name="android:background">#ffd1d1d1</item>
        <item name="android:padding">6dip</item>
    </style>

    <style name="onerow_option_style" parent="@style/onerow_option_height_style">
        <item name="android:focusable">true</item>
        <item name="android:clickable">true</item>
        <item name="android:layout_width">fill_parent</item>
        <item name="android:gravity">left|center_vertical</item>
    </style>
</resources>
```

主要的布局文件 main.xml 代码如下，其中 android:duplicateParentState 属性的作用是复制父节点的状态，也就是同时响应父标签的状态：

```xml
<?xml version="1.0" encoding="utf-8"?>
```

```xml
<ScrollView xmlns:android="http://schemas.android.com/apk/res/android"
    android:id="@+id/scrollView1" android:layout_width="wrap_content"
    android:layout_height="match_parent">
    <LinearLayout android:orientation="vertical"
        android:layout_width="fill_parent" android:layout_height="fill_parent"
        style="@style/MainStyle">
        <LinearLayout android:background="@drawable/onerow_option_selector"
            android:id="@+id/linearLayout1" android:orientation="horizontal"
            style="@style/onerow_option_style">
            <TextView android:text="个性签名" android:id="@+id/textView1"
                android:layout_width="wrap_content" android:layout_height="wrap_content"
                android:textColor="@color/blackText" android:layout_weight="1"></TextView>
            <ImageView android:id="@+id/imageView1" android:src="@drawable/arrow_selector"
                android:layout_width="wrap_content" android:layout_height="wrap_content"
                android:duplicateParentState="true"></ImageView>
        </LinearLayout>
        <View android:layout_width="fill_parent" android:layout_height="10dip"></View>
        <LinearLayout android:background="@drawable/onerow_option_selector"
            android:id="@+id/linearLayout1" android:orientation="horizontal"
            style="@style/onerow_option_style">
            <TextView android:text="个性签名" android:id="@+id/textView1"
                android:layout_width="wrap_content" android:layout_height="wrap_content"
                android:textColor="@color/blackText" android:layout_weight="1"></TextView>
            <ImageView android:id="@+id/imageView1" android:src="@drawable/arrow_selector"
                android:layout_width="wrap_content" android:layout_height="wrap_content"
                android:duplicateParentState="true"></ImageView>
        </LinearLayout>
        <LinearLayout android:background="@drawable/onerow_option_selector"
            android:id="@+id/linearLayout1" android:orientation="horizontal"
            style="@style/onerow_option_style">
            <TextView android:text="个性签名" android:id="@+id/textView1"
                android:layout_width="wrap_content" android:layout_height="wrap_content"
                android:textColor="@color/blackText" android:layout_weight="1"></TextView>
            <ImageView android:id="@+id/imageView1" android:src="@drawable/arrow_selector"
                android:layout_width="wrap_content" android:layout_height="wrap_content"
                android:duplicateParentState="true"></ImageView>
        </LinearLayout>
    </LinearLayout>
</ScrollView>
```

程序运行后的效果如图 9.32 所示。

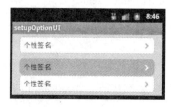

图 9.32 程序运行效果

在单击其中的条目时将改变背景图片及右边箭头的样式。

9.3.4　美化 ListView 控件

有了美化前面 Option 选项列表控件的经验后再美化 ListView 控件相对来讲就比较简单了，本节的实验运行效果如图 9.33 所示。

创建名称为 ListViewUI 的 Android 项目，完整的项目文件结构如图 9.34 所示。

图 9.33　ListView 美化后的效果

图 9.34　完整的项目文件结构

文件 ItemEntity.java 是代表每个列表条目内容信息的实体，代码如下：

```
package entity;

public class ItemEntity {

    private String title;

    public ItemEntity(String title) {
        super();
        this.title = title;
    }

    public String getTitle() {
        return title;
    }

    public void setTitle(String title) {
        this.title = title;
    }
}
```

}

由十本节 ListView 中的每一个列表条目的布局是自定义的，所以必须创建一个 Adapter 适配器类 ListViewAdapter.java，代码如下：

```java
public class ListViewAdapter extends BaseAdapter {

    private List<ItemEntity> itemList = new ArrayList();
    private Context context;

    public ListViewAdapter(Context context, List<ItemEntity> itemList) {
        super();
        this.itemList = itemList;
        this.context = context;
    }

    public int getCount() {
        return itemList.size();
    }

    public Object getItem(int arg0) {
        return null;
    }

    public long getItemId(int arg0) {
        return 0;
    }

    public View getView(int arg0, View arg1, ViewGroup arg2) {
        View view = ((Activity) context).getLayoutInflater().inflate(
                R.layout.listitem, null);
        LinearLayout linearLayout1 = (LinearLayout) view
                .findViewById(R.id.linearLayout1);
        linearLayout1.setBackgroundDrawable(context.getResources().getDrawable(
                R.drawable.preference_item));

        TextView textView1 = (TextView) view.findViewById(R.id.textView1);
        textView1.setText(itemList.get(arg0).getTitle());

        return view;
    }
}
```

有了条目信息的实体和自定义的 Adapter 适配器类，还需要在 Activity 类中进行组合，Main.java 的代码如下：

```java
public class Main extends Activity {
    private ListView listView1;

    @Override
```

```java
public void onCreate(Bundle savedInstanceState) {
    super.onCreate(savedInstanceState);
    setContentView(R.layout.main);

    List<ItemEntity> itemList = new ArrayList();
    for (int i = 0; i < 20; i++) {
        itemList.add(new ItemEntity("我是标题" + (i + 1)));
    }

    listView1 = (ListView) this.findViewById(R.id.listView1);
    listView1.setAdapter(new ListViewAdapter(this, itemList));
}
```

上面列举的是 Java 文件相关的代码，下面开始 XML 配置文件的代码。

在 res 文件夹中创建 drawable 子文件夹，在 drawable 文件夹中创建 XML 配置文件 arrow_selector.xml，功能是按下 ListView 中的条目时，将当前条目右边箭头改变样式，代码如下：

```xml
<?xml version="1.0" encoding="utf-8"?>
<selector xmlns:android="http://schemas.android.com/apk/res/android">
    <item android:state_pressed="true" android:drawable="@drawable/mm_submenu_pressed" />
    <item android:drawable="@drawable/mm_submenu_normal" />
</selector>
```

还要在 res/drawable 文件夹中创建 XML 配置文件 preference_item.xml，功能是按下 ListView 中的条目时改变当前条目的状态，代码如下：

```xml
<?xml version="1.0" encoding="utf-8"?>
<selector xmlns:android="http://schemas.android.com/apk/res/android">
    <item android:state_enabled="true" android:state_selected="true"
        android:drawable="@drawable/ic_preference_pressed" />
    <item android:state_enabled="true" android:state_pressed="true"
        android:drawable="@drawable/ic_preference_pressed" />
    <item android:drawable="@drawable/ic_preference_normal" />
</selector>
```

由于 ListView 中的每一个条目布局是自定义的，所以在 layout 文件夹中创建布局文件 listitem.xml，代码如下：

```xml
<?xml version="1.0" encoding="utf-8"?>
<LinearLayout android:id="@+id/linearLayout1"
    xmlns:android="http://schemas.android.com/apk/res/android"
    android:orientation="horizontal" android:layout_width="fill_parent"
    android:layout_height="wrap_content" android:padding="15dip">
    <TextView android:paddingLeft="8dip" android:layout_weight="1"
        android:text="TextView" android:id="@+id/textView1"
        android:layout_width="fill_parent" android:layout_height="50dip"
        android:gravity="left|center_vertical" android:textColor="#000000"
        android:textSize="16dip"></TextView>
    <ImageView android:id="@+id/imageView1"
```

```
            android:layout_height="fill_parent" android:layout_width="wrap_content"
            android:src="@drawable/arrow_selector" android:layout_gravity="center"
            android:duplicateParentState="true"> </ImageView>
</LinearLayout>
```

 提示　在自定义布局 XML 文件中想要实现改变 LinearLayout 的高度就可以改变 ListView 中条目的高度时，必须要在自定义的布局文件中以双层 LinearLayout 嵌套的方式来进行控件布局，并且在内部的 LinearLayout 控件设置高度即可。

主布局文件 main.xml 代码如下：

```
<?xml version="1.0" encoding="utf-8"?>
<LinearLayout xmlns:android="http://schemas.android.com/apk/res/android"
    android:orientation="vertical" android:layout_width="fill_parent"
    android:layout_height="fill_parent">
    <ListView android:id="@+id/listView1" android:fadingEdge="none"
        android:layout_width="fill_parent" android:layout_height="fill_parent"
        android:listSelector="#00000000" android:divider="#00000000"
        android:dividerHeight="0.0px"></ListView>
</LinearLayout>
```

代码 android:fadingEdge="none" 的功能是使 ListView 上面和下面的阴影去掉，而代码 android:listSelector="#00000000" 使 ListView 自带的黄色背景变为透明。

程序运行后就获得图 9.33 所示的效果。

9.3.5　美化 TabHost 控件

控件 TabHost 在开发项目时经常使用到，所以掌握美化 TabHost 是学习 Android 项目开发的必经之路。

新建名称为 tabhostTest 的 Android 项目。程序运行的外观效果如图 9.35 所示。其中 tab 导航条的效果如图 9.36 所示。

图 9.35　程序运行效果

图 9.36　tab 导航条

需要使用如图 9.37 所示的图片资源来实现美化。

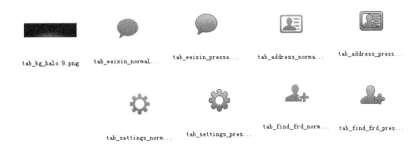

图 9.37 tab 导航需要的图片资源

单击每个 tab 导航中的按钮后有一个绿色椭圆背景，使用的图片资源如图 9.38 所示。
导航 tabHost 的背景是黑色的，使用的图片资源如图 9.39 所示。
全部的图片资源在 res/ drawable-hdpi 文件夹中，列表如图 9.40 所示。

图 9.38 绿色椭圆图片资源

图 9.39 背景为黑色的图片资源

图 9.40 全部的图片资源

有了这么多的图片资源，就需要有相对应的 selector 对象，在 res/drawable 文件夹中创建 5 个 selector 选择器 XML 配置文件。

文件 radiobutton1state.xml 的代码如下：

```
<?xml version="1.0" encoding="utf-8"?>
<selector xmlns:android="http://schemas.android.com/apk/res/android"
    android:constantSize="true" android:dither="true"
    android:variablePadding="true">
    <item android:state_checked="true" android:drawable="@drawable/tab_weixin_pressed" />
    <item android:drawable="@drawable/tab_weixin_normal" />
</selector>
```

文件 radiobutton2state.xml 的代码如下：

```
<?xml version="1.0" encoding="utf-8"?>
<selector xmlns:android="http://schemas.android.com/apk/res/android"
    android:constantSize="true" android:dither="true"
    android:variablePadding="true">
    <item android:state_checked="true" android:drawable="@drawable/tab_address_pressed" />
    <item android:drawable="@drawable/tab_address_normal" />
</selector>
```

文件 radiobutton3state.xml 的代码如下：

```xml
<?xml version="1.0" encoding="utf-8"?>
<selector xmlns:android="http://schemas.android.com/apk/res/android"
    android:constantSize="true" android:dither="true"
    android:variablePadding="true">
    <item android:state_checked="true" android:drawable="@drawable/tab_settings_pressed" />
    <item android:drawable="@drawable/tab_settings_normal" />
</selector>
```

文件 radiobutton4state.xml 的代码如下：

```xml
<?xml version="1.0" encoding="utf-8"?>
<selector xmlns:android="http://schemas.android.com/apk/res/android"
    android:constantSize="true" android:dither="true"
    android:variablePadding="true">
    <item android:state_checked="true" android:drawable="@drawable/tab_find_frd_pressed" />
    <item android:drawable="@drawable/tab_find_frd_normal" />
</selector>
```

文件 radiobuttonbackgroundstate.xml 的主要作用是单击 RadioButton 时切换背景效果，代码如下：

```xml
<?xml version="1.0" encoding="utf-8"?>
<selector xmlns:android="http://schemas.android.com/apk/res/android"
    android:constantSize="true" android:dither="true"
    android:variablePadding="true">
    <item android:state_checked="true" android:drawable="@drawable/tab_bg_halo" />
    <item android:drawable="@drawable/mm_trans" />
</selector>
```

在 values 目录下创建一个名称为 styles.xml 的样式文件，该文件主要是对 RadioButton 按钮添加外观样式，代码如下：

```xml
<?xml version="1.0" encoding="utf-8"?>
<resources>
    <style name="radiobuttonstyle">
        <item name="android:layout_weight">1</item>
        <item name="android:button">@null</item>
        <item name="android:textSize">13dip</item>
        <item name="android:gravity">center_horizontal</item>
        <item name="android:textColor">@color/radiobuttontextcolorstate</item>
    </style>
</resources>
```

控件 TabHost 每一页对应一个 XML 布局文件，代码如图 9.41 所示。

UI 控件的美化与动画 第 9 章

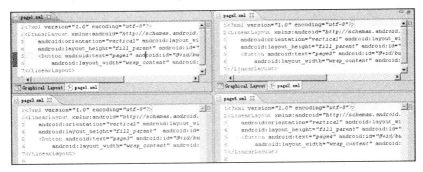

图 9.41 每个布局 XML 代码

最重要的 main.xml 布局文件的代码如下：

```xml
<?xml version="1.0" encoding="utf-8"?>
<LinearLayout xmlns:android="http://schemas.android.com/apk/res/android"
    android:orientation="vertical" android:layout_width="fill_parent"
    android:layout_height="fill_parent">
    <TabHost android:id="@android:id/tabhost" android:layout_width="match_parent"
        android:layout_height="match_parent">
        <LinearLayout android:layout_width="match_parent"
            android:id="@+id/linearLayout1" android:layout_height="match_parent"
            android:orientation="vertical">
            <TabWidget android:visibility="gone" android:layout_width="match_parent"
                android:layout_height="wrap_content" android:id="@android:id/tabs"></TabWidget>
            <FrameLayout android:layout_width="match_parent"
                android:layout_weight="1" android:layout_height="match_parent"
                android:id="@android:id/tabcontent">
                <LinearLayout android:layout_width="match_parent"
                    android:layout_height="match_parent"
                    android:id="@+id/contentlayout"></LinearLayout>
            </FrameLayout>
            <LinearLayout android:background="@drawable/mmfooter_bg"
                android:orientation="vertical" android:layout_width="fill_parent"
                android:layout_height="wrap_content">
                <RadioGroup android:orientation="horizontal"
                    android:layout_width="match_parent" android:id="@+id/radioGroup1"
                    android:layout_height="wrap_content">
                    <RadioButton style="@style/radiobuttonstyle"
                        android:background="@drawable/radiobuttonbackgroundstate"
                        android:drawableTop="@drawable/radiobutton1state"
                        android:layout_width="wrap_content" android:id="@+id/radio1"
                        android:layout_height="wrap_content" android:text="高洪岩1"
                        android:checked="true"></RadioButton>
                    <RadioButton style="@style/radiobuttonstyle"
                        android:background="@drawable/radiobuttonbackgroundstate"
                        android:drawableTop="@drawable/radiobutton2state"
                        android:layout_width="wrap_content" android:id="@+id/radio2"
                        android:layout_height="wrap_content"         android:text="高洪岩2"></RadioButton>
                    <RadioButton style="@style/radiobuttonstyle"
```

```xml
                        android:background="@drawable/radiobuttonbackgroundstate"
                        android:drawableTop="@drawable/radiobutton3state"
                        android:layout_width="wrap_content" android:id="@+id/radio3"
                        android:layout_height="wrap_content"         android:text=" 高 洪 岩 3"></RadioButton>
                    <RadioButton style="@style/radiobuttonstyle"
                        android:background="@drawable/radiobuttonbackgroundstate"
                        android:drawableTop="@drawable/radiobutton4state"
                        android:layout_width="wrap_content" android:id="@+id/radio4"
                        android:layout_height="wrap_content"         android:text=" 高 洪 岩 4"></RadioButton>
                </RadioGroup>
            </LinearLayout>
        </LinearLayout>
    </TabHost>
</LinearLayout>
```

还有名称为 Main.java 的 Activity 对象文件，代码如下：

```java
public class Main extends TabActivity {

    private TabHost tabHostRef;
    private RadioButton radioButton1;
    private RadioButton radioButton2;
    private RadioButton radioButton3;
    private RadioButton radioButton4;

    private RadioGroup radioGroup;

    @Override
    public void onCreate(Bundle savedInstanceState) {
        super.onCreate(savedInstanceState);
        setContentView(R.layout.main);

        radioGroup = (RadioGroup) this.findViewById(R.id.radioGroup1);
        radioGroup.setOnCheckedChangeListener(new OnCheckedChangeListener() {
            public void onCheckedChanged(RadioGroup arg0, int arg1) {
                switch (arg1) {
                case R.id.radio1:
                    tabHostRef.setCurrentTab(0);
                    break;
                case R.id.radio2:
                    tabHostRef.setCurrentTab(1);
                    break;
                case R.id.radio3:
                    tabHostRef.setCurrentTab(2);
                    break;
                case R.id.radio4:
                    tabHostRef.setCurrentTab(3);
                    break;
                }
```

```
        }
    });

    radioButton1 = (RadioButton) this.findViewById(R.id.radio1);
    radioButton2 = (RadioButton) this.findViewById(R.id.radio2);
    radioButton3 = (RadioButton) this.findViewById(R.id.radio3);
    radioButton4 = (RadioButton) this.findViewById(R.id.radio4);

    tabHostRef = this.getTabHost();

    LayoutInflater.from(this).inflate(R.layout.page1,
            tabHostRef.getTabContentView());
    LayoutInflater.from(this).inflate(R.layout.page2,
            tabHostRef.getTabContentView());
    LayoutInflater.from(this).inflate(R.layout.page3,
            tabHostRef.getTabContentView());
    LayoutInflater.from(this).inflate(R.layout.page4,
            tabHostRef.getTabContentView());

    tabHostRef.addTab(tabHostRef.newTabSpec("第一页").setContent(R.id.page1)
            .setIndicator("第一页"));
    tabHostRef.addTab(tabHostRef.newTabSpec("第二页").setContent(R.id.page2)
            .setIndicator("第二页"));
    tabHostRef.addTab(tabHostRef.newTabSpec("第三页").setContent(R.id.page3)
            .setIndicator("第三页"));
    tabHostRef.addTab(tabHostRef.newTabSpec("第四页").setContent(R.id.page4)
            .setIndicator("第四页"));

    tabHostRef.setCurrentTab(2);
    radioButton3.setChecked(true);

    }
}
```

程序运行的结果如图 9.35 所示。

9.3.6 美化 RadioGroup 组件

在 Android 项目中，导航的互斥效果大多使用 RadioGroup 组件来实现，本示例的运行效果如图 9.42 所示。

新建名称为 radioGroupUI 的 Android 项目，准备图片资源如图 9.43 所示。

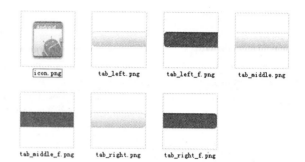

图 9.42 导航效果的示例　　　　　图 9.43 本示例使用的图片资源

由于导航控件在左右方向有圆角的效果，所以分别要创建左、中、右方向的 selector 文件。
在 res/ drawable 文件夹中创建文件 eachradiobuttonuileft.xml，代码如下：

```
<?xml version="1.0" encoding="utf-8"?>
<selector xmlns:android="http://schemas.android.com/apk/res/android"
    android:constantSize="true" android:dither="true"
    android:variablePadding="true">
    <item android:drawable="@drawable/tab_left_f"
        android:state_checked="true">
    </item>
    <item android:drawable="@drawable/tab_left" android:state_checked="false">
    </item>
    <item android:drawable="@drawable/tab_left">
    </item>
</selector>
```

在 res/ drawable 文件夹中创建文件 eachradiobuttonuiright.xml，代码如下：

```
<?xml version="1.0" encoding="utf-8"?>
<selector xmlns:android="http://schemas.android.com/apk/res/android"
    android:constantSize="true" android:dither="true"
    android:variablePadding="true">
    <item android:drawable="@drawable/tab_right_f"
        android:state_checked="true">
    </item>
    <item android:drawable="@drawable/tab_right"
        android:state_checked="false">
    </item>
    <item android:drawable="@drawable/tab_right">
    </item>
</selector>
```

在 res/ drawable 文件夹中创建文件 eachradiobuttonuimiddle.xml，代码如下：

```
<?xml version="1.0" encoding="utf-8"?>
<selector xmlns:android="http://schemas.android.com/apk/res/android"
    android:constantSize="true" android:dither="true"
    android:variablePadding="true">
    <item android:drawable="@drawable/tab_middle_f"
        android:state_checked="true">
```

```xml
        </item>
        <item android:drawable="@drawable/tab_middle"
            android:state_checked="false">
        </item>
        <item android:drawable="@drawable/tab_middle">
        </item>
</selector>
```

在 res/color 中创建文字 selector 效果配置文件 eachradiobuttontextcolor.xml，代码如下：

```xml
<?xml version="1.0" encoding="utf-8"?>
<selector xmlns:android="http://schemas.android.com/apk/res/android"
    android:constantSize="true" android:dither="true"
    android:variablePadding="true">
    <item android:color="#ffffff" android:state_checked="true">
    </item>
    <item android:color="#000000" android:state_checked="false">
    </item>
    <item android:color="#000000">
    </item>
</selector>
```

文件 main.xml 的代码如下：

```xml
<?xml version="1.0" encoding="utf-8"?>
<LinearLayout xmlns:android="http://schemas.android.com/apk/res/android"
    android:orientation="vertical" android:layout_width="fill_parent"
    android:layout_height="fill_parent">
    <RadioGroup android:id="@+id/radioGroup1"
        android:orientation="horizontal" android:layout_height="30dip"
        android:layout_width="match_parent">
        <RadioButton android:textSize="13dip"
            android:textColor="@color/eachradiobuttontextcolor"
            android:background="@drawable/eachradiobuttonuileft" android:button="@null"
            android:gravity="center" android:layout_width="fill_parent"
            android:layout_weight="1" android:id="@+id/radio1" android:text="数码频道"
            android:layout_height="fill_parent" android:checked="true"></RadioButton>
        <RadioButton android:textSize="13dip"
            android:textColor="@color/eachradiobuttontextcolor"
            android:background="@drawable/eachradiobuttonuimiddle"
            android:button="@null" android:gravity="center" android:layout_width="fill_parent"
            android:layout_weight="1" android:checked="true" android:id="@+id/radio2"
            android:text="汽车频道" android:layout_height="fill_parent"></RadioButton>
        <RadioButton android:textSize="13dip"
            android:textColor="@color/eachradiobuttontextcolor"
            android:background="@drawable/eachradiobuttonuimiddle"
            android:button="@null" android:gravity="center" android:layout_width="fill_parent"
            android:layout_weight="1" android:id="@+id/radio3" android:text="电器频道"
            android:layout_height="fill_parent"></RadioButton>
        <RadioButton android:textSize="13dip"
            android:textColor="@color/eachradiobuttontextcolor"
            android:background="@drawable/eachradiobuttonuiright" android:button="@null"
```

```
                    android:gravity="center" android:layout_width="fill_parent"
                    android:layout_weight="1" android:id="@+id/radio4" android:text="手机频道"
                    android:layout_height="fill_parent"> </RadioButton>
            </RadioGroup>
</LinearLayout>
```

程序运行的效果和图 9.32 一样。

9.3.7 美化 ExpandableListView 组件

本示例使用 ExpandableListView 控件并对其进行美化，最终的美化效果如图 9.44 所示。用到的图片资源如图 9.45 所示。

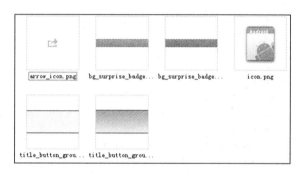

图 9.44　美化效果　　　　　　　　图 9.45　用到的图片资源

新建名称为 eList 的 Android 项目，再创建 Sheng.java 实体类，代码如下：

```
public class Sheng {

    private String id;
    private String shengName;
    private List<Shi> shiList = new ArrayList<Shi>();

    public Sheng(String id, String shengName, List<Shi> shiList) {
        super();
        this.id = id;
        this.shengName = shengName;
        this.shiList = shiList;
    }
    //get set 方法省略
}
```

创建 Shi.java 实体类，代码如下：

```
public class Shi {
```

```
    private String id;

    public Shi(String id, String shiName) {
        super();
        this.id = id;
        this.shiName = shiName;
    }
    //get set 方法省略
}
```

创建自定义适配器 MyBaseExpandableListAdapter.java，代码如下：

```
public class MyBaseExpandableListAdapter extends BaseExpandableListAdapter {

    private Context context;
    private List<Sheng> shengList = new ArrayList<Sheng>();

    public MyBaseExpandableListAdapter(Context context, List<Sheng> shengList) {
        super();
        this.context = context;
        this.shengList = shengList;
    }

    public Object getChild(int arg0, int arg1) {
        return null;
    }

    public long getChildId(int arg0, int arg1) {
        return 0;
    }

    public View getChildView(int arg0, int arg1, boolean arg2, View arg3,
            ViewGroup arg4) {
        View view = ((Activity) context).getLayoutInflater().inflate(
                R.layout.shilayout, null);
        TextView textView1 = (TextView) view.findViewById(R.id.textView1);
        textView1.setText(""
                + shengList.get(arg0).getShiList().get(arg1).getShiName());
        return view;
    }

    public int getChildrenCount(int arg0) {
        return shengList.get(arg0).getShiList().size();
    }

    public Object getGroup(int arg0) {
        return null;
    }

    public int getGroupCount() {
        return shengList.size();
```

```java
    }

    public long getGroupId(int arg0) {
        return 0;
    }

    public View getGroupView(int arg0, boolean arg1, View arg2, ViewGroup arg3) {
        View view = ((Activity) context).getLayoutInflater().inflate(
                R.layout.shenglayout, null);
        TextView textView1 = (TextView) view.findViewById(R.id.textView1);
        textView1.setText("" + shengList.get(arg0).getShengName());
        return view;
    }

    public boolean hasStableIds() {
        return false;
    }

    public boolean isChildSelectable(int arg0, int arg1) {
        // 返回真，让子元素可以选中
        return true;
    }

}
```

Activity 文件 Main.java 核心代码如下：

```java
public class Main extends Activity {
    private ExpandableListView expandableListView;

    @Override
    public void onCreate(Bundle savedInstanceState) {
        super.onCreate(savedInstanceState);
        this.requestWindowFeature(Window.FEATURE_NO_TITLE);
        setContentView(R.layout.main);

        expandableListView = (ExpandableListView) this
                .findViewById(R.id.expandableListView1);
        expandableListView.setGroupIndicator(null);

        // 省 1 及市列表--开始
        Shi sheng1_shi1 = new Shi("101", "省 1 市 1");
        Shi sheng1_shi2 = new Shi("102", "省 1 市 2");
        Shi sheng1_shi3 = new Shi("103", "省 1 市 3");
        Shi sheng1_shi4 = new Shi("104", "省 1 市 4");

        List<Shi> sheng1ShiList = new ArrayList<Shi>();
        sheng1ShiList.add(sheng1_shi1);
        sheng1ShiList.add(sheng1_shi2);
        sheng1ShiList.add(sheng1_shi3);
```

```java
sheng1ShiList.add(sheng1_shi4);

Sheng sheng1 = new Sheng("100", "省 1", sheng1ShiList);
// 省 1 及市列表--结束

// 省 2 及市列表--开始
Shi sheng2_shi1 = new Shi("201", "省 2 市 1");
Shi sheng2_shi2 = new Shi("202", "省 2 市 2");
Shi sheng2_shi3 = new Shi("203", "省 2 市 3");
Shi sheng2_shi4 = new Shi("204", "省 2 市 4");

List<Shi> sheng2ShiList = new ArrayList<Shi>();
sheng2ShiList.add(sheng2_shi1);
sheng2ShiList.add(sheng2_shi2);

Sheng sheng2 = new Sheng("200", "省 2", sheng2ShiList);
// 省 2 及市列表--结束

// 省 3 及市列表--开始
Shi sheng3_shi1 = new Shi("301", "省 3 市 1");
Shi sheng3_shi2 = new Shi("302", "省 3 市 2");
Shi sheng3_shi3 = new Shi("303", "省 3 市 3");
Shi sheng3_shi4 = new Shi("304", "省 3 市 4");

List<Shi> sheng3ShiList = new ArrayList<Shi>();
sheng3ShiList.add(sheng3_shi1);
sheng3ShiList.add(sheng3_shi2);
sheng3ShiList.add(sheng3_shi3);
sheng3ShiList.add(sheng3_shi4);

Sheng sheng3 = new Sheng("300", "省 3", sheng3ShiList);
// 省 3 及市列表--结束

// 省 4 及市列表--开始
Shi sheng4_shi1 = new Shi("401", "省 4 市 1");
Shi sheng4_shi2 = new Shi("402", "省 4 市 2");
Shi sheng4_shi3 = new Shi("403", "省 4 市 3");
Shi sheng4_shi4 = new Shi("404", "省 4 市 4");

List<Shi> sheng4ShiList = new ArrayList<Shi>();
sheng4ShiList.add(sheng4_shi1);
sheng4ShiList.add(sheng4_shi2);
sheng4ShiList.add(sheng4_shi3);
sheng4ShiList.add(sheng4_shi4);

Sheng sheng4 = new Sheng("400", "省 4", sheng4ShiList);
// 省 4 及市列表--结束

// 省 5 及市列表--开始
Shi sheng5_shi1 = new Shi("501", "省 5 市 1");
```

```java
            Shi sheng5_shi2 = new Shi("502", "省 5 市 2");
            Shi sheng5_shi3 = new Shi("503", "省 5 市 3");
            Shi sheng5_shi4 = new Shi("504", "省 5 市 4");
            Shi sheng5_shi5 = new Shi("505", "省 5 市 5");
            Shi sheng5_shi6 = new Shi("506", "省 5 市 6");
            Shi sheng5_shi7 = new Shi("507", "省 5 市 7");
            Shi sheng5_shi8 = new Shi("508", "省 5 市 8");

            List<Shi> sheng5ShiList = new ArrayList<Shi>();
            sheng5ShiList.add(sheng5_shi1);
            sheng5ShiList.add(sheng5_shi2);
            sheng5ShiList.add(sheng5_shi3);
            sheng5ShiList.add(sheng5_shi4);
            sheng5ShiList.add(sheng5_shi5);
            sheng5ShiList.add(sheng5_shi6);
            sheng5ShiList.add(sheng5_shi7);
            sheng5ShiList.add(sheng5_shi8);

            Sheng sheng5 = new Sheng("500", "省 5", sheng5ShiList);
            // 省 5 及市列表--结束

            final List<Sheng> shengList = new ArrayList<Sheng>();
            shengList.add(sheng1);
            shengList.add(sheng2);
            shengList.add(sheng3);
            shengList.add(sheng4);
            shengList.add(sheng5);

            MyBaseExpandableListAdapter adapter = new MyBaseExpandableListAdapter(
                    this, shengList);
            expandableListView.setAdapter(adapter);
            expandableListView.setOnChildClickListener(new OnChildClickListener() {
                public boolean onChildClick(ExpandableListView arg0, View arg1,
                        int arg2, int arg3, long arg4) {
                    Log.v("!", ""
                            + shengList.get(arg2).getShiList().get(arg3)
                                    .getShiName());
                    return false;
                }
            });

        }
    }
```

在 res/color 文件夹下创建"省"的文字状态配置文件 shengtextcolorstate.xml,代码如下:

```xml
<?xml version="1.0" encoding="utf-8"?>
<selector xmlns:android="http://schemas.android.com/apk/res/android"
    android:constantSize="false" android:dither="true"
    android:variablePadding="true">
    <item android:state_pressed="true" android:color="#000000" />
```

```xml
    <item android:color="#ffffff" />
</selector>
```

在 res/color 文件夹下创建"市"的文字状态配置文件 shitextcolorstate.xml，代码如下：

```xml
<?xml version="1.0" encoding="utf-8"?>
<selector xmlns:android="http://schemas.android.com/apk/res/android"
    android:constantSize="false" android:dither="true"
    android:variablePadding="true">
    <item android:state_pressed="true" android:color="#508ddb" />
    <item android:color="#508ddb" />
</selector>
```

在 res/drawable 文件夹下创建箭头状态列表文件 arrowdrawable.xml，代码如下：

```xml
<?xml version="1.0" encoding="utf-8"?>
<selector xmlns:android="http://schemas.android.com/apk/res/android"
    android:constantSize="false" android:dither="true"
    android:variablePadding="true">
    <item android:drawable="@drawable/arrow_icon" />
</selector>
```

在 res/drawable 文件夹下创建"省"状态列表文件 shengdrawable.xml，代码如下：

```xml
<?xml version="1.0" encoding="utf-8"?>
<selector xmlns:android="http://schemas.android.com/apk/res/android"
    android:constantSize="false" android:dither="true"
    android:variablePadding="true">
    <item android:state_pressed="true" android:drawable="@drawable/bg_surprise_badge_on" />
    <item android:drawable="@drawable/bg_surprise_badge_off" />
</selector>
```

在 res/drawable 文件夹下创建"市"状态列表文件 shidrawable.xml，代码如下：

```xml
<?xml version="1.0" encoding="utf-8"?>
<selector xmlns:android="http://schemas.android.com/apk/res/android"
    android:constantSize="false" android:dither="true"
    android:variablePadding="true">
    <item android:state_pressed="true" android:drawable="@drawable/title_button_group_middle_selected" />
    <item android:drawable="@drawable/title_button_group_middle_normal" />
</selector>
```

创建"省"布局文件 shenglayout.xml，代码如下：

```xml
<?xml version="1.0" encoding="utf-8"?>
<LinearLayout xmlns:android="http://schemas.android.com/apk/res/android"
    android:orientation="horizontal" android:layout_width="fill_parent"
    android:layout_height="30dip" android:background="@drawable/shengdrawable"
    android:padding="5dip" android:gravity="center_vertical">
    <ImageView android:layout_marginLeft="10dip"
        android:duplicateParentState="true" android:background="@drawable/arrowdrawable"
        android:layout_height="wrap_content" android:layout_width="wrap_content"
        android:id="@+id/imageView1"></ImageView>
    <TextView android:layout_marginLeft="10dip"
```

```
        android:duplicateParentState="true" android:textColor="@color/shengtextcolorstate"
        android:textSize="16dip" android:id="@+id/textView1"
        android:layout_width="wrap_content" android:layout_height="wrap_content"></TextView>
</LinearLayout>
```

创建"市"布局文件 shilayout.xml，代码如下：

```
<?xml version="1.0" encoding="utf-8"?>
<LinearLayout xmlns:android="http://schemas.android.com/apk/res/android"
    android:orientation="horizontal" android:layout_width="fill_parent"
    android:background="@drawable/shidrawable" android:layout_height="25dip"
    android:padding="5dip" android:gravity="center_vertical">
    <TextView android:duplicateParentState="true"
        android:textColor="@color/shitextcolorstate"
        android:layout_marginLeft="10dip" android:id="@+id/textView1"
        android:textSize="16dip" android:layout_width="wrap_content"
        android:layout_height="wrap_content"></TextView>
</LinearLayout>
```

到此代码介绍完毕，如果代码没有出错，就可以看到运行结果了。

9.4 动画

在 Android 中主要有两种动画表现方式：补间动画 Tween Animation 和逐帧动画 Frame Animation。由于逐帧动画主要用于游戏开发，而应用程序的开发用得比较少，所以本章主要介绍补间动画。

补间动画 Tween Animation 是 Android 中表现动画的主要方式，例如对控件添加动画、在 Activity 之间切换时添加动画等情况，Tween Animation 动画的 XML 使用语法如下：

```
<?xml version="1.0" encoding="utf-8"?>
<set xmlns:android="http://schemas.android.com/apk/res/android">
    android:interpolator="@[package:]anim/interpolator_resource"
    android:shareInterpolator=[" true " false ">
    <alpha android:fromAlpha="float" android:toAlpha="float" />
    <scale android:fromXScale="float" android:toXScale="float"
        android:fromYScale="float" android:toYScale="float" android:pivotX="float"
        android:pivotY="float" />
    <translate android:fromX="float" android:toX="float"
        android:fromY="float" android:toY="float" />
    <rotate android:fromDegrees="float" android:toDegrees="float"
        android:pivotX="float" android:pivotY="float" />
    <set> ...   </set>
</set>
```

从 XML 配置文件中可以看到，补间动画可以支持 alpha 透明、scale 缩放、translate 移动和 rotate 旋转等，后面的内容将会一一介绍它们的使用。

9.4.1 alpha 透明动画演示

新建名称为 anim_1 的 Android 项目，在 res/anim 文件夹下创建名称为 anim1.xml 的动画文件，它的作用是使 Button 按钮及名称为 Second 的 Activity 以渐现的方式显示出来，代码如下：

```xml
<?xml version="1.0" encoding="utf-8"?>
<set xmlns:android="http://schemas.android.com/apk/res/android">
    <alpha android:fromAlpha="0" android:toAlpha="1"
        android:duration="12000" />
</set>
```

属性 android:fromAlpha 的含义是 alpha 起始值，android:toAlpha 是终止值，它们的取值范围是 0.0 到 1.0 之间，属性 android:duration 是动画的持续时间，以毫秒为单位。

再创建名称为 exit.xml 的动画配置文件，它的作用是当名称为 Main 的 Activity 离开时播放的动画，代码如下：

```xml
<?xml version="1.0" encoding="utf-8"?>
<set xmlns:android="http://schemas.android.com/apk/res/android">
    <alpha android:fromAlpha="1" android:toAlpha="0"
        android:duration="10000" />
</set>
```

文件 Main.java 的代码如下：

```java
public class Main extends Activity {
    private Button button;

    @Override
    public void onCreate(Bundle savedInstanceState) {
        super.onCreate(savedInstanceState);
        setContentView(R.layout.main);

        button = (Button) this.findViewById(R.id.button1);
        button.startAnimation(AnimationUtils.loadAnimation(this, R.anim.anim1));

        button.setOnClickListener(new OnClickListener() {
            public void onClick(View arg0) {
                Intent intent = new Intent(Main.this, Second.class);
                Main.this.startActivity(intent);
                Main.this.overridePendingTransition(R.anim.anim1, R.anim.exit);
                //第 1 个参数：第二个 activity 进入时的动画
                //第 2 个参数：第一个 activity 退出时的动画
            }
        });
    }
}
```

程序运行后控件 Button 以 alpha 透明的方式呈现出由没有到有的效果，如图 9.46 所示。

再单击 Button 按钮时切换到 Second.java，这时显示 Activity 切换时的动画，如图 9.47 所示。

图 9.46　按钮从无到有的 alpha 效果

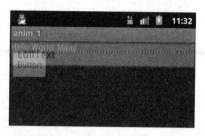
图 9.47　Activity 切换时播放动画

9.4.2　scale 缩放动画演示

新建名称为 anim_2 的 Android 项目，在 res/anim 文件夹下创建动画文件 anim1.xml，代码如下：

```xml
<?xml version="1.0" encoding="utf-8"?>
<set xmlns:android="http://schemas.android.com/apk/res/android">
    <scale android:fromXScale="0.0" android:toXScale="1.5"
        android:fromYScale="0" android:toYScale="1.5" android:pivotX="50%"
        android:pivotY="50%" android:duration="1000" />
</set>
```

属性的解释。

- android:fromXScale: 控件在动画执行开始的宽度起始值，0 为隐藏状态，1 为原始大小，大于 1 为放大的动画效果。
- android:toXScale: 控件在动画结束时的宽度终止值，0 为隐藏状态，1 为原始大小，大于 1 为放大的动画效果。
- android:fromYScale: 控件在动画执行开始的高度起始值，0 为隐藏状态，1 为原始大小，大于 1 为放大的动画效果。
- android:toYScale: 控件在动画结束时的高度终止值，0 为隐藏状态，1 为原始大小，大于 1 为放大的动画效果。
- android:pivotX="50%": 代表动画从控件宽度的 50%处开始。
- android:pivotY="50%": 代表动画从控件高度的 50%处开始。
- android:duration: 动画持续的时间，以毫秒为单位。

继续在 res/anim 文件夹下创建动画文件 input.xml，代表新的 Activity 显示时的切换动画，代码如下：

```xml
<?xml version="1.0" encoding="utf-8"?>
<set xmlns:android="http://schemas.android.com/apk/res/android">
    <scale android:fromXScale="0" android:toXScale="1"
        android:fromYScale="0" android:toYScale="1" android:pivotX="50%"
        android:pivotY="50%" android:duration="1000" />
</set>
```

还要在 res/anim 文件夹下创建动画文件 out.xml，代表旧的 Activity 退出时的切换动画，代码如下：

```xml
<?xml version="1.0" encoding="utf-8"?>
<set xmlns:android="http://schemas.android.com/apk/res/android">
    <scale android:fromXScale="1" android:toXScale="0"
        android:fromYScale="1" android:toYScale="0" android:pivotX="50%"
        android:pivotY="50%" android:duration="1000" />
</set>
```

核心 Activity 文件 Main.java 的代码如下：

```java
public class Main extends Activity {
    private Button button;

    @Override
    public void onCreate(Bundle savedInstanceState) {
        super.onCreate(savedInstanceState);
        setContentView(R.layout.main);

        button = (Button) this.findViewById(R.id.button1);
        Animation animRef = AnimationUtils.loadAnimation(this, R.anim.anim1);
        // animRef.setFillAfter(true);
        // 最终是否显示动画的最后一帧上
        // 这个参数要在 java 代码中进行设置
        button.startAnimation(animRef);
        button.setOnClickListener(new OnClickListener() {
            public void onClick(View arg0) {
                Intent intent = new Intent(Main.this, Second.class);
                Main.this.startActivity(intent);
                Main.this.overridePendingTransition(R.anim.input, R.anim.out);
            }
        });
    }
}
```

名称为 second.xml 的布局文件内容就是多个 EditText 控件，如图 9.48 所示。

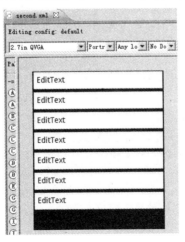

图 9.48　second.xml 布局中的内容

程序运行出来的效果就是 Button 由小到大，再还原到原始大小的动画效果，单击 Button 时，main.xml 布局在屏幕中间处由小变大，而 second.xml 布局在屏幕中间处由大变小的动画效果。

如果把 Main.java 文件中的代码注释去掉，即成为下面的情况：

animRef.setFillAfter(**true**);

则名称为 Main.java 的 Activity 显示出来时按钮的状态如图 9.49 所示。

图 9.49　fillAfter 为 true 时的效果

按钮停在了动画的最后一帧上，并没有还原原始的大小。

9.4.3　translate 移动动画演示

新建名称为 anim_3 的 Android 项目，在 res/anim 文件夹下创建名称为 anim1.xml 的动画文件，代码如下：

```
<?xml version="1.0" encoding="utf-8"?>
<set xmlns:android="http://schemas.android.com/apk/res/android">
    <translate android:fromXDelta="100%p" android:toXDelta="0%p"
        android:fromYDelta="100%p" android:toYDelta="0%p" android:duration="1000" />
</set>
```

属性解释。

- android:fromXDelta="100%p"：起始 x 轴座标，100%p 代表从屏幕最外面以 x 轴开始移动。
- android:toXDelta="0%p"：终止 x 轴座标，以 x 轴为座标，0%p 代表移动到屏幕最左边。
- android:fromYDelta="100%p"：起始 Y 轴座标，100%p 代表从屏幕最下面开始动画。
- android:toYDelta="0%p"：终止 Y 轴座标，0%p 代表屏幕最上面。
- android:duration="1000"：动画持续时间，毫秒为单位。

这个动画的效果是从屏幕的右下角移动到屏幕的左上角，再新创建一个 Activity 进入时的动画 XML 文件 input.xml，代码如下：

```
<?xml version="1.0" encoding="utf-8"?>
<set xmlns:android="http://schemas.android.com/apk/res/android">
    <translate android:fromXDelta="100%p" android:toXDelta="0%p"
        android:fromYDelta="0%p" android:toYDelta="0%p" android:duration="1000" />
</set>
```

再创建一个 Activity 退出时的动画 XML 文件 out.xml，代码如下：

```
<?xml version="1.0" encoding="utf-8"?>
<set xmlns:android="http://schemas.android.com/apk/res/android">
```

```xml
        <translate android:fromXDelta="0%p" android:toXDelta="100%p"
            android:fromYDelta="0%p" android:toYDelta="0%p" android:duration="1000" />
</set>
```

文件 Main.java 的代码如下：

```java
public class Main extends Activity {
    private Button button;

    @Override
    public void onCreate(Bundle savedInstanceState) {
        super.onCreate(savedInstanceState);
        setContentView(R.layout.main);
        button = (Button) this.findViewById(R.id.button1);
        Animation animRef = AnimationUtils.loadAnimation(this, R.anim.anim1);
        button.startAnimation(animRef);
        button.setOnClickListener(new OnClickListener() {
            public void onClick(View arg0) {
                Intent intent = new Intent(Main.this, Second.class);
                Main.this.startActivity(intent);
                Main.this.overridePendingTransition(R.anim.input, R.anim.out);
            }
        });
    }
}
```

程序运行后的效果就是 Button 按钮从右下角移动到左上角，单击 Button 按钮后，main.xml 布局以横向向右移出屏幕，而 second.xml 布局从右到左移进屏幕。

9.4.4 rotate 旋转动画演示

新建名称为 anim_4 的 Android 项目，在 res/anim 文件夹下创建名称为 anim1.xml 的动画文件，代码如下：

```xml
<?xml version="1.0" encoding="utf-8"?>
<set xmlns:android="http://schemas.android.com/apk/res/android">
    <rotate android:fromDegrees="0" android:toDegrees="720"
        android:pivotX="50%" android:pivotY="50%" android:duration="2000" />
</set>
```

属性解释。

- android:fromDegrees="0"：开始的角度。
- android:toDegrees="720"：结束的角度。
- android:pivotX="50%"：在控件宽度的百分比处开始动画。
- android:pivotY="50%"：在控件高度的百分比处开始动画。
- android:fromDegrees 值大于 android:toDegrees 就是逆时针旋转，反过来就是顺时针旋转。

继续创建 Activity 进入动画文件 input.xml，代码如下：

```xml
<?xml version="1.0" encoding="utf-8"?>
<set xmlns:android="http://schemas.android.com/apk/res/android">
    <rotate android:fromDegrees="180" android:toDegrees="0"
        android:pivotX="50%" android:pivotY="50%" android:duration="2000" />
</set>
```

动画的效果是从 180 度旋转为 0 度，是逆时针旋转。

继续创建 Activity 退出动画文件 out.xml，代码如下：

```xml
<?xml version="1.0" encoding="utf-8"?>
<set xmlns:android="http://schemas.android.com/apk/res/android">
    <rotate android:fromDegrees="0" android:toDegrees="90"
        android:pivotX="50%" android:pivotY="50%" android:duration="2000" />
</set>
```

动画的效果是用顺时针旋转 90º 进行动画退出。

9.4.5 动画中 Interpolators 的使用

Android 动画技术中的 Interpolators 指的是动画在某一种形态下的速率，也称为动画渲染器，比如在旋转状态下使用某一个 Interpolators 对象来设置这个旋转的方式为加速或减速，或实现一些具有反弹效果的旋转样式，这时就要使用 Interpolators 对象了。

使用它的方式是在动画配置文件 XML 中的<set>标签中设置 android:interpolator 属性值即可，示例如下：

```xml
<set xmlns:android="http://schemas.android.com/apk/res/android"
    android:interpolator="@android:anim/accelerate_decelerate_interpolator">
```

Android 中的 Interpolators 具有如图 9.50 所示的样式。

Interpolator class	Resource ID
AccelerateDecelerateInterpolator	@android:anim/accelerate_decelerate_interpolator
AccelerateInterpolator	@android:anim/accelerate_interpolator
AnticipateInterpolator	@android:anim/anticipate_interpolator
AnticipateOvershootInterpolator	@android:anim/anticipate_overshoot_interpolator
BounceInterpolator	@android:anim/bounce_interpolator
CycleInterpolator	@android:anim/cycle_interpolator
DecelerateInterpolator	@android:anim/decelerate_interpolator
LinearInterpolator	@android:anim/linear_interpolator
OvershootInterpolator	@android:anim/overshoot_interpolator

图 9.50　Android 系统自带的 Interpolators 对象

在名称为 anim_Interpolator 的项目中已经将这 9 个 Interpolators 的代码实现了，可以参看一下运行效果。

9.4.6 动画的混合应用演示

前面的示例都是将一种动画效果应用于 Button 控件上，本示例将把 4 种动画效果应用于 Button 控件上，实现的动画效果是 Button 按钮从右下角向左上角移动（translate），并且在移动的过程中旋转 3600°（rotate），到达指定位置后由小变大（scale），最后透明度变为 50%（alpha）。

新建名称为 anim_5 的 Android 动画实验项目，新建名称为 anim1.xml 的动画文件，效果是一边移动一边旋转，代码如下：

```xml
<?xml version="1.0" encoding="utf-8"?>
<set xmlns:android="http://schemas.android.com/apk/res/android">
    <rotate android:fromDegrees="3600" android:toDegrees="0"
        android:pivotX="50%" android:pivotY="50%" android:duration="5000" />
    <translate android:fromXDelta="100%p" android:toXDelta="0%p"
        android:fromYDelta="100%p" android:toYDelta="0%p" android:duration="5000" />
</set>
```

再新建动画文件 anim2.xml，效果是放大，代码如下：

```xml
<?xml version="1.0" encoding="utf-8"?>
<set xmlns:android="http://schemas.android.com/apk/res/android">
    <scale android:fromXScale="1" android:toXScale="1.5"
        android:fromYScale="1" android:toYScale="1.5" android:pivotX="50%"
        android:pivotY="50%" android:duration="1000" />
</set>
```

再新建动画文件 anim3.xml，效果是改变透明度为 50%，代码如下：

```xml
<?xml version="1.0" encoding="utf-8"?>
<set xmlns:android="http://schemas.android.com/apk/res/android">
    <alpha android:fromAlpha="1" android:toAlpha="0.5"
        android:duration="1000" />
</set>
```

文件 Main.java 的代码如下：

```java
public class Main extends Activity {
    private Button button;

    private AnimationListener animationListener1 = new AnimationListener() {
        public void onAnimationStart(Animation arg0) {
        }

        public void onAnimationRepeat(Animation arg0) {
        }

        public void onAnimationEnd(Animation arg0) {
            Animation animRef2 = AnimationUtils.loadAnimation(Main.this,
                    R.anim.anim2);
            animRef2.setAnimationListener(animationListener2);
            button.startAnimation(animRef2);
        }
    };

    private AnimationListener animationListener2 = new AnimationListener() {
        public void onAnimationStart(Animation arg0) {
        }
```

```java
            public void onAnimationRepeat(Animation arg0) {

            }

            public void onAnimationEnd(Animation arg0) {
                Animation animRef3 = AnimationUtils.loadAnimation(Main.this,
                        R.anim.anim3);
                animRef3.setFillAfter(true);
                button.startAnimation(animRef3);
            }
        };

        @Override
        public void onCreate(Bundle savedInstanceState) {
            super.onCreate(savedInstanceState);
            setContentView(R.layout.main);
            button = (Button) this.findViewById(R.id.button1);

            Animation animRef1 = AnimationUtils.loadAnimation(this, R.anim.anim1);
            animRef1.setAnimationListener(animationListener1);

            button.startAnimation(animRef1);

        }
    }
```

由于动画效果有执行的顺序问题，所以使用 AnimationListener 监听来判断动画是否结束，如果结束，则开始下一个动画效果。

第 10 章 Fragment 对象的使用

本章主要学习 Fragment 片段/碎片技术的使用，通过此技术可以掌握在 Android 中 UI 界面以模块组件式的开发，有利于软件结构的组织及软件后期升级的维护，现阶段越来越多的项目融入了 Fragment 技术，以创建结构更加合理的软件系统。

本章应该着重关注以下内容：

- 多个 Fragment 碎片之间的通信
- Fragment 与 Activity 之间的通信
- 如何用代码来切换不同的 Fragment 对象
- Fragment 的动态创建
- 使用 Fragment 实现 View 分页的效果

10.1 Fragment 对象简介

随着科技的发展，平板电脑以大屏幕、操作性优秀、硬件性能大幅提升等原因越来越受到市场的欢迎，那么不管是平板大屏幕电脑的开发，还是开发小屏幕的手机软件，几乎所有的软件公司都在遵循着 OOP 编程，目的是使业务代码能得到大幅的复用。但 Android 的知识直到学习到这里，仍然是一直用代码的方式来进行业务组件的复用，其实使 Android 界面组件复用也是非常重要的，比如开发一个点餐系统的界面，希望此界面也能在其他的项目中得到非常简单的复用，这时使用 Android3.0 中的 Fragment 技术就可以处理这种情况。

对象 Fragment 是 Android 3.0 新增的概念，它可以将界面 UI 进行分块，以"块"的方式组织 UI，完全可以达到界面 UI 组件的复用。Fragment 和 Activity 十分相似，Fragment 用来描述一些行为或一部分用户界面在一个 Activity 中，你可以合并多个 Fragment 在一个单独的 Activity 中建立选项卡面板，还可以同时重用 Fragment 在多个 Activity 中，可以把 Fragment 作为一个 Activity 中的一个模块。Fragment 有自己的生命周期，接收自己的输入事件，还可以在运行中的 Activity 对象中添加或移除它。一个 Fragment 必须总是嵌入在一个 Activity 中，同时 Fragment 的生命周期受 Activity 的影响。

10.2 Fragment 对象生命周期与事务

对象 Fragment 也有自己的生命周期与 UI 界面改变有关的事务 Fragment Transaction，在本节将介绍这两个非常重要的知识点。

10.2.1 Fragment 对象生命周期

需要说明的是，Fragment 是 Android 3.0 中的技术，那么如果使用 Android 2.x 版本就不能使用了吗？不会~!Google 已经开发出在 Android 1.6 到 Android 2.X 版本之间使用 Fragment 技术的兼容 jar 包，使用这个 jar 包就像在 Android SDK 3.x 中开发一样，这个 jar 包文件名称为 android-support-v4.jar，可以在 Android SDK 3.x 文件夹中找到它：

> android-sdk-windows\extras\android\support

在新建 Android 项目时把这个 jar 包添加到构建路径中就可以了。

在第 1 章学习过 Activity 对象的生命周期，关于生命周期的知识点在 Android 的学习中是非常重要的，因为学习它可以知道对象的生存状态，包括创建、运行、销毁各个阶段应该做哪些任务。

前面说过 Fragment 是放在 Activity 中运行的，是受 Activity 的后退栈管理，那么为了兼容小于 Android SDK 3.0 版本的 SDK，兼容包中提供了一个 Activity 的子类 FragmentActivity 来处理 Activity 对象中管理 Fragment 的问题。类结构如图 10.1 所示。

Fragment 对象的生命周期过程如图 10.2 所示。

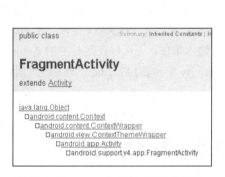

图 10.1 FragmentActivity 的继承关系

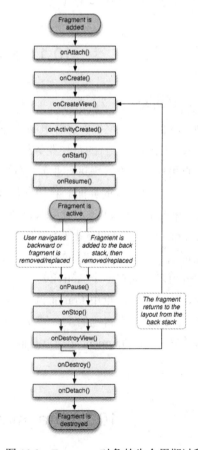

图 10.2 Fragment 对象的生命周期过程

从图 10.2 中可以看到，Fragment 对象有 11 个生命周期方法，这些回调方法的解释如下。

- onAttach 方法：当 1 个 Fragment 对象关联到 1 个 Activity 对象时调用。
- onCreate 方法：初始创建 Fragment 对象时调用。
- onCreateView 方法：创建与 Fragment 对象关联的 View 视图时调用。
- onActivityCreated 方法：当 Activity 对象完成自己的 onCreate()方法时调用。
- onStart 方法：Fragment 对象可以在 UI 可见时调用。
- onResume 方法：Fragment 对象的 UI 可以与用户交互时调用。
- onPause 方法：有控件以透明的方式遮挡，或 Activity 对象转为 onPause 状态时调用。
- onStop 方法：有控件以完全遮挡的方式，或 Activity 对象转为 onStop 状态时调用。
- onDestroyView 方法：Fragment 对象清理 View 资源时调用。
- onDestroy 方法：Fragment 对象完成对象清理 View 资源时调用。
- onDetach 方法：Fragment 对象没有与 Activity 对象关联时调用。

本节就用代码的方式来介绍 Activity 对象与 Fragment 对象生命周期的关系。新建名称为 fragmentlifecycle 的 Android 项目。新建布局文件 layout1.xml，代码如下：

```xml
<?xml version="1.0" encoding="utf-8"?>
<LinearLayout xmlns:android="http://schemas.android.com/apk/res/android"
    android:orientation="vertical" android:layout_width="fill_parent"
    android:layout_height="wrap_content" android:background="#ff0000">
    <Button android:text="Button" android:id="@+id/button1"
        android:layout_width="wrap_content" android:layout_height="wrap_content"></Button>
</LinearLayout>
```

新建布局文件 layout2.xml，代码如下：

```xml
<?xml version="1.0" encoding="utf-8"?>
<LinearLayout xmlns:android="http://schemas.android.com/apk/res/android"
    android:orientation="vertical" android:layout_width="fill_parent"
    android:layout_height="wrap_content" android:background="#00ff00">
    <Button android:text="Button" android:id="@+id/button1"
        android:layout_width="wrap_content" android:layout_height="wrap_content"></Button>
</LinearLayout>
```

再新建一个与 Activity 对象关联的布局文件 dialogactivitylayout.xml，代码如下：

```xml
<?xml version="1.0" encoding="utf-8"?>
<LinearLayout xmlns:android="http://schemas.android.com/apk/res/android"
    android:orientation="vertical" android:layout_width="200dip"
    android:layout_height="200dip" android:id="@+id/linearLayout1"
    android:gravity="center">
    <AnalogClock android:id="@+id/analogClock1"
        android:layout_width="wrap_content" android:layout_height="wrap_content"></AnalogClock>
</LinearLayout>
```

主布局文件 main.xml 的代码如下：

```xml
<?xml version="1.0" encoding="utf-8"?>
```

```xml
<LinearLayout xmlns:android="http://schemas.android.com/apk/res/android"
    android:orientation="vertical" android:layout_width="fill_parent"
    android:layout_height="fill_parent" android:id="@+id/linearLayout1">
    <Button android:text="我是 main.xml 中的按钮，点我显示半透明的对话框" android:id="@+id/button1"
        android:layout_width="wrap_content" android:layout_height="wrap_content"></Button>
</LinearLayout>
```

在 src 路径下创建新的包 myfragment，在该包中创建两个 Fragment 对象。

文件 MyFragment1.java 的核心代码如下：

```java
public class MyFragment1 extends Fragment {
    //其他 10 个回调方法代码省略……
    @Override
    public View onCreateView(LayoutInflater inflater, ViewGroup container,
            Bundle savedInstanceState) {
        Log.v("MyFragment1", "onCreateView");
        return inflater.inflate(R.layout.layout1, container, false);
    }
    //……
}
```

上面代码中回调方法 onCreateView()的主要作用是创建一个 View 对象，使 Fragment 关联这个 View 对象在 Activity 界面中显示出来，其中代码：

```java
inflater.inflate(R.layout.layout1, container, false);
```

第 1 个参数是要把某一个布局文件转成 View 对象，第 2 个参数是把这个 View 对象放入 Activity 关联的布局文件中，第 3 个参数表示是否把这个 View 对象追加到 container 参数的后面，有关这个知识点，在后面有专门的介绍。

文件 MyFragment2.java 的核心代码如下：

```java
public class MyFragment1 extends Fragment {
    //其他 10 个回调方法代码省略……
    @Override
    public View onCreateView(LayoutInflater inflater, ViewGroup container,
            Bundle savedInstanceState) {
        Log.v("MyFragment2", "onCreateView");
        return inflater.inflate(R.layout.layout2, container, false);
    }
    //……
}
```

新建具有对话框 Dialog 样式的 Activity 文件 DialogActivity.java，核心代码如下：

```java
public class DialogActivity extends Activity {
    //注意：其他的 Activity 的生命周期方法省略……
    @Override
    public void onCreate(Bundle savedInstanceState) {
        super.onCreate(savedInstanceState);
        Log.v("DialogActivity", "onCreate");
        setContentView(R.layout.dialogactivitylayout);
```

```
        }
    //......
}
```

文件 Main.java 的核心代码如下：

```java
public class Main extends FragmentActivity {
    //注意：Main.java 其他回调方法省略……
    private Button button1;

    @Override
    public void onCreate(Bundle savedInstanceState) {
        super.onCreate(savedInstanceState);
        setContentView(R.layout.main);

        Log.v("Main", "onCreate");

        //创建 2 个 Fragment 对象
        MyFragment1 myf1 = new MyFragment1();
        MyFragment2 myf2 = new MyFragment2();
        //通过 getSupportFragmentManager()方法取得 FragmentManager 对象
        FragmentManager fm = this.getSupportFragmentManager();
        //开启对 Fragment 操作的事务
        FragmentTransaction ft = fm.beginTransaction();
        //添加 2 个 Fragment 到 linearLayout1 布局对象中
        ft.add(R.id.linearLayout1, myf1, "myf1");
        ft.add(R.id.linearLayout1, myf2, "myf2");
        //提交事务
        ft.commit();

        button1 = (Button) this.findViewById(R.id.button1);
        button1.setOnClickListener(new OnClickListener() {
            public void onClick(View arg0) {
                Intent intent = new Intent(Main.this, DialogActivity.class);
                Main.this.startActivity(intent);
            }
        });
    }
    //......
}
```

程序初始运行效果如图 10.3 所示。

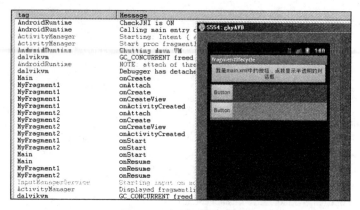

图 10.3　初始运行效果

从图 10.3 中可以看到，Activity 对象执行完 onCreate()方法后就开始创建 Fragment 对象，并且分别经过 onAttach、onCreate、onCreateView 和 onActivityCreated 这 4 个回调方法，再执行 Fragment 对象的 onStart()回调方法。Fragment 对象都创建完毕后再执行 Activity 对象的 onStart()和 onResume()方法，最后再执行 Fragment 对象的 onResume()方法。

当按下 Back 按钮完全退出当前的项目时，LogCat 面板追加打印日志，如图 10.4 所示。

MyFragment1	onPause
MyFragment2	onPause
Main	onPause
MyFragment1	onStop
MyFragment2	onStop
Main	onStop
MyFragment1	onDestroyView
MyFragment1	onDestroy
MyFragment1	onDetach
MyFragment2	onDestroyView
MyFragment2	onDestroy
MyFragment2	onDetach
Main	onDestroy
dalvikvm	GC_EXPLICIT free

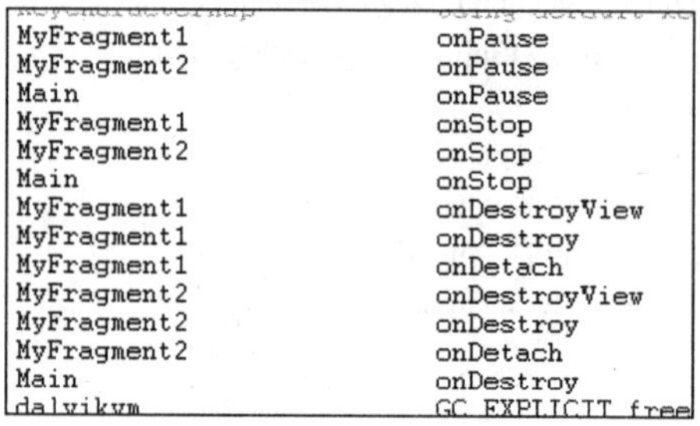

图 10.4　完全退出应用程序的追加日志

从图 10.4 中可以看到，完全退出应用程序前，先执行 Fragment 对象的 onPause()方法，再执行 Main 的 onPause()方法，再执行 Fragment 对象的 onStop()方法和 Main 对象的 onStop()方法，然后依次执行 Fragment 对象的回调方法 onDestroyView、onDestroy()和 onDetach()方法，将 Fragment 对象进行销毁，最后执行 Main 对象的 onDestroy()方法把 Activity 对象销毁。

上面的过程是，Activity 对象与 Fragment 对象创建及销毁时全部的生命周期过程，可以发现 Activity 对象的生命周期完全影响到 Fragment 对象的生命周期，也就是说 Fragment 对象的宿主是 Activity 对象，宿主的生命周期可以影响到自己（Fragment）。

再继续测试，重新进入项目，当单击 Button 按钮时以半透明的方式弹出一个对话框样式的 Activity 对象，生命周期也发生了一些变化，如图 10.5 所示。

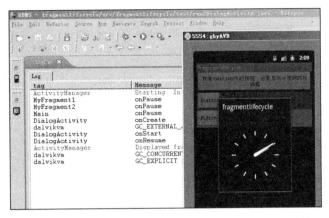

图 10.5　弹出半透明 Dialog 对话框样式的 Activity 日志效果

从图 10.5 中可以看到，Fragment 对象先 onPause 暂停，然后 Main 再 onPause 暂停，DialogActivity 对象执行 onCreate()、onStart()和 onResume()回调，这时再按下 Back 按钮，让 DialogActivity 对象销毁，再看看 LogCat 打印的日志，如图 10.6 所示。

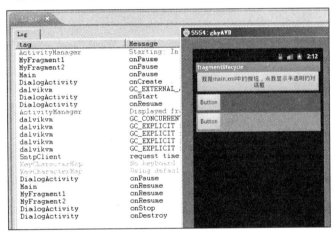

图 10.6　销毁 DialogActivity 对象的生命周期过程

通过这个实验可以发现，Fragment 的生命周期是随着 Activity 的改变而改变的。

10.2.2　Fragment 对象的事务

所谓的 FragmentTransaction 类可以理解成是在 Activity 界面中用于增删改查 Fragment 对象的工具类，此对象也可以把历史的 UI 界面放入一个"back stack"后退栈中，这个后退栈由 Activity 管理，使用这个栈的目的是用户按下 Back 按钮后还能看见以前曾经操作过的界面，有些类似于 Ctrl+Z 还原的功能。

FragmentTransaction 对象中的 addToBackStack()方法来将当前未 commit()的 UI 界面添加进 "back stack"后退栈中，以便按下 Back 按钮进行 UI 界面的恢复。

新建名称为 fragmentTransactionTest 的 Android 项目，更改 main.xml 的代码如下：

```
<?xml version="1.0" encoding="utf-8"?>
```

```xml
<LinearLayout xmlns:android="http://schemas.android.com/apk/res/android"
    android:orientation="horizontal" android:layout_width="fill_parent"
    android:layout_height="fill_parent">
    <LinearLayout android:id="@+id/leftLayout"
        android:orientation="vertical" android:layout_width="80dip"
        android:layout_height="fill_parent">
    </LinearLayout>
    <LinearLayout android:id="@+id/rightLayout"
        android:orientation="vertical" android:layout_width="fill_parent"
        android:layout_height="fill_parent" android:layout_weight="1">
    </LinearLayout>
</LinearLayout>
```

创建一个名称为 leftlayout.xml 的布局文件,此布局文件的主要作用是关联一个 Fragment 对象,这个 Fragment 对象要添加进 main.xml 中 id 为 leftLayout 的<LinearLayout>标签中,代码如下:

```xml
<?xml version="1.0" encoding="utf-8"?>
<LinearLayout xmlns:android="http://schemas.android.com/apk/res/android"
    android:orientation="vertical" android:layout_width="fill_parent"
    android:layout_height="fill_parent" android:background="#ff0000">
    <Button android:text="Button" android:id="@+id/button1"
        android:layout_width="wrap_content" android:layout_height="wrap_content"></Button>
    <Button android:text="Button" android:id="@+id/button2"
        android:layout_width="wrap_content" android:layout_height="wrap_content"></Button>
    <Button android:text="Button" android:id="@+id/button3"
        android:layout_width="wrap_content" android:layout_height="wrap_content"></Button>
    <Button android:text="Button" android:id="@+id/button4"
        android:layout_width="wrap_content" android:layout_height="wrap_content"></Button>
</LinearLayout>
```

再创建一个 Fragment 对象 LeftFragment.java,此文件关联 leftlayout.xml 文件,代码如下:

```java
public class LeftFragment extends Fragment {

    private Button button1;
    private Button button2;
    private Button button3;
    private Button button4;

    private View view;

    private OnClickListener buttonOnClickListener = new OnClickListener() {
        public void onClick(View arg0) {
            FragmentManager fm = LeftFragment.this.getActivity()
                    .getSupportFragmentManager();
            FragmentTransaction ft = fm.beginTransaction();
            switch (arg0.getId()) {
            case R.id.button1:
                ft.addToBackStack("" + arg0.getId());
                MyFragment1 myFragment1 = new MyFragment1();
                ft.replace(R.id.rightLayout, myFragment1, "right1");
```

```
                        break;
                case R.id.button2:
                        ft.addToBackStack("" + arg0.getId());
                        MyFragment2 myFragment2 = new MyFragment2();
                        ft.replace(R.id.rightLayout, myFragment2, "right2");
                        break;
                case R.id.button3:
                        ft.addToBackStack("" + arg0.getId());
                        MyFragment3 myFragment3 = new MyFragment3();
                        ft.replace(R.id.rightLayout, myFragment3, "right3");
                        break;
                case R.id.button4:
                        Fragment rightFragment1 = LeftFragment.this.getActivity()
                                        .getSupportFragmentManager()
                                        .findFragmentByTag("right1");
                        ft.remove(rightFragment1);
                        break;
                }
                ft.commit();
        }
    };

    @Override
    public View onCreateView(LayoutInflater inflater, ViewGroup container,
            Bundle savedInstanceState) {
        view = inflater.inflate(R.layout.leftlayout, container, false);

        button1 = (Button) view.findViewById(R.id.button1);
        button2 = (Button) view.findViewById(R.id.button2);
        button3 = (Button) view.findViewById(R.id.button3);
        button4 = (Button) view.findViewById(R.id.button4);

        button1.setOnClickListener(buttonOnClickListener);
        button2.setOnClickListener(buttonOnClickListener);
        button3.setOnClickListener(buttonOnClickListener);
        button4.setOnClickListener(buttonOnClickListener);

        return view;

    }
}
```

还要创建 main.xml 布局中右边待切换的 Fragment 对象关联的 3 个 XML 文件,如图 10.7 所示。

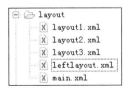

图 10.7　创建 layout1.xml、layout2.xml 和 layout3.xml 布局文件

文件 layout1.xml 中有 1 个 EditText 控件，文件 layout2.xml 中有 2 个 EditText 控件，文件 layout3.xml 中有 3 个 EditText 控件。

再创建与这 3 个 XML 布局文件关联的 3 个 Fragment 对象，其中 MyFragment1.java 的核心代码如下：

```java
public class MyFragment1 extends Fragment {

    @Override
    public View onCreateView(LayoutInflater inflater, ViewGroup container,
            Bundle savedInstanceState) {
        Log.v("MyFragment1", "onCreateView");
        return inflater.inflate(R.layout.layout1, container, false);
    }
}
```

其他两个 Fragment 代码基本雷同，只是关联的布局文件不一样。

```java
return inflater.inflate(R.layout.layout2, container, false);
return inflater.inflate(R.layout.layout3, container, false);
```

非常重要的 Main.java 的代码如下：

```java
public class Main extends FragmentActivity {

    @Override
    public void onCreate(Bundle savedInstanceState) {
        super.onCreate(savedInstanceState);
        setContentView(R.layout.main);

        LeftFragment leftFragmentRef = new LeftFragment();
        MyFragment1 myFragment1 = new MyFragment1();

        FragmentManager fm = this.getSupportFragmentManager();
        FragmentTransaction ft = fm.beginTransaction();
        ft.add(R.id.leftLayout, leftFragmentRef, "left");
        ft.add(R.id.rightLayout, myFragment1, "right1");
        ft.commit();

    }
}
```

程序运行后的 UI 效果如图 10.8 所示。

面板 LogCat 打印出的日志信息，如图 10.9 所示。

图 10.8　初始运行效果

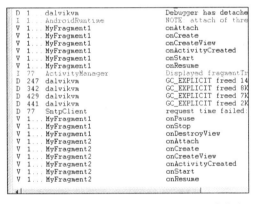

图 10.9　初始运行效果的日志信息

此界面是由 Main.java 代码初始化的，可以发现在 Main.java 文件中的 onCreate()回调函数中并没有使用 addToBackStack()方法，所以初始运行的 UI 界面不纳入"back stack"后退栈中，这时单击左边第 2 个 Button 按钮，出现的界面如图 10.10 所示。

此时日志信息如图 10.11 所示。

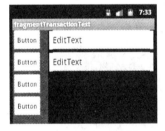

图 10.10　显示 layout2.xml 布局文件　　图 10.11　显示 layout2.xml 布局文件的日志信息

单击第 2 个按钮后执行的是 LeftFragment.java 文件中的代码：

```
case R.id.button2:
    ft.addToBackStack("" + arg0.getId());
    MyFragment2 myFragment2 = new MyFragment2();
    ft.replace(R.id.rightLayout, myFragment2, "right2");
    break;
```

从上面的代码中可以看到，单击第 2 个 Button 按钮执行了 addToBackStack()方法，所以将当前的 UI 界面添加到"back stack"栈中（此界面是具有一个 EditText 状态的界面），这时从图 10.11 中可以发现，MyFragment1 并没有执行 onDestroy()销毁方法，因为它已经被纳入到"back stack"栈中了，这时按下 Back 按钮就回到了图 10.8 的界面效果了。

到此，Fragment 事务的基本使用就告一段落。

如果不把 MyFragment1 放入"back stack"栈中呢，情况又如何呢？我们更改 LeftFragment.java

文件中的代码如下：

```
case R.id.button2:
    // ft.addToBackStack("" + arg0.getId());
    MyFragment2 myFragment2 = new MyFragment2();
    ft.replace(R.id.rightLayout, myFragment2, "right2");
    break;
```

把 addToBackStack()方法屏蔽掉，再次重新运行项目，出现的界面和日志如图 10.12 所示。

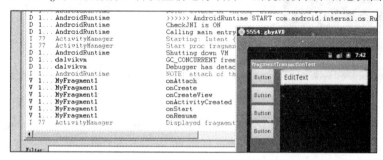

图 10.12　屏蔽掉 addToBackStack 方法的初始运行效果

从图 10.12 中可以看到并没有什么新奇的效果，只是按 Fragment 生命周期的顺序执行回调函数，这时单击第 2 个按钮，再看 UI 和日志，如图 10.13 所示。

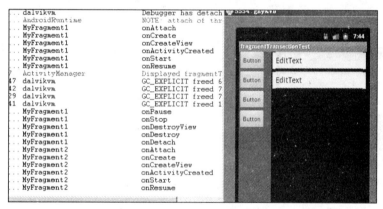

图 10.13　继续单击第 2 个按钮的 UI 和日志信息

从图 10.13 中可以看到，MyFragment1 被销毁掉了，因为执行了 onDestroy()方法，并没有把 UI 界面状态添加到事务中。

最后一个要实验的就是删除 Fragment 了，重新运行项目，直接单击第 4 个按钮，UI 界面及日志信息如图 10.14 所示。

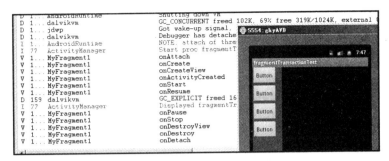

图 10.14 单击第 4 个按钮的效果

由于此过程从未使用过"back stack",所以这时按下 Back 按钮完成退出了应用程序。

10.3 Fragment 对象使用案例

学习完 Fragment 对象的生命周期和事务后,现在来学习与 Fragment 对象有关的使用案例,Fragment 在 Android 中是非常重要的知识点,使用它可以实现很多实用的功能。

10.3.1 Fragment 对象的初步使用与 inflate 方法参数的解析

新建名称为 fragmentTest1 的 Android 项目。布局文件 main.xml 的代码如下:

```
<?xml version="1.0" encoding="utf-8"?>
<LinearLayout xmlns:android="http://schemas.android.com/apk/res/android"
    android:orientation="vertical" android:layout_width="fill_parent"
    android:layout_height="fill_parent" android:id="@+id/linearLayout1">
    <Button android:text="我是main.xml 中的按钮" android:id="@+id/button1"
        android:layout_width="wrap_content" android:layout_height="wrap_content"></Button>
</LinearLayout>
```

新建布局文件 layout1.xml,代码如下:

```
<?xml version="1.0" encoding="utf-8"?>
<LinearLayout xmlns:android="http://schemas.android.com/apk/res/android"
    android:orientation="vertical" android:layout_width="fill_parent"
    android:layout_height="wrap_content" android:background="#ff0000">
    <Button android:text="Button" android:id="@+id/button1"
        android:layout_width="wrap_content" android:layout_height="wrap_content"></Button>
</LinearLayout>
```

新建布局文件 layout2.xml,代码如下:

```
<?xml version="1.0" encoding="utf-8"?>
<LinearLayout xmlns:android="http://schemas.android.com/apk/res/android"
    android:orientation="vertical" android:layout_width="fill_parent"
    android:layout_height="wrap_content" android:background="#00ff00">
    <Button android:text="Button" android:id="@+id/button1"
        android:layout_width="wrap_content" android:layout_height="wrap_content"></Button>
</LinearLayout>
```

新建两个自定义的 Fragment 对象 MyFragment1.java 和 MyFragment2.java，其中在 public View onCreateView(LayoutInflater inflater, ViewGroup container,Bundle savedInstanceState)回调方法中的核心代码分别为：

```java
MyFragment1.java:
@Override
public View onCreateView(LayoutInflater inflater, ViewGroup container,
        Bundle savedInstanceState) {
    Log.v("MyFragment1", "onCreateView");
    return inflater.inflate(R.layout.layout1, container, false);
}

MyFragment2.java:
@Override
public View onCreateView(LayoutInflater inflater, ViewGroup container,
        Bundle savedInstanceState) {
    Log.v("MyFragment2", "onCreateView");
    return inflater.inflate(R.layout.layout2, container, false);
}
```

这两个 Fragment 对象的代码仅仅是关联的布局文件不同。

文件 Main.java 的核心代码如下：

```java
public class Main extends FragmentActivity {

    @Override
    public void onCreate(Bundle savedInstanceState) {
        super.onCreate(savedInstanceState);
        setContentView(R.layout.main);

        MyFragment1 myf1 = new MyFragment1();
        MyFragment2 myf2 = new MyFragment2();

        FragmentManager fm = this.getSupportFragmentManager();
        FragmentTransaction ft = fm.beginTransaction();
        ft.add(R.id.linearLayout1, myf1, "myf1");
        ft.add(R.id.linearLayout1, myf2, "myf2");
        ft.commit();

    }
}
```

程序运行结果如图 10.15 所示。打印日志信息如图 10.16 所示。

图 10.15　运行结果　　　　图 10.16　初始日志记录

这段代码并没有什么异常与特殊，仅仅作为一种使用 Fragment 对象的复习吧，运行结果可以把 Fragment 对象显示到 Activity 对象中。但本示例要讨论的问题并不仅仅在于此。下面继续实验如下代码中的知识点：

```
@Override
public View onCreateView(LayoutInflater inflater, ViewGroup container,
        Bundle savedInstanceState) {
    Log.v("MyFragment1", "onCreateView");
    return inflater.inflate(R.layout.layout1, container, false);
}
```

方法 inflate() 有 3 个参数，第 1 个参数表示 Fragment 对象关联的是哪个 XML 布局文件，此参数非常容易理解，第 2 个参数是把这个 XML 布局文件转成的 View 对象放入 container 容器中，那么这个 container 容器到底是谁呢？一起来测试一下。

更改 main.xml 的代码如下：

```
<?xml version="1.0" encoding="utf-8"?>
<LinearLayout xmlns:android="http://schemas.android.com/apk/res/android"
    android:orientation="vertical" android:layout_width="fill_parent"
    android:layout_height="fill_parent" android:id="@+id/linearLayout1"
    android:tag="my name is linearLayout1">
    <Button android:text="我是main.xml 中的按钮" android:id="@+id/button1"
        android:layout_width="wrap_content" android:layout_height="wrap_content"></Button>
</LinearLayout>
```

对 LinearLayout 标签添加 android:tag 属性值，更改 2 个 Fragment 的对象核心代码如下：

```
@Override
public View onCreateView(LayoutInflater inflater, ViewGroup container,
        Bundle savedInstanceState) {
    Log.v("MyFragment1", "onCreateView");
    Log.v("tag=", "" + container.getTag());
    return inflater.inflate(R.layout.layout1, container, false);
}
```

程序重新运行后打印信息如图 10.17 所示。

图 10.17　打印出了 android:tag 属性值

通过这个实验可以发现，这个 container 容器其实就是 main.xml 文件中的根结点 <LinearLayout>，至此第 2 个参数的介绍结束了，第 3 个参数呢？

第 3 个参数代表这个 View 对象是否添加到 container 容器内部，在以前的实验中第 3 个参数都是 false，也就是不把这个 View 添加到 container 容器内部，因为是手动添加，就要手写代码来实现，所以就会在 Main.java 文件中出现这样的代码：

```
ft.add(R.id.linearLayout1, myf1, "myf1");
ft.add(R.id.linearLayout1, myf2, "myf2");
```

那如果第 3 个参数为 true 是什么含义呢，这是表示将第 1 个参数转成的 View 对象放入第 2 个参数的内部了，Fragment 的代码更改如下：

```
@Override
public View onCreateView(LayoutInflater inflater, ViewGroup container,
        Bundle savedInstanceState) {
    Log.v("MyFragment1", "onCreateView");
    Log.v("tag=", "" + container.getTag());
    inflater.inflate(R.layout.layout1, container, true);
    return super.onCreateView(inflater, container, savedInstanceState);
}
```

运行项目，也出现了正确的界面，如图 10.18 所示。

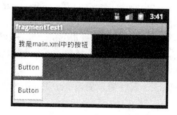

图 10.18　出现正确界面

在实际开发中，为了代码容易阅读及更新维护的便利，经常使用如下代码的方式添加 Fragment 对象：

```
@Override
public View onCreateView(LayoutInflater inflater, ViewGroup container,
        Bundle savedInstanceState) {
    Log.v("MyFragment2", "onCreateView");
    return inflater.inflate(R.layout.layout2, container, false);
}
```

也就是第 3 个参数为 false 的形式。

10.3.2　FragmentActivity 与 Fragment 对象交互

有时候需要从 FragmentActivity 传递参数到 Fragment 对象，而且这种情况是非常常见的，使用上也非常简单。

新建名称为 fragmentTest2 的 Android 项目，创建名称为 layout1.xml 的布局文件，代码如下：

```
<?xml version="1.0" encoding="utf-8"?>
<LinearLayout xmlns:android="http://schemas.android.com/apk/res/android"
    android:layout_width="fill_parent" android:layout_height="fill_parent"
```

```xml
        android:orientation="vertical" android:id="@+id/fragment_layout1">
    <Button android:id="@+id/button1" android:layout_width="wrap_content"
        android:layout_height="wrap_content" android:text="我是页1" />
    <Button android:id="@+id/button2" android:layout_width="wrap_content"
        android:layout_height="wrap_content" android:text="取得main_button1 的text" />
</LinearLayout>
```

继续创建名称为 layout2.xml 的布局文件，代码如下：

```xml
<?xml version="1.0" encoding="utf-8"?>
<LinearLayout xmlns:android="http://schemas.android.com/apk/res/android"
    android:layout_width="fill_parent" android:layout_height="fill_parent"
    android:orientation="vertical" android:id="@+id/fragment_layout2">
    <Button android:id="@+id/button1" android:layout_width="wrap_content"
        android:layout_height="wrap_content" android:text="我是页2" />
</LinearLayout>
```

再创建与这两个布局文件 layout1.xml 和 layout2.xml 相关联的两个 Fragment 对象，其中 MyFragment1.java 的核心代码如下：

```java
public class MyFragment1 extends Fragment {

    private Button button2;

    // 初始创建 Fragment 对象时调用
    @Override
    public void onCreate(Bundle savedInstanceState) {
        super.onCreate(savedInstanceState);
        //以下代码取得从 FragmentActivity 对象传递过来的数据
        Log.v("!", "MyFragment1 onCreate param value="
                + this.getArguments().getString("username"));

    }

    // 创建与 Fragment 对象关联的 View 视图时调用
    @Override
    public View onCreateView(LayoutInflater inflater, ViewGroup container,
            Bundle savedInstanceState) {
        Log.v("!", "MyFragment1 onCreateView");

        View view = inflater.inflate(R.layout.layout1, container, false);
        button2 = (Button) view.findViewById(R.id.button2);
        button2.setOnClickListener(new OnClickListener() {
            public void onClick(View v) {
                //以下代码用于从 Fragment 对象中取得 FragmentActivity
                //对象中控件的属性值
                Button main_button1 = (Button) MyFragment1.this.getActivity()
                        .findViewById(R.id.main_button1);
                Log.v("!", "main_button1 text=" + main_button1.getText());
            }
        });
```

```
        return view;
    }

    //其他代码省略
```

文件 MyFragment2.java 的代码没有什么特别之处，只是实现了全部 11 个生命周期方法，并用 Log.v()打印日志就可以了。

主布局文件 main.xml 的代码如下：

```xml
<?xml version="1.0" encoding="utf-8"?>
<LinearLayout xmlns:android="http://schemas.android.com/apk/res/android"
    android:layout_width="fill_parent" android:layout_height="fill_parent"
    android:orientation="horizontal">

    <LinearLayout android:layout_width="200dip"
        android:layout_height="fill_parent" android:orientation="vertical">
        <Button android:id="@+id/main_button1" android:layout_width="wrap_content"
            android:layout_height="wrap_content" android:text="显示第 1 页" />

        <Button android:id="@+id/main_button2" android:layout_width="wrap_content"
            android:layout_height="wrap_content" android:text="显示第 2 页" />
    </LinearLayout>

    <LinearLayout android:id="@+id/linearLayout1"
        android:layout_width="fill_parent" android:layout_height="fill_parent"
        android:layout_weight="1" android:background="#ff0000"
        android:orientation="vertical">
    </LinearLayout>

</LinearLayout>
```

文件 Main.java 的代码如下：

```java
public class Main extends FragmentActivity {

    private Button main_button1;
    private Button main_button2;

    @Override
    public void onCreate(Bundle savedInstanceState) {
        super.onCreate(savedInstanceState);
        setContentView(R.layout.main);

        main_button1 = (Button) this.findViewById(R.id.main_button1);
        main_button2 = (Button) this.findViewById(R.id.main_button2);

        main_button1.setOnClickListener(new OnClickListener() {
            public void onClick(View v) {
                FragmentManager fragmentManager = Main.this
                        .getSupportFragmentManager();
```

```java
                FragmentTransaction fragmentTransaction = fragmentManager
                            .beginTransaction();
                MyFragment1 fragment1 = new MyFragment1();
                Bundle bundle = new Bundle();
                bundle.putString("username", "gotoPage1");
                fragment1.setArguments(bundle);
                fragmentTransaction.replace(R.id.linearLayout1, fragment1);
                fragmentTransaction.commit();
            }
        });

        main_button2.setOnClickListener(new OnClickListener() {
            public void onClick(View v) {
                FragmentManager fragmentManager = Main.this
                            .getSupportFragmentManager();
                FragmentTransaction fragmentTransaction = fragmentManager
                            .beginTransaction();
                MyFragment2 fragment2 = new MyFragment2();
                Bundle bundle = new Bundle();
                bundle.putString("username", "gotoPage2");
                fragment2.setArguments(bundle);
                fragmentTransaction.replace(R.id.linearLayout1, fragment2);
                fragmentTransaction.commit();
            }
        });

        // 初始化界面
        FragmentManager fragmentManager = Main.this.getSupportFragmentManager();
        FragmentTransaction fragmentTransaction = fragmentManager
                    .beginTransaction();
        // 将 MyFragment1 对象默认显示在界面中
        MyFragment1 fragment1 = new MyFragment1();
        Bundle bundle = new Bundle();
        bundle.putString("username", "gotoPage1");
        // 并给 MyFragment1 对象传递 Bundle 参数
        fragment1.setArguments(bundle);
        // 将 fragment1 对象添加到 R.id.linearLayout1 内部
        fragmentTransaction.add(R.id.linearLayout1, fragment1);
        fragmentTransaction.commit();
    }
}
```

程序运行后的日志和界面如图 10.19 所示。

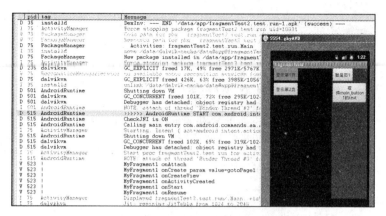

图 10.19　初始运行界面及日志

单击图 10.19 界面中的"取得 main_button1 的 text"按钮，打印日志如图 10.20 所示。

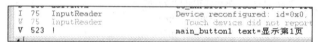

图 10.20　取得 FragmentActivity 界面中的控件属性值

通过这个实验，Fragment 与 FragmentActivity 就可以互相通信了。

10.3.3　Fragment 对象之间的交互

对象 Fragment 之间也可以实现交互。新建名称为 fragmentTest3 的 Android 项目，创建 layout 布局文件 XML 及对应的 Fragment 对象，整个项目结构如图 10.21 所示。

图 10.21　整个项目结构

文件 main.xml 的代码如下：

```
<?xml version="1.0" encoding="utf-8"?>
<LinearLayout xmlns:android="http://schemas.android.com/apk/res/android"
    android:layout_width="fill_parent" android:layout_height="fill_parent"
    android:orientation="horizontal">
    <LinearLayout android:id="@+id/leftLinearLayout"
```

```
        android:layout_width="200dip" android:layout_height="fill_parent"
        android:background="#ff0000" android:orientation="vertical">
    </LinearLayout>
    <LinearLayout android:id="@+id/rightLinearLayout"
        android:layout_width="fill_parent" android:layout_height="fill_parent"
        android:layout_weight="1" android:orientation="vertical">
    </LinearLayout>
</LinearLayout>
```

文件 Main.java 的代码如下：

```
public class Main extends FragmentActivity {
    @Override
    public void onCreate(Bundle savedInstanceState) {
        super.onCreate(savedInstanceState);
        setContentView(R.layout.main);

        MyFragment1 MyFragment1Ref = new MyFragment1();
        MyFragment2 MyFragment2Ref = new MyFragment2();

        FragmentManager fm = this.getSupportFragmentManager();
        FragmentTransaction transaction = fm.beginTransaction();
        transaction.add(R.id.leftLinearLayout, new MyFragment1(),
                "leftfragment");
        transaction.add(R.id.rightLinearLayout, new MyFragment2(),
                "rightfragment");
        transaction.commit();

    }
}
```

其他文件源代码省略，但从程序运行初始界面如图 10.22 所示的效果来看，也能大致分析出 Fragment 及对应 XML 布局的代码。

单击"去第一项"按钮右边的 EditText 控件改变内部文本，如图 10.23 所示。

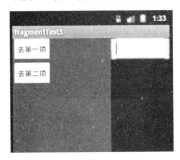

图 10.22　程序初始运行效果　　　　图 10.23　单击去第一项按钮

单击"去第二项"按钮右边的 EditText 控件改变内部文本，如图 10.24 所示。

图 10.24　单击去第二项按钮

通过这个实验，多个 Fragment 对象之间就可以互相通信了。

10.3.4　在 DialogFragment 对象中使用 onCreateView 回调函数生成对话框

对象 DialogFragment 的继承关系如图 10.25 所示。

图 10.25　DialogFragment 的继承关系

从图 10.25 中可以看到，DialogFragment 继承 Fragment 对象，但 DialogFragment 对象具有的外观样式和它的名称一样，所以可以在 Fragment 对象中使用 DialogFragment 类来实现对话框的效果。但 DialogFragment 类还具有非常重要的特性，它可以被放入 Activity 对象的 back stack 后退栈中，因为它的父类 Fragment 也具有同样的特性。但在本示例中并不打算演示将 DialogFragment 放入 back stack 后退栈的效果，仅仅以一个简单的案例演示如何使用 DialogFragment 对象。

新建名称为 fragmentTest4 的 Android 项目，创建名称为 MyDialogFragment.java 的 DialogFragment 对象，代码如下：

```java
public class MyDialogFragment extends DialogFragment {

    private Button button1;
    private EditText usernameEditText;
    private EditText passwordEditText;

    @Override
    public View onCreateView(LayoutInflater inflater, ViewGroup container,
            Bundle savedInstanceState) {

        View view = inflater.inflate(R.layout.dialogfragmentlayout, container,
                false);

        button1 = (Button) view.findViewById(R.id.button1);
        usernameEditText = (EditText) view.findViewById(R.id.editText1);
        passwordEditText = (EditText) view.findViewById(R.id.editText2);

        button1.setOnClickListener(new OnClickListener() {
```

```java
            public void onClick(View arg0) {
                Log.v("!", "username value="
                        + usernameEditText.getText().toString()
                        + " password value="
                        + passwordEditText.getText().toString());
                MyDialogFragment.this.dismiss();
            }
        });

        return view;
    }
}
```

与这个 Fragment 关联的布局 XML 文件 dialogfragmentlayout.xml 代码如下：

```xml
<?xml version="1.0" encoding="utf-8"?>
<LinearLayout xmlns:android="http://schemas.android.com/apk/res/android"
    android:orientation="vertical" android:layout_width="fill_parent"
    android:layout_height="fill_parent">
    <LinearLayout android:orientation="vertical"
        android:layout_width="300dip" android:layout_height="200dip"
        android:gravity="center">
        <EditText android:layout_height="wrap_content" android:id="@+id/editText1"
            android:text="EditText" android:layout_width="match_parent"></EditText>
        <EditText android:layout_height="wrap_content" android:id="@+id/editText2"
            android:text="EditText" android:layout_width="match_parent"></EditText>
        <Button android:text="Button" android:id="@+id/button1"
            android:layout_width="wrap_content" android:layout_height="wrap_content"></Button>
    </LinearLayout>
</LinearLayout>
```

对象 FragmentActivity 文件 Main.java 的代码如下：

```java
public class Main extends FragmentActivity {
    private Button button1;

    private MyDialogFragment dialog1 = new MyDialogFragment();

    @Override
    public void onCreate(Bundle savedInstanceState) {
        super.onCreate(savedInstanceState);
        setContentView(R.layout.main);

        button1 = (Button) this.findViewById(R.id.button1);
        button1.setOnClickListener(new OnClickListener() {
            public void onClick(View arg0) {
                dialog1.show(Main.this.getSupportFragmentManager(), "dialog1");
            }
        });
    }
}
```

程序运行效果如图 10.26 所示。

图 10.26　程序运行效果

单击 Button 按钮取出 EditText 中的文本内容，如图 10.27 所示。

```
V 1...!                    username value=EditText password value=EditText
```

图 10.27　取出 EditText 中的文本

再使用 public View onCreateView(LayoutInflater inflater, ViewGroup container, Bundle savedInstanceState)回调方法中的 inflate 对象来生成 View 对象，即可以在 DialogFragment 上显示出任何的 UI 控件，比如 ListView 等，并且 ListView 也支持自定义的 Adapter 适配器，可以自行写代码进行实验。

10.3.5　将 DialogFragment 对象放入 back stack 后退栈中

前面说过 DialogFragment 继承自 Fragment 对象，所以也可以将 DialogFragment 对象放入 back stack 栈中。

新建名称为 fragmentTest5 的 Android 项目，创建两个 DialogFragment 对象，第 1 个 MyDialogFragment1.java 的核心代码如下：

```java
public class MyDialogFragment1 extends DialogFragment {

    private Button button1;

    private MyDialogFragment2 dialog2 = new MyDialogFragment2();

    @Override
    public void onDestroy() {
        super.onDestroy();
        Log.v("!", "MyDialogFragment1 onDestroy");
    }

    @Override
    public void onDestroyView() {
        super.onDestroyView();
        Log.v("!", "MyDialogFragment1 onDestroyView");
    }
```

```java
@Override
public View onCreateView(LayoutInflater inflater, ViewGroup container,
        Bundle savedInstanceState) {
    View view = inflater.inflate(R.layout.dialogfragmentlayout1, container,
            false);
    button1 = (Button) view.findViewById(R.id.button1);
    button1.setOnClickListener(new OnClickListener() {
        public void onClick(View arg0) {
            // FragmentTransaction 中的 remove()方法是删除容器中的 Fragment 对象，
            // 如果不执行 remove()方法，则 MyDialogFragment1 以半透明的方式在底层显示，
            // 影响美观。
            //
            // 如果方法 ft.addToBackStack("");参与执行，
            // MyDialogFragment1 生命周期只到 onDestroyView()，
            // 因为放入 back stack 栈中，以备还原恢复。
            //
            // 如果方法 ft.addToBackStack("");不参与执行
            // MyDialogFragment1 生命周期经过 onDestroyView()到 onDestroy()方法，
            // 因为不需要还原所以直接销毁了
            FragmentManager fm = MyDialogFragment1.this
                    .getFragmentManager();
            FragmentTransaction ft = fm.beginTransaction();
            ft.remove(MyDialogFragment1.this);
            ft.addToBackStack("");
            dialog2.show(ft, "dialog2");
        }
    });

    return view;

}
}
```

第 2 个文件 MyDialogFragment2.java 的核心代码如下：

```java
public class MyDialogFragment2 extends DialogFragment {

    @Override
    public View onCreateView(LayoutInflater inflater, ViewGroup container,
            Bundle savedInstanceState) {

        return inflater.inflate(R.layout.dialogfragmentlayout2, container,
                false);
    }
}
```

还要创建与两个 DialogFragment 对应的 XML 布局文件，dialogfragmentlayout1.xml 的代码如下：

```xml
<?xml version="1.0" encoding="utf-8"?>
<LinearLayout xmlns:android="http://schemas.android.com/apk/res/android"
    android:orientation="horizontal" android:layout_width="fill_parent"
    android:layout_height="fill_parent">
    <Button android:text="Button" android:id="@+id/button1"
        android:layout_width="300dip" android:layout_height="300dip"></Button>
</LinearLayout>
```

文件 dialogfragmentlayout2.xml 的代码如下：

```xml
<?xml version="1.0" encoding="utf-8"?>
<LinearLayout xmlns:android="http://schemas.android.com/apk/res/android"
    android:orientation="horizontal" android:layout_width="fill_parent"
    android:layout_height="fill_parent">
    <EditText android:text="EditText" android:id="@+id/editText1"
        android:layout_width="fill_parent" android:layout_height="wrap_content"></EditText>
</LinearLayout>
```

名称为 Main.java 的 FragmentActivity 对象的代码如下：

```java
public class Main extends FragmentActivity {
    private Button button1;
    private MyDialogFragment1 dialog1 = new MyDialogFragment1();

    @Override
    public void onCreate(Bundle savedInstanceState) {
        super.onCreate(savedInstanceState);
        setContentView(R.layout.main);

        button1 = (Button) this.findViewById(R.id.button1);
        button1.setOnClickListener(new OnClickListener() {
            public void onClick(View arg0) {
                dialog1.show(Main.this.getSupportFragmentManager(), "dialog1");
            }
        });
    }
}
```

程序运行后如图 10.28 所示。

图 10.28　初始运行效果

单击图 10.28 中的 Button 按钮，弹出 DialogFragment 对话框，如图 10.29 所示。

再单击图 10.29 中的大按钮，出现如图 10.30 所示的效果。

此时单击 AVD 中的 Back 按钮，前一个 DialogFragment 被还原，界面如图 10.31 所示。

图 10.29　单击 Button 后的效果　　图 10.30　出现 EditText 界面　　图 10.31　被还原的 DialogFragment 界面

10.3.6　在 DialogFragment 对象中使用 onCreateDialog 回调函数生成对话框

新建名称为 fragmentTest6 的 Android 项目，创建 DialogFragment 对象文件 MyDialogFragment.java，代码如下：

```java
public class MyDialogFragment extends DialogFragment {

    @Override
    public Dialog onCreateDialog(Bundle savedInstanceState) {

        AlertDialog dialog = new AlertDialog.Builder(this.getActivity())
                .setView(
                        this.getActivity().getLayoutInflater().inflate(
                                R.layout.dialogfragmentlayout, null)).setTitle(
                "我是标题").create();
        dialog.setButton("关闭", new OnClickListener() {
            public void onClick(DialogInterface arg0, int arg1) {
            }
        });

        return dialog;
    }

}
```

此 DialogFragment 对应的布局 XML 文件 dialogfragmentlayout.xml 的代码如下：

```xml
<?xml version="1.0" encoding="utf-8"?>
<LinearLayout xmlns:android="http://schemas.android.com/apk/res/android"
    android:orientation="horizontal" android:layout_width="fill_parent"
    android:layout_height="fill_parent" android:gravity="center">
    <AnalogClock android:id="@+id/analogClock1"
        android:layout_width="wrap_content" android:layout_height="wrap_content"></AnalogClock>
</LinearLayout>
```

文件 Main.java 的核心代码如下：

```java
public class Main extends FragmentActivity {
    private Button button1;
    private MyDialogFragment dialog = new MyDialogFragment();

    @Override
    public void onCreate(Bundle savedInstanceState) {
        super.onCreate(savedInstanceState);
        setContentView(R.layout.main);

        button1 = (Button) this.findViewById(R.id.button1);
        button1.setOnClickListener(new OnClickListener() {
            public void onClick(View arg0) {
                dialog.show(Main.this.getSupportFragmentManager(), "dialog5");
            }
        });
    }
}
```

程序运行后单击 main.xml 中的按钮弹出对话框，如图 10.32 所示。

图 10.32 弹出对话框

10.3.7 切换 Fragment 添加动画效果

创建名称为 fragmentTest7 的 Android 项目，添加两个与 Fragment 对象关联的布局 XML 文件，在这两个布局文件中各添加一个控件，如图 10.33 所示。

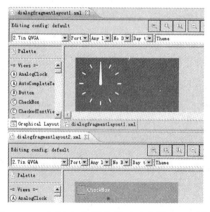

图 10.33 在每个布局文件中各添加一个控件

主布局文件 main.xml 的代码如下:

```xml
<?xml version="1.0" encoding="utf-8"?>
<LinearLayout xmlns:android="http://schemas.android.com/apk/res/android"
    android:orientation="vertical" android:layout_width="fill_parent"
    android:layout_height="fill_parent">
    <LinearLayout android:orientation="vertical"
        android:id="@+id/inner_linearLayout" android:layout_weight="1"
        android:layout_width="fill_parent" android:layout_height="fill_parent">
    </LinearLayout>
    <Button android:text="Button" android:id="@+id/button1"
        android:layout_width="wrap_content" android:layout_height="wrap_content"></Button>
</LinearLayout>
```

对应的 FragmentActivity 文件 Main.java 的代码如下:

```java
public class Main extends FragmentActivity {
    private Button button1;

    private MyFragment1 myFragment1Ref = new MyFragment1();
    private MyFragment2 myFragment2Ref = new MyFragment2();

    @Override
    protected void onCreate(Bundle arg0) {
        super.onCreate(arg0);
        this.setContentView(R.layout.main);

        FragmentManager fm = this.getSupportFragmentManager();
        FragmentTransaction ft = fm.beginTransaction();
        ft.add(R.id.inner_linearLayout, myFragment1Ref, "myFragment6_1Ref");
        ft.commit();

        button1 = (Button) this.findViewById(R.id.button1);
        button1.setOnClickListener(new OnClickListener() {
            public void onClick(View arg0) {

                FragmentManager fm = Main.this.getSupportFragmentManager();
                FragmentTransaction ft = fm.beginTransaction();
                ft.setCustomAnimations(R.anim.fragment_slide_left_enter,
                        R.anim.fragment_slide_left_exit,
                        R.anim.fragment_slide_right_enter,
                        R.anim.fragment_slide_right_exit);
                ft.replace(R.id.inner_linearLayout, myFragment2Ref,
                        "myFragment6_2Ref");
                ft.addToBackStack(null);
                ft.commit();

            }
        });

    }
```

}

定义 decelerateInterpolator 动画配置文件 decelerate_quint.xml，代码如下：

```xml
<?xml version="1.0" encoding="utf-8"?>
<decelerateInterpolator xmlns:android="http://schemas.android.com/apk/res/android"
    android:factor="2.5" />
```

动画配置文件 decelerate_quint.xml 是自定义的 Interpolator，属性 android:factor="2.5" 定义减速的幅度。

动画文件 fragment_slide_left_enter.xml 的代码如下：

```xml
<?xml version="1.0" encoding="utf-8"?>
<set xmlns:android="http://schemas.android.com/apk/res/android"
    android:interpolator="@anim/decelerate_quint">
    <translate android:fromXDelta="33%" android:toXDelta="0%p"
        android:duration="@android:integer/config_mediumAnimTime" />
    <alpha android:fromAlpha="0.0" android:toAlpha="1.0"
        android:duration="@android:integer/config_mediumAnimTime" />
</set>
```

动画文件 fragment_slide_left_exit.xml 的代码如下：

```xml
<?xml version="1.0" encoding="utf-8"?>
<set xmlns:android="http://schemas.android.com/apk/res/android"
    android:interpolator="@anim/decelerate_quint">
    <translate android:fromXDelta="0%" android:toXDelta="-33%p"
        android:duration="@android:integer/config_mediumAnimTime" />
    <alpha android:fromAlpha="1.0" android:toAlpha="0.0"
        android:duration="@android:integer/config_mediumAnimTime" />
</set>
```

动画文件 fragment_slide_right_enter.xml 的代码如下：

```xml
<?xml version="1.0" encoding="utf-8"?>
<set xmlns:android="http://schemas.android.com/apk/res/android"
    android:interpolator="@anim/decelerate_quint">
    <translate android:fromXDelta="-33%" android:toXDelta="0%p"
        android:duration="@android:integer/config_mediumAnimTime" />
    <alpha android:fromAlpha="0.0" android:toAlpha="1.0"
        android:duration="@android:integer/config_mediumAnimTime" />
</set>
```

动画文件 fragment_slide_right_exit.xml 的代码如下：

```xml
<?xml version="1.0" encoding="utf-8"?>
<set xmlns:android="http://schemas.android.com/apk/res/android"
    android:interpolator="@anim/decelerate_quint">
    <translate android:fromXDelta="0%" android:toXDelta="33%p"
        android:duration="@android:integer/config_mediumAnimTime" />
    <alpha android:fromAlpha="1.0" android:toAlpha="0.0"
        android:duration="@android:integer/config_mediumAnimTime" />
</set>
```

程序运行的效果就是切换 Fragment 时有切换的动画。

10.3.8 Fragment 的显示和隐藏

对象 Fragment 也可以隐藏或显示，本示例代码相对简单，不提供演示项目，Demo 代码如下：

```java
public class Button7_FragmentActivity extends FragmentActivity {

    private Button button7_fragmentactivity_layout_button1;
    private Button button7_fragmentactivity_layout_button2;

    private MyFragment7_1 myFragment7_1Ref = new MyFragment7_1();
    private MyFragment7_2 myFragment7_2Ref = new MyFragment7_2();

    @Override
    protected void onCreate(Bundle arg0) {
        super.onCreate(arg0);
        this.setContentView(R.layout.button7_fragmentactivity_layout);

        FragmentManager fm = this.getSupportFragmentManager();
        FragmentTransaction ft = fm.beginTransaction();
        ft.add(R.id.inner_linearLayout, myFragment7_2Ref, "myFragment7_2Ref");
        ft.add(R.id.inner_linearLayout, myFragment7_1Ref, "myFragment7_1Rcf");
        ft.commit();

        addInit(R.id.button7_fragmentactivity_layout_button1, myFragment7_1Ref);
        addInit(R.id.button7_fragmentactivity_layout_button2, myFragment7_2Ref);

    }

    private void addInit(int buttonId, final Fragment fragment) {

        Button button = (Button) this.findViewById(buttonId);
        button.setOnClickListener(new OnClickListener() {
            public void onClick(View arg0) {

                // 注意：show()和hide()方法不参与生命周期
                // 如果使用replace()方法则经过生命周期：
                // onPause onStop onDestroyView onDestroy
                FragmentManager fm = Button7_FragmentActivity.this
                        .getSupportFragmentManager();
                FragmentTransaction ft = fm.beginTransaction();
                if (fragment.isHidden() == true) {
                    ft.show(fragment);
                } else {
                    ft.hide(fragment);
                }
                ft.commit();

            }
```

            });
        }
    }

10.3.9　ListFragment 对象的使用

ListFragment 对象的主要作用就是在 Fragment 中生成一个列表，它的继承关系如图 10.34 所示。

图 10.34　ListFragment 继承关系

创建名称为 fragmentTest9 的 Android 项目，新建 ListFragment 对象文件 MyListFragment.java，代码如下：

```java
public class MyListFragment extends ListFragment {

    private List usernameList = new ArrayList();

    @Override
    public void onActivityCreated(Bundle savedInstanceState) {
        super.onActivityCreated(savedInstanceState);

        for (int i = 0; i < 100; i++) {
            usernameList.add("username" + (i + 1));
        }

        ArrayAdapter adapter = new ArrayAdapter(this.getActivity(),
                R.layout.simple_list_item_1, usernameList);
        this.setListAdapter(adapter);

    }

    @Override
    public void onListItemClick(ListView l, View v, int position, long id) {
        super.onListItemClick(l, v, position, id);
        Log.v("!", "" + usernameList.get(position));
    }

}
```

注意，使用 ListFragment 对象时不需要布局 XML 文件。

FragmentActivity 的对象文件 Main.java 的代码如下：

```java
public class Main extends FragmentActivity {
    private Button button1;
    private MyListFragment fragment = new MyListFragment();

    @Override
    public void onCreate(Bundle savedInstanceState) {
        super.onCreate(savedInstanceState);
        setContentView(R.layout.main);
        button1 = (Button) this.findViewById(R.id.button1);
        button1.setOnClickListener(new OnClickListener() {
            public void onClick(View arg0) {
                FragmentManager fm = Main.this.getSupportFragmentManager();
                FragmentTransaction ft = fm.beginTransaction();
                ft.add(R.id.linearLayout1, fragment, "fragment");
                ft.commit();
            }
        });

    }
}
```

程序运行后单击按钮出现如图 10.35 所示的界面。

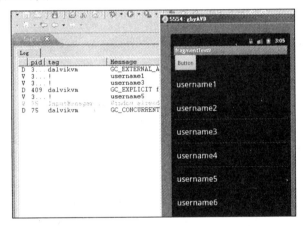

图 10.35　出现 ListFragment 列表界面

10.3.10　Fragment 对象的分页处理方式 1

一个 Activity 可以只显示一个 Fragment 对象，也可以显示多个 Fragment 对象，但在只显示 1 个 Fragment 对象时还要切换到其他的 Fragment 对象，这时就有必要在当前的 Activity 中处理 Fragment 的分页了。

创建名称为 fragmentTest10 的 Android 项目，再创建 5 个 Fragment 对象，如图 10.36 所示。

再创建 5 个 Fragment 对应的 5 个 XML 布局文件，如图 10.37 所示。

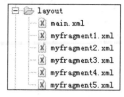

图 10.36　创建 5 个 Fragment 对象　　图 10.37　5 个 XML 布局文件

由于在界面中需要分页，所以在 main.xml 中需要添加如下代码：

```xml
<?xml version="1.0" encoding="utf-8"?>
<LinearLayout xmlns:android="http://schemas.android.com/apk/res/android"
    android:orientation="vertical" android:layout_width="fill_parent"
    android:layout_height="fill_parent">
    <android.support.v4.view.ViewPager
        android:id="@+id/pager" android:layout_width="match_parent"
        android:layout_height="0px" android:layout_weight="1">
    </android.support.v4.view.ViewPager>
    <LinearLayout xmlns:android="http://schemas.android.com/apk/res/android"
        android:orientation="horizontal" android:layout_width="fill_parent"
        android:layout_height="wrap_content">
        <Button android:text="上一页" android:id="@+id/button1"
            android:layout_width="wrap_content" android:layout_height="wrap_content"></Button>
        <Button android:text="下一页" android:id="@+id/button2"
            android:layout_width="wrap_content" android:layout_height="wrap_content"></Button>
    </LinearLayout>
</LinearLayout>
```

创建自定义的分页适配器类 MyPageAdapter.java，代码如下：

```java
public class MyPageAdapter extends FragmentPagerAdapter {

    private List<Fragment> listFragment = new ArrayList<Fragment>();

    public MyPageAdapter(FragmentManager fm) {
        super(fm);
    }

    public void setListFragment(List<Fragment> listFragment) {
        this.listFragment = listFragment;
    }

    @Override
    public Fragment getItem(int arg0) {
        return listFragment.get(arg0);
    }

    @Override
    public int getCount() {
        return listFragment.size();
```

        }
    }

在前面讲解使用 Adapter 对象时，都是通过方法 getView()返回 View 对象，但在此示例中通过 FragmentPagerAdapter 对象的 getItem()方法返回的却是 Fragment 对象，其实 Adapter 和 FragmentPagerAdapter 的目的是一样的，都是返回 UI 界面。

FragmentPagerAdapter 的主要作用是实现分页的功能，它里面的 Fragment 虽然不显示，但还是占用内存空间，相当于 Fragment 是静态的，不被销毁的，生命周期只执行到 onDestroyView()回调函数，所以它只适合于分页比较少的情况。

文件 Main.java 的代码如下：

```
public class Main extends FragmentActivity {

    private ViewPager viewPager;
    private MyPageAdapter adapter;
    private List<Fragment> listFragment = new ArrayList<Fragment>();

    private Button button1;
    private Button button2;

    private int currentPage = 0;

    @Override
    public void onCreate(Bundle savedInstanceState) {
        super.onCreate(savedInstanceState);
        setContentView(R.layout.main);

        MyFragment1 myFragment1 = new MyFragment1();
        MyFragment2 myFragment2 = new MyFragment2();
        MyFragment3 myFragment3 = new MyFragment3();
        MyFragment4 myFragment4 = new MyFragment4();
        MyFragment5 myFragment5 = new MyFragment5();

        listFragment.add(myFragment1);
        listFragment.add(myFragment2);
        listFragment.add(myFragment3);
        listFragment.add(myFragment4);
        listFragment.add(myFragment5);

        adapter = new MyPageAdapter(this.getSupportFragmentManager());
        adapter.setListFragment(listFragment);

        viewPager = (ViewPager) this.findViewById(R.id.pager);

        viewPager.setAdapter(adapter);

        button1 = (Button) this.findViewById(R.id.button1);
        button2 = (Button) this.findViewById(R.id.button2);
```

```java
            button1.setOnClickListener(new OnClickListener() {
                public void onClick(View arg0) {
                    if (currentPage > 0) {
                        currentPage--;
                        viewPager.setCurrentItem(currentPage);
                    } else {
                        currentPage = 0;
                        viewPager.setCurrentItem(currentPage);
                    }
                }
            });

            button2.setOnClickListener(new OnClickListener() {
                public void onClick(View arg0) {
                    if (currentPage < listFragment.size() - 1) {
                        currentPage++;
                        viewPager.setCurrentItem(currentPage);
                    } else {
                        currentPage = listFragment.size() - 1;
                        viewPager.setCurrentItem(currentPage);
                    }
                }
            });
        }
    }
```

程序运行结果如图 10.38 所示。

但 Fragment 对象从未销毁过，日志如图 10.39 所示。

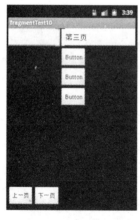

图 10.38　正在拖动分页操作　　　　图 10.39　Fragment 对象从未销毁

10.3.11　Fragment 对象的分页处理方式 2

上一节使用的是 FragmentPagerAdapter 类来实现 Fragment 分页处理，但从性能和内存使用率上来看，并不是理想的解决方案，因为创建出来的 Fragment 从来不销毁，这样在分页比较多的时候内存很快就会溢出,所以使用本节的 FragmentStatePagerAdapter 对象来解决这样的问题,只需要将项目 fragmentTest10 中的所有文件复制到 fragmentTest11 项目中,并更改核心代码如下:

public class MyPageAdapter **extends** FragmentStatePagerAdapter

运行后在 LogCat 打印出了销毁 Fragment 的日志,如图 10.40 所示。

```
MyFragment1 onDestroyView
MyFragment1 onDestroy
MyFragment1 onDetach
```

图 10.40 FragmentStatePagerAdapter 销毁 Fragment

10.3.12 使用 Fragment 对象实现 TabHost 样式的分页及滑动

新建名称为 fragmentTest12 的 Android 项目。创建 4 个 Fragment 对象,如图 10.41 所示。
创建 4 个 Fragment 对应的 XML 布局文件,如图 10.42 所示。

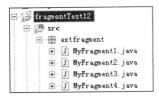
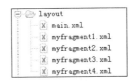

图 10.41 4 个 Fragment 对象　　　　图 10.42 4 个 Fragment 对应的布局文件

创建 FragmentStatePagerAdapter 类的子类 MyPageAdapter.java,代码如下:

```java
public class MyPageAdapter extends FragmentStatePagerAdapter {

    private List<Fragment> listFragment = new ArrayList<Fragment>();

    public MyPageAdapter(FragmentManager fm) {
        super(fm);
    }

    public void setListFragment(List<Fragment> listFragment) {
        this.listFragment = listFragment;
    }

    @Override
    public Fragment getItem(int arg0) {
        return listFragment.get(arg0);
    }

    @Override
    public int getCount() {
        return listFragment.size();
    }

}
```

创建 stateList 配置文件,因篇幅所限,代码不再显示,如图 10.43 所示。

使用到的图片资源如图 10.44 所示。

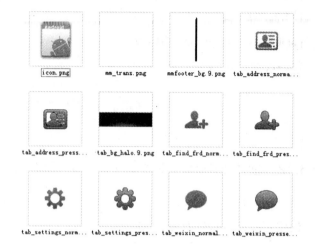

图 10.43　stateList 配置文件　　　　　图 10.44　使用到的图片资源

核心 Activity 文件 Main.java 的代码如下：

```java
public class Main extends FragmentActivity {
    private ViewPager viewPager;
    private MyPageAdapter adapter;
    private List<Fragment> listFragment = new ArrayList<Fragment>();

    private int currentPage = 0;

    private RadioButton radioButton1;
    private RadioButton radioButton2;
    private RadioButton radioButton3;
    private RadioButton radioButton4;

    private ArrayList radioButtonIdList = new ArrayList();

    OnClickListener radioButtonOnClick = new OnClickListener() {
        public void onClick(View arg0) {

            switch (arg0.getId()) {
            case R.id.radio1:
                viewPager.setCurrentItem(0);
                break;
            case R.id.radio2:
                viewPager.setCurrentItem(1);
                break;
            case R.id.radio3:
                viewPager.setCurrentItem(2);
                break;
            case R.id.radio4:
                viewPager.setCurrentItem(3);
                break;
```

```java
            }
        }
    };

    @Override
    public void onCreate(Bundle savedInstanceState) {
        super.onCreate(savedInstanceState);
        setContentView(R.layout.main);

        radioButton1 = (RadioButton) this.findViewById(R.id.radio1);
        radioButton2 = (RadioButton) this.findViewById(R.id.radio2);
        radioButton3 = (RadioButton) this.findViewById(R.id.radio3);
        radioButton4 = (RadioButton) this.findViewById(R.id.radio4);

        radioButton1.setOnClickListener(radioButtonOnClick);
        radioButton2.setOnClickListener(radioButtonOnClick);
        radioButton3.setOnClickListener(radioButtonOnClick);
        radioButton4.setOnClickListener(radioButtonOnClick);

        radioButtonIdList.add(radioButton1.getId());
        radioButtonIdList.add(radioButton2.getId());
        radioButtonIdList.add(radioButton3.getId());
        radioButtonIdList.add(radioButton4.getId());

        MyFragment1 myFragment1 = new MyFragment1();
        MyFragment2 myFragment2 = new MyFragment2();
        MyFragment3 myFragment3 = new MyFragment3();
        MyFragment4 myFragment4 = new MyFragment4();

        listFragment.add(myFragment1);
        listFragment.add(myFragment2);
        listFragment.add(myFragment3);
        listFragment.add(myFragment4);

        adapter = new MyPageAdapter(this.getSupportFragmentManager());
        adapter.setListFragment(listFragment);

        viewPager = (ViewPager) this.findViewById(R.id.pager);

        viewPager.setAdapter(adapter);

        viewPager.setOnPageChangeListener(new OnPageChangeListener() {

            public void onPageSelected(int arg0) {
                Log.v("!", "onPageSelected");
            }

            public void onPageScrolled(int arg0, float arg1, int arg2) {
                Log.v("!", "onPageScrolled");
```

```
                    ((RadioButton) Main.this
                            .findViewById((Integer) radioButtonIdList.get(arg0)))
                            .setChecked(true);
                }

                public void onPageScrollStateChanged(int arg0) {
                    Log.v("!", "onPageScrollStateChanged");

                }
            });

        }
    }
```

程序运行结果如图 10.45 所示。

图 10.45　程序运行结果

在图 10.45 的界面中，既可以手动滑动改变 Fragment，还可以单击下面的 RadioButton 来实现切换 Fragment 界面。